ADVANCES IN NEUROLOGY

Volume 93

ADVANCES IN NEUROLOGY

Volume 93

The Parietal Lobes

Editors

Adrian M. Siegel, M.D.
Department of Neurology
University of Zürich
Zürich, Switzerland

Richard A. Andersen, Ph.D.
Division of Biology
California Institute of Technology
Pasadena, California

Hans-Joachim Freund, M.D.
Department of Neurology
University of Düsseldorf
Düsseldorf, Germany

Dennis D. Spencer, M.D.
Department of Neurosurgery
Yale University
New Haven, Connecticut

LIPPINCOTT WILLIAMS & WILKINS
A **Wolters Kluwer** Company
Philadelphia • Baltimore • New York • London
Buenos Aires • Hong Kong • Sydney • Tokyo

Acquisitions Editor: Anne M. Sydor
Developmental Editor: Jenny Kim
Production Editor: Jeff Somers
Manufacturing Manager: Colin Warnock
Cover Designer: Patricia Gast
Compositor: Lippincott Williams & Wilkins Desktop Division
Printer: Maple-Press

530 Walnut Street
Philadelphia, PA 19106 USA
LWW.com

Printed in the USA

Library of Congress Cataloging-in-Publication Data

ISBN: 0-7817-3625-0
ISSN: 0091-3952

---

10 9 8 7 6 5 4 3 2 1

Advances in Neurology Series

Vol. 92: Ischemic Stroke. *H.J.M. Barnett, J Bogousslavsky, and H. Meldrum, editors.* 496 pp., 2003.
Vol. 91: Parkinson's Disease. *A. Gordin, S. Kaakkola,H. Teräväinen, editors.* 426 pp., 2002.
Vol. 90: Neurological Complications of Pregnancy. Second Edition: *B. Hainline and O. Devinsky, editors.* 368 pp., 2002.
Vol. 89: Myoclonus and Paroxysmal Dyskinesias: *S. Fahn, S.J. Frucht, M. Hallett, D.D. Truong, editors.* 528 pp., 2002.
Vol. 88: Neuromuscular Disorders: *R. Pourmand and Y. Harati, editors.* 368 pp., 2001.
Vol. 87: Gait Disorders: *E. Ruzicka, M. Hallett, and J. Jankovic, editors.* 432 pp., 2001.
Vol. 86: Parkinson's Disease: *D. Calne and S. Calne, editors.* 512 pp., 2001.
Vol. 85: Tourette Syndrome: *D. J. Cohen, C. Goetz, and J. Jankovic, editors.* 432 pp., 2001.
Vol. 84: Neocortical Epilepsies: *P. D. Williamson, A. M. Siegel, D. W. Roberts, V. M. Thadani, and M. S. Gazzaniga, editors.* 688 pp., 2000.
Vol. 83: Functional Imaging in the Epilepsies: *T. R. Henry, J. S. Duncan, and S. F. Berkovic, editors.* 348 pp., 2000.
Vol. 82: Corticobasal Degeneration and Related Disorders: *I. Litvan, C .G. Goetz, and A. E. Lang, editors.* 280 pp., 2000.
Vol. 81: Plasticity and Epilepsy: *H. Stefan, P. Chauvel, F. Anderman, and S.D. Shorvon, editors.* 396 pp., 1999.
Vol. 80: Parkinson's Disease: *Gerald M. Stern, editor.* 704 pp., 1999.
Vol. 79: Jasper's Basic Mechanisms of the Epilepsies. Third Edition: *A. Delgado-Escueta, W. Wilson, R. Olsen, and R. Porter, editors.* 1,132 pp., 1999.
Vol. 78: Dystonia 3:*S. Fahn, C. D. Marsden, and M. R. DeLong, editors.* 374 pp., 1998.
Vol. 77: Consciousness: At the Frontiers of Neuroscience: *H. H. Jasper, L. Descarries, V. F. Castellucci, S. Rossignol, editors.* 300 pp., 1998.
Vol. 76: Antiepileptic Drug Development: *J. A. French, I. E. Leppik, and M.A. Dichter, editors.* 276 pp., 1998.
Vol. 75: Reflex Epilepsies and Reflex Seizures: *B. G. Zifkin, F. Andermann, A. Beaumanoir, and A. J. Rowan, editors.* 310 pp., 1998.
Vol. 74: Basal Ganglia and New Surgical Approaches for Parkinson's Disease: *J. A. Obeso, M. R. DeLong, C. Ohye, and C. D. Marsden, editors.* 286 pp., 1997.
Vol. 73: Brain Plasticity: *H. J. Freund, B. A. Sabel, and O. W. Witte, editors.* 448 pp., 1997.
Vol. 72: Neuronal Regeneration, Reorganization, and Repair: *F. J. Seil, editor.* 416 pp., 1996.
Vol. 71: Cellular and Molecular Mechanisms of Ischemic Brain Damage: *B. K. Siesjö and T. Wieloch, editors.* 560 pp., 1996.
Vol. 70: Supplementary Sensorimotor Area: *H. O. Lüders, editor.* 544 pp., 1996.
Vol. 69: Parkinson's Disease: *L. Battistin, G. Scarlato, T. Caraceni, and S. Ruggieri, editors.* 752 pp., 1996.
Vol. 68: Pathogenesis and Therapy of Amyotrophic Lateral Sclerosis: *G. Serratrice and T. L. Munsat, editors.* 352 pp., 1995.
Vol. 67: Negative Motor Phenomena: *S. Fahn, M. Hallett, H. O. Lüders, and C. D. Marsden, editors.* 416 pp., 1995.
Vol. 66: Epilepsy and the Functional Anatomy of the Frontal Lobe: *H. H. Jasper, S. Riggio, and P. S. Goldman-Rakic, editors.* 400 pp., 1995.
Vol. 65: Behavioral Neurology of Movement Disorders: *W. J. Weiner and A. E. Lang, editors.* 368 pp., 1995.
Vol. 64: Neurological Complications of Pregnancy: *O. Devinsky, E. Feldmann, and B. Hainline, editors.* 288 pp., 1994.
Vol. 63: Electrical and Magnetic Stimulation of the Brain and Spinal Cord: *O. Devinsky, A. Beric, and M. Dogali, editors.* 352 pp., 1993.
Vol. 62: Cerebral Small Artery Disease: *P. M. Pullicino, L. R. Caplan, and M. Hommel, editors.* 256 pp., 1993.
Vol. 61: Inherited Ataxias: *A. E. Harding and T. Deufel, editors.* 240 pp., 1993.
Vol. 60: Parkinson's Disease: From Basic Research to Treatment: *H. Narabayashi, T. Nagatsu, N. Yanagisawa, and Y. Mizuno, editors.* 800 pp., 1993.
Vol. 59: Neural Injury and Regeneration: *F. J. Seil, editor.* 384 pp., 1993.
Vol. 58: Tourette Syndrome: Genetics, Neurobiology, and Treatment: *T. N. Chase, A. J. Friedhoff, and D. J. Cohen, editors.* 400 pp., 1992.
Vol. 57: Frontal Lobe Seizures and Epilepsies: *P. Chauvel, A. V. Delgado-Escueta, E. Halgren, and J. Bancaud, editors.* 752 pp., 1992.

Contents

Part III: Human Studies

Contributing Authors

Marie-Claire Albanese, M.Sc.
Department of Physiology
McGill University
1205 Dr. Penfield Avenue
Montreal, Quebec H3A 1B1
Canada

Richard A. Andersen, Ph.D.
James G. Boswell Professor of Neuroscience
Division of Biology
Mail Code 216-76
California Institute of Technology
Pasadena, California
USA

Alain Berthoz, Ph.D.
Professor and Chair
Laboratory of Physiology
College de France-LPPA
11 place Marcellin Berthelot
Paris, France

Ferdinand Binkofski, M.D.
Department of Neurology
University Hospital Schleswig-Holstein
Campus Lübeck
Ratzeburger Allee 160
23538 Lübeck
Germany

James W. Bisley, Ph.D.
Post-doctoral Research Fellow
David Mahoney Center of Brain and Behavior
Center for Neurobiology and Behavior
Columbia University College of Physicians and Surgeons
1051 Riverside Drive
Unit 87, Rm 5-06
New York, New York
USA

Gabriella Bottini, M.D., Ph.D.
Department of Psychology
University of Pavia
P.zza Botta, 6
Pavia 27100
Italy

Christopher A. Buneo, Ph.D.
Senior Research Fellow
Division of Biology
California Institute of Technology
Mail Code 216-76
Pasadena, California
USA

Paul Cisek, Ph.D.
Adjunct Investigator
Section of Neurophysiology
Laboratory of Systems Neuroscience
National Institute of Mental Health
Bldg. 49, Room B1EE17 MSC 4401
49 Convent Drive
Bethesda, Maryland
USA

Christine E. Collins, Ph.D.
Research Associate
Department of Psychology
Vanderbilt University
301 Wilson Hall
111 21st Avenue South
Nashville, Tennessee
USA

Mary K. Colvin
Department of Psychological and Brain Sciences
Center for Cognitive Neuroscience
6207 Moore Hall
Dartmouth College
Hanover, New Hampshire
USA

Jon Driver, B.A., D. Phil.
Professor of Neuroscience
Institute of Cognitive Neuroscience
University College London
17 Queen Square
London WC1N 3AR
United Kingdom

Gary H. Duncan, D.D.S., Ph.D.
Professeur Titulaire
Faculté de Médecin Dentaire
Centre de Recherche en Sciences Neurologiques
Université de Montréal
Adjunct Professor
Department of Neurology and Neurosurgery
Montréal Neurologic Institute
McGill University
Montréal, Quebec
Canada

Simon Eickhoff
Institute of Medicine
Research Center Jülich
Leo-Brandt-Str.
52428 Jülich
Germany

Alessandro Farnè, Ph.D.
Department of Psychology
University of Bologna
Viale Berti Pitchat
Bologna, Italy

Herta Flor, Ph.D.
Professor of Neuropsychology
Department of Neuropsychology
University of Heidelberg
Central Institute of Mental Health
J5, 68159 Mannheim
Germany

Hans-Joachim Freund, M.D., FRCP
Department of Neurology
University of Düsseldorf
Moorenstrasse 5
Düsseldorf, Germany

Michael S. Gazzaniga, Ph.D.
Department of Psychological and Brain Sciences
Director, Center for Cognitive Neuroscience
6162 Moore Hall
Dartmouth College
Hanover, New Hampshire
USA

Mitchell Glickstein, Ph.D.
Professor Emeritus of Neuroscience
Department of Anatomy
University College London
Gower Street
London, United Kingdom

Michael E. Goldberg, M.D.
David Mahoney Professor of Brain and Behavior
Departments of Neurology and Psychiatry
Center for Neurobiology and Behavior
Columbia University College of Physicians and Surgeons
1051 Riverside Drive, Unit #87
New York, New York
USA

Melvyn A. Goodale, Ph.D., F.R.S.C.
Canada Research Chair in Visual Neuroscience
CIHR Group on Action and Perception
Department of Psychology
University of Western Ontario
London, Ontario
Canada

Nadia Gosselin-Kessiby
Doctoral Candidate
Départment de Physiologie
Pavillion Paul-G-Desmarais
Université de Montréal
C.P. 6128, Succursale Centre-ville
Montréal, Canada

Marie-Hélène Grosbras
Cognitive Neuroscience Unit
Montreal Neurological Institute
3801 University
Montreal, Quebec
Canada

Angela M. Haffenden, Ph.D.
Clinical Neuropsychologist
Psychology Department
Foothills Medical Centre
1403 29th Street NW
Calgary, Alberta
Canada

Todd C. Handy, Ph.D.
Department of Psychological and Brain Sciences
Center for Cognitive Neuroscience
6162 Moore Hall
Dartmouth College
Hanover, New Hampshire
USA

Marc Jeannerod, M.D.
Institut des Sciences Cognitives
67 Boulevard Pinel
69675 Bron, France

Jon H. Kaas, Ph.D.
Distinguished Professor
Department of Psychology
Vanderbilt University
301 David K. Wilson Hall
111 21st Avenue South
Nashville, Tennessee
USA

John F. Kalaska, Ph.D.
Département de Physiologie
Pavillon Paul-G-Desmarais
Université de Montréal
C.P. 6128, Succursale Centre-ville
Montréal, Québec H3C 3J7
Canada

Hahnah J. Kasowski
Department of Neurosurgery
Yale University
333 Cedar Street
P. O. Box 208082
New Haven, Connecticut
USA

Ramón C. Leiguarda, M.D.
Department of Neurology
Rául Carrea Institute of Neurological Research
FLENI Montaneses 2325
Buenos Aires, Argentina

Emiliano Macaluso, Ph.D.
Research Fellow
Institute of Cognitive Neuroscience
University College London
Alexandra House
17 Queen Square
London, WC1N 3AR
United Kingdom

Pattanasak Mongkolwat, Ph.D.
Research Assistant and Professor of Radiology
Department of Radiology
Northwestern University Medical School
448 E. Ontario St., Ste 300
Chicago, Illinois
USA

Nicola Palomero-Gallagher, Dr. rer. Nat.
Scientist
Institut für Medizin
Forschungszentrum Jülich
52425 Jülich
Germany

Eraldo Paulesu, M.D.
Professor of Psychobiology and Physiological Psychology
Department of Psychology
University of Milano-Bicocca
Piazza dell'Ateneo Nuovo 1
Milano 20126
Italy

Eric J. Russell, M.D.
Chairman
Department of Radiology
Northwestern University Medical School
676 N. St. Clair, Suite #800
Chicago, Illinois
USA

Hideo Sakata, M.D.
Professor
Departments of of Anatomy and Physiology
Seitoku Junior College of Nutrition
1-4-6 Nishi-shinkoiwa
Katsushika-ku
Tokyo, Japan

Georges Salamon, M.D.
Visiting Researcher
Department of Radiology
The David Geffen School of Medicine at UCLA
10833 Le Conte Avenue – BL 428 CHS
Box 951721
Los Angeles, California
USA

Noriko Salamon-Murayama, M.D.
Assistant Professor
Department of Radiology
The David Geffen School of Medicine at UCLA
10833 Le Conte Avenue – BL 428 CHS
Box 951721
Los Angeles, California
USA

Rüdiger J. Seitz, M.D.
Department of Neurology
University Hospital Düsseldorf
Moorenstrasse 5
40225 Düsseldorf
Germany

Adrian M. Siegel, M.D.
Neurologist FMH
Attending Neurologist
Department of Neurology
University Hospital of Zürich
8091 Zürich
Switzerland

Dennis D. Spencer, M.D.
Department of Neurosurgery
Yale University
333 Cedar Street
P. O. Box 208082
New Haven, Connecticut
USA

Susan S. Spencer, M.D.
Departments of Neurosurgery and Neurology
Lab of Clinical Investigation
Yale University
333 Cedar Street, P. O. Box 208018
New Haven, Connecticut
USA

Michael R. Stoffman, M.D.
Department of Neurosurgery
Yale University
333 Cedar Street
P. O. Box 208082
New Haven, Connecticut
USA

Giuseppe Vallar, M.D.
Professor of Physiological Psychology
Department of Psychology
University of Milano-Bicocca
Piazza dell'Ateneo Nuovo 1
Milano 20126
Italy

Karl Zilles, M.D.
Professor of Brain Research
Institute of Medicine
Research Center Jülich
Leo Brandt Str.
52425 Jülich
Germany

Preface

In his seminal book on the parietal lobe Critchley contemplated that "the parietal lobes are empirical conceptions rather than autonomous entities in the anatomical and physiological sense," and that "there is something essentially artificial about a full-dress discussion limited to a sector of the cerebral hemisphere." Nevertheless he wrote a masterpiece on this issue that still represents the gold standard in terms of clarity and completeness of the classical observations on the phenomenology of parietal lobe disorder.

Since that time, our views on parietal lobe function have been further advanced by neurophysiological investigation. This new avenue of parietal research began in the 1970s with the seminal studies by Vernon Mountcastle and his colleagues at the Johns Hopkins University School of Medicine. They recorded the activity of single neurons from the posterior parietal cortex in behaving monkeys, a relatively new experimental approach at the time. They showed activities related to sensation and behaviour, implicating the posterior parietal cortex in transformations between the two. Following these groundbreaking studies, a number of laboratories divided the posterior parietal cortex using functional and anatomical criteria, finding sensory-motor areas for grasping, reaching, and eye movements. Mechanisms not only for the spatial representation of goals, but also for the transformation between coordinate frames required for sensory-motor integration, have been elucidated in the course of the last 20 years. Moreover, a number of experiments have better delineated how attention, learning, prediction, and other higher cognitive functions operate in the context of sensory-motor processing in the parietal lobes.

These new findings, elaborated in several of the chapters by leading investigators in parietal neurophysiology, help us to understand the neural hardware whose destruction accounts for the deficits Critchley so eloquently described a half a century ago. Moreover, a number of these experimental studies have been inspired by insights derived from Critchley's reviews of the neurological literature.

Experimental studies employ microlesions to complement the information gained from neurophysiological and neuroanatomical data and to probe selective dysfunctions of the respective neuronal task groups. In the human a similar approach is possible—though on a coarser scale—by comparing the effects of lesions with functional activation studies. Accordingly, most of the chapters on human data in this book employ both approaches for their respective topic. They show that a variety of specific dysfunctions ensuing parietal lobe damage are mirrored by neuroimaging data revealing parietal lobe involvement in tasks that reflect a positive image of the deficits seen after lesions. But functional imaging also demonstrates how the respective activations are nested in widely distributed functional networks. Such a comparative evaluation of parietal lobe function is not always straightforward because of some principal differences between the information provided by either approach.

The wealth of new information from this side and from the refinement of the behavioural analysis allows the elaboration of more complex cognitive functions including the fascinating aspects of hemispheric specialisation. Here, the role of the parietal lobe for language and calculus, attention and learning, spatial functions and the conception of motor behaviour are particularly challenging issues.

As we look to the future, we can see that non-invasive imaging techniques such as functional magnetic resonance imaging and magneto encephalography will further bridge the gap between experiments in humans and non-human primates. This approach, combined with techniques of simultaneous recording of multiple cells and the recording of electrical activity from the human cortex in the course of clinical evaluations for diseases such as epilepsy, will lead to a deeper understanding of the parietal circuitry.

Adrian M. Siegel, M.D.
Richard A. Andersen, Ph.D.
Hans-Joachim Freund, M.D.
Dennis D. Spencer, M.D.

Foreword

It is now half a century since Critchley's monograph *The Parietal Lobes* was published, in 1953, and republished a decade later (1). Scarcely before or since has a single authored monograph in systems neuroscience had such an immediate and lasting impact in an important area of clinical and experimental neurology. Critchley achieved in his own mind, and expressed between two covers, a critical synthesis of the then-current knowledge in the field, gathered from all relevant disciplines. Moreover, he was well aware of and cogently expressed the idea that "the parietal lobe" is an artificial abstraction useful as an ordering principle, and that the several areas with different functions, already discernible then, were nodes in distributed systems involving other cortical areas and subcortical structures. Critchley achieved this synthesis using his knowledge and experience in clinical neurology, with little help from what we would now regard as the relatively primitive technical aids available to him. I read this monograph avidly in the late 1960's when a serendipitous experimental observation led me into the unknowns of the neurophysiology of the parietal cortex. I suspect that many investigators in what has become a very large field have their own, well-read copies.

A comparison of the contents of the present volume with Critchley's book suggests two generalities. The first is that direct study of the higher functions of the brain at the level of neural operations is now routine in brain research. This remarkable advance is due to the development of a psychophysics for monkeys and the combined experiment in them, improved methods for observing and measuring human behavior, and imaging methods in humans. Any behavior that an animal can be trained to emit, or that a human subject will willingly display in the laboratory setting, can be studied directly in terms of the relevant neural activity. The further development of the imaging methods to provide information in the dynamic time domain of cerebral function will be an important advance.

The second generality is the "globalization" of neuroscience. The 22 papers in this book were generated in laboratories located in many countries, from Japan to Argentina, to Germany, England, Italy, Switzerland, and France in Europe, to Canada and the United States. With this diversity has come an international fellowship of individuals with common interests, marked by good will and a ready exchange of ideas. No single individual, laboratory, or country dominates the field.

Knowledge of posterior parietal lobe cortical function in the 1950's was based almost exclusively on studies of patients with cortical lesions. By contrast, somatic sensory areas of the anterior parietal cortex, the postcentral gyrus, were under intensive study with a variety of electrophysiological methods. Woolsey and his colleagues had used the evoked potential method to produce a map of the postcentral gyrus in monkeys compatible with the necessarily less precise maps obtained by electrical stimulation of the exposed postcentral gyrus in waking humans, by Cushing, Foerster, and Penfield. The first single neuron studies of the postcentral areas in monkeys were completed, and initiated a series that has continued in the last decades in the form of the combined experiment in which behavioral and neural events are recorded simultaneously in waking monkeys. Use of this latter method in monkeys working in tasks requiring parietal lobe function for their execution revealed sets of neurons with functional properties that are positive images of several of the defects produced in monkeys by parietal lobe lesions. Moreover, these experiments showed the powerful effect of directed attention on the excitability of parietal lobe neurons, and uncovered sets of parietal visual neurons with receptive fields and directional selectivities appropriate for signaling the movement of the visual flow fields. For many neurophysiologists of that era, experienced in the stimulus-response paradigm of experiments on sensory systems, each day's recording in the early experiments on the parietal lobe was an adventure, for it was possible to see the direct correlation between the recorded neuronal activity and an item of behavior—in some cases rather high order behavior. Correlations of course do not lead directly to causality, and there was not then, and still is not, evidence in this class of experiments that some other, unobserved neuronal activity is critical for the behavior under study. But, for the moment, correlation was enough.

The intensive, continuing study of the parietal lobe in waking monkeys in several laboratories has produced a rich and expanding knowledge of the function of the parietal systems, extending from the somatic sensory functions of the anterior parietal fields in the detection of and discrimination between somatic sensory stimuli, to the role of the posterior parietal fields in visuospatial perception, visual attention, and visuomotor transformations leading to reaching and grasping, and other complex functions, described by several authors in the present volume.

The chapters in this book are both reviews of particular aspects of parietal lobe function and direct descriptions of recent research results. The first part contains three anatomical chapters. Chapter 1 gives an overview of the cytoarchitecture of the parietal cortex. It is based on an observer-independent method developed by Zilles and his colleagues, and used by them in studies of the cytoarchitecture of the postcentral fields in humans. When combined with receptor autoradiography, the method promises to define in cytoarchitectural terms the several areas of the posterior parietal cortex discovered in electrophysiological studies in waking monkeys, and outlined with increasing accuracy in studies of humans with small lesions of the parietal cortex. This new method appears to be the first major contribution to cytoarchitectural studies in several decades.

Salamon and his colleagues summarize what is of great value in imaging studies: A precise delineation of cortical gyri and sulci in the human brain. Chapter 3 by Mitchell Glickstein calls attention to what is sometimes neglected, that the parietal lobe is an efferent structure, sending descending projections to thalamus, basal ganglia, brain stem, cerebellum, and spinal cord, and that these efferents must be taken into account in syntheses of parietal lobe function, particularly in consideration of the apraxias.

The structure and connectivities of the postcentral and intrasylvian areas that compose the first steps to the distributed cortical system serving somatic sensibilities are described in Chapter 4, written by Jon Kaas, a major contributor to the field. Knowledge produced over many years with classical neuroanatomical methods, and greatly incremented by use of the method of microelectrode mapping in a large number of mammalian species, has produced a coherent description of the evolution of these areas of the cerebral cortex, so far as this is possible by study of living species. Chapter 6, by the same author, contains a description of the plasticity which appears immediately after partial lesions or denervations of the somatic system. The evidence summarized supports the idea that these early changes are produced by an uncovering, or imbalance, of normal functional properties of the system. The delayed and long-lasting plastic changes that follow require other explanations. They are discussed in Chapter 12 by Herta Flor with reference to the phantoms and phantom pain experienced by limb amputees.

Chapter 5 contains an elegant summary of present knowledge of the cortical system serving pain and temperature in humans. It comes from the group at the Université de Montreal, who are major contributors to the field. There can be little doubt that this continually enlarging fund of new information about the pain system is one of the major additions to knowledge produced by the use of methods for imaging the human brain. The pleomorphic and variable nature of pain is matched by a widely distributed cortical pain system in which different nodes contribute different attributes to the overall pain experience.

The sequence then turns to what are regarded as the higher order functions of the posterior parietal lobe systems, those between afferent input and motor output. Sakata and Kalaska deal with the sensory-motor transition functions generating signals for the movement of the arm and grasping by the hand, and describe results obtained in combined experiments in monkeys working in appropriate reaching and grasping tasks. Each paper describes the generation and nature of these signals. Kalaska discusses the difficult problem of the frames of reference that intervene between the visual and manual grasping of objects. Bisley and Goldberg summarize the evidence for two hypotheses concerning our visual interface with the world about us. The first is that the lateral intraparietal area (LIP) contains a salience map for the direction of attention in the visual fields, instead of or in addition to its putative function in generating saccades. The second is that LIP contains a mechanism for updating the visual field during saccadic eye movements, and thus contributes to maintenance of a stable visual world in spite of moving eyes. This chapter might appropriately be read in parallel with that by Grosbras and Berthoz, which summarizes studies of the human oculomotor control system, made with imaging methods. The human parieto-frontal circuits appear more complex than those presently known in monkeys, and homologies of relevant cortical areas in the two primates are still uncertain.

Andersen and Buneo focus first on the specific functions of posterior parietal areas LIP and the parietal reach region (PRR), respectively operative in the sensory-motor transitions leading to eye movements and to projection of the arm and grasping by the hand, which are now understood directly in terms of parietal neuron activity, knowledge to which Andersen and his colleagues have made a series of important contributions. The review then generalizes considering several candidate hypotheses concerning where these parietal areas are located along the continuum from sensory inputs to outputs from the frontal motor system. The general conclusion is that they process specific signals for eye movements and arm projections, but that they are also parts of widely distributed cortical systems through which they are influenced by attention and affective set, and through which they play a role in learning and memory of the events they generate specifically. They are located in the large in-between area.

The chapters in Part III document that knowledge of parietal lobe functions in normal humans and in patients with parietal lobe lesions has reached a new level of insight and sophistication. This is due to new methods for measuring human performance and deficiencies in performance, and to the steadily improving imaging methods. Beyond that, several chapters reveal an intrepid confidence that, with the new methods, new vistas are opening in the field of human neurology, and perhaps nowhere with more promise than in studies of the parietal lobe system.

Hans-Joachim Freund provides a scholarly overview of the field of neurology around which this book is organized. The author covers the full range of defects produced by lesions of the parietal lobe cortex. Meticulous study of patients with small lesions combined with improved methods for determining the location and extent of those lesions provides evidence that the human parietal cortex, like that of the monkey, is organized in a modular manner. Seitz and Binkofski provide a meta-analysis of 50 activation studies made between 1991 and 2002 in humans as they executed tasks relevant to parietal lobe function. These studies converge to confirm that the human parietal lobe cortex is composed of several modular regions with specific functions and specific extrinsic connections. These two reviews might be read alongside that of Leiguarda, who reviews a number of clinical studies, beginning with that of Liepmann, which focus on the role of the left-parietal cortex in right-handed individuals in the control of complex motor skills. It is clear, however, that in many of these studies a small but significant number of patients with apraxias had right hemisphere lesions.

Vallar, Bottini, and Paulesu present a meta-analysis of the neglect syndromes, particularly unilateral spatial neglect, summarizing results obtained in the large number of patients with parietal lobe lesions, and inactivation studies in normal humans. A consensus in the field is that unilateral spatial neglect is a high order cognitive defect, and that the continuous maintenance of awareness in the immediately surrounding spatial field is essential for spatial attention and spatial representation. Unilateral spatial neglect is most often produced by lesions of the inferior parietal lobule, but a significant number of such patients have lesions of the frontal premotor cortex or of the connections between the two. This leads to the inference that the maintenance of awareness depends upon the dynamic ongoing activity in this distributed system, and is not "localized" to one or the other area alone.

Goodale and Haffenden address the question whether the dorsal and ventral streams of the trans-cortical projecting visual system function as independently as has hitherto been supposed. They summarize evidence from a variety of experiments that lead to the conclusion that the two systems are tightly integrated in normal function, although each is primarily concerned with different aspects of the recognition and identification of objects in the visual world, and the direction of action upon them. The integration appears complete although the two can be dissociated in humans by lesions of either the parietal or the temporal lobes.

Jeannerod and Farne first describe the dissociations between reaching and grasping produced in monkeys by small lesions or reversible lesions in activations of MIT or AIT, respectively. They then center their discussion on the disorders that follow lesions of the parietal lobe in humans, particularly optic ataxia, and the dissociations between such disorders and disturbances of visual perception produced by lesions of the ventral stream in the inferotemporal cortex. They describe an intensively studied patient with defects in reaching and grasping who could achieve almost normal performance in reaching-grasping when presented with familiar objects. Semantic information about those objects is thought to have been obtained from the ventral stream. They suggest, as do Goodale and Haffenden, that the two major streams are integrated in normal function.

Macaluso and Driver summarize the results of imaging studies in humans which show that regions of cortex in the anterior and dorsal regions of the intraparietal sulcus are activated by both visual and tactile stimuli, from similar locations in the space of the contralateral hemifield. Other regions in the inferior parietal lobule are similarly activated in this multimodal but space location-specific way, in either hemifield. It is implied that parietal lobe mechanisms build a panoramic spatial representation independently of the specific modalities of sensory inputs upon which it depends.

Chapter 20 contains a summarizing review of the hemispheric specialization in parietal lobe function, a striking example of differences in function of the two hemispheres in hominoids, and particularly in man, a subject to which Michael Gazzaniga has contributed perhaps more than any other researcher in a long series of studies of split-brain humans. It is written, as the author has put it, "in light of modern cognitive neuroscience." The left parietal cortex is specialized for motor and temporal processing, the right for visuospatial processing. These high-level cognitive operations are embedded in the dynamic activity in the reciprocal connections between the posterior parietal lobe areas and the frontal lobe. It is uncertain whether the functionally specialized posterior parietal areas of the two hemispheres differ in size or area distributions. If not, the different functions may be attributed to differences in their extrinsic connectivities, or in their intrinsic operations, or both.

Adrian M. Siegel deals with a difficult clinical entity, parietal lobe epilepsy. This is difficult because localizing symptoms may become apparent only after spread of the ictal focus to adjacent cortical regions. In many such cases the focus produces no unusual experiences at all, but in others a mixture of experiences related to parietal lobe function. The epileptic focus is not constrained by modular boundaries, so that the resulting clinical picture may include a mixture of quite different symptoms. Chapter 22, from researchers at the Yale University Department of Neurosurgery, shows that with careful and detailed preoperative study, the condition can be treated successfully by surgical means in a majority of patients.

The future development in this field of neuroscience is predictably unpredictable. For the combined experiment in waking monkeys working in tasks requiring parietal lobe actions, however, the future is now! This is so because laboratories in several parts of the world have perfected methods for chronically implanting large numbers of microelectrodes in the cerebral cortices of monkeys and recording from them for many months. This work will allow direct study of what we all rather glibly call the "population signal". Imagine, for example, the reach-to-grasp experiment performed in a monkey with large numbers of electrodes implanted in the parietal, prefrontal, and motor cortices. With such experimental objectives, neurophysiology moves from little to big science, with all the problems of funding, of assuring adequate rewards for the numbers of experienced scientists the experiment will require, and the like. All of these problems will be difficult; all can be solved, and the rewards in advancement of our understanding of how the waking, active brain performs will be great. The history of neuroscience is filled with instances in which the most significant advances have depended on new methods or improvement of old ones. This will certainly occur in the field of imaging, in which the technical advances move with lightning speed. Its practitioners know the needs: To solve the problem of the relation of the second order events recorded in imaging experiments to the specific details of the neuronal activities that produce them; to improve spatial resolution by an order of magnitude; to decrease collect times so that the dynamic flow of activity of the cerebral cortex can be followed in real time. This book contains a series of scholarly essays in which almost all aspects of parietal lobe function and its disorder by lesions are considered. It is not a book for browsing. Its attentive study will be rewarding to all systems neuroscientists.

REFERENCE

1. Critchley. The Parietal Lobes, 1953. London: Arnold. Reprint, New York: Habner, 1969.

Vernon B. Mountcastle, M.D.
Professor of Neuroscience
Johns Hopkins University

Acknowledgments

We wish to acknowledge Professor Vernon B. Mountcastle, whose groundbreaking research on the posterior parietal cortex has formed the cornerstone for much of the work described in this book. His research gave us our first glimpses into the neural circuits that are the basis of parietal lobe function. It was a great honor for us that Professor Mountcastle agreed to write the foreword to this volume.

We thank all of the contributors for their superb chapters. We would also like to thank the students and postdoctoral fellows who have worked with us through the years for their insights and enthusiasm in the course of their studies of this most interesting cortical area. We acknowledge the generous support of our work by the Boswell Foundation, the National Eye Institute, the Defense Advanced Research Agency, and Office of Naval Research. Finally, the Editors would like to thank Anne Sydor, Jenny Kim, and all the staff at Lippincott Williams & Wilkins for their assistance in preparing this book.

1

The Human Parietal Cortex: A Novel Approach to Its Architectonic Mapping

Karl Zilles,*† Simon Eickhoff,*† and Nicola Palomero-Gallagher*

*Institute of Medicine, Research Center Jülich, 52425 Jülich, Germany;
†C.&O. Vogt-Institute of Brain Research, University of Düsseldorf, 40001 Düsseldorf, Germany

INTRODUCTION

The parietal cortex occupies approximately one fourth of the human brain. On the medial surface, it is delimited rostrally by the central sulcus, ventrally by the subparietal sulcus, and caudally by the parieto-occipital sulcus. On the lateral surface, it is separated from the frontal lobe by the central sulcus, but no clear macroscopic landmarks enable precise differentiation from the occipital or temporal lobes.

Numerous cytoarchitectonic and myeloarchitectonic maps of the human brain have been published since the beginning of the last century (1–11). And yet the parcellation of the parietal cortex remains the subject of discussion, because the existing maps differ considerably concerning the number and size of individual areas (Figs. 1-1 and 1-2).These architectonic observations are also riddled with methodologic drawbacks: (a) The definition of areal borders is based on a highly observer-dependent method, and thus the significance of borders could not be tested. (b) They lack information concerning interhemispheric and intersubject variability of location and size of cortical areas. (c) They are simplified schematic drawings, and thus lack a registration of these borders in a spatial reference system. Furthermore, the parcellation schemes depicted fail to explain the much more detailed areal pattern of the parietal cortex in general, and of the intraparietal sulcus in particular, which has recently been revealed by functional imaging studies (12–19).

In the present chapter we review the parcellation schemes of the classical cytoarchitectonic and myeloarchitectonic maps and complement them with our own cytoarchitectonic and Receptor architectonic studies. Receptors for classical neurotransmitters are heterogeneously distributed throughout the human brain, and unveil hitherto unknown parcellations of the human parietal cortex on the basis of the distribution pattern of molecules functionally relevant for neurotransmission. Finally, the connectivity between the parietal and particularly the frontal cortex is reviewed briefly.

CYTOARCHITECTONIC AND MYELOARCHITECTONIC MAPS

To date, the most widely used cytoarchitectonic map is that of Brodmann (5), since it serves, via the atlas of Talairach and Tournoux (20), as an anatomic reference in numerous functional imaging studies. Therefore, Brodmann's parcellation concept (5,6) is used in the present chapter as the reference point for comparison with the maps of Campbell (7), Elliot Smith (4), Vogt (11), von Economo and Koskinas (8), Gerhardt (9), Sarkissov (10),

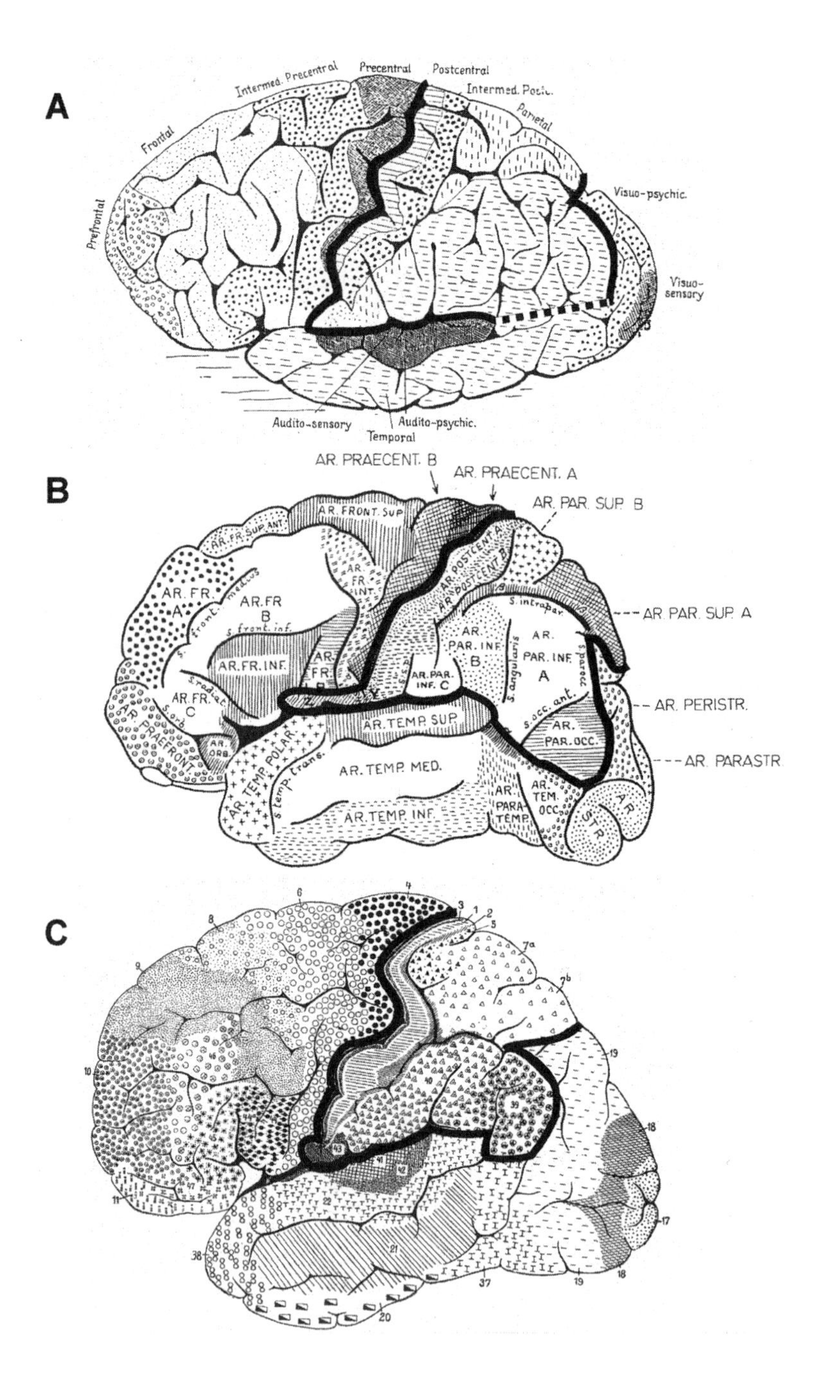

FIG. 1-1. Lateral view of the human parietal cortex, adapted from the classical maps of **(A)** Campbell (7), **(B)** Elliot Smith (4), **(C)** Brodmann (6), **(D)** Vogt (11), **(E)** von Economo and Koskinas (8), and **(F)** Sarkissov et al. (10). Cytoarchitectonic areas are marked by different hatches, and the parietal cortex is surrounded by a thick black line. For a comparison of the different nomenclatural systems see Table 1-1. Note differences in sulcal patterns and variations in the shape, number, and size of the areas depicted. Campbell (7) did not define a border between the inferior parietal lobule and the temporal lobe (thus the dotted line in the figure). The basal parietal region PH of von Economo and Koskinas (8) is not included here, because this area probably belongs to the visual cortex.

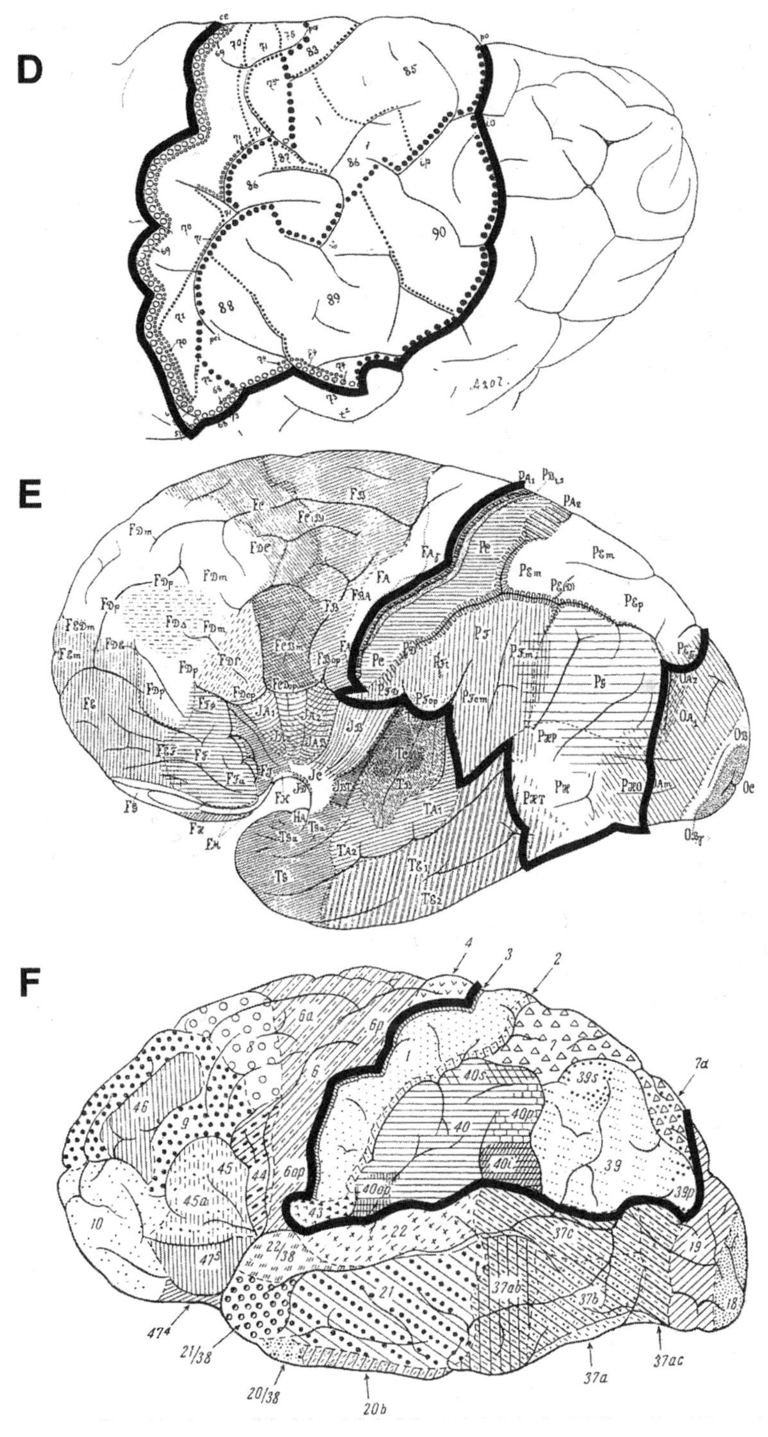

FIG. 1-1. *Continued.*

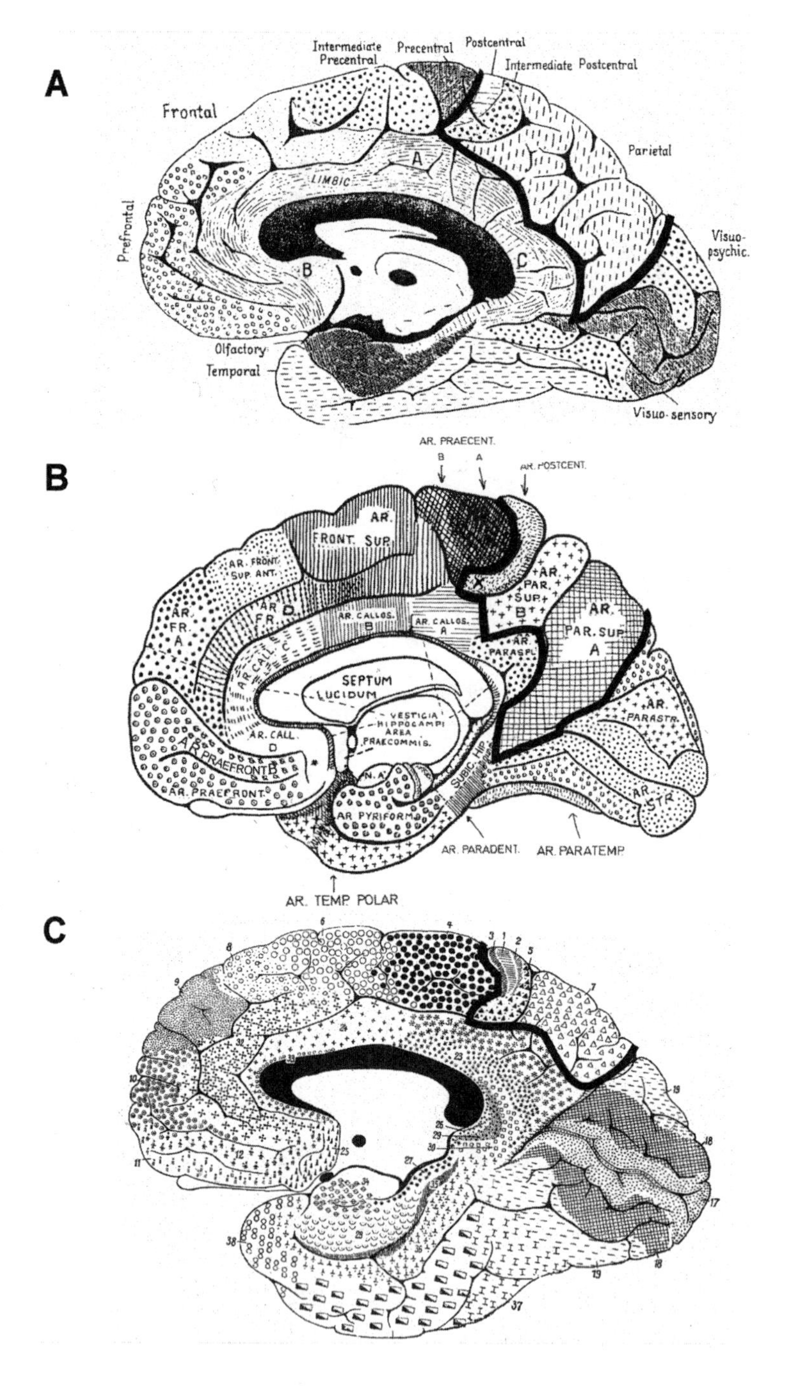

FIG. 1-2. Mesial view of the human parietal cortex, adapted from the classical maps of **(A)** Campbell (7), **(B)** Elliot Smith (4), **(C)** Brodmann (6), **(D)** Vogt (11), **(E)** von Economo and Koskinas (8), and **(F)** Sarkissov (10). Cytoarchitectonic areas are marked by different hatches, and the parietal cortex is surrounded by a thick black line. For a comparison of the different nomenclatural systems see Table 1-1. Note differences in sulcal patterns and variations in the shape, number, and size of the areas depicted.

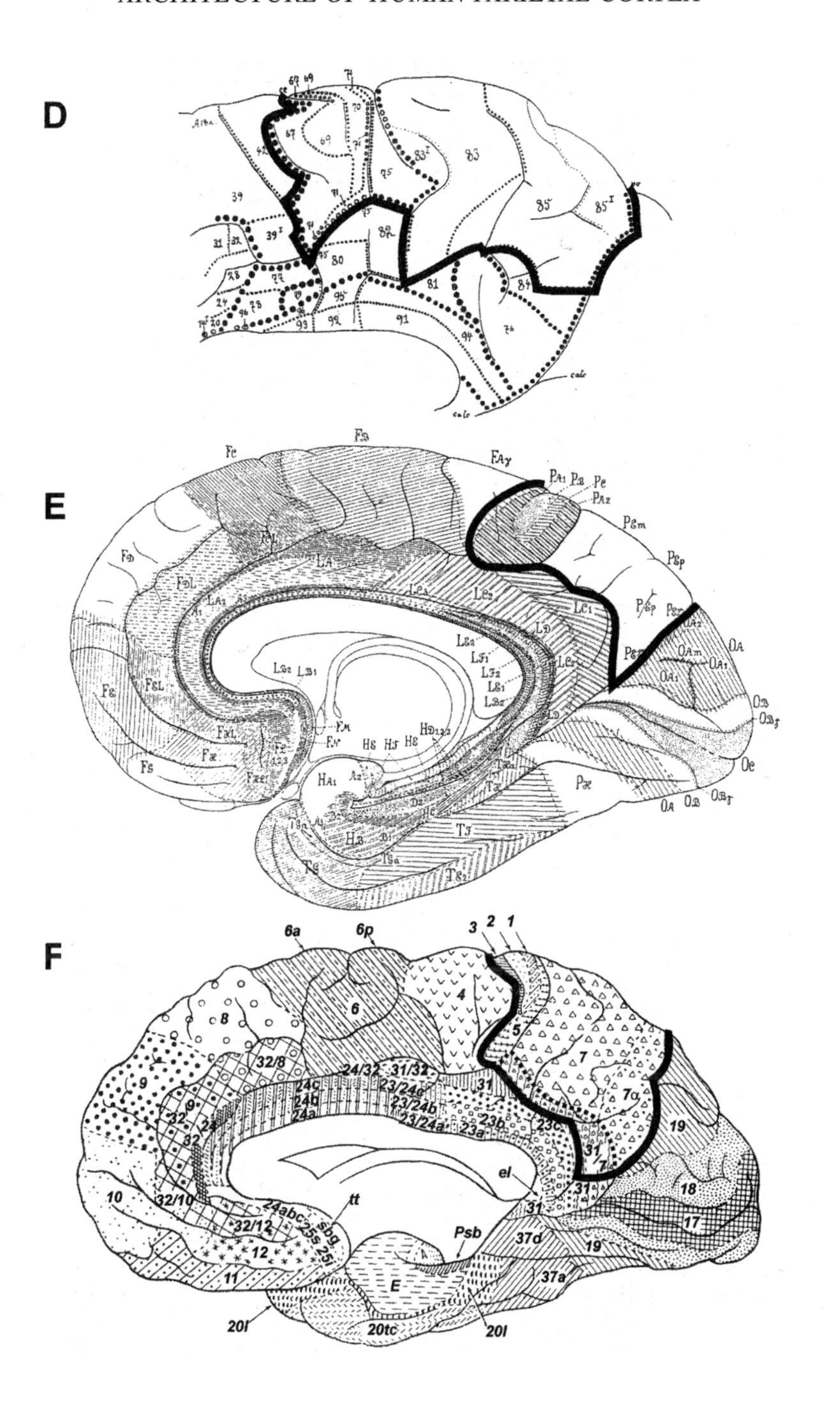

FIG. 1-2. *Continued.*

TABLE 1-1. *Comparison of Parcellation Schemes of the Parietal Cortex*

Campbell (1905)	Elliott Smith (1907)	Brodmann (1909)	Vogt (1911)	Von Economo and Koskinas (1925)	Sarkissov (1955)
Postcentral region					
Postcentral	AR postcentralis A	**3** (3a)[a]	67	PA_1	3
		(3b)	69	PB_1, PB_2	
		1	70	PC	1
	AR postcentralis B	**2**	71	PD	2
Intermediate postcental	Visuosensory band β			PDE	
Transition zone	Area Z	**43**	72	PFD	43
Parietal region					
Parietal	AR parietalis sup. B	**5**	75	PA_2	5
		7a	83 (86)	PE_m	7
	AR parietalis sup. A	**7b**	85	PE_p, PE_y	7α
Parietotemporal region	AR parietalis inf. A	**39**	90	PG	39, 39s, 39p
	AR parietalis inf. B	**40**	89 (86)	PF, PF_m, PF_{cm}	40, 40s, 40i, 40p
	AR parietalis inf. C		88	PF_t	
				PF_{op}	40op

Comparison of Brodmann's parcellation scheme of the parietal cortex, and that of Campbell, Elliott Smith, Vogt, von Economo and Koskinas, and Sakissov et al. The subdivision of the parietal cortex into a postcentral and a parietal region is based on Brodmann's concept. Note that von Economo and Koskinas consider their area PA_2 to be a postcentral region, whereas Brodmann regards his (equivalent) BA 5 as a parietal region. Italics indicate areas that Campbell delineated, but did not name.

[a]Brodmann never coined the terms 3a and 3b; they have been included here for comprehensive reasons. See text for further details.

From: Elliot Smith G. A new topographical survey of the human cerebral cortex, being an account of the distribution of the anatomically distinct cortical areas and their relationship to the cerebral sulci. *J Anat Physiol* 1907:41:237–254; Brodmann K. *Vergleichende Lokalisationslehre der Grosshirnrinde in ihren Prinzipien dargestellt auf Grund des Zellbaues.* Leipzig: Barth, 1909. [English translation: Garey LJ. *Brodmann's "Localisation in the cerebral cortex."* London: Smith-Gordon, 1994]; Campbell AW. *Histological studies on the localisation of cerebral function.* Cambridge, UK: Cambridge University Press, 1905; von Economo C, Koskinas GN. *Die Cytoarchitektonik der Hirnrinde des erwaschsenen Menschen.* Berlin: Springer, 1925; Sarkissov SA, Filimonoff IN, Preobrashenskaya NS. *Cytoarchitecture of the human cortex cerebri.* Moscow: Medgiz, 1949 [in Russian]; Vogt O. Die Myeloarchitektonik des Isocortex parietalis. *J Psychol Neurol* 1911;18:379–396, with permission.

and Batsch (2) (Figs. 1-1 and 1-2), as well as with our most recent cytoarchitectonic results. For a comparison of the different nomenclatural systems presently used, see Table 1-1.

Brodmann (5) subdivided the human parietal cortex into a postcentral and a parietal region. The postcentral region comprises the postcentral gyrus and parts of the paracentral lobule and the operculum Rolandi. It can be further subdivided into four architectonically related areas: BA 3, BA 1, BA 2 (according to their anterior-posterior sequence, located on the mesial and lateral surfaces) and BA 43 (on the operculum Rolandi). The parietal region is formed by the superior and inferior parietal lobules, the most posterior part of the paracentral lobule, and the mesial extent of the superior parietal lobule down to the subparietal sulcus and to the parietooccipital sulcus. According to Brodmann (5), this region comprises four architectonically related areas: BA 5, BA 7, BA 39, and BA 40.

Brodmann's Area 1

BA 1, the intermediate postcentral area, is located between BA 2 and BA 3. It occupies a cortical band along the apex of the postcentral gyrus, and can be distinguished cytoarchitectonically from both neighboring areas (see Brodmann's Area 2).

BA 1 corresponds to von Economo and Koskinas' area PC (8), and to area 70 of Vogt

(11), Gerhardt (9), and Batsch (2). The cytoarchitecture of BA 1 can be classified as a homotypical six-layered isocortex with large pyramidal cells in deeper layer III and a well-developed layer IV. Thus, the cytoarchitecture and location of BA 1 fit into a functional interpretation as a unimodal somatosensory region.

von Economo and Koskinas (8) defined a subarea of their Pc, that is, Pc$_\gamma$ (8), a narrow strip located on the caudal part of the paracentral lobule on the mesial hemispheric surface. Subarea Pc$_\gamma$ differs from Pc in that it contains giant pyramidal cells in layer V. Furthermore, Pc$_\gamma$ differs from the adjacent preparietal area PA$_2$ (see Brodmann's area 5), which also contains some giant pyramidal cells in layer V, in that its layer IV is better developed.

Brodmann's Area 2

BA 2, the caudal postcentral area of Brodmann (5), is a narrow band located mainly on the anterior wall of the postcentral gyrus. Its borders are neither constant in their location nor sharp. It is found only at times on the apex of the postcentral gyrus. More often, it is located completely within the postcentral sulcus, and even crosses the fundus of the postcentral sulcus caudally, thus encroaching on the superior parietal lobule. According to Brodmann (5), BA 2 extends fairly far caudally within the intraparietal sulcus.

It is difficult to delineate BA 2 from BA 1 by pure visual inspection in Nissl-stained sections, because both areas belong to the homotypical six-layered isocortex. Although this border was defined in most of the classical architectonic maps, Elliot Smith (4) and von Economo and Koskinas (8) expressed some doubt concerning its existence. However, we recently demonstrated the reliability of the border between BA 2 and BA 1 using an observer-independent cytoarchitectonic technique and receptor architectonic analysis (21–26) in a sample of human brains.

BA 2 corresponds to von Economo and Koskinas' area PD (8) and to area 71 of Vogt (11), Gerhard (9), and Batsch (2). Elliot Smith (4) defined the part of BA 2 located within the intraparietal sulcus as his "sensory band β," and clearly separated it from his area postcentralis B.

BA 2 and area PD differ considerably regarding their extent. Brodmann (5) described this region as occasionally encroaching the posterior part of the paracentral lobule, whereas von Economo and Koskinas (8) did not describe such a mesial expansion of PD. We were able to confirm von Economo and Koskinas' parcellation (8) by means of a quantitative analysis of cortical areas based on a multivariate statistical analysis of their cytoarchitecture (21).

von Economo and Koskinas (8) defined a further postcentral region, their area PDE (or PED), a region of transition between PD and PE (BA 7). PDE continues for a considerable distance on the upper and lower bank of the intraparietal sulcus. This intraparietal part of PDE can be compared with Brodmann's (5) caudal part of BA 2 in the intraparietal sulcus, and with the visuosensory band B of Elliot Smith (4).

BA 2, as BA 1, can be described as a homotypical six-layered isocortex. It contains large pyramidal cells in deeper layer III and a clearly visible layer IV. Thus, the cytoarchitecture and location of BA 2 would also fit into a functional interpretation as a unimodal somatosensory region. Indeed, it has been demonstrated using functional imaging studies that BA 2 is involved in somatosensory tasks of varying complexity (27–30).

Brodmann's Area 3

BA 3, the rostral postcentral area of Brodmann (5), is located on the posterior bank of the central sulcus, where it borders the primary motor cortex, that is, BA 4. However, it is important to note that Brodmann emphasized the fact that this border does not consistently coincide with the fundus of the central sulcus but can encroach on the anterior or posterior walls of the central sulcus to locally and intersubjectively variable extents. Furthermore, the border between these two areas is not always striking, but a small region displaying cytoarchitectonic characters of areas 3 and 4 can be found at the bottom of the central sulcus, the anterior part of the paracentral lobule, and the posterior part of the operculum Rolandi. Brodmann (5) did not delineate this small intermediate cortical region as an independent cortical unit.

Campbell (7) and Elliot Smith (4) did not consider the above-mentioned intermediate cortical region a separate entity either. Indeed, Campbell (7) only differentiates two fields within the postcentral region, which he terms "postcentral area" and "intermediate postcentral area," whereas Elliot Smith (4) leaves open the question as to whether two or three postcentral regions can be defined. Conversely, von Economo and Koskinas (8) and Vogt (11) considered this transitional region as an independent area, and named it PA_1 (von Economo and Koskinas) or area 67 (Vogt), respectively. This transition zone is now defined as area 3a (for a review see 31).

The much larger, caudal, part of BA 3, which is now known as area 3b, corresponds to area 69 of Vogt (11), Gerhardt (9), and Batsch (2), and to von Economo and Koskinas' (8) areas PB_1 and PB_2. Their subdivision of area 3b into two areas was based on the stronger invasion of layer III of PB_1 by small granular cells (thus obscuring the borders between layers II and IV) compared to the reduced granularization of layer III in PB_2.

Area 3b is characterized by a conspicuous layer IV with small granular neurons invading layer III; therefore, it can be classified as koniocortex (heterotypical six-layered isocortex with an overexpression of layer IV), an architectonic type that is found in all primary sensory areas of the human brain (32). Thus, the cytoarchitecture of area 3b supports the functional interpretation as primary somatosensory area.

Brodmann's Area 5

BA 5, the preparietal area, is cytoarchitectonically clearly distinguishable from neighboring areas owing to a thick inner granular layer and the presence of extraordinarily large pyramidal cells in layer V, which at times attain the size of Betz giant pyramidal cells. According to Brodmann (5), although the size of the pyramidal cells in this region can vary considerably in individual cases, its position is essentially constant. It covers an area extending from the posterior part of the paracentral lobule to the rostral bank of the callosmarginal sulcus, and continues laterally between the superior part of the postcentral sulcus and the anterior border of BA 7.

BA 5 is a controversial brain region since it was classified by Brodmann as part of his parietal region (5), but von Economo and Koskinas (8) identified their comparable PA_2 (containing a conspicuous layer IV and extraordinarily large pyramidal cells in layer V) as being part of the postcentral region. Furthermore, von Economo and Koskinas conceived PA_2 as being a mere extension of their area PA_1 (area 3a) and did not consider the declaration of PA_2 as an individual area necessary, an opinion shared by Campbell (7) and Elliot Smith (4).

Brodmann's Area 7

BA 7, the superior parietal region, occupies most of the lateral superior parietal lobule (except for the part occupied by BA 5) and of the medial precuneus. Brodmann (5,6) described gradual rostrocaudal architectonic changes within area 7; thus, he defined the existence of two subdivisions, which he named 7a and 7b, although he did not define a clear border between them.

BA 7, as BA 5, also is a highly controversial brain region. In monkeys, Brodmann (5) identified area 5 as an area representing the complete superior parietal lobule, and area 7 as covering the complete inferior parietal lobule. In the human brain, both BA 5 and BA 7 are located on the superior parietal lobe, whereas the inferior parietal lobule is occupied by what he termed "human-specific" areas BA 39 and BA 40. According to Brodmann (5), this interspecies difference could be explained by the hypothesis that monkey area 7 represents an undifferentiated precursor zone for all human parietal areas (i.e., BA 7, BA 39, and BA 40), with the exception of BA 5. This explanation implies an extremely fast cortical evolution of BA 39 and BA 40 and a complete architectonic reorganization of the parietal lobe during brain evolution from monkeys to humans. However, there are no available findings that support this hypothesis. Consequently, area 7 of monkeys should not be considered equivalent to area 7 of humans (33).

Since Brodmann (5,6) did not provide a cytoarchitectonic description or a micrograph of BA 7, comparisons with the maps of other authors can be performed only on the basis of topography. von Economo and Koskinas (8) described a practically identical location for their area PE, which is characterized by a sharply delineated band with a conspicuously low cell packing density and corresponds to deeper layer V (layer Vb). The borders of this band are blurred in the adjacent BA 39 and BA 40. PE was subdivided by von Economo and Koskinas (8) into the anterior area PE_m, with a more pronounced magnocellular appearance compared to the posterior relatively smaller-celled area PE_p. The border between these areas is marked approximately by the superior parietal sulcus. PE_m and PE_p are probably equivalent to Brodmann's subdivisions 7a and 7b, respectively. This rostrocaudal subdivision of BA 7 also was reported by Sarkissov et al. (10), although with a differing nomenclature.

A further subdivision was found by von Economo and Koskinas (8) in the most posterior part of PE_p, the gigantopyramidal area PE_γ, characterized by widely spaced, very large, but slim pyramidal cells in layers IIIc and V. PE_γ is located mainly on the anterior wall of the parieto-occipital sulcus.

Brodmann's Area 39

BA 39, the angular area, is located on the inferior parietal lobule, and "... corresponds broadly to the angular gyrus" (5; translated by Garey LJ, 1994, p. 119). Brodmann did not describe the cytoarchitectonic features of BA 39, although he did mention the lack of sharp borders between this area and BA 19 as well as BA 37. He defined the intraparietal sulcus as being the approximate border between BA 39 and BA 7.

von Economo and Koskinas' (8) area PG is their equivalent of BA 39. PG is located posteriorly of the sulcus Jensen, below the intraparietal sulcus, above area PH (BA 37), and rostrally to the occipital cortex. It can be distinguished from BA 7 by its generally smaller cells and a blurred layer Vb, and from the occipital cortex by smaller pyramidal cells in lower layer IIIc, a conspicuously darker and wider layer V, and a wider but cell-poorer layer VI. The cytoarchitecture of PG takes an intermediate position between PE (BA 7) and PF (BA 40), since: (a) layer III of PG is smaller than in PE; (b) layer V of PG is lighter than layer VI, therefore enabling an easier discrimination of both layers than in PF; and (c) its overall cell size is smaller than in PE and larger than in PF. Area PG shows a similar columnar arrangement of its cell bodies as area PF, but the columns extending from layer III to VI are somewhat broader and more widely spaced in PG.

Sarkissov and coworkers (10) further subdivided BA 39. They defined three subregions: area 39, area 39s (superior), and area 39p (posterior). Area 39 constitutes the "core" region, and areas 39s and 39p were defined as transitional subareas. Thus, the cytoarchitecture of area 39s takes an intermediate position between BA 39 and BA 7, whereas area 39p is a combination of the architectonic features of BA 39 and the occipital cortex.

Brodmann's Area 40

BA 40, the supramarginal area of Brodmann (5), corresponds approximately to the supramarginal gyrus. It is separated from the postcentral area by the inferior postcentral and the posterior subcentral sulci, and from BA 39 by the sulcus of Jensen, but has no clear border to BA 22.

von Economo and Koskinas (8) described the absence of sharp borders between PF, their equivalent of BA 40, and adjacent areas. They described the existence of transitional areas, that is, PF_t, PF_{op}, and PF_{cm}, which are highly variable in their cytoarchitecture and size. However, they left this question open: Should they be recognized as separate areas or as interindividually variable expressions of transition fields at the border to surrounding areas? Sarkissov and coworkers (10) also defined several transition zones within their area 40: area 40s (superior supramarginal), area 40p (posterior), area 40i (inferior), and area 40op (opercular). However, these transition zones were merely enumerated, without any further explanation. Vogt (11) defined at least two regions within BA 40: area 88 (located mainly on the anterior branch of the supramarginal gyrus) and area 89 (located more caudally). Additionally, parts of their area 86 could be located below the intraparietal sulcus. We were able to find a similarly located border to that described between areas 88 and 89 by means of a receptor architectonic study (34).

The major part of PF can be distinguished from the surrounding parietal cortex due to: (a) its generally smaller cells in all layers, including layers III and V; (b) layers V and VI with a similar low density of cell bodies; (c) wide layers II and IV; and (d) a conspicuous, fine columnar arrangement of cells. The delineation to the temporal areas is based on the higher cell density in layers V and VI of these areas in comparison to PF.

von Economo and Koskinas' (8) subareas PF_{cm}, PF_t, and PF_{op} contain even smaller cell bodies than the "core" of PF. Area PF_{cm} is characterized by a striking columnar arrangement in layers III and IV and large cells in deeper layer III. Layers III and V of areas PF_t and PF_{op} are relatively cell sparse, and their layer III is very small. von Economo and Koskinas (8) did not define a specific border between these two areas. The border between PF_t and PED can be defined on the basis of a more conspicuous expression of layers IV and VI in the former area. According to von Economo and Koskinas (8), area PF_{op}, extends into the depth of the Sylvian fissure, reaching their insular and retroinsular fields.

However, the delineation of these border areas of PF remains difficult without systematic, observer-independent measurements of the cytoarchitecture of these areas.

Brodmann's Area 43 and the Parietal Operculum

BA 43, the subcentral postcentral region of Brodmann (5)

> . . . is formed by the union of the pre- and postcentral gyri at the inferior end of the central sulcus and thus lies on the Rolandic operculum. From its architecture, this area belongs to the postcentral cortex. Its anterior border is quite sharp and coincides approximately with the anterior subcentral sulcus; posteriorly it disappears gradually around the posterior subcentral sulcus in the retrocentral transition zone and in the anterior portion of the supramarginal area 40. It extends widely over the inner surface of the operculum, that is to say in the depths of the Sylvian fissure; in this region it has a distinct boundary with the insular cortex (5; translated by Garey LJ, 1994, p. 112).

Unfortunately, this is as far as Brodmann's description of area 43 goes. Furthermore, he did not accompany it with a sufficiently detailed cytoarchitectonic description or a figure. Therefore, any comparisons with the maps of other authors can be performed only on the basis of topography.

Brodmann (5) explicitly referred to Elliot Smith's area Z (4) and to the small region that Campbell (7) described on the Rolandic operculum, but considered more a mixed zone than a special area, as being equivalents of his area 43. The most probable candidate for a comparable area in the maps of Vogt (11), Gerhardt (9), and Batsch (2) is their area 72, and in von Economo and Koskinas' map (8), their area PFD, although they doubted whether it could be classified as an independent cortical unit.

We recently were able to show (22,23) that the parietal operculum, contrary to classical parcellation schemes, contains at least four different microstructural areas: OP 1, OP 2,

OP 3, and OP 4 (35). Considering their location, areas OP 3 and OP 4 could be the opercular part of BA 43, whereas areas OP 1 and OP 2 would represent the opercular part of BA 40. However, the border defined by von Economo and Koskinas (8) between their areas PFD (BA 43) and PF_{op} (BA 40) would lead to the grouping of areas OP 1, OP 2, and OP 3 (equivalent to PF_{op}), and would separate them from area OP 4.

We have chosen a neutral nomenclature because of this nomenclatural confusion in the literature and the resulting uncertainty in identifying homolog structures: OP (for operculum) and an ascending numeration in the caudal-rostral direction. Although the region as a whole can be characterized as a fairly well laminated granular cortex, the four areas defined differ clearly in the size and number of pyramidal cells located in lower layer III and upper layer V as well as in the packing density of the cells located in layers IV and VI.

Two of the defined areas are located close to the free surface of the parietal cortex, reaching it on a regular basis (area OP 4 rostrally, area OP 1 caudally), whereas the other two lie in the depth of the Sylvian fissure, bordering the insula (rostrally area OP 3 and caudally area OP 4). The rostral ends of areas OP 4 and OP 3 reach the level of the central sulcus. However, we never observed areas OP 1 and OP 2 to reach the caudal end of the Sylvian fissure, as they were always followed by retroinsular and inferior parietal fields, which stretch on to the upper bank of the Sylvian fissure.

OP 1 (Fig. 1-3) is characterized by the arrangement of cells in wide, pronounced

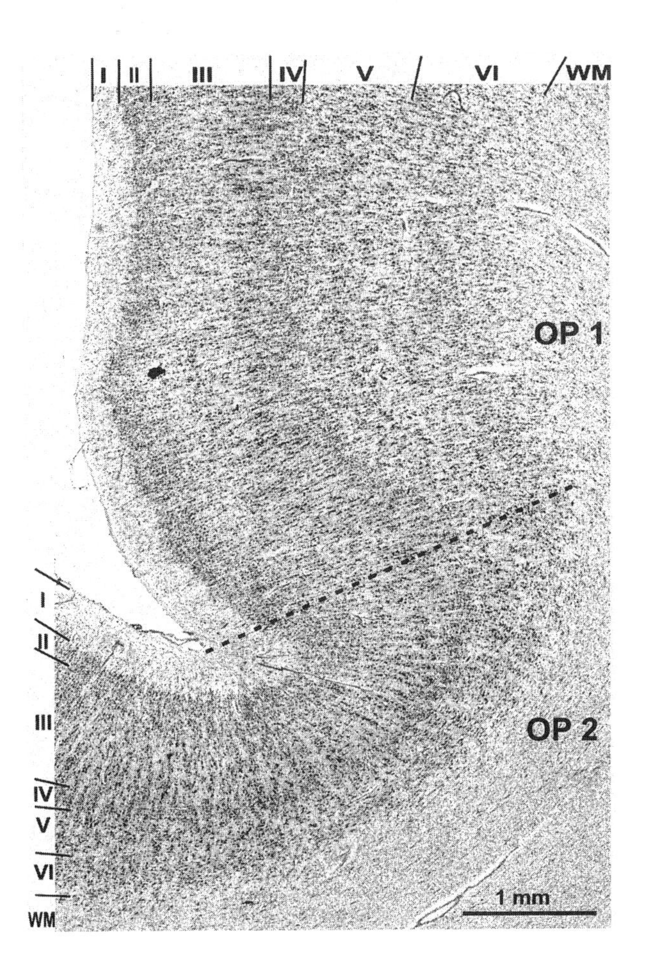

FIG. 1-3. Coronal section through the human parietal operculum, stained for cell bodies. Area OP 1 is clearly separable from OP 2 due to the well developed, densely packed, inner granular layer of the latter area and the large pyramidal cells in layer IIIc of OP 1. Cortical layers are indicated by roman numerals. WM, white matter.

columns, thick infragranular layers equaling the supragranular ones in width, and a very blurred border between lamina VI and the underlying white matter. Other main features are the larger pyramidal cells found in sublaminar IIIc and large granular cells in layer VI.

The thin cortex of OP 2 (Fig. 1-3) shows only an inconspicuous columnar arrangement, small infragranular layers, and a distinct white matter border. The inner granular layer is well developed and contains a high density of granular cells, whereas prominent pyramidal cells are rarely found.

OP 3 is characterized by a thin, barely columnated cortex, with low overall cell density, and a lack of prominent pyramidal cells. The infragranular layers are particularly poorly developed and cell sparse. Although this area belongs to the granular cortex, its lamina IV contains very loosely packed granular cells, almost resembling dysgranular areas.

OP 4 shows a striking columnar arrangement, with fine cell columns reaching up to upper layer III, wide infragranular layers, and a poor demarcation of the white matter border. Other distinct characteristics are the many medium-sized pyramidal cells found in lower layer III and upper layer V.

It remains to be determined whether this cytoarchitectonic heterogeneity reflects somatotopy or areas of different modalities (e.g., secondary somatosensory, gustatory, or vestibular areas), and whether further opercular areas rostrally and caudally of OP 1 to OP 4 can be identified. In order to do so, probability maps of the defined areas will be transformed into a standardized stereotaxic space, thus enabling direct comparison with activations derived from future functional imaging studies.

The Intraparietal Sulcus

Although the intraparietal sulcus has been intensely studied in the macaque brain (see also the following for comparisons), neither Brodmann (5,6) nor von Economo and Koskinas (8) provide a distinct cytoarchitectonic parcellation of the cortex located within the human intraparietal sulcus which would enable a direct correlation with the detailed areal pattern revealed by functional imaging studies (e.g., AIP, VIP, LIP, MIP, and PIP) (12–19).

This complex sulcus consists of numerous inner branches and contains many areas of similar cytoarchitecture. Furthermore, within a given region, cytoarchitectonic features are relatively variable between subjects, making the determination of areal borders particularly difficult. To date, we have characterized two areas located on the anterior ventral bank of the intraparietal sulcus (36) in a sample of 10 human brains by means of a quantitative analysis of cortical areas (and their borders) and multivariate statistical analysis of their cytoarchitecture (22,23). As already mentioned (see Brodmann's Area 43), we decided to create a neutral nomenclature when defining areas lacking a suitable historical denomination; therefore, we have named these two regions ip1 and ip2 (for intraparietal). Area ip1 is located in the depth of the sulcus, whereas the neighboring area ip2 is found on the lower bank of the intraparietal sulcus and extends toward the free surface of the inferior parietal lobule, although it generally does not reach it. Area ip2 starts more rostrally than ip1; that is, at the intersection of the intraparietal sulcus with the postcentral sulcus.

Both ip1 and ip2 differ from neighboring areas owing to their prominent layer III (which contains pyramidal cells of varying sizes) and their higher density of pyramidal cells in layer V. In spite of their cytoarchitectonic similarity, ip1 and ip2 can be distinguished owing to the fact that: (a) lower layer III of area ip2 has a higher density of larger pyramidal cells than ip1; (b) layer V of area ip2 is broader, and therefore layer III appears to be narrower than in ip1; and (c) layer V in area ip2 has a higher density of pyramidal cells and displays a better visible subdivision in Va and Vb, since Vb is brighter than Va.

The stereotaxic location of areas ip1 and ip2, and their relation to each other and to the neighboring parietal areas suggest that they are candidates for the functionally defined areas VIP (12) and AIP (37), respectively.

General Aspects of Cytoarchitectonic Maps of the Parietal Cortex

A comparison of the described cortical maps (1–11) shows a similarity of certain areal patterns located within the postcentral area, with practically identical architectonic descriptions and comparable gross relations of the areas to macroscopic landmarks. However, differences are found when the sizes and borders of equivalent areas are compared between the studies. This type of differences is not surprising, because the size of a given cortical area generally shows a considerable degree of intersubject variability (21,24–26,38–41). Most important, striking differences become evident when comparing the parcellation schemes of the posterior part of parietal cortex. Therefore, the architecture of the posterior parietal cortex of the human brain is a relatively unexplored topic of brain mapping. The classical cytoarchitectonic and myeloarchitectonic studies can only be considered guidelines for future multimodal and observer-independent quantitative architectonic analyses (22,23).

Finally, microstructural brain mapping requires a far more refined presentation of the data (e.g., three-dimensional [3D] representations in computerized brain atlases, flat maps, etc.) than just a simple two-dimensional schematic drawing. Therefore, we have established a method by which microscopic data can be integrated into a 3D representation of a reference brain located in stereotaxic space, in order to create 3D probabilistic maps of cortical areas (for a comprehensive review, see refs. 42 and 43), thus enabling a combined analysis of architectonic maps and functional imaging data (28–30,40,41,43–52).

RECEPTOR ARCHITECTONIC MAPPING

Receptors for classical neurotransmitters are heterogeneously distributed throughout the parietal cortex, showing clear regional differences both in their mean densities and laminar distribution patterns. Quantitative *in vitro* receptor autoradiography has become a powerful tool in the field of brain mapping (23,53) since it enables visualization of the architectonic organization of the cerebral cortex and computation of mean receptor densities not only at a regional, but also at a laminar level (24,33,40,54–57).

The complex regional codistribution patterns of receptors for classical neurotransmitters in the human brain induced the introduction of so-called "receptor fingerprints" (33). They are the representation of the mean densities of several different receptors over all cortical layers within a single brain region in the form of a polar coordinate plot. They demonstrate the balance between different receptor types and transmitter systems and may differ in shape and/or size between different cortical areas. Thus, the receptor fingerprints reflect multimodal aspects of the organization of cortical areas. The largest difference is found between the primary motor and somatosensory cortices, an intermediate difference is found between the primary somatosensory and superior parietal association cortex, and the smallest difference results from the comparison between the superior and inferior parietal association cortices (33). Thus, the degree of similarity in shape and/or size between different fingerprints seems to reflect the degree of similarity both in architecture and function.

Within the postcentral region, the regional and laminar distribution patterns of the muscarinic cholinergic M_2 receptors delineate the known cytoarchitectonic subdivisions of the somatosensory cortex into BA 1, BA 2, and BA 3 (Fig. 1-4). Furthermore, they also reveal the subdivision of BA 3 into two areas (Figs. 1-3 and 1-4). Area 3b is characterized by an exceptionally high density of M_2 receptors in all layers, clearly delineating it from the adjacent BA 1 and area 3a (Figs. 1-1 through 1-5). Conspicuously high M_2 receptor densities are not only found in area 3b, the primary somatosensory cortex, but also in the human primary auditory (Fig. 1-4) (33) and the human and monkey visual cortices (58). The exceptional high density of the muscarinic M_2 receptor generally characterizes primary sen-

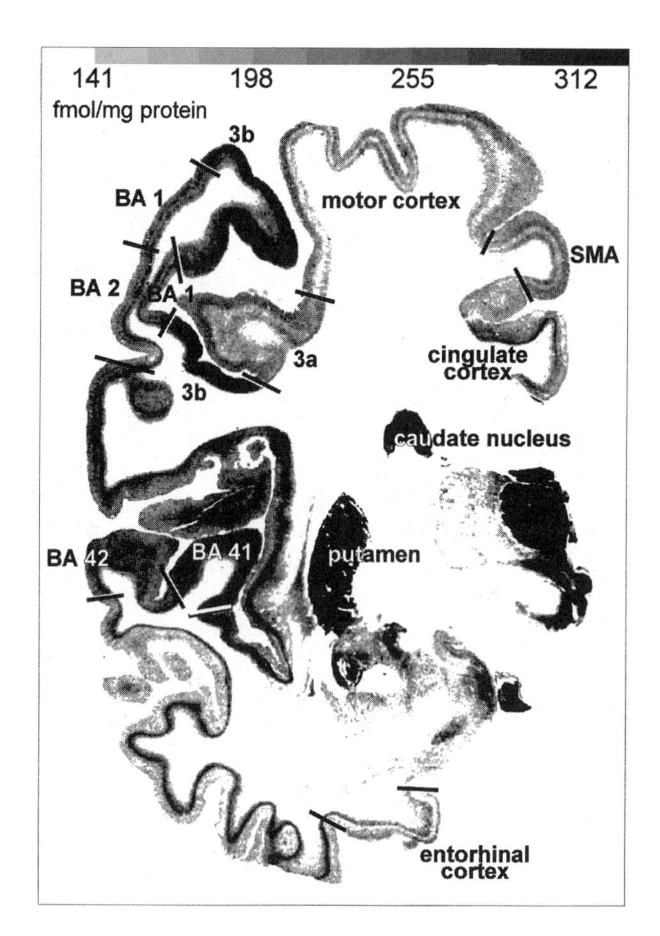

FIG. 1-4. Coronal section through a whole human hemisphere, in which the cholinergic muscarinic M_2 receptors were visualized by means of $[^3H]$oxotremorine-M. The scale indicates the gray intensity coding of binding site densities in fmol/mg protein. The primary somatosensory (area 3b) and primary auditory (BA 41) cortices are clearly distinguishable from surrounding areas due to their conspicuously high M_2 receptor densities. BA, Brodmann area; BA 1, BA 2, area 3a, nonprimary unimodal sensory areas; BA 42, secondary auditory cortex; SMA, supplementary motor cortex.

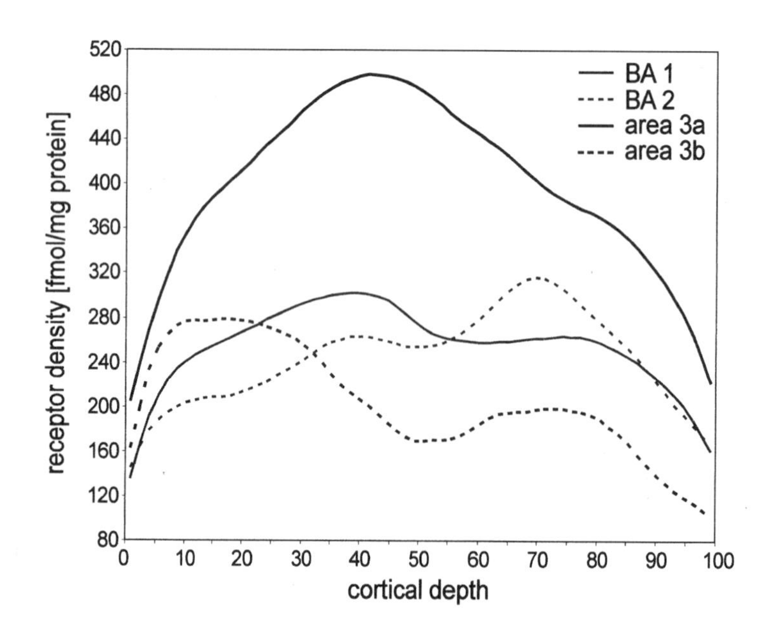

FIG. 1-5. Laminar distribution pattern of M_2 receptors (visualized by means of $[^3H]$oxotremorine-M) in BA 1, BA 2, area 3a, and area 3b. The normalized cortical depth extending from the pial surface (0%) to the border between layer VI and the white matter (100%) is represented on the x-axis, and the densities of M_2 receptors (in fmol/mg protein) are represented on the y-axis.

sory areas of the cortex not only in humans but also in nonhuman primates (33,58). BA 1 contains a bilaminar distribution of intermediate M_2 receptor densities, whereas BA 2 shows an overall low receptor density, with intermediate values only in the deepest cortical layer (Figs. 1-4 and 1-5). BA 2 contains the lowest M_2 receptor densities within the somatosensory cortex, further supporting the idea that BA 2, as BA 1, should be considered a nonprimary somatosensory area (27–30,55).

The distribution patterns of [^{3}H]kainate binding sites support Sarkissov's (10) idea of a heterogeneous BA 39. The distribution pattern of [^{3}H]kainate binding sites enabled the definition of three receptor-architectonically different areas, which we named (in their dorsal-ventral order of appearance) 90d, 90i, and 90v (33). They differ from each other mainly by changes of receptor densities in the central cortical layers, with area 90i showing the relatively lowest density. The exact relationship between our areas 90d, 90i, and 90v and areas 39, 39s, and 39p of Sarkissov et al. (10) remains to be determined.

The distribution patterns of AMPA, kainate, NMDA, $GABA_A$, α_1, M_2, and 5-HT_{1A} receptors enable the subdivision of superficial BA 40 into at least two receptor architectonic areas (34), which correspond approximately to Vogt's areas 88 and 89 (11). Conversely, the M_1 and 5-HT_2 receptors did not reveal further areal subdivisions within the supramarginal gyrus, thus confirming the existence of an area 40 as defined by Brodmann (5).

Following the trend set by our most recent cytoarchitectonic studies of the anterior intraparietal sulcus, the distribution patterns of receptors for classical neurotransmitters also disclose a much finer parcellation scheme of the posterior intraparietal sulcus than that described in any of the classical brain maps. The laminar and regional densities of the glutamatergic AMPA and kainate receptors (Fig. 1-6) enable

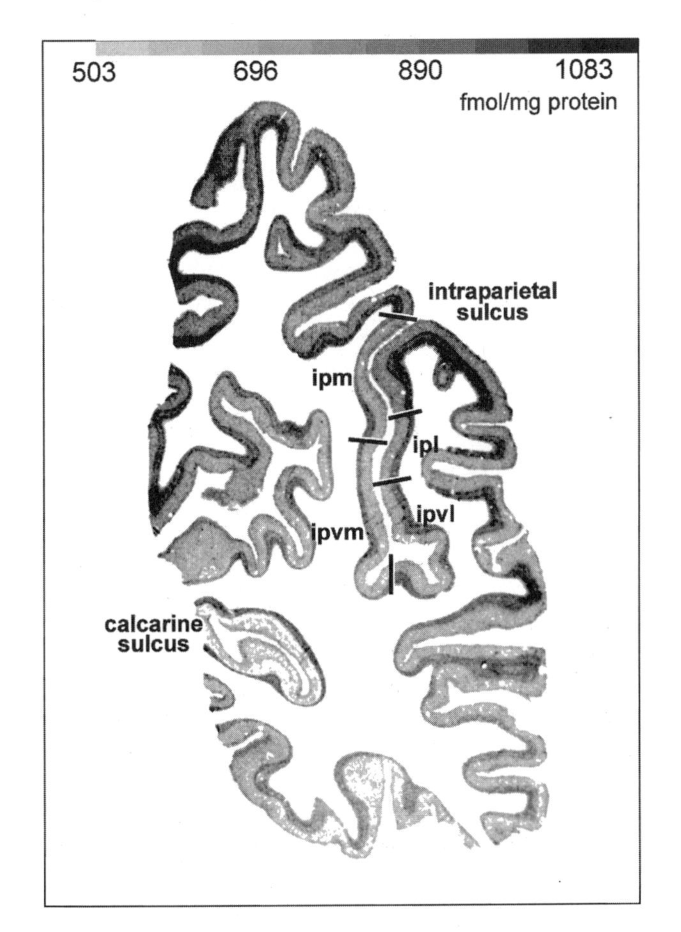

FIG. 1-6. Coronal section through a whole human hemisphere, showing the posterior intraparietal sulcus. The glutamatergic kainate receptors were visualized by means of [^{3}H]kainate, and the scale indicates the gray intensity coding of binding site densities in fmol/mg protein, ipm, ipvm, ipvl, and ipl: medial, ventro-medial, ventro-lateral, and lateral intraparietal areas, respectively. The comparability of these receptor architectonically identified areas with functionally identified areas in the intraparietal sulcus is presently under analysis.

detection of hitherto unidentified areas in the posterior intraparietal sulcus; that is, area ipm, ipv, and ipl (for intraparietal sulcus; medial, ventral, and lateral parts) (33). Area ipv may be further subdivided into an area ipvm, located on the superior (medial) wall, and an area ipvl, located on the inferior (lateral) wall of the intraparietal sulcus, by the higher receptor density in layer VI of ipvl. The functional meaning of these parcellations remains to be elucidated.

CONNECTIVITY OF THE PARIETAL CORTEX AND THE HUMAN–MONKEY COMPARISON

The precise structural connectivity of the human parietal cortex is largely unknown, since axonal tracing techniques cannot be applied in the human brain for obvious reasons. The recently introduced *in vivo* fiber tracking based on magnetic resonance diffusion tensor imaging is still not powerful enough to demonstrate the anatomic connectivity at the required level of spatial resolution. Thus, our knowledge about the connectivity of the human parietal lobe is based mainly on axonal tracing studies in the macaque brain.

This procedure, however, requires the comparability of the architectonic organization between monkey and human parietal cortices, in order to identify equivalent (homolog) sites in the brains of both species. This condition seemed not to be fulfilled until recently, since the classical cytoarchitectonic maps of Brodmann (5,6) display a superior parietal lobule represented by area 5 (and only in its most posterior region a very small portion of area 7), and an inferior parietal lobule represented by area 7 in the monkey brain, whereas the same author found both areas 5 and 7 in the superior lobule of the human brain, and described the "human-specific" areas 39 and 40 in the inferior parietal lobule. Moreover, various recently identified areas in the intraparietal sulcus of the monkey brain (see the preceding and Fig. 1-7) are not depicted in Brodmann's map of the human cortex.

Thus, the intraparietal sulcus turned out to be the crucial landmark for any valid comparison between the architectonic and connectivity structures of the monkey and human parietal cortices. As described in the preceding in more detail, the equivalents of the monkey areas AIP and VIP have recently been identified in the human intraparietal sulcus using functional imaging (12,37) and cytoarchitectonic (33,35,36) techniques. These results are convincing arguments for the hypothesis that the structural organization in the superior and inferior parietal lobules of humans and monkeys are comparable, although we have to keep in mind the functional specializations in the inferior parietal lobule of the human brain, particularly those associated with language and reading abilities (59–63).

The parietal cortex of the macaque monkey, which receives sensory input from the visual and somatosensory cortices (64–66), projects to the posterior motor areas F1 to F5 (Figs. 1-7 and 1-8), whereas the more rostral motor areas F6 and F7 receive major input from the prefrontal cortex (67,68). Since areas F1 to F5 give rise to the corticospinal tract, one of the major functions of the parieto-frontal cortical system is sensory-motor transformations necessary for motor actions.

Both the superior and inferior parietal lobules of the monkey parietal cortex receive input from the somatosensory and visual cortices. The intraparietal sulcus contains multimodal areas integrating both modalities, and additionally the input from the auditory cortex (12,13). *The circuit between area PE,* which receives somatosensory but not visual input, and the *primary motor cortex* (BA 4 or F1) provides information on the location of body parts necessary for the control of limb movements (67,68).

The *circuit between VIP,* which receives a major input from the areas of the dorsal visual stream and from the areas PEc and PFG, *and F4,* which contains representations of movement of the arm, neck, face and mouth, is active during encoding of peripersonal space and transforming object locations into movements toward them (67,68).

The *AIP-F5b circuit* is crucial for the capacity to transform 3D properties of the ob-

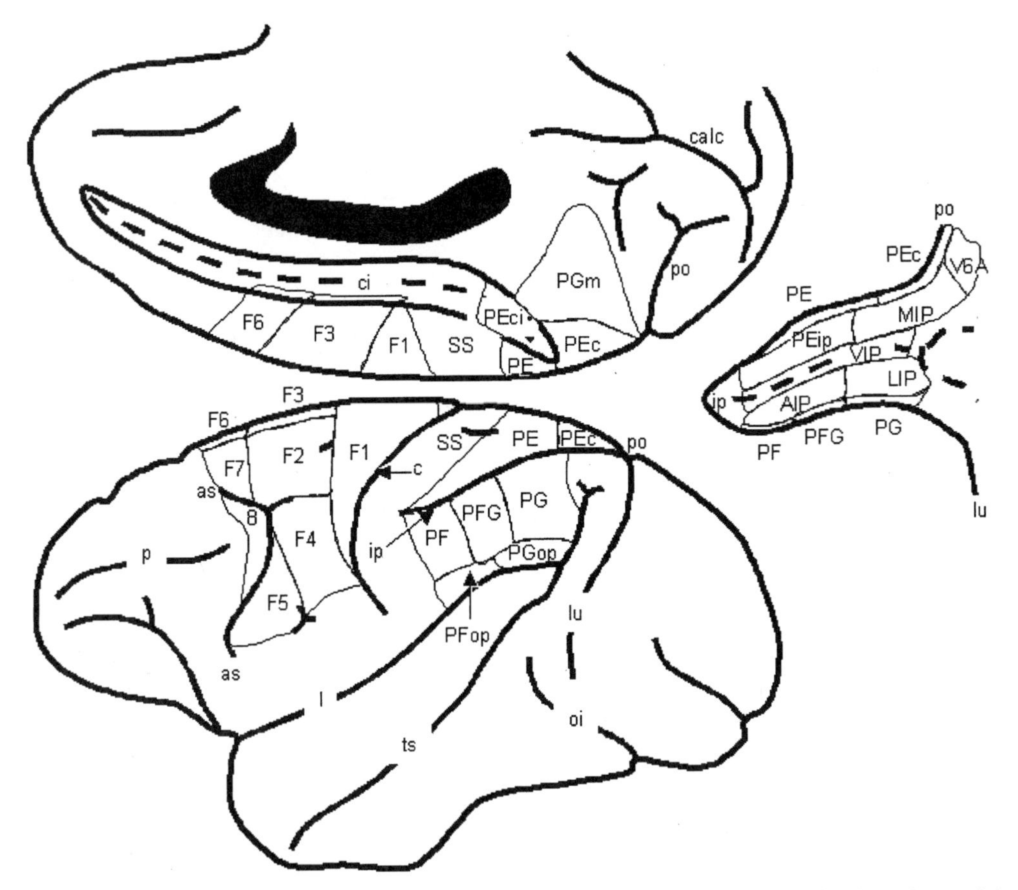

FIG. 1-7. Map of the cytoarchitectonic areas in the motor (F1 to F7) and parietal cortices of the macaque monkey (modified after refs. 67 and 68). The motor area F1 is the equivalent of Brodmann's area 4, the areas F2 to F7 are equivalent to Brodmann's area 6. F3 is the supplementary motor area (SMA) of functional studies, and F6 is the presupplementary motor area (pre-SMA). The parietal areas PE and PE$_{ci}$ are found in the position of Brodmann's area 5, and the areas PE$_{c}$, PF, PFG, PF$_{op}$, PG, PG$_{m}$, are PG$_{op}$ are subdivisions of area 7 in Brodmann's map of the monkey cortex (5). AIP, anterior intraparietal area; as, arcuate sulcus; c, central sulcus; calc, calcarine sulcus; ci, cingulated sulcus; ip, intraparietal sulcus; l, lateral fissure; LIP, lateral intraparietal area; lu, lunate sulcus; MIP, medial intraparietal area; oi, inferior occipital sulcus; p, principal sulcus; po, parieto-occipital sulcus; SS, somatosensory cortex; ts, superior temporal sulcus; VIP, ventral intraparietal area; V6A, visual area V6A; 8, Brodmann area 8.

ject into hand movements (67,68). AIP is active during grasping of specific objects independent of the object position in space, whereas F5b performs goal-directed movements of the hand and/or mouth.

The *PF-F5c circuit* with the latter area containing the mirror neurons (69) is involved in an execution/observation matching system when the monkey observes grasping, placing, and manipulating of objects (67,68).

Area F2d (Fig. 1-8) receives input from PEc and PEip (Fig. 1-7), which contain neurons involved in the somatosensory control of movements, whereas area F2vr (Fig. 1-7) is connected with V6A and MIP, which contains neurons responding to visual (V6A) or both visual and somatosensory stimuli. Thus, the *PEc/PEip-F2d circuit* seems to be active in planning and controlling arm and leg movements, whereas the *MIP/V6A-F2vr circuit* is

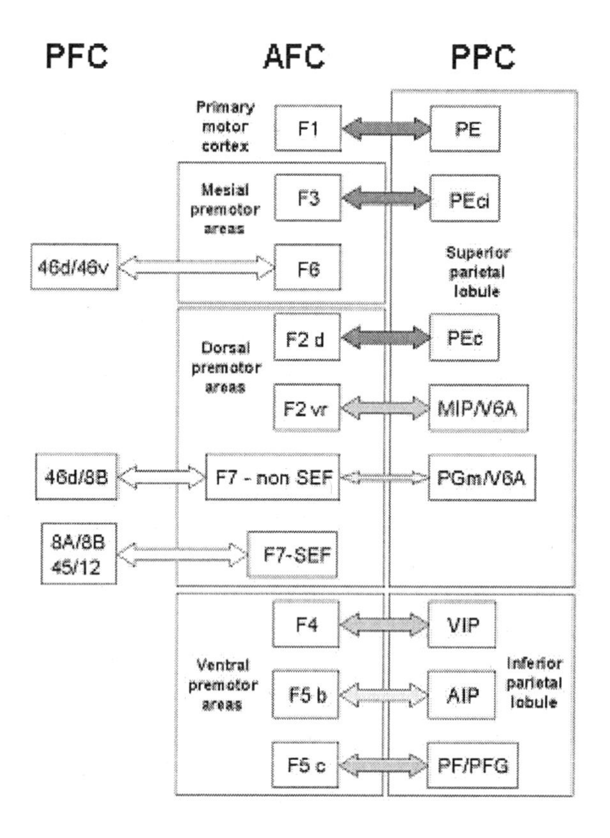

FIG. 1-8. Connectivity between parietal and frontal motor areas in the macaque brain (modified after an unpublished scheme courtesy of M. Matelli, Parma, Italy). Arrows indicate the connections. *Dark gray arrows* correspond to somatosensory, *light gray arrows* to visual, and *medium gray arrows* to somatosensory and visual connections. *White arrows* indicate connections between the prefrontal and rostral motor areas. AFC, agranular frontal cortex; F2d, F2 area around the frontal dimple; F2vr, ventrorostral part of F2; F5b, F5 area of the arcuate bank; F5c, F5 area of the cortical convexity; PFC, prefrontal cortex; PPC, posterior parietal cortex; SEF, supplementary eye field; 8A/8B, subdivisions of Brodmann's area 8; 45/12, Brodmann's areas 45 and 12; 46d/46v, dorsal and ventral parts of Brodmann's area 46. For all the other abbreviations see legend of Fig. 1-7.

monitoring and controlling arm position during the transport phase. The latter circuit seems to represent the neural system producing optic ataxia symptoms after impairment (67,68).

The *PGm-F7 circuit* plays an important role in conditional movement selection (67,68). Lesions of the human dorsal premotor cortex lead to similar symptoms (70,71) as reported after lesions in the monkey F7 area (72,73); that is, arbitrary goal-directed movement in response to visual stimuli cannot be executed although no obvious motor deficit is present.

The supplementary motor area *F3,* which contains a complete somatotopical representation, receives its major parietal input from *PEci.* F3 responds predominantly to somatosensory, but also to visual stimuli. The F3 neurons discharge during active movement in a time-locked manner. In contrast to the primary motor cortex F1, where stimulation leads to movements of a single joint, F3 stimulation evokes movement of two or more joints. Thus, the *PEci-F3 circuit* is involved in motor functions by controlling motor activity in a global way (68).

These monkey data together with the comparability of key landmarks in the human and monkey parietal (see the preceding) and frontal (68) cortices suggest the existence of similar parieto-frontal networks in the human brain. Recent functional imaging studies also support the concept of functionally and somatotopically specific parieto-frontal circuits (74–76) in the human brain that are organized in a way comparable to those in the monkey brain. Also, the analysis of the different roles of parietal and frontal areas in one circuit now can be studied with functional imaging techniques (77).

ACKNOWLEDGMENTS

We thank Hi-Jae Choi, Christian Grefkes, and Filip Scheperjans for productive discussions and support. This work was supported by grants of the DFG (SFB 194/A6; Klinische Forschergruppe), the Volkswagen-Stiftung, and the Human Brain Project/Neuroinformatics Research (funded by the National Institute of Mental Health, National Institute of Neurological Disorders and Stroke, National Institute on Drug Abuse, and the National Cancer Institute).

REFERENCES

1. Bailey P, von Bonin G. *The isocortex of man.* Urbana, IL: University of Illinois Press, 1951.
2. Batsch E-G. Die myeloarchitektonische Untergliederung des Isocortex parietalis beim Menschen. *J Brain Res* 1956;2:225–258.
3. Eidelberg D, Galaburda AM. Inferior parietal lobule. Divergent architectonic asymmetries in the human brain. *Arch Neurol* 1984;41:843–852.
4. Elliot Smith G. A new topographical survey of the human cerebral cortex, being an account of the distribution of the anatomically distinct cortical areas and their relationship to the cerebral sulci. *J Anat Physiol* 1907;41:237–254.
5. Brodmann K. *Vergleichende Lokalisationslehre der Grosshirnrinde in ihren Prinzipien dargestellt auf Grund des Zellbaues.* Leipzig: Barth, 1909. [English translation: Garey LJ. *Brodmann's "Localisation in the cerebral cortex."* London: Smith-Gordon, 1994].
6. Brodmann K. Physiologie des Gehirns. *Neue Dtsch Chir* 1914;11:85–426.
7. Campbell AW. *Histological studies on the localisation of cerebral function.* Cambridge, UK: Cambridge University Press, 1905.
8. von Economo C, Koskinas GN. *Die Cytoarchitektonik der Hirnrinde des erwachsenen Menschen.* Berlin: Springer, 1925.
9. Gerhardt E. Die Cytoarchitektonik des Isocortex parietalis beim Menschen. *J Psychol Neurol* 1940;49: 367–419.
10. Sarkissov SA, Filimonoff IN, Preobrashenskaya NS. *Cytoarchitecture of the human cortex cerebri.* Moskow: Medgiz, 1949 [in Russian].
11. Vogt O. Die Myeloarchitektonik des Isocortex parietalis. *J Psychol Neurol* 1911;18:379–396.
12. Bremmer F, Schlack A, Shah NJ, et al. Polymodal motion processing in posterior parietal and premotor cortex: a human fMRI study strongly implies equivalencies between humans and monkeys. *Neuron* 2001;29: 287–296.
13. Bremmer F, Schlack A, Duhamel JR, et al. Space coding in primate posterior parietal cortex. *NeuroImage* 2001;14:S46–S51.
14. Dong Y, Fukuyama H, Honda M, et al. Essential role of the right superior parietal cortex in Japanese kana mirror reading: an fMRI study. *Brain* 2000;123:790–799.
15. Fink GR, Marshall JC, Weiss PH, et al. "Where" depends on "what": a differential functional anatomy for position discrimination in one- versus two-dimensions. *Neuropsychologia* 2000;38:1741–1748.
16. Fink GR, Marshall JC, Shah NJ, et al. Line bisection judgments implicate right parietal cortex and cerebellum as assessed by fMRI. *Neurology* 2000;54: 1324–1331.
17. Fink GR, Marshall JC, Gurd J, et al. Deriving numerosity and shape from identical visual displays. *NeuroImage* 2001;13:46–55.
18. Fink GR, Marshall JC, Weiss PH, et al. The neuronal basis of vertical and horizontal line bisection judgments: an fMRI study of normal volunteers. *NeuroImage* 2001;14:S59–S67.
19. Weiss PH, Marshall JC, Wunderlich G, et al. Neural consequences of acting in near versus far space: a physiological basis for clinical dissociations. *Brain* 2000;123:2531–2541.
20. Talairach J, Tournoux P. *Co-planar stereotaxic atlas of the human brain.* Stuttgart: Thieme, 1988.
21. Grefkes C, Geyer S, Schormann T, et al. Human somatosensory area 2: observer-independent cytoarchitectonic mapping, interindividual variability, and population map. *NeuroImage* 2001;14:617–631.
22. Schleicher A, Amunts K, Geyer S, et al. Observer-independent method for microstructural parcellation of cerebral cortex: a quantitative approach to cytoarchitectonics. *NeuroImage* 1999;9:165–177.
23. Schleicher A, Amunts K, Geyer S, et al. A stereological approach to human cortical architecture: identification and delineation of cortical areas. *J Chem Neuroanat* 2000;20:31–47.
24. Geyer S, Schleicher A, Zilles K. The somatosensory cortex of human: cytoarchitecture and regional distributions of receptor-binding sites. *NeuroImage* 1997;6:27–45.
25. Geyer S, Schleicher A, Zilles K. Areas 3a, 3b, and 1 of human primary somatosensory cortex. 1. Microstructural organization and interindividual variability. *NeuroImage* 1999;10:63–83.
26. Geyer S, Schormann T, Mohlberg H., et al. Areas 3a, 3b, and 1 of human primary somatosensory cortex. 2. Spatial normalization to standard anatomical space. *NeuroImage* 2000;11:684–696.
27. Bodegård A, Geyer S, Grefkes C, et al. Hierarchical processing of tactile shape in the human brain. *Neuron* 2001;31:317–328.
28. Bodegård A, Geyer S, Naito E, et all. Somatosensory areas in man activated by moving stimuli: cytoarchitectonic mapping and PET. *NeuroReport* 2000;11: 187–191.
29. Bodegård A, Ledberg A, Geyer S, et al. Object shape differences reflected by somatosensory cortical activation. *J Neurosci* 2000;20:RC51.
30. Naito E, Ehrsson HH, Geyer S, et al. Illusory arm movements activate cortical motor areas: a positron emission tomography study. *J Neurosci* 1999;19:6134–6144.
31. Jones EG, Porter R. What is area 3a? *Brain Res Rev* 1980;2:1–43.
32. Zilles K. Cortex. In: Paxinos G, ed. *The human nervous system.* San Diego: Academic Press, 1990:757–802.
33. Zilles K, Palomero-Gallagher N. Cyto-, myelo-, and receptor architectonics of the human parietal cortex. *NeuroImage* 2001;14:8–20.
34. Palomero-Gallagher N, Amunts K, Mazziotta J, et al. Brodmann's Area 40 as revealed by quantitative receptor autoradiography. *NeuroImage* 2001;13:S583.
35. Eickhoff S, Geyer S, Amunts K, et al. Cytoarchitectonic

analysis and stereotaxic map of the human secondary somatosensory cortex region. Eighth International Conference on Functional Mapping of the Human Brain, June 2–6, 2002, Sendai, Japan. Available on CD-Rom in *NeuroImage* 2002;16/2:792. *http://www.academicpress.com/journals/hbm2002/13871.html*

36. Choi H-J, Amunts K, Mohlberg H, et al. Cytoarchitectonic mapping of the anterior ventral bank of the intraparietal sulcus in humans. Eighth International Conference on Functional Mapping of the Human Brain, June 2–6, 2002, Sendai, Japan. Available on CD-Rom in *NeuroImage* 2002;16/2:591. *http://www.academicpress.com/journals/hbm2002/14097.html*
37. Grefkes C, Weiss PH, Zilles K, et al. Crossmodal processing of object features in human anterior intraparietal cortex: an fMRI study strongly implies equivalencies between humans and monkey. *Neuron* 2002;35: 173–184.
38. Amunts K, Schleicher A, Bürgel U, et al. Broca's region revisited: cytoarchitecture and intersubject variability. *J Comp Neurol* 1999;412:319–341.
39. Amunts K, Malikovic A, Mohlberg H, et al. Brodmann's areas 17 and 18 brought into stereotaxic space-where and how variable? *NeuroImage* 2000;11:66–84.
40. Geyer S, Ledberg A, Schleicher A, et al. Two different areas within the primary motor cortex of man. *Nature* 1996;382:805–807.
41. Roland PE, Geyer S, Amunts K, et al. Cytoarchitectural maps of the human brain in standard anatomic space. *Hum Brain Map* 1997;5:222–227.
42. Amunts K, Zilles K. Advances in cytoarchitectonic mapping of the human cerebral cortex. In: Naidich TP, Yousry TA, Mathews VP, eds. *Neuroimag Clin North Am* 2001;11:151–169.
43. Roland PE, Zilles K. Structural divisions and functional fields in the human cerebral cortex. *Brain Res Rev* 1998;26:87–105.
44. Naito E, Kinomura S, Geyer S, et al. Fast reaction to different sensory modalities activates common fields in the motor areas, but the anterior cingulate cortex is involved in the speed of reaction. *J Neurophysiol* 2000;83:1701–1709.
45. Mazziotta J, Toga A, Evans A, et al. A four-dimensional probabilistic atlas of the human brain. *J Am Med Inform Assoc* 2001;8:401–430.
46. Mazziotta J, Toga A, Evans A, et al. A probabilistic atlas and reference system for the human brain: International Consortium for Brain Mapping (ICBM). *Phil Trans R Soc Lond B Biol Sci* 2001;356:1293–1322.
47. Roland PE, Zilles K. Brain atlases: a new research tool. *Trends Neurosci* 1994;17:458–467.
48. Roland PE, Zilles K. The developing European computerized human brain database for all imaging modalities. *NeuroImage* 1996;4:S39–S47.
49. Zilles K, Schlaug G, Matelli M, et al. Mapping of human and macaque sensorimotor areas by integrating architectonic, transmitter receptor, MRI and PET data. *J Anat* 1995;187:515–537.
50. Zilles K, Schleicher A, Langemann C, et al. Quantitative analysis of sulci in the human cerebral cortex: development, regional heterogeneity, gender difference, asymmetry, intersubject variability and cortical architecture. *Hum Brain Map* 1997;5:218–221.
51. Ehrsson HH, Naito E, Geyer S, et al. Simultaneous movements of upper and lower limbs are coordinated by motor representations that are shared by both limbs: a PET study. *Eur J Neurosci* 2000;12:3385–3398.
52. Larsson J, Amunts K, Gulyás B, et al. Neuronal correlates of real and illusory contour perception: functional anatomy with PET. *Eur J Neurosci* 1999;11:4024–4036.
53. Zilles K, Schleicher A. Correlative imaging of transmitter receptor distributions in human cortex. In: Stumpf WE, Solomon HF, eds. *Autoradiography and correlative imaging.* San Diego: Academic Press, 1995:277–307.
54. Geyer S, Matelli M, Lupino G, et al. Receptor autoradiographic mapping of the mesial motor and premotor cortex of the macaque monkey. *J Comp Neurol* 1998;397:231–250.
55. Grefkes C, Scheperjans F, Palomero-Gallagher N, et al. Which areas represent the human primary somatosensory cortex? New insights from receptor architectonics. Eighth International Conference on Functional Mapping of the Human Brain, June 2–6, 2002, Sendai, Japan. Available on CD-Rom in *NeuroImage* 2002;16/2:783. *http://www.academicpress.com/journals/hbm2002/13870.html*
56. Scheperjans F, Grefkes C, Palomero-Gallagher N, et al. Human area 5 in the hierarchy of sensorimotor processing: conclusions from quantitative receptor autoradiography. Eighth International Conference on Functional Mapping of the Human Brain, June 2–6, 2002, Sendai, Japan. Available on CD-Rom in *NeuroImage* 2002;16/2: 791. *http://www.apnet.com/journals/hbm2002/13870.html#13870*
57. Zilles K, Schlaug G, Geyer S, et al. Anatomy and transmitter receptors of the supplementary motor areas in the human and nonhuman primate brain. In Lüders HO, ed. Advances in Neurology, Vol. 70: *Supplementary sensorimotor area.* Philadelphia: Lippincott-Raven, 1996:29–43.
58. Zilles K, Clarke S. Architecture, connectivity, and transmitter receptors of human extrastriate visual cortex. Comparison with nonhuman primates. *Cereb Cortex* 1997;12:673–742.
59. Rumsey JM, Andreason P, Zemetkin AJ et al. Failure to activate the left temporoparietal cortex in dyslexia. An oxygen-15 positron emission tomographic study. *Arch Neurol* 1992;49:527–534.
60. Rumsey JM, Nace K, Donohue B, et al. A positron emission tomographic study of impaired word recognition and phonological processing in dyslexic men. *Arch Neurol* 1997;54:562–573.
61. Paulesu E, Frith U, Snowling M, et al. Is developmental dyslexia a disconnection syndrome? Evidence from PET scanning. *Brain* 1996;119:143–157.
62. Shaywitz SE, Shaywitz PA, Pugh KR, et al. Functional disruption in the organization of the brain for reading in dyslexia. *Proc Natl Acad Sci USA* 1998;95:2636–2641.
63. Klingberg T, Hedehus M, Temple E, et al. Microstructure of temporo-parietal white matter as a basis for reading ability: evidence from diffusion tensor magnetic resonance imaging. *Neuron* 2000;25:493–500.
64. Cavada C, Goldman-Rakic PS. Posterior parietal cortex in rhesus monkey: I. Parcellation of areas based on distinctive limbic and sensory corticocortical connections. *J Comp Neurol* 1989;287:393–421.
65. Cavada C., Goldman-Rakic PS. Posterior parietal cortex in rhesus monkey: II. Evidence for segregated corticocortical networks linking sensory and limbic areas with the frontal lobe. *J Comp Neurol* 1989;287:422–445.
66. Cavada C., Goldman-Rakic PS. Multiple visual areas in the posterior parietal cortex of primates. In: Hicks TP,

Molotchnikoff S, Ono T, eds. *Progress in brain research,* vol 95. Amsterdam: Elsevier, 1993:123–137.
67. Rizzolatti G, Luppino G. The cortical motor system. *Neuron* 2001;31:889–901.
68. Rizzolatti G, Luppino G, Matelli M. The organization of the cortical motor system: new concepts. *Electroencephalogr Clin Neurophysiol* 1998;106:283–296.
69. Rizzolatti G, Fadiga L, Gallese V, et al. Premotor cortex and the recognition of motor actions. *Cogn Brain Res* 1996;3:131–141.
70. Halsband U, Freund H-J. Premotor cortex and conditional motor learning in man. *Brain* 1990;113:207–222.
71. Petrides M. Deficits in conditional associative-learning tasks after frontal and temporal lesions in man. *Neuropsychology* 1985;23:601–614.
72. Halsband U, Passingham R. The role of premotor and parietal cortex in the direction of action. *Brain Res* 1982;240:368–372.
73. Petrides M. Motor conditional associative-learning after selective prefrontal lesions in the monkey. *Behav Brain Res* 1982;5:407–413.
74. Binkofski F, Buccino G, Stephan KM, et al. A parieto-premotor network for object manipulation: evidence from neuroimaging. *Exp Brain Res* 1999;128:210–213.
75. Buccino G, Binkofski F, Fink GR, et al. Action observation activates premotor and parietal areas in a somatotopic manner: an fMRI study. *Eur J Neurosci* 2001;13:400–404.
76. Binkofski F, Fink GR, Geyer S, et al. Neural activity in human primary motor cortex areas 4a and 4p is modulated differentially by attention to action. *J Neurophysiol* 2002;88:514–519.
77. Thoenissen D, Zilles K, Toni I. Differential involvement of parietal and precentral regions in movement preparation and motor intention. *J Neurosci* 2002;22:9024–9034.

2

Magnetic Resonance Imaging Study of the Parietal Lobe: Anatomic and Radiologic Correlations

Georges Salamon*, Noriko Salamon-Murayama*, Pattanasak Mongkolwat**, and Eric J. Russell**

**Department of Radiology, The David Geffen School of Medicine at UCLA, Los Angeles, California; **Department of Radiology, Northwestern University Medical School, Chicago, Illinois*

INTRODUCTION

Until recent years, dissections and tissue sections have been the only precise tools for gross anatomic analysis of the brain. Brodmann classification, the division of cortex into primary, unimodal, heteromodal, or limbic areas, and the most sophisticated methods used in neuroanatomy always refer to the division of the brain based on sulcal and gyral nomenclature (1).

Technical advances continue to secure the role of magnetic resonance (MR) imaging as the best technique for morphologic study of the brain. High-resolution MR studies of the brain can be made with great precision in multiple planes in normal subjects from the prenatal period to advanced age. Subjects may be selected according to sex and age, and evaluated with neuropsychological tests.

Precise delineation of the cerebral cortex and deep nuclei of the brain is important for functional neurosurgery and the diagnosis of neurocognitive disorders. For precise localization of functional data in functional imaging with positron emission tomography (PET) or functional MRI (fMRI), a detailed anatomic MRI study is required to allow excellent correlations with gyral projections.

METHODS

Several methods of MRI may be used. In practice, for routine clinical examination, the plane of the examination is often considered less important than the sequence protocol, which may include the following pulse sequences: T1, T2, proton density, fluid attenuation inversion recovery (FLAIR), diffusion, perfusion, T2*-weighted gradient echo, sequences for the MR examination of vessels (MR angiography [MRA]), and MR spectroscopy (MRS). The intrinsic nature of the lesion, delineated by these sophisticated methods, may be more important than its exact topographic projection for pathologic diagnosis. However, the location is extremely important for therapeutic planning and research.

MR images are routinely acquired in the commissural (AC-PC) plane derived from Talairach (2), and then matched with an atlas (3,4). Orthogonal reconstructions and three-dimensional (3D) images that can depict the surface of the brain also may be made from a volumetric acquisition. Voxel-based analysis even can be used to prepare an atlas of *in vivo* normal human brain structure, which can be matched with MR results in patients with diseases such as schizophrenia or Alzheimer's disease (5–8).

For patients with cognitive disorders and memory deficits, it is mandatory to have the best structural images, which also may be matched with functional imaging in some cases. The standard protocol we used is the following: T1 spin echo (SE) sagittal, proton density and T2 axial whole brain, T1 3D fast field echo (FFE) coronal, T2 turbo spin echo (TSE) coronal, two-dimensional (2D) gradient echo, and FLAIR sagittal images. The most important rule for this examination is to always use the line crossing the anterior and posterior white commissures (AC-PC line) as a plane of reference. This plane may be used for comparison with images from the main atlas or textbooks on the anatomy of the brain (2,9–12). This plane may be used easily by any technician in an MRI laboratory; the two commissures are easily visualized on sagittal T1 sections. Horizontal sections are parallel to the commissural plane and coronal sections are made perpendicular to it. For the evaluation of normal sulci, their variations are compared to anatomic brain specimens; and for patients with memory disorders, we use a program that allows the observer to simultaneously visualize any point in the brain in sagittal, coronal, and horizontal planes. This is important for accurate labeling of the cerebral cortex and a good appreciation of corticonuclear connections. R. Mullick developed this program (Etdips) at the National Institutes of Health (NIH) (13).

LIMITS AND DIVISIONS OF THE PARIETAL LOBE

Neuroanatomy and neurology textbooks (10,14–16) describe the same classical anatomic landmarks (Figs. 2-1 and 2-2).

Externally, the parietal lobe is limited anteriorly by the central sulcus. Posteriorly, the border with the occipital lobe is a virtual line from the superior limit of the parieto-occipital sulcus to the preoccipital notch at the inferior surface of the parieto-occipital area. Inferiorly, there is not a clear delineation between parietal and temporal lobes at the end of the lateral sulcus and superior temporal sulcus (Fig. 2-1). The supramarginal gyrus is located surrounding the termination of the ascending ramus of the lateral sulcus. The angular gyrus is found at the posterior margin of the superior temporal sulcus.

Internally (Fig. 2-2), the limits of the parietal lobe are marked anteriorly by the central sulcus, posteriorly by the deep parieto-occipital fissure, and inferiorly by the subparietal sulcus. Several sulci delineate different gyri inside the parietal lobe. The postcentral sulcus (posteriorly) and the central sulcus (anteriorly) are the limits of the primary sensory area. The parietal operculum is the inferior part of this gyrus, corresponding to the lateral sulcus. At the upper part arising from the postcentral sulcus, a deep transversal sulcus (the interparietal sulcus) divides the corresponding parietal area in two lobules, the superior and inferior parietal lobules. The surface of the inferior parietal lobule is different on the right and left sides in right-handed subjects. The termination of the sylvian fissure on the right side is located more anteriorly than on the left side, where the planum temporale is larger (17). This explains why the central sulcus (18) is more often anteriorly located (58%) on the right side than on the left (mean difference, 3 mm) on horizontal sections in these subjects. The inferior parietal lobule sometimes is divided by the small sulcus of Jensen (Dejerine) or angular sulcus (19), which courses between the supramarginal gyrus and angular gyrus.

The primary sensory and motor areas form the paracentral lobule on the internal aspect of the brain. It is limited anteriorly by the paracentral sulcus and posteriorly by the marginal sulcus. The part of the paracentral lobule devoted to the parietal lobe is smaller than the part corresponding to the primary motor area (frontal lobe). The remainder of the internal part of the parietal lobe, the precuneus, is limited anteriorly by the marginal sulcus, posteri-

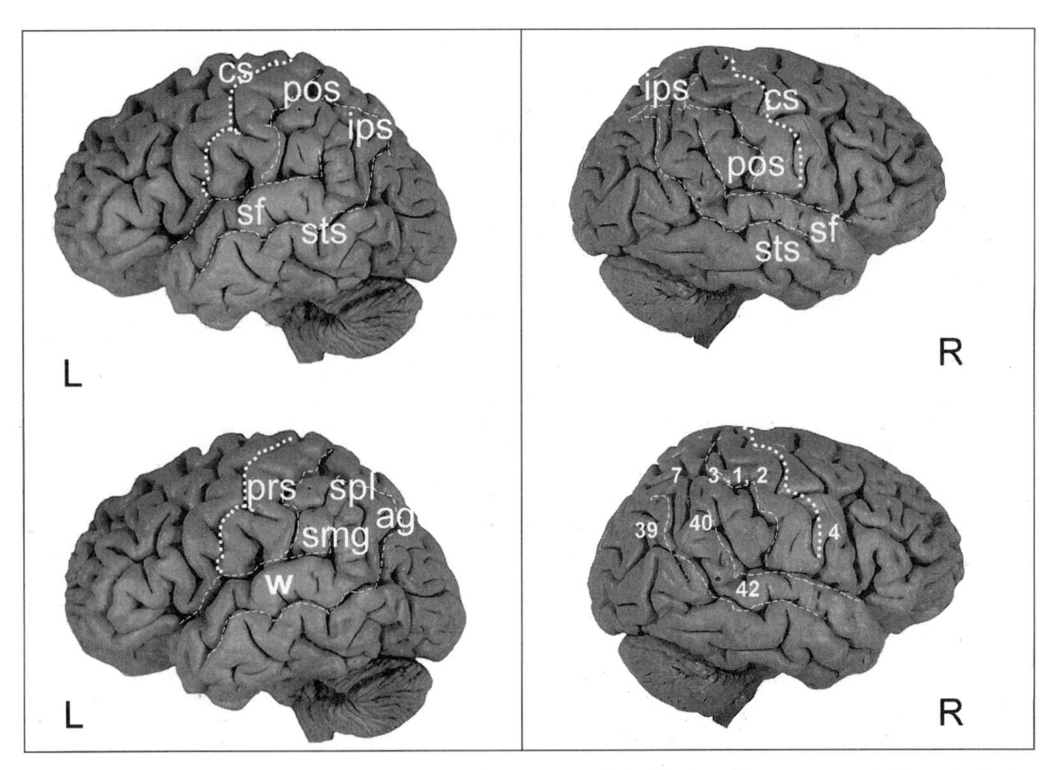

FIG. 2-1. External view of the parietal lobe. Dissection of the brain with an external view of the two hemispheres. On the upper row on the left and right sides are labeled the main sulci, which delineate the parietal lobe and divide this lobe in different gyri (cs, central sulcus; ips, interparietal sulcus; pos, postcentral sulcus; sf, sylvian fissure; sts, superior temporal sulcus). On the lower row are presented on the left the different gyri inside the parietal lobe on the external aspect and on the right are labeled the corresponding Brodmann areas (ag, angular gyrus; prs, primary sensory area; smg, supramarginal gyrus; spl, superior parietal lobule). Wernicke area adjacent to supramarginal gyrus is also labeled (W).

orly by the parieto-occipital fissure, and inferiorly by the subparietal sulcus.

The different gyri of the parietal lobe follow Brodmann classification (Figs. 2-1 and 2-2): Areas 3a, 3b, 1, and 2 correspond to the primary sensory area. More posteriorly is the small area 5. The major part of the parietal cortex located above the interparietal sulcus, the superior parietal lobule, corresponds to area 7. Area 19 lies more posteriorly. The angular gyrus corresponds to Brodmann area 39, and the supramarginal gyrus to area 40. At the internal aspect of the brain, the precuneus is labeled as Brodmann area 7. The correspondence between different functional types of the cerebral cortex (heteromodal, unimodal, primary, limbic) and its Brodmann area (Fig. 2-3) is clearly presented by Mesulam (20). Many secondary sulci may be observed in all of these parts of the parietal lobe (21). Asymmetric variations also are observed at the parietal and temporal lobe levels (22–24), and these may have an important relationship with various functions of the right parietal lobe (space recognition) and left parietal lobe (body image and language). All these differences in sulcal and gyral patterns are analyzed herein.

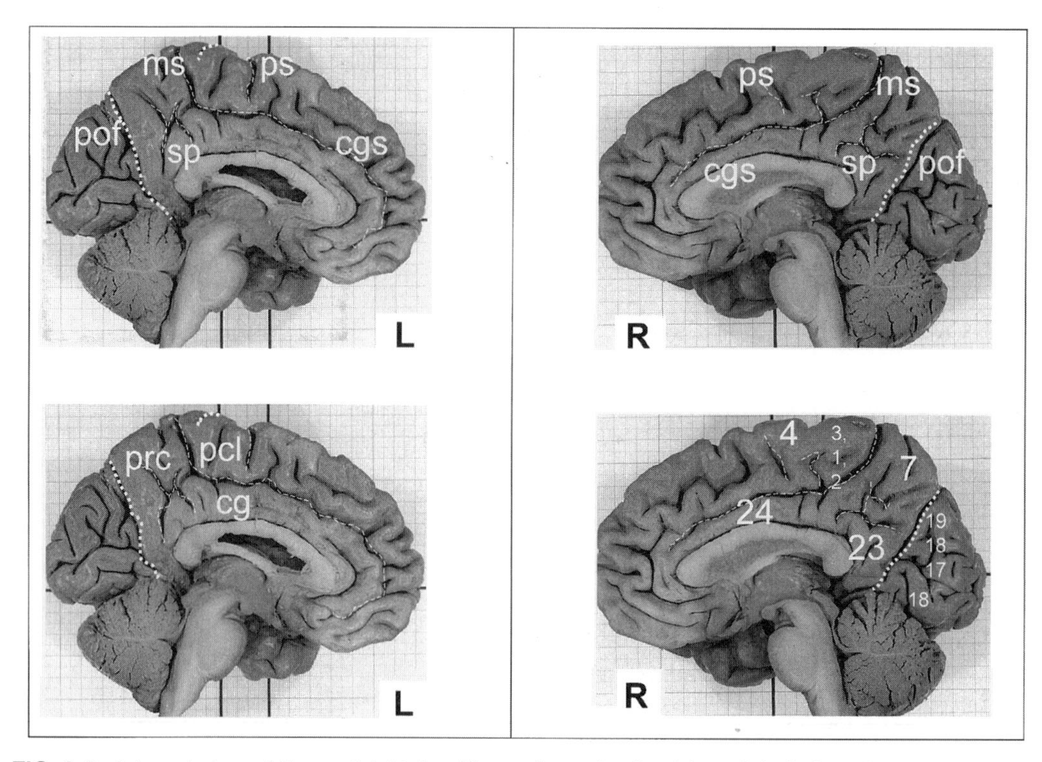

FIG. 2-2. Internal view of the parietal lobe. The main parietal sulci are labeled on the upper row on the left and right side (cgs, cingulated sulcus; ms, marginal sulcus; ps, paracentral sulcus; pof, parieto-occipital fissure; sp, subparietal sulcus). The different gyri are labeled on the inferior row on the left side and the corresponding Brodmann areas on the right side (pcl, paracentral lobule; prc, precuneus). The cingulated gyrus (cg) is also labeled, being at the parietal lobe level located close to the precuneus.

MAGNETIC RESONANCE IMAGING OF THE PRIMARY SENSORY AREA (EXTERNAL SURFACE) (FIGS. 2-3, 2-4, 2-5, AND 2-6)

On the external aspect of the brain this area is limited anteriorly by the central sulcus, posteriorly by the postcentral sulcus, and inferiorly at the sylvian fissure at opercular level. On the medial aspect this area is defined by the posterior part of the paracentral lobule (Fig. 2-2). On sagittal sections these limits are very clear, but there are several anatomic variations that may be visible by dissection or MRI. The central sulcus has an S shape with two curves (genu) visible when the external part of the brain is exposed after dissection (18,21,30). This is visible on sagittal MR sections also. The superior curve of the genu, convex anteriorly, corresponds to the hand area. Correlations have been made at this level with functional studies (18,26,27). The area corresponding to the projection of the face is close to the opercular level. Different secondary patterns also may be seen at the level of the central sulcus depending on its shape, dimension, depth, and termination. The inferior part of the central sulcus does not end at the sylvian fissure in 84% of cases, and it is visible at the internal aspect of the brain in 88% of cases. It is seen as a continuous sulcus in 92% of cases.

Side branches of the central sulcus have been found by dissection, under surgical examination, and with MRI. One or two anterior side branches are the most common pattern and go over the precentral sulcus. One to three branches go over the postcentral gyrus. The

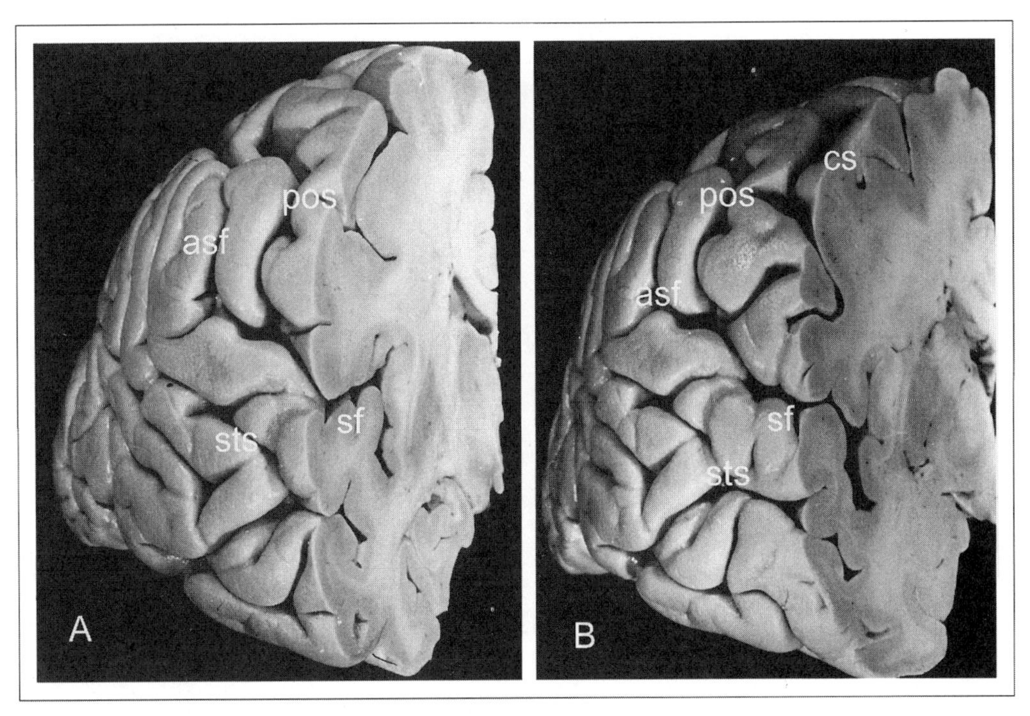

FIG. 2-3. Oblique view of the primary sensory area **(A)** and of the central sulcus **(B).** This photograph is reprinted with the courtesy of H. Duvernoy. Note the location of the postcentral sulcus (pos) anterior to the terminal ramus of the sylvian fissure (asf) and the depth of the central sulcus (cs), which does not reach the sylvian fissure (sf) (sts, superior temporal sulcus).

motor cortex, which is much thicker than the sensory area, can be measured with high-resolution MRI (28). Asymmetry of the central sulcus is observed commonly in right-handed subjects, as confirmed with standard MRI (65%) and 3D MRI (18,24,29). The depth of the central sulcus has a mean value of 16.6 mm. Comparison of functional MRI and magnetoencephalography can help with anatomic localization also (25,27,30,31). A more detailed discussion of MRI can be found in the chapter on functional imaging of the parietal

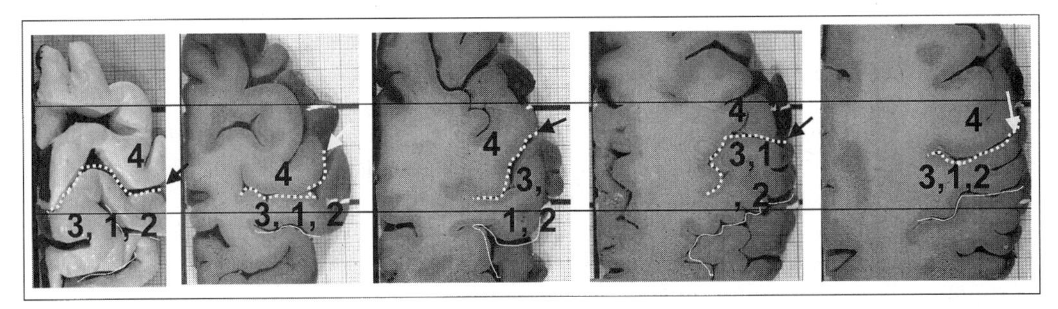

FIG. 2-4. Central sulcus. Shape of the central sulcus on horizontal sections of the same brain made every 5 mm, starting from the superior margin of the brain *(left).* MRI is certainly the most accurate tool (see Fig. 2-5) for such anatomic study, with T1 3D FFE coronal sections and reconstruction in three orthogonal planes.

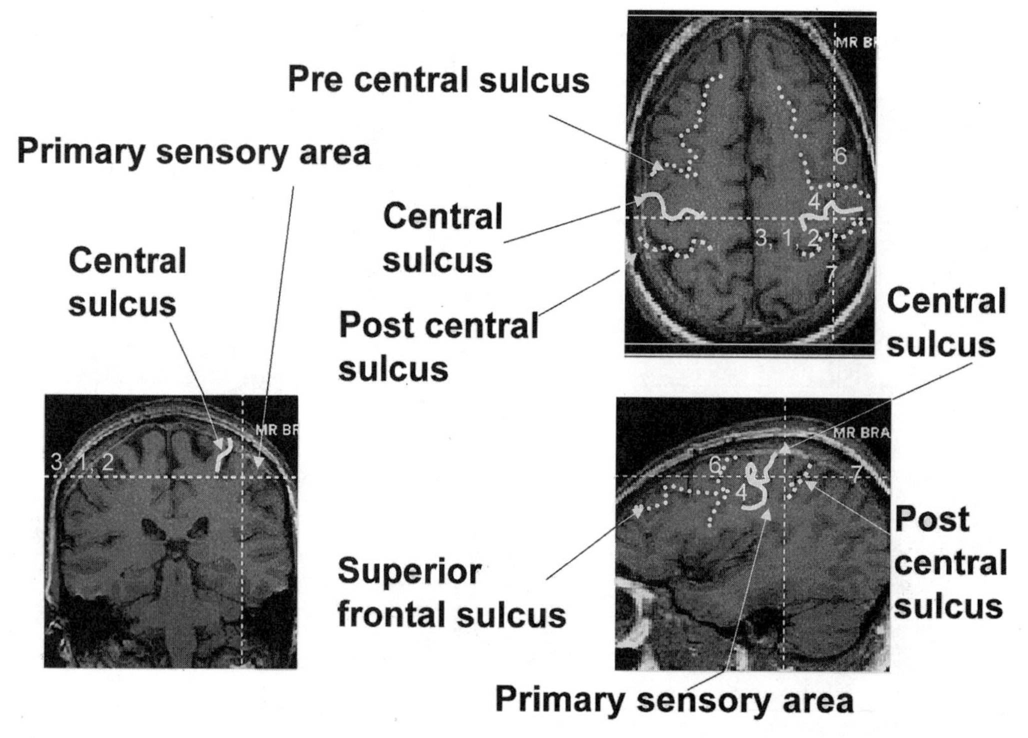

FIG. 2-5. Projection of the primary sensory area. This section corresponds to the superficial aspect of the sensory area at the hand level. The postcentral gyrus may be followed on the horizontal, sagittal, or coronal plane.

lobe. Imaging of arterial branches can aid in localization. For example, identification of the branches of the central artery, using either angiography or MRA, can aid in determining the location and direction of the central sulcus (32). Mathematical algorithms have been proposed for automatic recognition of the central sulcus as well (33).

The posterior limit of the primary sensory area is the postcentral sulcus. It presents more variations than the precentral sulcus and it is often interrupted. Ono (21) found that only 48% of postcentral sulci were continuous on dissection. A similar pattern can be observed on sagittal MRI. Generally, true interruption of the postcentral gyrus is visible in the middle part of its course at the lateral aspect of the brain. A real secondary postcentral sulcus is found in 12% of cases. The largest connection on the posterior side of the postcentral sulcus is the interparietal sulcus (Figs. 2-7, 2-8, and 2-9). Several smaller secondary transverse sulci can be observed on the surface of the postcentral gyrus, or over the superior or inferior parietal lobules. The inferior termination of the postcentral gyrus forms the first large sulcus parallel to the ascending ramus of the sylvian fissure (Fig. 2-3). Steinmetz (34) has provided a detailed anatomic study of the inferior part of the postcentral gyrus—the parietal operculum—along with other personal studies comparing MRI and brain specimens. Several asymmetries or variations have been described according to the side, sex, and handedness of the subject (34,35). The distance between the

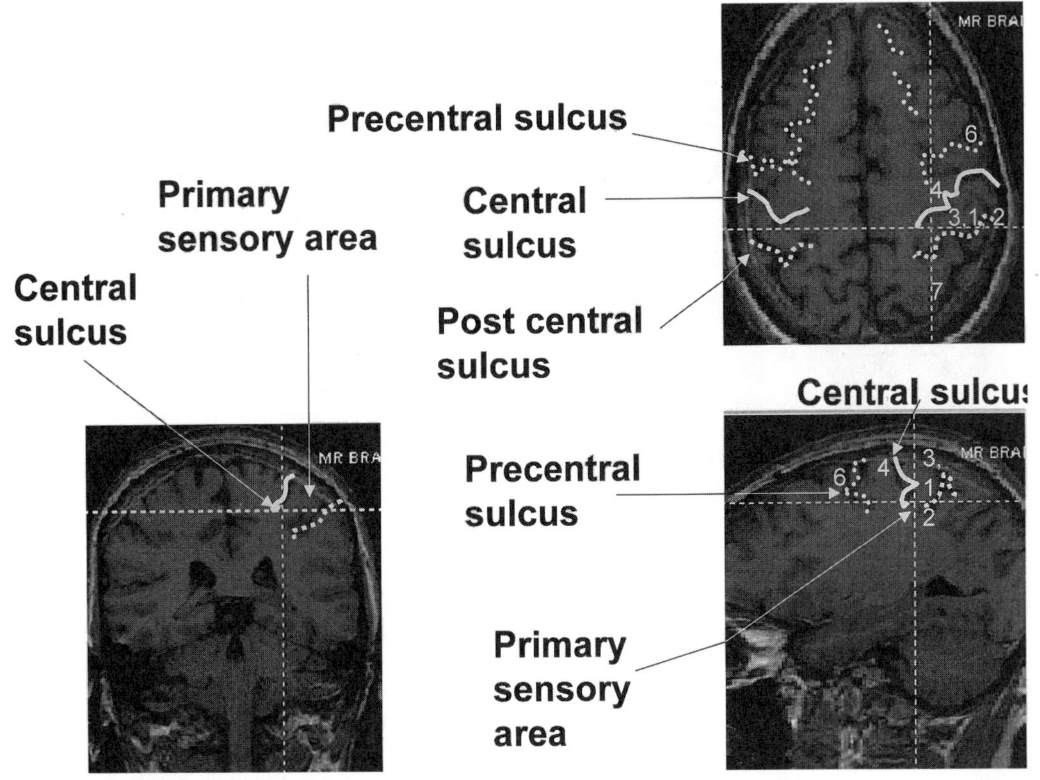

FIG. 2-6. Projection of the deep part of the sensory area (hand level). Because the depth and orientation of the postcentral sulcus, the coronal plane is more posterior than on Fig. 2-6 and the sagittal projection more medial. The superficial and the deep parts of this area may be analyzed on this horizontal section.

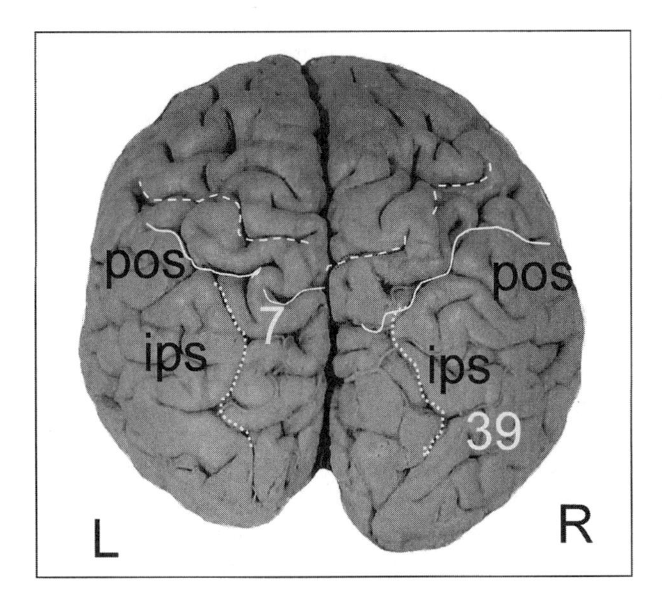

FIG. 2-7. Posterior view of the parietal lobe. Location of the superior parietal lobule (Brodmann area 7) and of the angular gyrus (Brodmann area 39) behind the postcentral sulcus (pos) separated by the interparietal sulcus (ips).

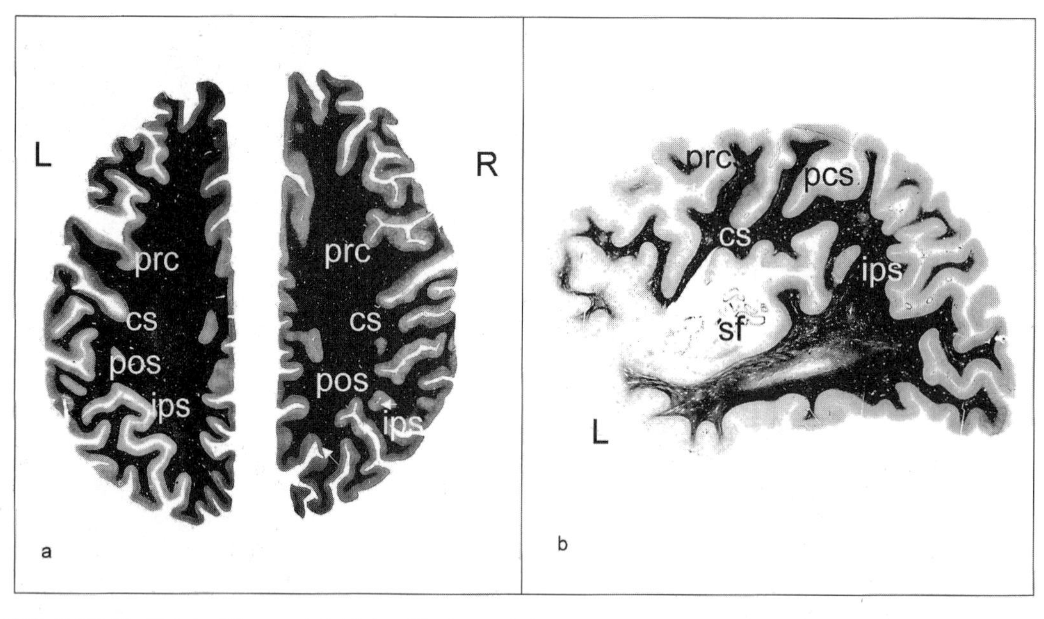

FIG. 2-8A,B. Sulci of the parietal lobe. Horizontal *(a)* and sagittal *(b)* sections of the brain. Note on the left side the continuous course of the interparietal sulcus (ips) with its shape of corrugated iron. On the right side this sulcus is more irregular and interrupted. The images of central sulcus (cs), between the precentral (prc) and postcentral (pos) sulcus are very helpful in these planes to trace the location of the primary sensory motor area, superior parietal lobule, and angular gyrus.

central sulcus and postcentral sulcus averages 11.5 mm (21).

More than 50% of the parietal cortex lies in the depths of the central sulcus. A large percentage is found at the level of hand area (18). On MRI, the primary sensory cortex is easily located with attention to the sylvian fissure and the central and postcentral sulci (Fig. 2-5, 2-6, and 2-7). Horizontal sections above the level of ventricles and corpus callosum (40 to 60 mm above commissural plane) clearly show the characteristic view of the central sulcus. Its position, depth, and S shape are the best keys to its recognition (Fig. 2-5); this can be confirmed by functional studies, intraoperative stimulation, or recordings (18,30). The central sulcus is located between the precentral sulcus (with an end-to-side connection to the superior frontal sulcus) and the postcentral sulcus (with a similar connection with the interparietal sulcus). Localization of the central sulcus helps in the recognition of the postcentral sulcus if there is some variation; the latter is located 11 mm more posteriorly. With the help of the Etdips or similar program, the corresponding sulcal location can be recognized on the two other orthogonal plane views once one of these sulci is recognized.

In the sagittal plane, the three sulci (precentral, central, and postcentral) have specific shapes (Figs. 2-6, 2-7, and 2-8). The precentral and superior frontal sulci form a right angle. Posteriorly, the postcentral and interparietal sulci form a similar image. The direction, shape, and mean projection of the central sulcus are unique. Another landmark is Heschl's gyrus on a more medial sagittal image: The projection of the central sulcus in the horizontal plane corresponds to the direction and location of the transverse temporal gyrus. If the MRI is acquired in the commissural (AC-PC) plane, a perpendicular line drawn from the middle part of Heschl's gyrus will cross the central sulcus. The same landmark can be used on coronal images, where the transverse gyrus is well outlined. A line perpendicular to

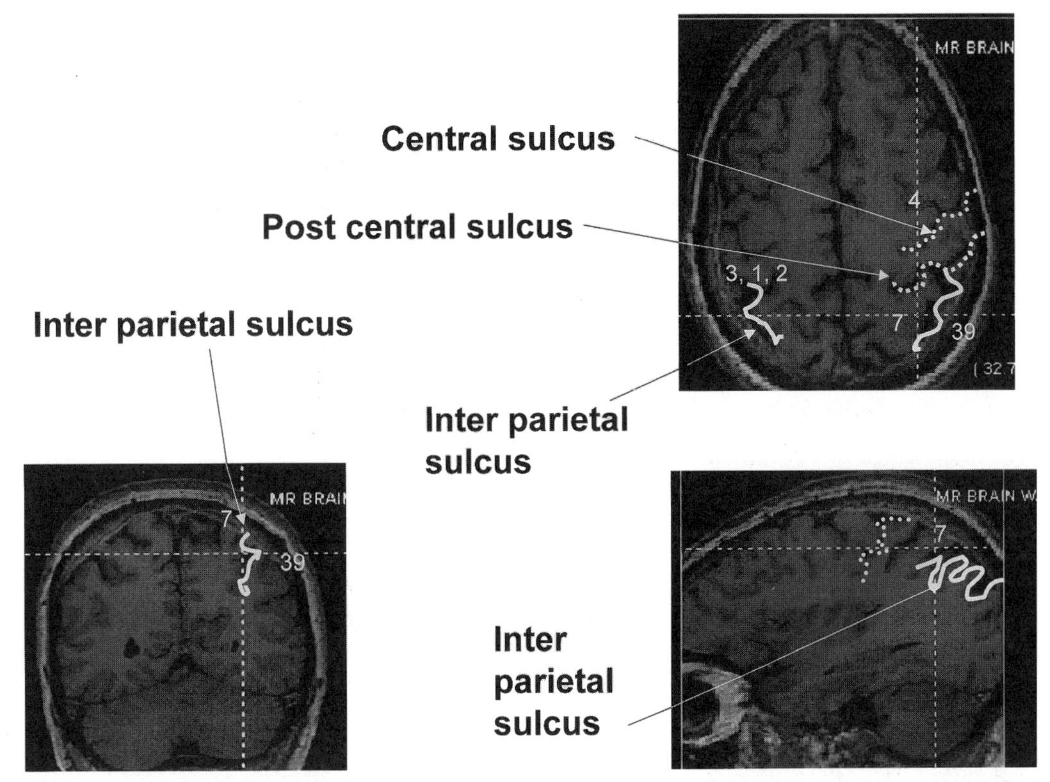

FIG. 2-9. Projection of the interparietal sulcus. This sulcus has a typical shape on sagittal view and is always different from the right to left side. The limits between the superior parietal lobule and the angular gyrus are well outlined by comparing coronal or horizontal sections.

Heschl's gyrus crosses the central sulcus. The primary sensory cortex is lateral to the sulcus on this plane.

Functional imaging can be helpful and complementary to conventional MRI for preoperative localization of brain tumors and other lesions associated with edema when the sulci are not visible or distorted. However, in the case of a small tumor without edema, a small vascular infarct, or degenerative disease, a detailed topographic MR report may be achieved easily with the help of the Etdips program or any other similar program, permitting simultaneous examination of the three planes of space (36,37).

Serial MRI images in the coronal plane are most useful for the study of the corresponding thalamic relay, ventroposterior lateral nucleus (VPL), and the projections of corticospinal fibers corresponding to the primary sensory area (38,39).

MAGNETIC RESONANCE IMAGING OF THE SUPERIOR PARIETAL LOBULE (FIGS. 2-9 AND 2-10)

The connections and subdivisions of the superior parietal lobule (7a, 7b, tip, 7m) have been well studied in monkeys (40–43). Research supports the role of the superior parietal lobe in attention (20,43,44), space recognition, and visual approach, indicating the great importance of this part of the parietal lobe. At the external part of the brain, the superior limit of the superior parietal lobule is the upper margin of the parietal lobe, whereas its inferior margin is the interparietal sulcus. The anterior limit is formed by the postcentral

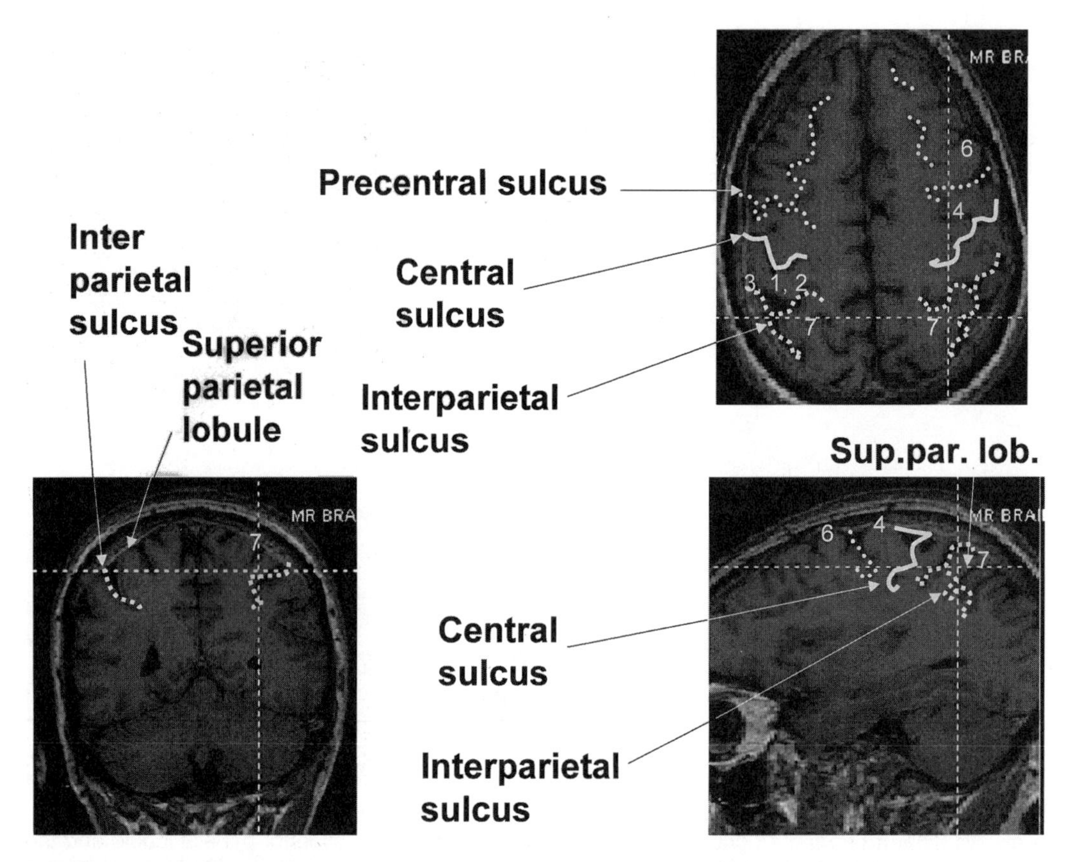

FIG. 2-10. Projection of the superior parietal lobule. On coronal and horizontal sections the superior parietal lobule is located medially to the projection of the deep interparietal sulcus. This interparietal sulcus is always visible by using this method, which projects simultaneously the same point in sagittal, coronal, or horizontal planes. On the sagittal plane area 7 is located above the level of the interparietal sulcus.

sulcus. There is not a clear posterior delineation with the occipital lobe. The line joining the upper part of the parieto-occipital fissure and preoccipital notch is considered a good landmark (14).

The interparietal sulcus is connected to the postcentral sulcus in 70% of cases (21). Many different patterns that also vary from right to left may be observed. On the right, a continuous interparietal sulcus occurs in 28%, whereas it occurs in 72% on the left. Three interruptions may be observed on the right side but never the left. Two interruptions occur in 28% on the left and 68% on the right side (21). Side branches upward or downward are common; however, their average number, from one to three, is not significantly different on the right or left side. Connections of the interparietal sulcus with adjacent sulci, the superior temporal sulcus, or the parieto-occipital fissure are observed in less than 20% of cases.

The interparietal sulcus is deep (19.5 mm); it is one of the deepest brain levels, with the exception of the lateral sulcus and parieto-occipital fissure. MRI done in a tangential parasagittal plane and compared with brain specimens allows for easy recognition (45) by looking at the junction of the postcentral and interparietal sulci. In our experience, recognition of the interparietal sulcus is easily achieved in all normal cases by using 3D analysis. On transverse axial MRI, the junction of the postcentral and interparietal sulci is the best landmark, especially on the left side, where the interparietal sulcus may be followed for the greatest part of its course. With attention to the more interrupted pattern observed on the right side, and by comparison with the left side, the two interparietal sulci are clearly outlined on this plane (Fig. 2-8).

On a coronal section (perpendicular to the commissural plane) made at the level of the atrium or more posteriorly, the interparietal sulcus is the deepest sulcus visible on the convexity on the two sides when the examiner follows the surface of the brain starting from interhemispheric fissure. The shape asymmetry of the two sulci is important. Comparison of coronal, horizontal, and sagittal sections of the same point is very helpful. On sagittal sections, recognition of the left interparietal sulcus is very easy if this sulcus is continuous. It has the characteristic "corrugated iron" shape (Fig. 2-9). The interparietal sulcus is always located away from the midline, in a sagittal plane where the hippocampus is visible as well, just above the projection of optic radiations (2). The surface and projection of the superior parietal lobule may be followed easily when the interparietal sulcus is localized. The superior parietal lobule is medial to this sulcus on the coronal plane, superior on the sagittal plane, and medial on the horizontal plane. The variations from left to right side are very important in this area; therefore, the individual analysis of MRI is more accurate than the use of a computed analysis program alone (46).

MAGNETIC RESONANCE IMAGING OF THE ANGULAR GYRUS (FIG. 2-11)

The presence of finger agnosia, right–left disorientation, acalculia, and agraphia of the Gerstmann syndrome have drawn attention to the importance of this area of the left hemisphere (15). The same interest is attached to the right side for left spatial neglect. With the development of functional studies, this area has received greater attention from practitioners of such fields as language, visual memory, and spatial recognition (47,48).

The angular gyrus is part of the inferior parietal lobule, located below the interparietal sulcus and around the posterior termination of the superior temporal sulcus. The main landmarks on MRI of the angular gyrus are the interparietal sulcus and the termination of the superior temporal sulcus. The former relationship has been described with the description of the superior parietal lobule. The superior temporal sulcus may be continuous or divided into two or three segments during its course (21). It ends as a single or double line. The angular gyrus is outlined anteriorly (84% on right, 92% on left) by an angular sulcus. In 20% of cases, this angular sulcus joins the interparietal sulcus on the right side (never on the left). The depth of this sulcus has a mean value of 18 mm on the right and 17.4 mm on the left.

MRI analysis of the angular gyrus is best performed by using sagittal MR sections. Even if the superior temporal sulcus is interrupted in its course parallel to the sylvian fissure, it is easy to follow. The cortex around this area may be outlined using the 3D method of mapping, whereas this is impossible on a single horizontal section, even if the examination is made along the commissural plane. On coronal sections used to study the projection of the superior parietal lobule, the image of the interparietal sulcus is also a good landmark; the angular gyrus is located on the lateral part of this sulcus. In our experience, the landmark provided by the superior temporal sulcus projection is the easiest with which to trace the projection and limits of the angular gyrus. The presence of an angular sulcus must be carefully differentiated from the termination of the superior temporal sulcus itself, to avoid confusion between the projection of the angular gyrus and the posterior part of the supramarginal gyrus.

MAGNETIC RESONANCE IMAGING OF THE SUPRAMARGINAL GYRUS (FIGS. 2-11 AND 2-12)

The location and surface of the supramarginal gyrus are in close relationship with the

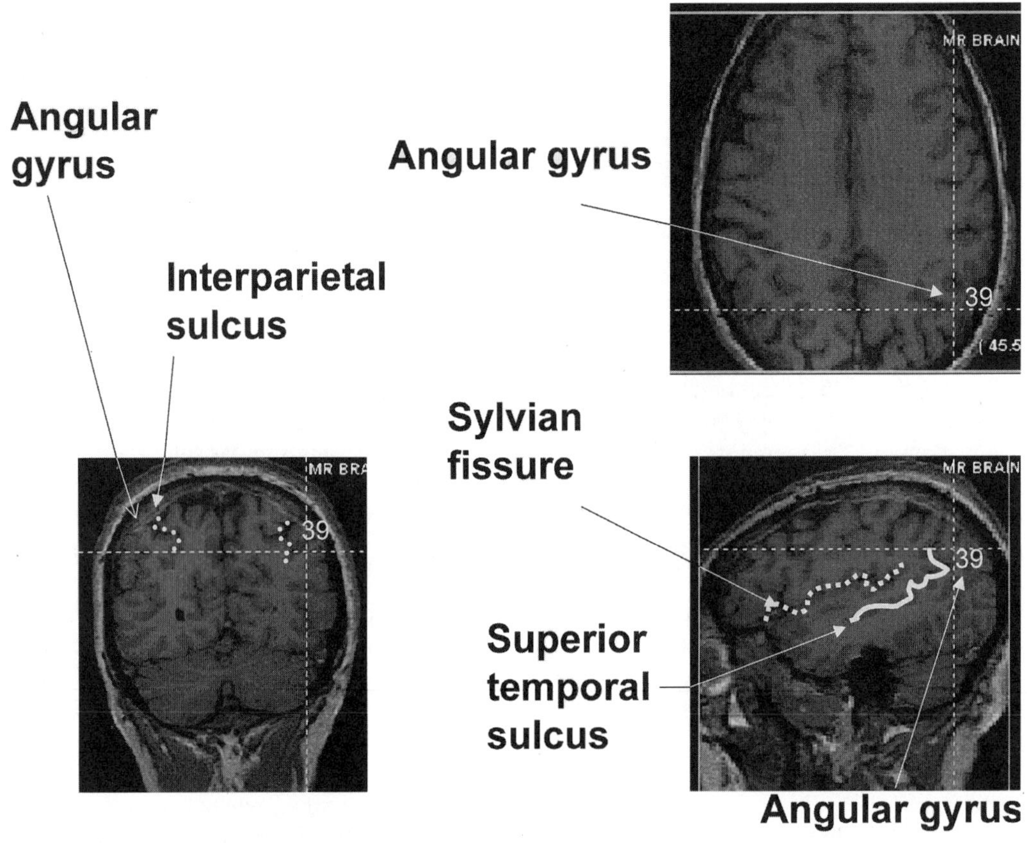

FIG. 2-11. Projection of the angular gyrus. The angular gyrus (area 39) is easily visible on a sagittal section by following the superior temporal sulcus. With the help of 3D navigation, the location of the angular gyrus is visible on horizontal sections and the coronal plane.

terminal portion of the sylvian fissure. The sylvian fissure (21,35) often is asymmetric. The terminal ascending ramus at the posterior bifurcation of the sylvian fissure represents its terminal branch, ascending with some variations. This branch is different from the right to the left side. There are two patterns on the right side: a vertical (96%) and a more horizontal (4%) one. On the left side the vertical pattern is less frequent (44%), whereas a more horizontal similar pattern is visible in 56%. There also are variations in the size of the lateral sulcus. The horizontal segment is twice as large on the left side, and the vertical segment is twice as large on the right. The depth of the sylvian fissure is greater at its posterior part on the right than on the left (mean difference, 3 to 5 mm). These variations are related to the development of the planum temporal (17). It is important to know that the terminal segment of the sylvian fissure has side branches directed superiorly in 12% on the right side and 16% on the left side (16). The delineation of the supramarginal gyrus is best made with reference to the sylvian fissure, mainly employing sagittal or coronal sections. The sagittal sections best depict the ascending terminal part of the sylvian fissure: The supramarginal gyrus is found surrounding this termination, with an anterior limit formed by the postcentral sulcus. Inferiorly, the delineation between the area of the supramarginal gyrus and the Wernicke area (areas 40 and 42) cannot be drawn with precision because these two areas are very close, without any macroscopically distinct landmark.

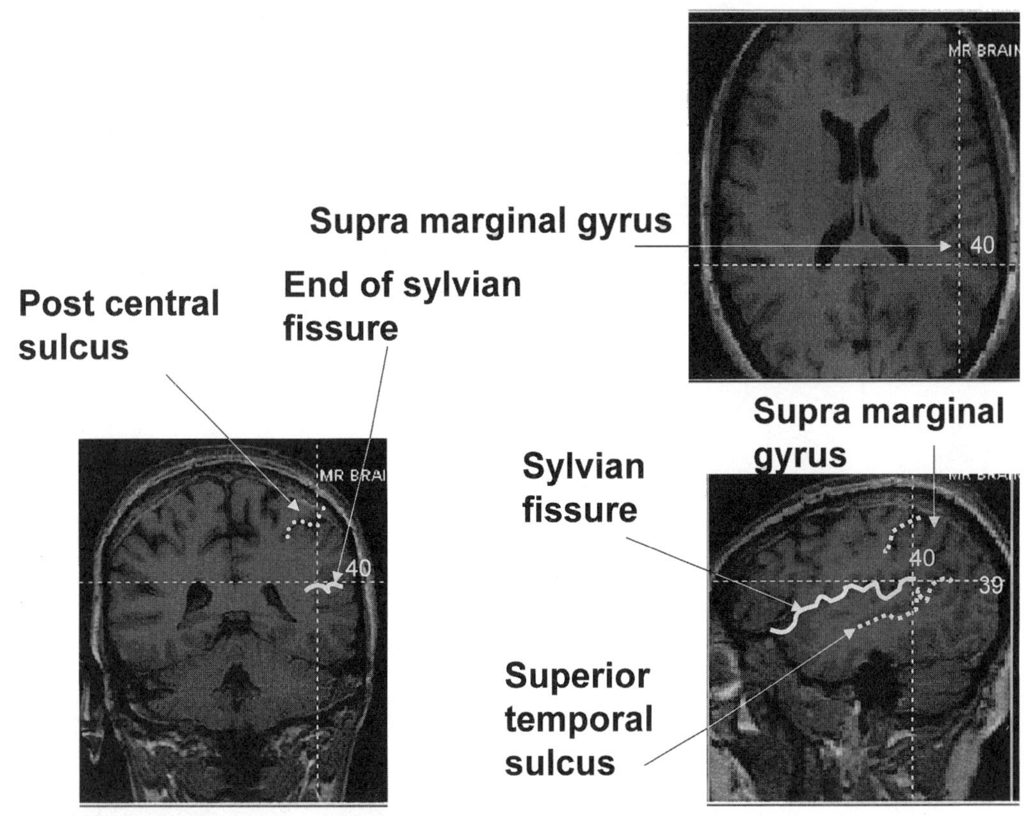

FIG. 2-12. Projection of the supramarginal gyrus. The termination of the lateral sulcus is helpful to delineate the supramarginal gyrus all around its ascending terminal branch. Another good landmark on coronal section is given by looking at the image of the transverse temporal gyrus. The sections posterior to this gyrus correspond to Wernicke's area and then to the supramarginal gyrus.

The lateral sulcus on coronal sections, and the typical transverse (Heschl's) gyrus at its middle level are excellent landmarks. Just posterior to Heschl's gyrus on coronal section is Wernicke's area. The supramarginal gyrus is more posterior, and is located on the two banks of the terminal end of the sylvian fissure. It is more difficult to determine the projection of the supramarginal gyrus on horizontal sections. If sagittal and coronal sections are analyzed simultaneously, the image of the supramarginal gyrus is traced very easily and helps to locate this area on horizontal sections. A simultaneous delineation in all three planes of the main landmarks of the supramarginal gyrus and adjacent cortical areas is important for patients with speech disorders or other cognitive deficits involving the right, left, or both hemispheres; this has been demonstrated on functional studies as well (49).

MAGNETIC RESONANCE IMAGING OF THE PRECUNEUS (FIGS. 2-13, 2-14, AND 2-15)

The main landmarks of the precuneus are the parieto-occipital fissure posteriorly, the subparietal sulcus inferiorly, and the marginal sulcus and upper part the superior margin of the brain superiorly. These landmarks are important for all recent studies that have elucidated the role of the precuneus as a component of a network in such different activities

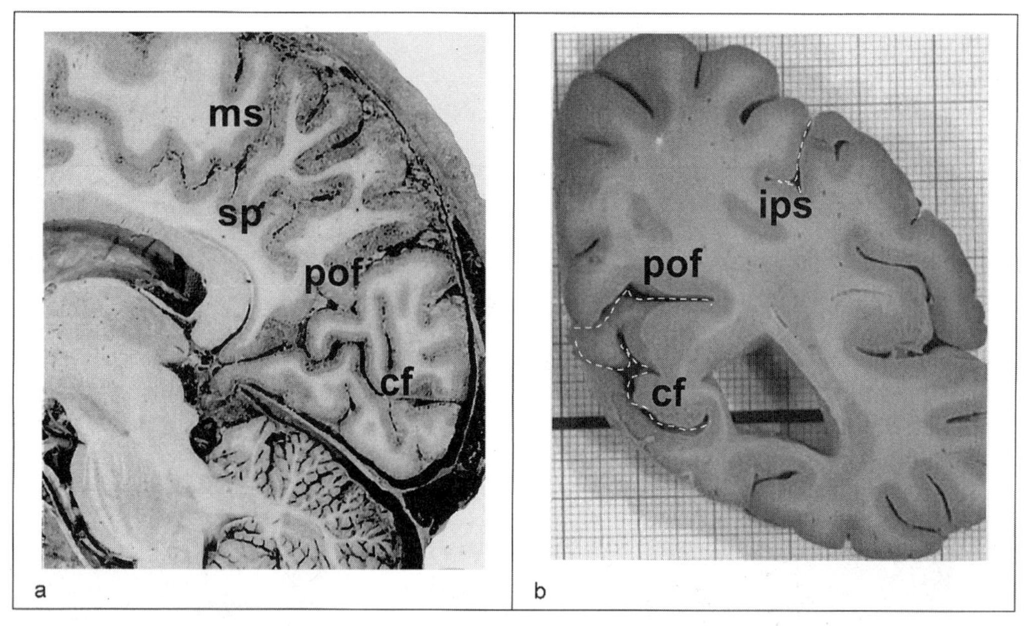

FIG. 2-13. Anatomic view of the parieto-occipital fissure. The parieto-occipital fissure (pof) is visible on the sagittal *(a)* and coronal *(b)* planes. On this coronal image, the black line corresponds to the plane of the commissures (all coronal anatomic sections were made perpendicular to this plane) (cf, calcarine fissure; ips, interparietal sulcus; ms, marginal sulcus; sp, subparietal sulcus).

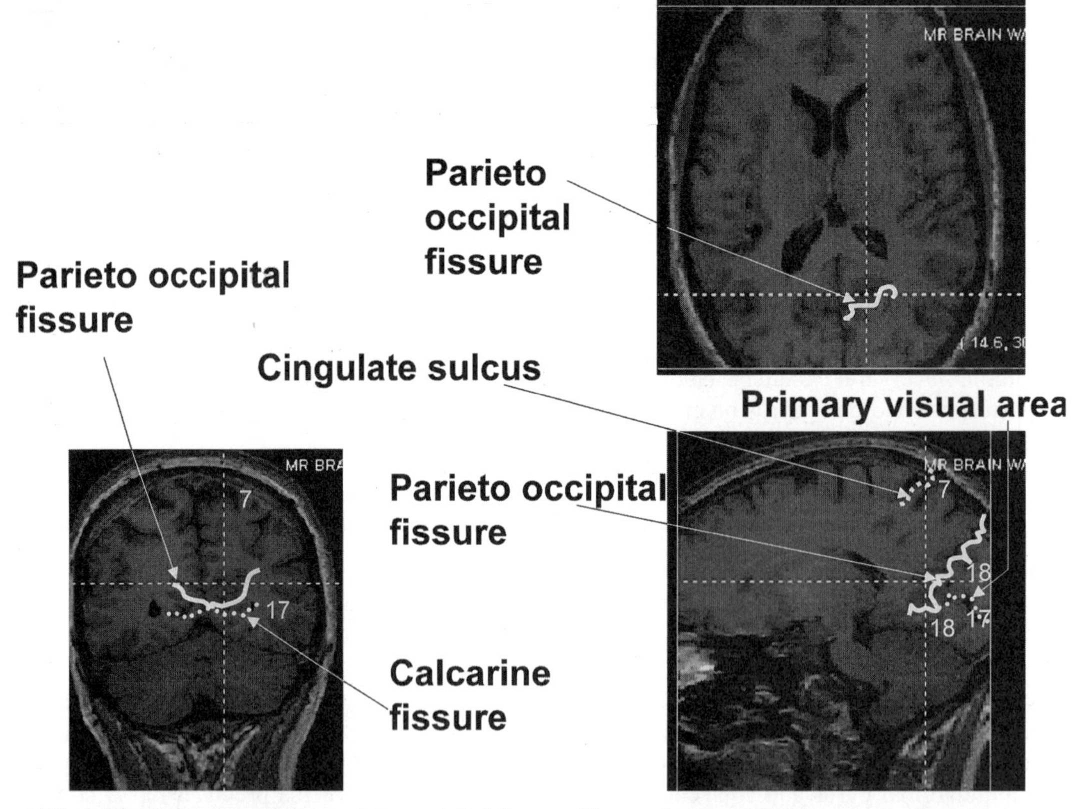

FIG. 2-14. Projection of the parieto-occipital fissure. The parieto-occipital fissure is always recognized on sagittal and coronal sections. It is then possible to trace its position on horizontal sections.

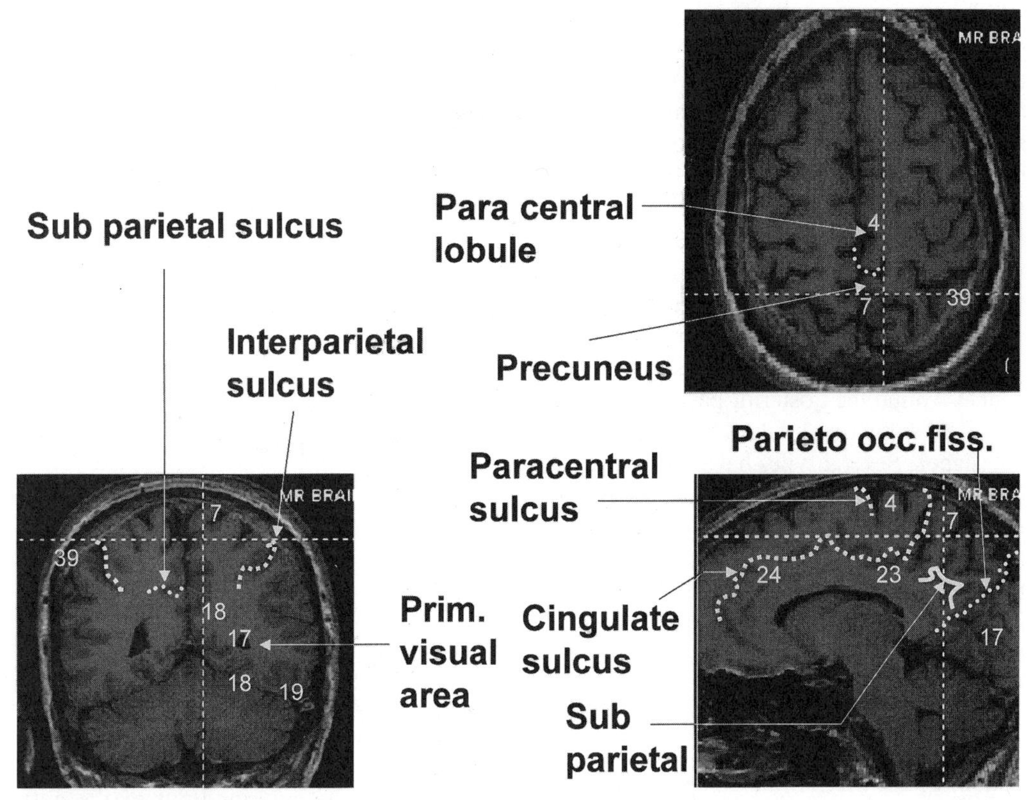

FIG. 2-15. Projection of the precuneus and of the para central lobule. The delineation of the precuneus on the lateral section is very easy. The projection on coronal or horizontal planes is done automatically at any level as soon as its limits are recognized. On coronal section, the entire cortex above the subparietal sulcus corresponds to the precuneus. On horizontal section, the precuneus is the cortex corresponding at the medial surface to the territory back to the projection of subparietal sulcus. The same sagittal plane may be used for the paracentral lobule delineation, with the drawing of the cingulate sulcus (its marginal part) and the termination of the central and paracentral sulcus. On coronal view, the paracentral lobule is found between the marginal sulcus and paracentral sulcus. The plane must be strictly perpendicular to the commissures.

as mental navigation, retrieval of auditory-verbal memory, and structural components of musical perception (50–52).

The parieto-occipital fissure has a limited extension on the upper part of the brain in all cases on the left and in 96% on the right (16). Its shape is variable: straight (20%), T-shaped (24%), or more complex with three branches. The three-branch pattern is observed more often on the right (48%) than the left (24%). Different side branches are visible, but its end point and depth characterize the main fissure. It ends at the level of the calcarine fissure, either directly or in very close proximity, with pseudoconnections in all cases on the two sides. Its end is always located at the level of the upper bend of the calcarine fissure, or slightly more anteriorly. Its depth has a mean value of 23 mm, greater than the adjacent calcarine fissure (16 mm). These findings are easily seen on right or left parasagittal MR sections.

The divergent courses of the calcarine and parieto-occipital fissures are characteristic on coronal sections. When the interparietal sulcus and the parieto-occipital fissure are visible along their entire course, the division between parietal and occipital lobe is very striking. In every normal case, coronal and sagittal sections may achieve recognition of the parieto-occipital fissure. As for many other divisions of the brain, this cannot be done on horizontal sections alone, and it is best done with a simultaneous 3D navigation system.

The inferior margin of the precuneus is the subparietal sulcus (Fig. 2-15). Although the cingulate sulcus turns upward and ends as a marginal branch at the anterior level of the precuneus, the subparietal sulcus continues its course around the posterior part of the cingulum. The subparietal sulcus may have different shapes. Several types have been described (21) and are found on MRI easily. Most often, it is not represented as a single line but as a complete or incomplete H shape (64%). One to three ascending branches may be observed along its course. Its terminal part is directed toward the splenium in 86% of cases. It has a mean depth of 7 mm. It may be traced best on sagittal MRI. On coronal sections at the middle part of the precuneus, starting from the bottom of the section the calcarine fissure is the first sulcus visible at the internal aspect of the brain; the parieto-occipital fissure is visible just above this. The subparietal sulcus is the next adjacent superior sulcus. Nevertheless, the best view for recognition of the precuneus is provided by sagittal section and orthogonal 3D analysis.

The last anterior landmark of the precuneus is the marginal branch of the cingulate sulcus. It marks the division at the inner part of the brain between the precuneus and paracentral lobule areas. The cingulate sulcus ends upward with a marginal branch, which extends over the lateral surface of the brain, or at the upper part of the brain in 58% of cases. For this reason, it can be seen at the upper part of horizontal MR sections just posterior to the central sulcus (53). The sagittal images are most useful to draw its precise location, even if the cingulate sulcus is difficult to follow in its entire course when it is interrupted (40%) or if it is represented by a double sulcus (24%). The entire cortical area between the marginal branch, parieto-occipital sulcus, subparietal sulcus, and superior margin of the parietal lobe comprises the precuneus.

MAGNETIC RESONANCE IMAGING OF THE PRIMARY SENSORY AREA OF THE PARACENTRAL LOBULE (FIG. 2-15)

The representation of the inferior limb at the sensory level is located posterior to the central sulcus on the internal aspect of the brain. This area represents one fourth of the entire area of the paracentral lobule. Its landmarks are best seen on sagittal sections in correlation with horizontal sections, which show the course of the central sulcus. The central sulcus ends at the internal aspect of the brain in 65% of cases, and is a continuous sulcus from operculum to the superior margin of the brain in 92%. It defines on upper horizontal sections the anterior limit of the sensory area on the paracentral lobule. The posterior margin of this area is the marginal sulcus. On sagittal MR, the anterior limit of the entire paracentral lobule is the paracentral sulcus, either a side branch of the cingulate sulcus or an extension of the posterior part of the cingulate sulcus (21). This sulcus has no connection with the cingulate sulcus in only 12% of cases, and represents a small sulcus from the lateral surface of the brain.

CONCLUSION

The entire surface of the parietal lobe and all its division may be followed on MRI by recognition of the configuration of its main sulci on sagittal, horizontal, or coronal plane MR images. When the examination is made according to the commissural (AC-PC) plane, this task is easier because this permits correlations with a standard reference atlas (2,9,11). Any type of computed program using a navigation system simultaneously following these different parts of the parietal lobe is extremely useful.

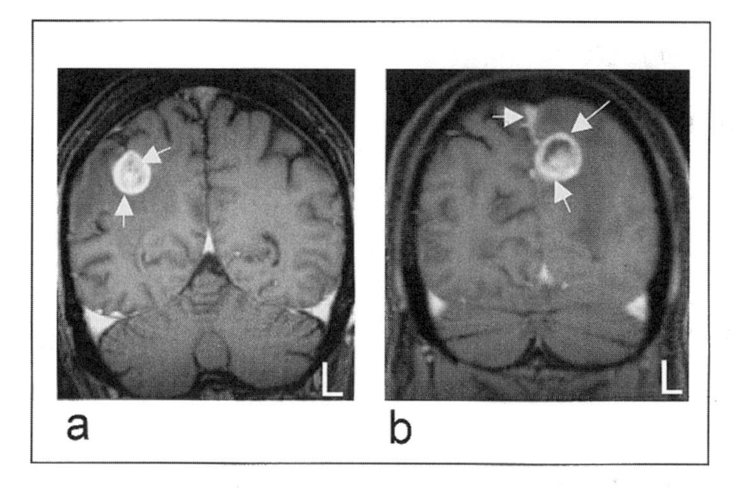

FIG. 2-16. Parietal tumors. Two locations of metastatic brain tumors on the parietal lobe. **A:** Metastasis (breast) in the depth of interparietal sulcus *(right side).* **B:** Metastasis (lung) with edema located at the precuneus *(left side).* A small contrast enhancement at the falx level reveals also a meningeal involvement.

MRI examination of the parietal lobe cannot be limited to its cortical surface. Parietal cortex is not only sensory, but also is an important heteromodal area of the brain with connections with other frontal, temporal heteromodal, and limbic areas. Some of the large connections represented by major white matter bundles may be followed on MRI. The projections of arcuate, cingulated, and superior longitudinal fasciculus fibers, as well as those of the optic radiations and callosal and corticospinal fibers may be analyzed on MRI (38,39). The main parts of the thalamus in relation to the parietal lobes, such as the ventro-posteroro-lateral (VPL) nucleus or the pulvinar may be analyzed on coronal or sagittal sections also (2,9).

The development of the cortical sulci in children can be followed on MRI (54,55). There is strong evidence that this development (56) is closely related to the development of the thalamus and basal ganglia. Several pediatric disorders with gyral malformations are very well documented with MRI (37,57). We have illustrated cases of parietal cortex anatomy in cases of tumors, vascular infarcts, and dementias (Figs. 2-16 to 2-21). They show (by comparison with normal analysis of the parietal lobe) the great importance of an in-depth knowledge of normal and variant parietal lobe gyral and sulcal anatomy for optimal analysis of pathologic parietal lobe imaging.

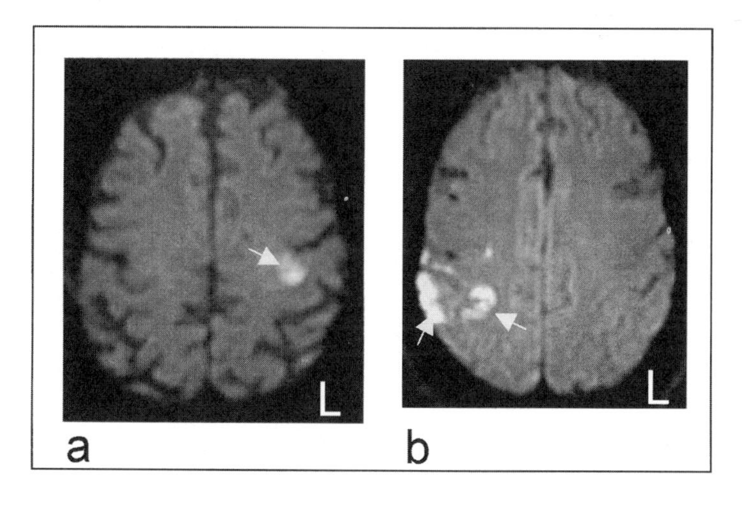

FIG. 2-17. Recent parietal infarcts (diffusion images). **A:** Location of infarct corresponds to the depth of the central sulcus and the most anterior part of the primary sensory area (rolandic artery). **B:** Infarct at inferior lobule level and in the depth of interparietal sulcus (angular gyrus artery).

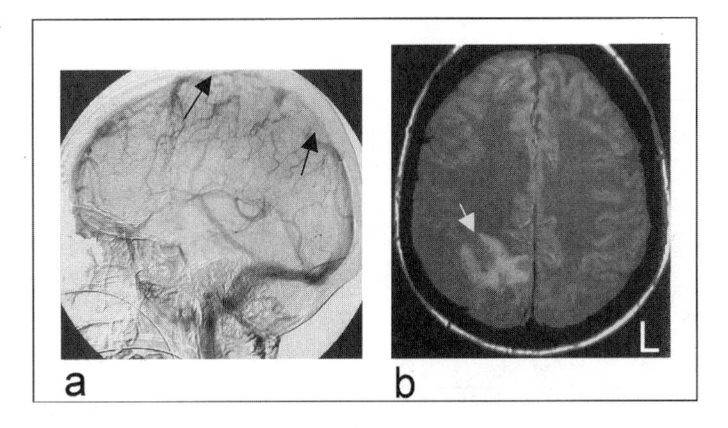

FIG. 2-18. Thrombosis of the superior sagittal sinus. **A:** Venous phase of carotid angiography with a defect at superior sagittal sinus level *(arrows).* **B:** Hemorrhagic infarct in the depth of parietal lobe between the superior parietal lobule and angular gyrus.

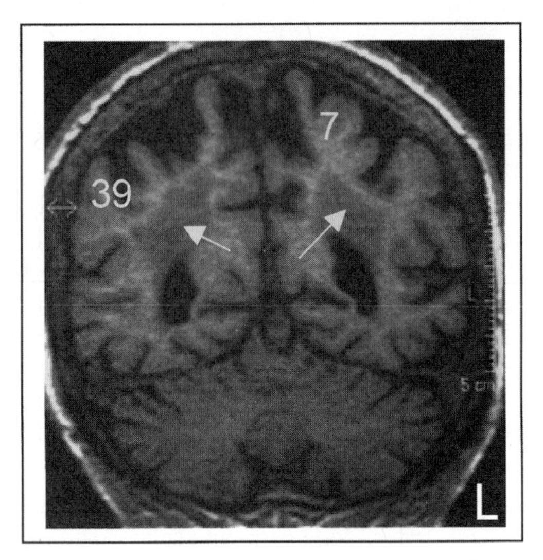

FIG. 2-19. Watershed vascular insufficiency. Female 76yo with hypertension. Bilateral images of parietal cortical atrophy (superior parietal lobule and angular gyrus) and deep images at white matter level, suggestive of vascular insufficiency *(arrows).*

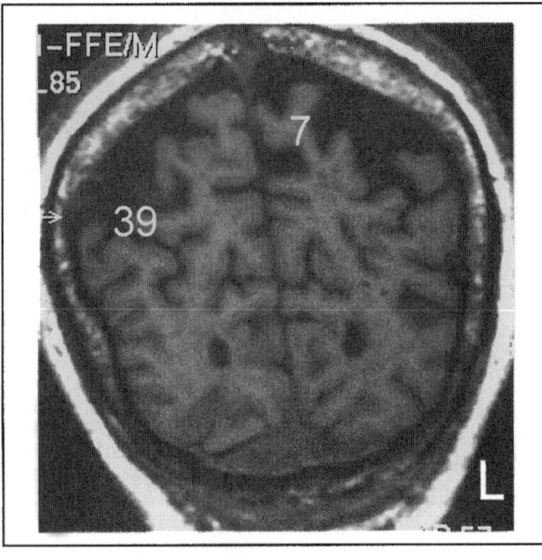

FIG. 2-20. Probable Alzheimer's disease. (76yo.mms:26). This is a patient with hippocampal atrophy. At the parietal lobe level, atrophy of the superior parietal lobule (7) is more important on the left side. Angular gyrus (39) atrophy is more marked on the right hemisphere.

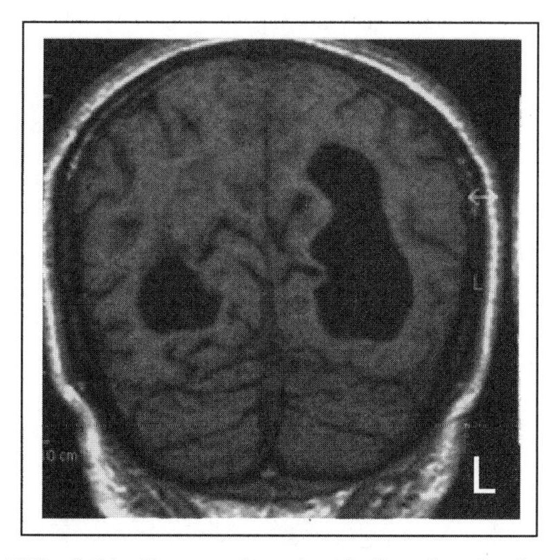

FIG. 2-21. Progressive visual disorder. At the time of the magnetic resonance image, this patient (76yo, female) had Balint syndrome with a history of progressive visual disorders for 5 years. Note important bilateral atrophy of the white matter on two sides at the parieto-occipital level with an important enlargement of the posterior horns of lateral ventricle (more on left).

REFERENCES

1. Chiavaras MM, Petrides M. Orbito frontal sulci of the human and macaque monkey brain. *J Comp Neurol* 2000;422:35–54.
2. Talairach J, Tournoux P. *Co-planar stereotaxic atlas of the human brain: 3-dimensional proportional system.* Stuttgart: Thieme, 1988.
3. Mazziotta JC, Toga A, Evans A, Fox, P, Lancaster J, Woods R. A probabilistic approach for mapping the human brain. In: Toga AW, Mazziotta JC, eds. Brain Mapping the Systems. San Diego: Academic Press, 2000.
4. Frackowiak RSJ, Friston KJ, Frith CD, Dolan RJ, Mazziotta JC. *Human brain function.* San Diego: Academic Press, 1997.
5. MacDonald JD, Avis D, Evans AC. Multiple surface identification and matching with magnetic resonance images. *Proc Soc Vis Biomed Comput* 1994:160–169.
6. Lohman G, von Cramon Y. Automatic labeling of the human cortical surface using sulcal basins. *Med Image Anal* 1999;2:1–12.
7. Friston KJ. Voxel-based morphometry: the methods. *NeuroImage* 2000;11(6):805–821.
8. Narr KL, Thompson PM, Sharma T, et al. Three-dimensional mapping of gyral shape and cortical surface asymmetries in schizophrenia: gender effects. *Am J Psychiatr* 2001;158:244–255.
9. Schaltenbrand G, Bailey P. *Introduction to stereotaxis with an atlas of the human brain.* Stuttgart: Thieme, 1959.
10. Nieuwenhuys R, Voogd J, van Huijzen Chr. *The human central nervous system.* Berlin: Springer, 1983.
11. Duvernoy HM. *The human hippocampus: an atlas of applied anatomy.* Munich: Bergmann, 1988.
12. Duvernoy HM. *The human brain surface: three-dimensional sectional anatomy and MRI.* Vienna: Springer-Verlag, 1991.
13. Mullick R. National Institutes of Health. Clinical image processing, 1999. *http://www.ccnih.gov/cip/software/etdips*
14. Parent A. *Carpenter's human neuroanatomy.* Baltimore: Williams & Wilkins, 1996.
15. Critchley M. *The parietal lobes.* London: Edward Arnold, 1953.
16. Williams PL, Warwick R, Dyson M, et al. *Gray's anatomy.* Edinburgh: Churchill Livingstone, 1989.
17. Geschwind N, Levitsky W. Human brain: left right asymmetries in temporal speech region. *Science* 1968;161:186–187.
18. Rumeau CL, Tzourio N, Murayama N, et al. Location of hand function in the sensorimotor cortex: MR and functional correlation. *Am J Neuroradiol* 1994;15(3):567–572.
19. Dejerine J. *Anatomie des centres nerveux.* Paris: Rueff, 1898 (republished, Paris: Masson, 1980).
20. Mesulam MM. *Principles of behavioral and cognitive neurology,* 2nd ed. New York: Oxford University Press, 2000.
21. Ono M, Kubik S, Abernathey CD. *Atlas of the cerebral sulci.* Stuttgart: Thieme, 1990.
22. Le May M. Morphological cerebral asymmetries of modern man, fossil man and non human primate. *Ann NY Acad Sci* 1976;280:349–366.
23. Habib M, Robichon F, Levrier O, et al. Diverging asymmetries of temporo-parietal cortical areas: a reappraisal of Geschwind-Galaburda theory. *Brain Language* 1995; 48(2):238–258.
24. Falk D, Hildebold C, Cheverud J, et al. Human cortical asymmetries determined with 3D MR technology. *J Neurosci Meth* 1991;39(2):185–191.
25. Sobel DF, Gallen CC, Schwartz BJ, et al. Locating the central sulcus: comparison of MR anatomic and magnetoencephalographic functional methods. *Am J Neuroradiol* 1993;14(4):915–925.
26. Yetzkin FZ, Papke RA, Mark LP, et al. Location of the sensorimotor cortex: functional and conventional MR compared. *Am J Neuroradiol* 1995;16(10):2109–2113.
27. Puce A, Constable RT, Luby ML, et al. Functional MRI of sensory and motor cortex: comparison with electrophysiological localization. *J Neurosurg* 1995;83(2): 262–270.
28. Meyer JR, Roychowdhury S, Russell EJ, et al. Location of the central sulcus via cortical thickness of the precentral and postcentral gyri on MR. *Am J Neuroradiol* 1996;17(9):1699–1706.
29. Amunts K, Schlaug G, Schleicher A, et al. Asymmetry in the human motor cortex and handedness. *NeuroImage* 1996;4(3):216–222.
30. Yousry TA, Schmid UD, Alkhadi H, et al. Localization of the motor hand area to a knob on the precentral gyrus. A new landmark. *Brain* 1997;120(10):141–157.
31. Lee CC, Jack CR, Riederer SJ. Mapping of the central sulcus with fMRI: active versus passive activation tasks. *Am J Neuroradiol* 1998;19(5):847–852.
32. Salamon G, Huang YP. *Radiologic anatomy of the brain.* New York: Springer, 1975.
33. Mangin JF, Frouin V, Bloch I, et al. From 3D MRI to structural representation of the cortex topography using

topology-preserving deformations. *J Math Imag Vision* 1995;5:297–318.
34. Steinmetz H, Ebeling U, Huang YX, et al. Sulcus topography of the parieto opercular region: an anatomic and MR study. *Brain Language* 1990;38(4):515–533.
35. Witelson SF, Kigar DL. Sylvian fissure morphology and asymmetry in men and women :bilateral differences in relation to handedness in men. *J Comp Neurol* 1992;323(3):326–340.
36. Damasio H. *Human brain anatomy in computerized images.* New York: Oxford University Press, 1995.
37. Tamraz JC, Outin C, Soussi B, et al. *Principles of MRI of the head, base of skull and spine, with clinical correlations.* Berlin: Springer, (in press).
38. Kasai K, Salamon-Murayama N, Levrier O, et al. Theoretical situation of brain white matter tracts evaluated by three-dimensional MRI. *Surg Radiol Anat* 1996;18(4): 295–302.
39. Peretti-Viton P, Azulay JP, Trefouret S, et al. MRI of the intracranial corticospinal tracts in amyotrophic and primary lateral sclerosis. *Neuroradiology* 1999;41(10):744–749.
40. Petrides M, Pandya DN. Projections to the frontal cortex from the posterior parietal region in rhesus monkey. *J Comp Neurol* 1984;228(1):105–116.
41. Seltzer B, Pandya DN. Posterior parietal projections to the intra parietal sulcus of the rhesus monkey. *Exp Brain Res* 1986;62(3):459–469.
42. Cavada C, Goldman-Rakic PS. Posterior parietal cortex in rhesus monkey: evidence for segregated corticocortical networks linking sensory and limbic areas with the frontal lobe. *J Comp Anat* 1989;287(4):422–445.
43. Morecraft RJ, Geula C, Mesulam MM. Architecture of connectivity within a cingulo-fronto-parietal neurocognitive network for directed attention. *Arch Neurol* 1993;50(30):279–284.
44. Gitelman DR, Nobre AC, Parrish TB, et al. A large-scale distributed network for covert spatial attention :further anatomic delineation based on stringent behavioural and cognitive controls. *Brain* 1999;122(6):1093–1106.
45. Ebeling U, Steinmetz H. Anatomy of the parietal lobe:mapping the individual pattern. *Acta Neurochir* 1995;136(1–2):8–11.
46. Nowinski WL, Bryan RN, Raghavan R. *The electronic clinical brain atlas. Multiplanar navigation of the human brain.* (based on the classic Talairach-Tournoux and Schaltenbrand atlases) New York: Thieme, 1997.
47. Roland PE, Gulyas B. Visual memory, visual imagery and visual recognition of large field patterns by the human brain: functional anatomy by PET. *Cereb Cortex* 1995;5(1):79–93.
48. Demonet JF, Price C, Wise R, et al. Differential activation of right and left posterior sylvian regions by semantic and phonological tasks: a PET study in normal human subjects. *Neurosci Lett* 1994;182(1):25–28.
49. Mazoyer BM, Dehaene S, Tzourio N, et al. The cortical representation of speech. *J Cog Neurosci* 1993;5: 457–479.
50. Ghaem O, Mellet E, Crivello F, et al Mental navigation along memorized routes activates the hippocampus, precuneus and insula. *Neuroreport* 1997;8(3): 739–744.
51. Fletcher PC, Frith CD, Grasby PM, et al. Brain systems for encoding and retrieval of auditory-verbal memory. An vivo study in humans. *Brain* 1995;118(2): 401–416.
52. Platel H, Price C, Baron JC, et al. The structural components of music perception. A functional anatomic study. *Brain* 1997;120(2):229–243.
53. Naidich TP, Blum JT, Firestone MI. The parasagittal line: an anatomic landmark for axial imaging. *Am J Neuroradiol* 2001;22(5):885–895.
54. Salamon G, Raynaud M, Regis J, et al. *Magnetic resonance imaging of the pediatric brain. An anatomic atlas.* New York: Raven Press, 1990.
55. Naidich TP, Grant JL, Altman N, et al. The developing cerebral surface. Preliminary report on the patterns of sulcal and gyral maturation-anatomy, ultrasound and MRI. *Neuroimag Clin North Am* 1994;4(2):201–240.
56. Walker W. Why does cerebral cortex fissure and fold? A review of the determinants of gyri and sulci. In: Jones EG, Peters A. *Cerebral cortex.* New York: Plenum, 1990:3–110.
57. Barkovich AJ. *Pediatric neuroimaging,* 3rd ed. Philadelphia: Lippincott Williams & Wilkins, 2001.

3

Subcortical Projections of the Parietal Lobes

Mitchell Glickstein

Department of Anatomy, University College London, London, England

INTRODUCTION

The focus of this chapter is on the subcortical projections of the parietal lobe, with an emphasis on its role in the sensory guidance of movement. I begin by briefly outlining the history of cortical localization with special reference to the parietal lobes. Early clues as to their functions came from study of monkeys and patients who sustained similar brain lesions. One of the cardinal symptoms of parietal lobe lesions in monkeys and man is a deficit in the visual guidance of movement. Anatomic studies on the connections from parietal lobe visual areas to motor areas of the brain have emphasized the role of corticocortical pathways. Here I review some of the subcortical targets, organizational principles, and possible functions of the subcortical connections. I briefly describe and contrast the projections to three of these; the superior colliculus, basal ganglia, and pontine nuclei. All three are probably involved in one or another aspect of visuomotor control. Finally, I discuss possible functions of reciprocal connections from cerebellum back to the parietal lobe visual areas.

Two principles about the structure and functions of the nervous system were established in the last half of the 19th century. Both remain fundamental to our thinking about the brain and spinal cord. One of these is the neuron doctrine; the idea that the brain and spinal cord are made up of individual elements, called neurons and their supporting elements. The processes of neurons may touch one another, but they do not fuse. The neuron doctrine is an easy history to trace. It is based almost entirely on the staining method of Camillo Golgi, and the brilliant anatomic studies and synthesis of Santiago Ramon y Cajal. The other principle is that of cerebral localization; the idea that different parts of the brain, especially the cerebral cortex, are involved in different functions. Cortical localization has a more complex history. Fritsch and Hitzig (1) first demonstrated that electrical stimulation of a specific region of the frontal lobe of a dog produces movement of a limb or the face on the opposite side of the body. Eleven years later Hermann Munk (2) reported that unilateral ablation of the occipital lobe of a monkey causes blindness in the opposite half of the visual field. By 1900, movement, vision, hearing, and somatic sensation were ascribed to specific locations on the cerebral cortex. Despite progress in the study of cortical localization, the functions of large areas outside of primary sensory and motor cortex, including much of the parietal lobe, remained less well understood.

Various suggestions had been put forward about the possible functions of the cortex between the primary visual, somatosensory, and auditory areas. One idea was that this parietal lobe region might serve as an "association area" in which input from the three senses could be blended. Alternatively, the parietal lobe might be a locus of memory storage; a place to store visual, auditory, or somatosensory memories. However, evidence for any of these proposed functions remained vague and indirect.

In the early part of the 20th century, study of the symptoms suffered by patients with parietal lobe lesions began to provide evidence about its possible functions. One of the most important contributions was by Rudolf Balint, who described the symptoms of a patient who had suffered bilateral damage to the parietal lobes. Figure 3-1 is from Balint's paper (3), illustrating the location of the lesions in the parietal lobes. Both parietal lobes were massively damaged. The damage to the left primary motor and somatosensory cortex occurred much later, shortly before the patient died (Fig. 3-1).

Balint studied his patient over a period of two and a half years, and described in detail the observed deficits (3). The patient seldom moved his eyes spontaneously, although he could shift his gaze when he was encouraged to do so. He also had great difficulty in estimating the distance of objects in his visual field and responding appropriately to them.

Balint wrote:

> . . . when he hears the sound of a wagon or of an electric tram, he can not estimate the distance, and so he is afraid that a misfortune might come to him. Because of this, for some time he has hardly ever gone out on the street.

Balint's patient was severely impaired in the visual guidance of his arms and hands.

> In the movement of his right hand however, a definite anomaly is obvious. He himself says that among his difficulties is that when he lights a cigar, he lights it in the middle rather than at the end. It often happened that when he was cutting a piece of meat on his plate, holding a fork in his left hand, the knife, which he held in his right hand, landed outside the plate. . . . Also, while searching in the room he made great mistakes. If he was instructed to take hold of an object in front of him with his right hand he typically reached beside it, and only succeeded in finding it when it bumped against his hand.

Balint also emphasized that his patient had a marked deficit in attention, an observation that was confirmed by subsequent authors (4,5). In severe cases, following unilateral lesions there may be neglect of the opposite side of the body as well as contralateral visual space (6).

However, although the symptom of neglect generally has been attributed to damage to the parietal lobe, the critical locus of the lesion that causes spatial neglect is in some dispute. Karnath and his colleagues (7), for example, have argued that it is a lesion in the temporal, not the parietal lobe that is associated with neglect.

The most universally agreed class of symptoms caused by parietal lobe lesions is sensorimotor disturbance. People and monkeys are impaired in their ability to guide the eyes or limbs under visual control. The ini-

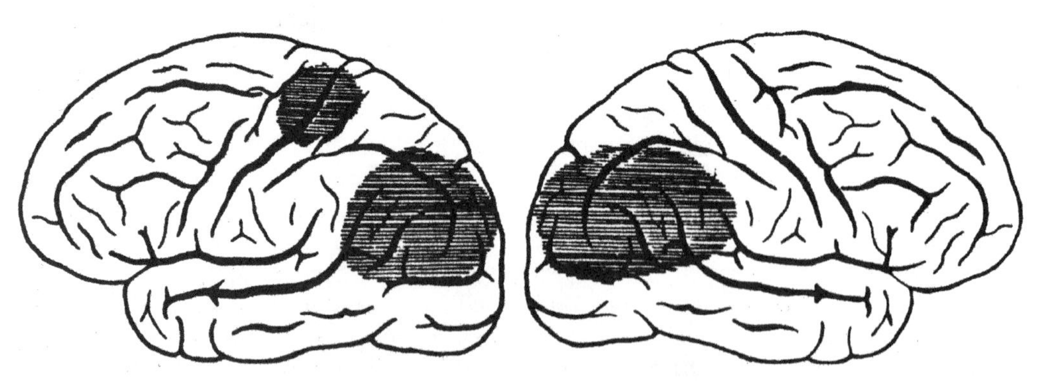

FIG. 3-1. Balint's (1909) drawing of the postmortem findings of bilateral parietal lobe lesions in his patient. The unilateral lesion of pericentral cortex occurred much later than the lesions to the parietal lobes (3).

tial symptoms caused by the lesion often are severe. David Ferrier (8) ablated the angular gyrus of the parietal lobes in monkeys and mistakenly believed that he had identified the primary visual cortex, because the animal appeared to be blind after the surgery. In his first experiments, Ferrier observed his animals for only 3 days after he operated on them, because if left alive any longer, they would inevitably develop fatal infections. Ferrier's conclusions were sharply criticized by Hermann Munk, who reported that unilateral occipital lesions produce hemianopia; bilateral lesions, blindness (2). Munk argued that Ferrier had made at best only speculative conclusions based on inadequate observation.

Ferrier never acknowledged the validity of Munk's critique, although in later experiments he adopted sterile surgical procedures and his animals could now live for days, weeks, or years. With Gerald Yeo (9), Ferrier replicated his earlier studies of the effects of angular gyrus lesions. Ferrier's protocols now made it clear that the animals recovered to a great extent from the severity of the initial symptoms. It was now clear that the lesions had not produced blindness, but a profound deficit in the visual guidance of movement. Ferrier wrote:

> The cherry was laid on the floor in front of it, but it was unable to find it, though looking eagerly for it. The animal enjoyed its food which it found by groping about with its hand in the cage.
>
> On the fourth day there was some indication of returning vision. A piece of orange was held before it, whereupon it came forward in a groping manner and tried to lay hold but missed repeatedly. When the piece of orange was laid on the floor it stretched out its hand over it, short of it and round about it before it succeeded in securing it.
>
> On the fifth day the animal came out of its cage spontaneously and walked about. It never knocked its head. It was evidently able to see its food, but constantly missed laying hold of it, at first putting its hand beyond it or short of it.
>
> On the sixth day the animal walked about freely avoiding obstacles, but vision was evidently defective, as on several occasions it was seen as if about to climb before it had come sufficiently near the ledge which it wished to mount. It was, however, able to pick up grains of rice scattered on the floor, but always with uncertainty as to the exact position.

Ferrier, Balint, and subsequent authors interpreted the deficits caused by parietal lobe lesions as visual. Ferrier's monkeys were aware of the food, and were eager to get it, but made errors in reaching toward it. What Ferrier and Balint actually observed after parietal lobe lesions was not a visual loss, but a severe deficit in the visual guidance of movement, a deficit so severe that Ferrier had misinterpreted it as blindness.

Ferrier's monkeys misdirected their arms when attempting to grasp a bit of food. Parietal lobe lesions also impair performance in a task in which monkeys are required to orient their wrists and fingers under visual guidance (10). Kuypers and his colleagues (11) devised a task in which a monkey had to align its thumb and forefinger along a narrow groove in a disc order to retrieve a bit of food. Figure 3-2 illustrates the disc, and Figure 3-3 shows the performance by a monkey after successive parietal lobe lesions.

In addition to its role in visual guidance of the limbs, the parietal lobe has connections to brain structures involved in oculomotor control. Balint's patient seldom initiated voluntary eye movements. In monkeys and people the lesions cause an increase in the latency and a decrease in the accuracy of saccadic eye movements (12).

Parietal lobe lesions impair visual guidance of movements of the eyes, limbs, or the whole body. How are parietal lobe visual areas connected to brain structures that control movement? In principle the connection to motor areas of the brain might be by a series of corticocortical connections. Alternatively, the link could be by way of a subcortical pathway.

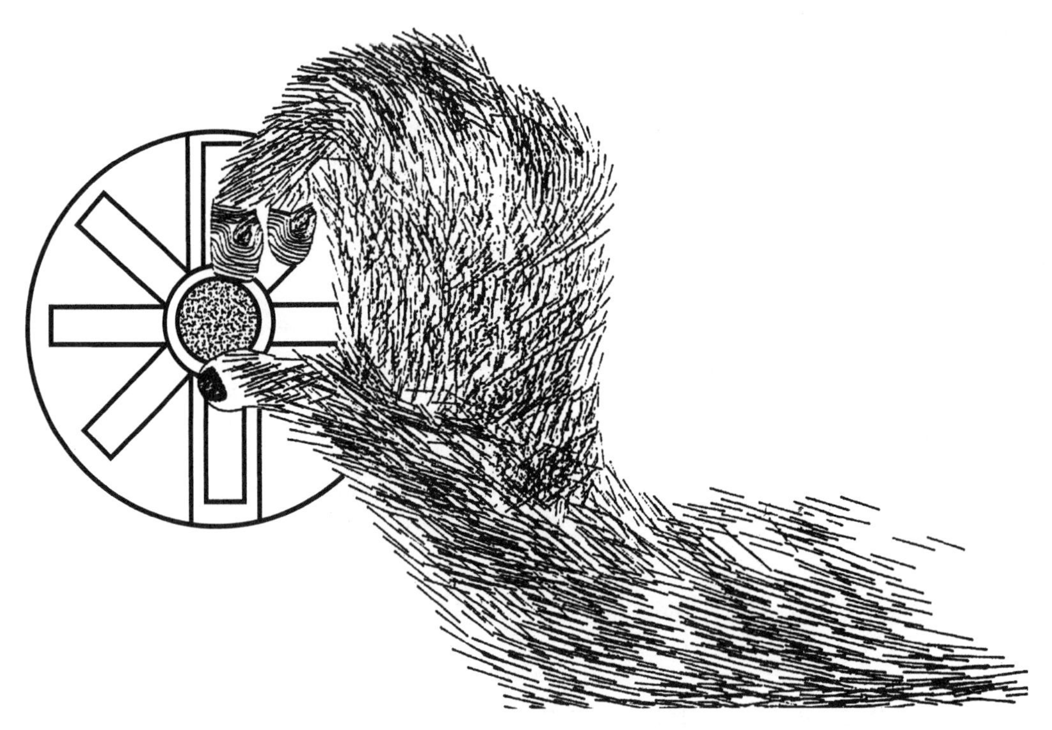

FIG. 3-2. The drawing illustrates the "Kuypers disc." The monkey had to orient its fingers and wrist properly to retrieve the reward. (Redrawn from Ref. 10, with permission.)

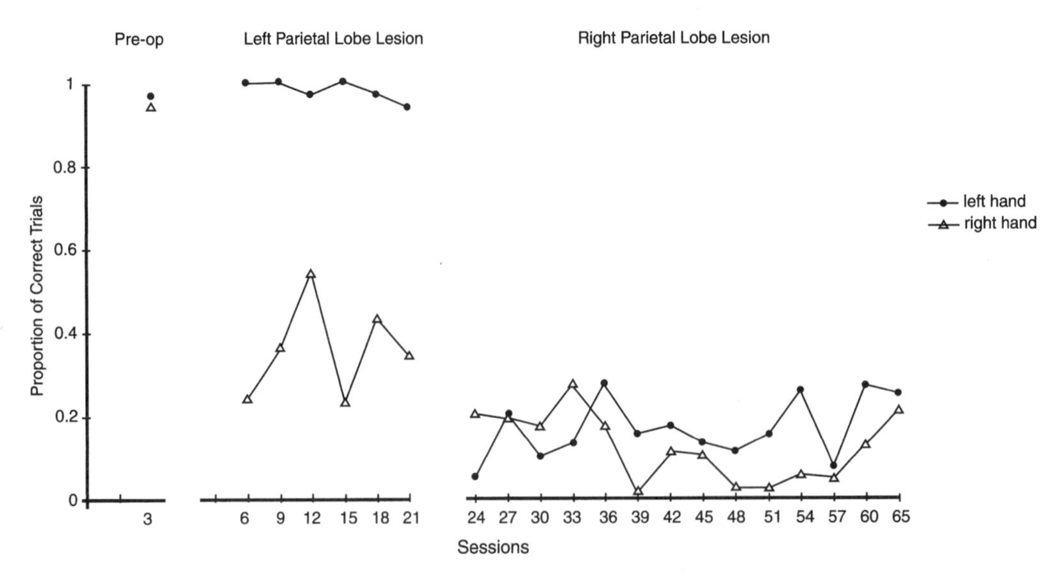

FIG. 3-3. Performance of a monkey on the Kuypers disc after successive ablation of the parietal lobes. (From Ref. 10, with permission.)

THE "OBVIOUS" IMPORTANCE OF CORTICOCORTICAL CIRCUITS

Anyone who has dissected fixed human brain tissue is soon aware that the cortex contains an elaborate network of corticocortical connections. U-fibers link adjacent gyri, and long corticocortical association bundles join remote cortical areas with one another. It seemed only natural to many authors to assume that sensory guidance of movement involves a series of connections from primary through secondary sensory cortical areas, ultimately to motor or premotor cortex. However, although they are prominent, corticocortical circuits cannot account for all instances of the sensory guidance of movement. If vision is restricted to one hemisphere, and all of the forebrain commissures are cut in the midline, monkeys can still guide either arm rapidly toward a visual target using either eye (13). They can also reach accurately after all intracortical fibers have been severed by making deep cuts in the white matter between the parietal lobe visual areas and the frontal motor areas. Brinkman and Kuypers (14) suggested that when vision is restricted to one hemisphere, and the corpus callosum is cut, the monkey uses the surviving corticocortical links from visual cortex to the ipsilateral motor cortex. The reach is then directed by a nondecussating descending corticospinal pathway that connects to neurons in the cord that control proximal muscles. However, although this pathway may play a role, as will become clear, there are alternative subcortical routes, by way of the pons and cerebellum, whereby visual cortical areas on one side of the brain can access the motor areas of the opposite hemisphere.

Three subcortical targets of the parietal lobe visual areas probably play a role in various aspects of visually guided movement. The parietal lobes project to the basal ganglia, the superior colliculus and, by way of a relay in the pontine nuclei, the cerebellum. I discuss here some broad questions about these pathways. Do all of the extrastriate cortical visual areas, including those in the temporal as well as the parietal lobe, project to all of the subcortical targets? Within a given cortical area, is there segregation of cells that project to different subcortical targets? Do descending axons bifurcate so that the same visual information is sent to more than one target? Although the focus of this chapter is on the role of the parietal cortex in visuomotor guidance, I begin with a discussion of efferent fibers from the somatosensory cortex of rats, because the pattern of subcortical projection is particularly clear in these animals.

PROJECTIONS FROM THE RAT SOMATOSENSORY CORTEX ARISE FROM INDEPENDENT GROUPS OF CORTICAL CELLS

Study of subcortical targets of the parietal lobe of rats reveals a virtually complete laminar segregation of the cells that project to subcortical targets. Most of the projections to subcortical targets, other than reciprocal connections to the thalamus, are pyramidal cells located in lamina V. This lamina of the rat somatosensory cortex is divisible into two obvious sublayers; a superficial lamina Va, which is pale in cytochrome oxidase preparations, and a deep lamina Vb, which is much darker.

Injection of a retrograde tracer substance into the pontine nuclei or the basal ganglia of rats reveals an almost complete segregation of cells that project to the two targets (15). Figure 3-4 shows the labeled cells in the somatosensory cortex of a rat that had been retrogradely labeled following a WGA-HRP injection into the pontine nuclei.

In the area illustrated every pyramidal cell in lamina Vb was retrogradely labeled following a large injection of WGA-HRP into the pontine nuclei. When the tracer was injected into the basal ganglia, nearly all of the retrogradely labeled cells were in layer Va. A narrow tier of cells at the border between lamina Va and Vb probably project to both targets. Layer Vb cells receive a direct input from the thalamus and they are activated at a short latency following natural stimulation (16). Layer Va cells are activated at a longer latency

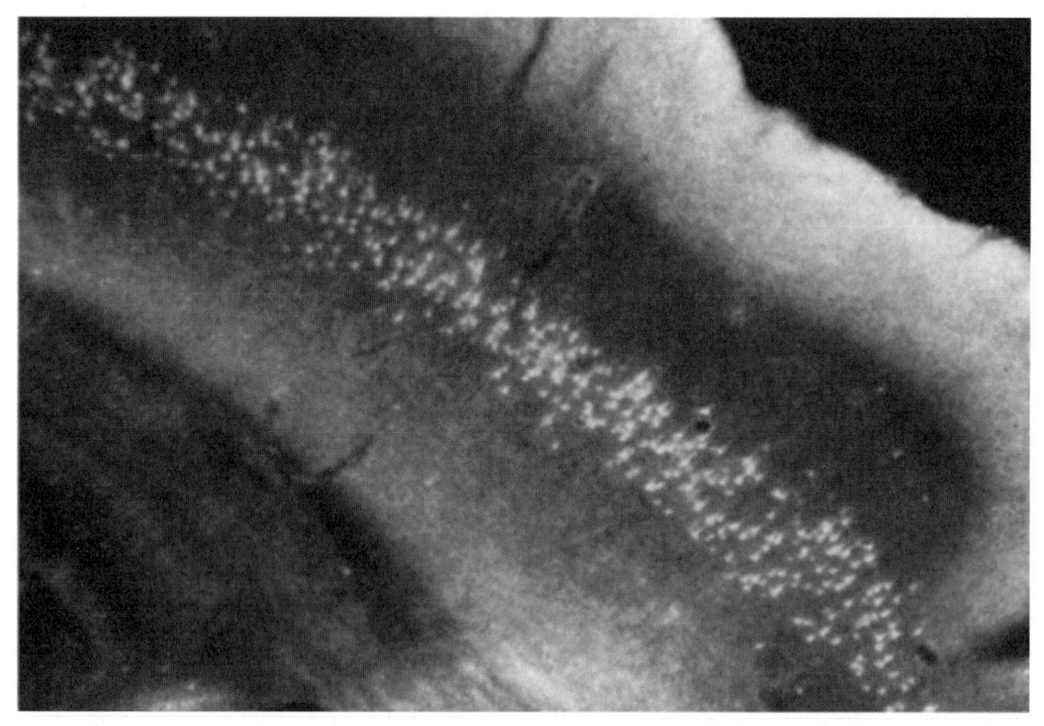

FIG. 3-4. Darkfield photomicrograph of labeled cells in layer Vb of rat somatosensory cortex.

via an intracortical relay. The rapid relay of information to the cerebellum by cells in layer Vb, and behavioral deficits in animals in which this pathway is interrupted at the level of the cerebral peduncles (17), suggest that one function of the pathway to the cerebellum is the sensory control of ongoing movements.

CONNECTIONS TO THE SUPERIOR COLLICULUS

Ferrier first suspected a visual role for the angular gyrus, because electrical stimulation in this parietal lobe area caused eye movements (8). The eye movements are probably evoked by way of connections to the superior colliculus. Saccadic eye movements elicited by stimulation of the parietal lobe visual areas are abolished by ablation of the superior colliculus (18). Parietal lobe visual areas project densely to intermediate and deep layers of the superior colliculus (19). The projection is particularly dense from the lateral intraparietal area (LIP), a region of the parietal lobes whose cells are active prior to saccadic eye movements (20). Cells in area LIP that are activated antidromically following electrical stimulation of the superior colliculus are active prior to saccadic eye movements (21). There is a systematic difference in the average latency of cell activity in LIP and the colliculus relative to the eye movement. Parietal lobe visual cells are activated earlier, and they are more closely tied to the sensory stimulus than to the execution of the saccade. Collicular neurons are activated later, and they are more closely tied to the eye movement. Paré and Wurtz (21) suggested that the circuit from parietal lobe to the colliculus constitutes "a progressive evolution in the neuronal processing for saccades."

Because of the complex shape and folding of cortical fissures, it is often difficult or impossible to specify the laminar depth of a particular neuron whose activity is being recorded. Because cells in different cortical layers have markedly different patterns of connectivity and response properties, recording studies may pool data from totally different classes of neurons. The Paré and Wurtz (21) recordings were from cells antidromically activated from the colliculus, consequently almost certainly in lamina V of the parietal lobe cortex. The technique they used offers a way to establish the target and location of a given cortical cell, and is helpful in working out the pattern of neuronal discharge in relation to a given movement.

SOME LAMINA V CELLS PROJECT TO TWO OR MORE SUBCORTICAL TARGETS

In the rat somatosensory cortex most of the projections to the pontine nuclei and basal ganglia arise from different cells. In other cases axons may bifurcate to innervate two or more subcortical targets. If two different retrograde tracers are placed into the superior colliculus and the pontine nuclei of cats, many cells in an extrastriate visual area are double labeled (22). The same class of visual cells may be activated antidromically by electrical stimulation both from the superior colliculus and the pontine nuclei (23). The latency of firing following electrical stimulation at the two subcortical sites demonstrated that the activation was by way of a bifurcated axon collateral; one branch projecting to the superior colliculus, the other to the pontine nuclei. These layer V cells are particularly sensitive to targets moving at a preferred speed and in a preferred direction Thus, visual motion information is relayed simultaneously to two independent targets; the colliculus for guiding movements of the eyes and neck, the other to the cerebellum, perhaps for guiding the limbs.

PROJECTIONS TO THE BASAL GANGLIA

All of the extrastriate visual areas, including those in the temporal as well as parietal lobes, project to the basal ganglia (24,25). Although there are few or no projections from the inferotemporal cortex to the colliculus or pons, there is an orderly projection from inferotemporal as well as parietal visual areas to the basal ganglia. Parietal lobe visual areas project to a midregion in the body of the caudate nucleus. Inferotemporal visual areas project to a nonoverlapping region in the tail of the caudate nucleus. The projections are to the region of the caudate nucleus that is nearest to the cortical site. In addition to this orderly projection from extrastriate visual areas, there is a more diffuse system in which there is a widespread projection to these same areas of the caudate nucleus from the anterior cingulate cortex.

These findings reflect major differences in the outputs of two classes of extrastriate visual areas. All extrastriate visual areas, including the inferotemporal cortex, project to the basal ganglia. There are few or no projections to the pontine nuclei from inferotemporal visual areas. Nearly all of the cortical visual input to the pons and superior colliculus comes from the parietal lobe. These differences in the pattern of projection to subcortical targets must reflect a major difference in the function of the two pathways. The projection to the colliculus and pontine nuclei probably represents a pathway that is involved directly in the sensory guidance of movement. The projection to the basal ganglia may carry more highly processed visual cues to trigger rather than guide complete movements, but it remains poorly understood.

PROJECTIONS FROM THE PARIETAL LOBE TO THE PONTINE NUCLEI

The parietal lobe can be broadly divided into two major divisions; one of which is dominated by its somatosensory, the other by its visual input. Rostral subdivisions of the parietal lobe include the primary somatosensory cortex, areas 3, 3a, 1 and 2, and the adjacent area 5. The more caudal areas of the parietal lobe receive direct and indirect input from the primary visual cortex. The focus of this review is on these, more caudal regions and their role in the visuomotor control.

ON THE ROUTE FROM CEREBRAL CORTEX TO THE PONTINE NUCLEI

All of the efferent fibers from the cerebral cortex travel in the internal capsule. Descending fibers are collected at the level of the midbrain into the large crura cerebri at the base of the cerebral peduncles. Fibers in this great efferent pathway maintain a topographic relationship to the cortical area from which they arise. If a small amount of tracer substance is injected into a specific area of the rat cerebral cortex, the course of the fibers can be followed in the cerebral peduncles (26). Fibers that arise from the frontal cortex travel in the most ventromedial region of the peduncle. Fibers from occipital and temporal cortex travel in the dorsolateral corner. Fibers from somatosensory cortex travel between these two areas. A similar arrangement is present in the internal capsule and cerebral peduncles of the human brain (27). The evidence in humans is sparse, because it is based on a limited number of postmortem cases in which degenerating fibers could be detected in the peduncle after a prior cortical lesion.

The arrangement of fibers in the efferent pathway from the parietal lobe through the internal capsule and cerebral peduncles can help to interpret clinical observations. Classen et al. (28) studied a patient who had suffered a major disturbance in visuomotor performance following a subcortical infarct. The lesion invaded the ventral posterior thalamus, and also interrupted the fibers traveling in the most caudal limb of the internal capsule. The lesion thus interrupted fibers en route from the parietal lobe visual areas to the pontine nuclei, thereby blocking the relay of visual information to the cerebellum. The observed visuomotor deficits probably followed from this loss of visual input to the cerebellum.

CORTICOPONTINE PROJECTIONS FROM THE PARIETAL LOBE

In all mammals studied there is a major projection from cortical visual areas to the pontine nuclei. Details of this pathway vary among different mammals. In rats, for example, all known visual areas project to the pontine nuclei (29). In cats there also are projections from all known visual areas (30,31), but there is great variability in the density of the projections. In monkeys, nearly all of the visual input is supplied by extrastriate areas, although there is a very small projection to the pontine nuclei from a restricted area of the striate cortex itself (32,33). If a retrograde tracer substance is injected so as to fill the pontine nuclei, labeled cells are found in about half of the cerebral cortex. Figure 3-5 illustrates the results in one such case.

Labeled cells are seen in extrastriate areas beginning with area MT (V5) and extending rostrally to include visual areas on the banks of the superior temporal and intraparietal sulci and the adjacent gyri. The visual areas had a higher density of labeled cells than the primary and secondary somatosensory areas. Few or no labeled cells were found in the inferotemporal cortex. The projection to the pontine nuclei is dominated by cortical areas whose cells respond to moving, directionally specific targets, making them particularly appropriate for guiding movement (34,35).

The pattern of termination of visual fibers within the pontine nuclei varies among species. In monkeys and rats, the major target of visual efferent fibers is to a region in the dorsolateral corner of the pontine nuclei, although some axon terminals extend well beyond this focus. In cats, pontine projections from extrastriate visual areas are distributed to a more central location within the pontine nuclei (36). Area 18 of the cat visual cortex projects to a central region, close to the corticospinal fibers as they traverse the pons. Fibers arising from cells in the lateral suprasylvian visual area are distributed in an annular arrangement outside the area 18 targets. The pattern of anatomic termination suggests that the two cortical visual areas may be specialized for two different sorts of movement. The pontine cells that receive their input from area 18 send their axons principally to the vermis. Fibers that arise from cells in the lateral suprasylvian areas terminate on pontine cells

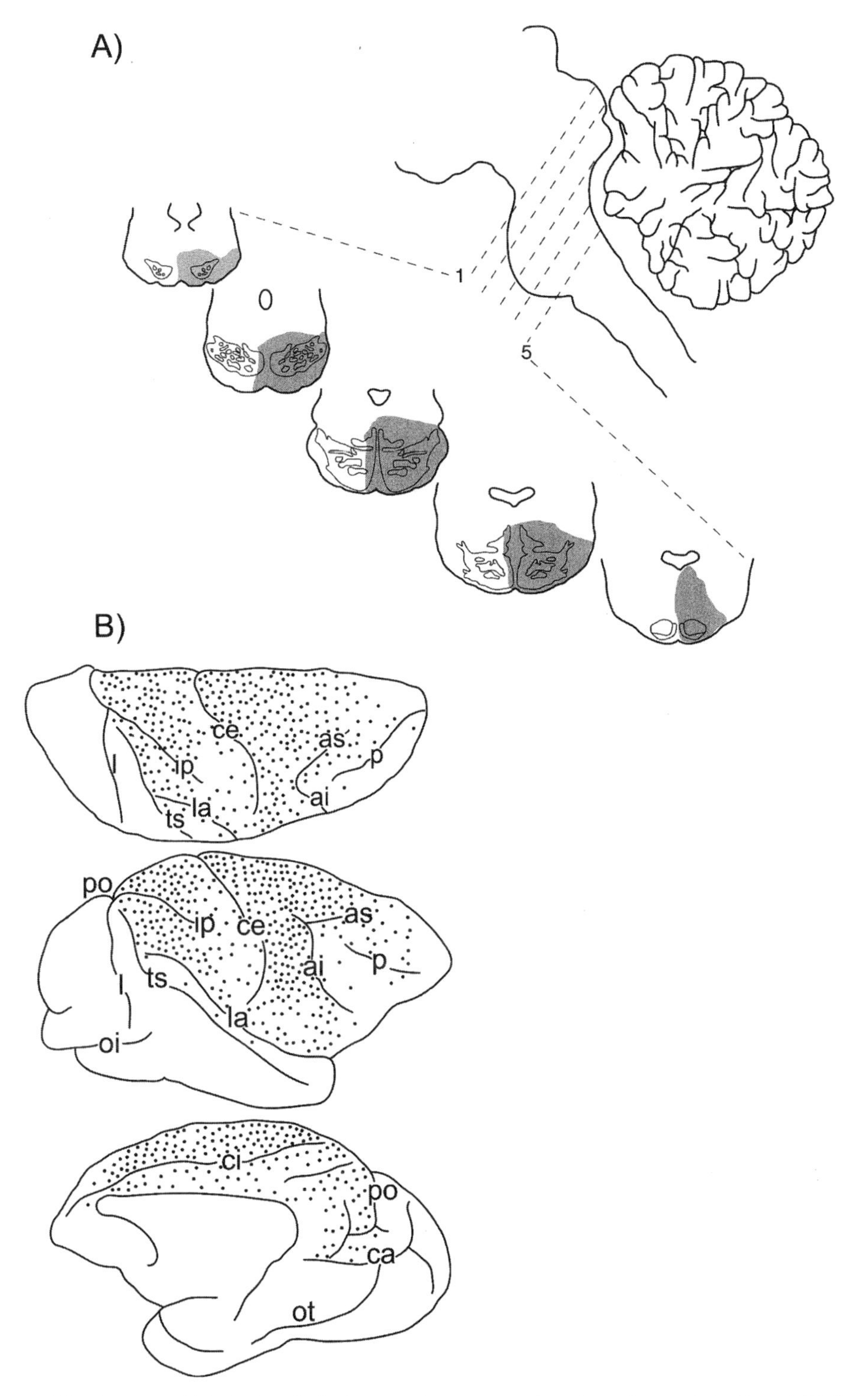

FIG. 3-5. Distribution of labeled cells in the monkey cerebral cortex after an injection of WGA-HRP into the pontine nuclei. (Redrawn from Ref. 33, with permission.) **A:** Diagram illustrating the size and rostrocaudal extent of the pontine injection. **B:** Dorsal, ventral and medial views of the right hemisphere with a reconstruction of the location of labeled cells. (ai, inferior ramus of arcuate sulcus; as, superior ramus of arcuate sulcus; ca, calcarine fissure; ip, intraparietal sulcus; l, lunate sulcus; la, lateral (sylvian) fissure; oi, inferior occipital sulcus; ot, occipitotemporal sulcus; p, sulcus principalis; po, parieto-occipital incisure; st, superior temporal sulcus)

that project to the cerebellar hemispheres. These data suggest that the visual input from area 18 to the vermis may be specialized for controlling whole body movements. The visual input from the suprasylvian cortex may be used for controlling movement of individual limbs.

DEMAGNIFICATION IN THE CORTICOPONTINE PATHWAY

Cells in the pontine nuclei can be activated by appropriate visual targets (30). In cats, the receptive fields of pontine visual cells are typically much larger than those of the cortical cells that project to them. A typical receptive field of a cell in lamina V in area 18 or the suprasylvian area of the cat cortex might be on the order of 3×4 degrees. Cells in the pontine nuclei often have large receptive fields that may extend to include an entire hemifield. In most mammals, the central visual fields are overrepresented in visual cortical areas. If pontine cells were to sample visual input uniformly from the cortex, overrepresentation of the central visual fields would present a challenge. Consider the cortical representation of a target moving uniformly across the visual field. The image of the target would move slowly across the region of cortex which maps the central visual field, and increasingly rapidly as the target moved toward the periphery. There must be connections from a wide area of cortex to single cells in the pontine nuclei to construct the large pontine receptive fields. If a pontine cell received an equal projection from central and peripheral field representation, it would respond optimally to different speeds as the target moved across the visual field. The projection from cortex to pons corrects for this by demagnifying the cortical projection. One source of evidence for demagnification comes from a study of the projections of the lateral suprasylvian visual area of a cat (37). Throughout much of its extent, the periphery of the visual field is represented at the lip of the suprasylvian fissure (labeled 1 on the inset in Figure 3-6), and the centre of the visual field at its depth (labeled 5 on the inset in Figure 3-6). When HRP was injected into the pontine nuclei, the number of cells that were retrogradely filled varied systematically from the lip to the depth of the fissure. We divided the cortex along the fissure into five equal areas, and counted the number of retrogradely filled cells in each. As Figure 3-6 illustrates, there was consistently a greater density of retrogradely labeled cells in the region corresponding to the peripheral visual field. The projection demagnifies the cortical projection.

The superior colliculus, like the cerebral cortex, maintains an overrepresentation of the central visual fields. Control injections into the superior colliculus in three other cats showed that the retrogradely filed cells were uniformly distributed in the suprasylvian cortex.

A similar demagnification is present in the projection to the pontine nuclei from the striate cortex of monkeys. Although there is only a very small projection to the pontine nuclei from area 17 itself, there are a few small labeled pyramidal cells in the cortex of the upper bank of the calcarine fissure. These labeled cells are restricted to the representation of the lower peripheral visual field. Figure 3-7 shows the number of cells found on the banks of the calcarine fissure at successive distances from the occipital pole to its extreme anterior end.

Here again, there is clear evidence that this projection systematically demagnifies the corticopontine projection.

PONTOCEREBELLAR PROJECTIONS

Pontine cells project to both cerebellar hemispheres (38). Although most pontocerebellar fibers terminate in the contralateral hemisphere, some fibers terminate in the cerebellum on the ipsilateral side. A bilateral corticopontocerebellar projection would still be present in the absence of corticocortical or commissural connections between the hemispheres. Savaki and her colleagues studied a monkey in which vision was restricted to one hemisphere, and all of the commissural links

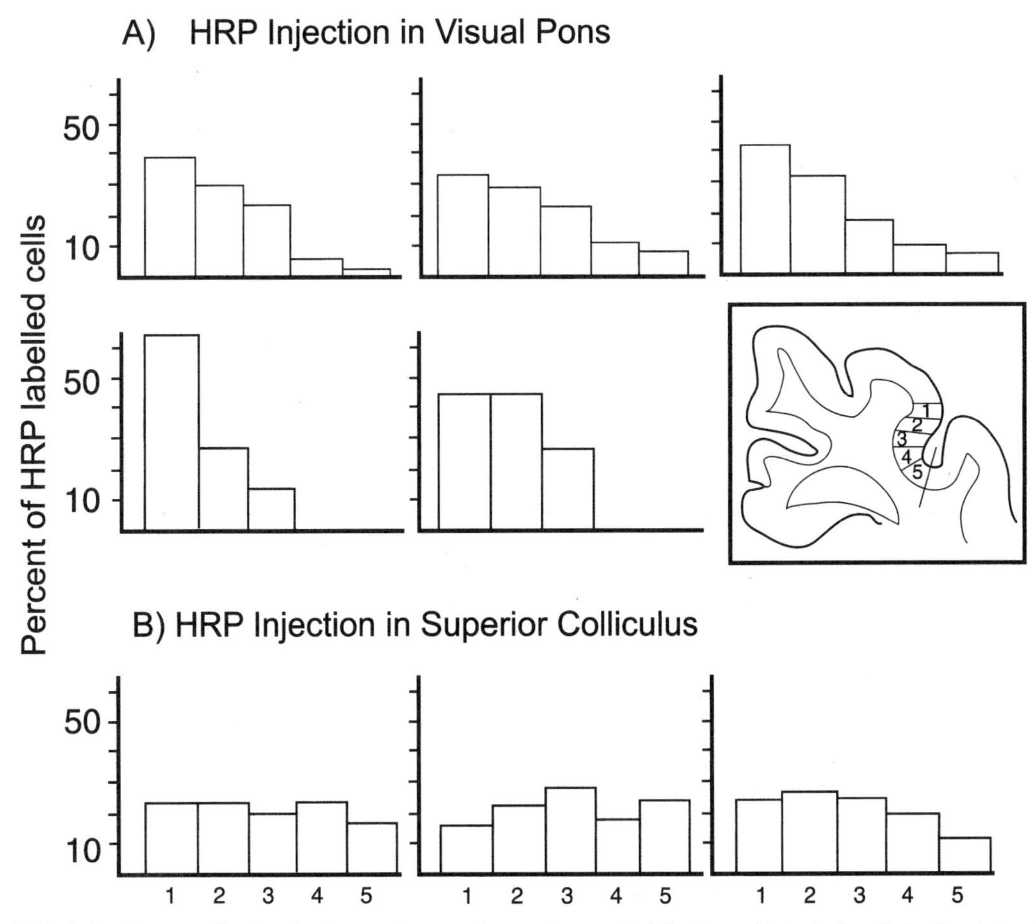

FIG. 3-6. Demagnification in the corticopontine pathway. Distribution of labeled cells in the lateral suprasylvian area of cats after an injection of WGA-HRP into the pontine nuclei or superior colliculi. (Redrawn from Ref. 37, with permission.)

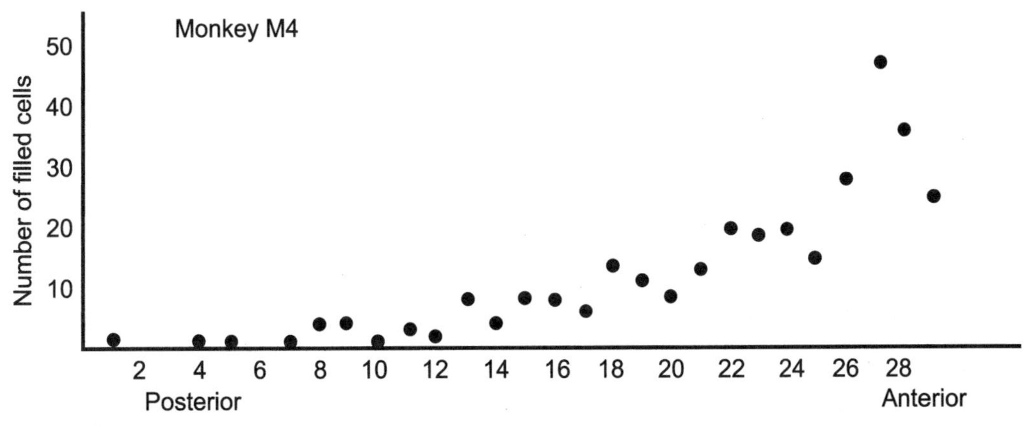

FIG. 3-7. Demagnification in the corticopontine pathway. The distribution of labeled cells in the calcarine fissure after an injection of WGA-HRP into the pontine nuclei. (From Ref. 32, with permission.)

between the two sides of the brain had been cut (39). The animal had been trained to perform a rapid visuomotor task using the arm ipsilateral to the visual input, and was tested while performing this task after receiving an injection of 2-deoxy glucose (2-DG). Although all corticocortical links had been cut, the pattern of 2-DG activation showed that the motor cortex contralateral to the arm used was active during the task. Further analysis of the 2-DG results (40) showed that the cerebellum was active during this task.

THE RECIPROCAL CONNECTIONS FROM THE CEREBELLUM TO THE PARIETAL LOBE VISUAL AREAS: TOWARD REMAPPING FROM RETINAL TO HEAD-CENTERED COORDINATES

There are reciprocal connections from the cerebellum back to the parietal lobe (41). Using transneuronal tracing techniques, Clower et al. (42) found a projection from cerebellar dentate nucleus to area 7b of the parietal lobe. This input could influence the receptive field properties of parietal lobe visual cells. In early stages of visual processing by the brain the visual fields are represented in retinotopic coordinates. There must be a way in which the retinotopic location can be related to the position of external objects, taking into account the position of the eye, head, and body. Andersen and his colleagues (20,43) have demonstrated that the visual responsiveness of parietal lobe neurons, the gain fields, are modified by changes in eye position. One possible function for a reciprocal connection, from cerebellum, via a thalamic relay, to the parietal lobe cortex would be to provide a signal of eye and head position used to modify receptive field properties.

SUMMARY AND CONCLUSIONS

Parietal lobe visual areas project to several subcortical targets. Three of the most prominent of these are to the basal ganglia, superior colliculus, and pontine nuclei. All three are probably involved in some aspect of visually guided movement. One of the largest systems of fiber connections in the human brain arises in the cerebral cortex, and relays to the cerebellum via the pontine nuclei. Pontocerebellar axons terminate within the cerebellar cortex as mossy fibers. Cells in the cerebral cortex that project to the pontine nuclei are all pyramidal in shape, and all are located in lamina V of the cerebral cortex. Subcortical projections from the somatosensory cortex of rats shows that cells that project to the basal ganglia and cerebellum occupy different sublaminae of layer V. Some cortical pyramidal cells' axons may divide, with one branch projecting to the superior colliculus, the other to the pontine nuclei. Temporal lobe visual areas, which are involved in form recognition and learning, do not project to the cerebellum. Reciprocal connections from the cerebellum back to the parietal lobe may play a role in converting visual information from retinotopic to head-centered coordinates.

ACKNOWLEDGMENTS

I thank Dr. Ines Kralj-Hans and Professor John Stein for critical discussion of the manuscript.

REFERENCES

1. Fritsch G, Hitzig E. *Über die elektrische Erregbarkeit des Grosshirns. Arch f Anat. physiol. und Wissenschaftl Mediz.* Leipzig: 1870:300–332.
2. Munk H. *Über die Functionen der Grosshirnrinde.* Berlin: Hirschwald, 1881.
3. Balint R. Seelenlähmung des 'Schauens', optische Ataxie, räumliche Störung der Aufmerksamkeit. *Monatschr für Psychiatrie und Neurologie* 1909;25: 51–81.
4. Pinéas H. Ein Fall von räumlicher Orientierungstörungen mit Dyschirie Z Ges. *Neurol Psychiatr* 1931;133: 180–195.
5. Zingerle H. Ueber Stoerungen der Wahrnehmung des eigenen Koerpers bei organischen Gehirnkrankungen. *Monatschr. Psychiatr Neurol* 1913;34:13–36.
6. Stein JF. The representation of egocentric space in the posterior parietal cortex. *Behav Brain Sci* 1992;72: 691–700.
7. Karnath HO, Ferber S, Himmelbach M. Spatial awareness is a function of the temporal, not the posterior parietal lobe. *Nature* 2001;411:950–953.
8. Ferrier D. *The functions of the brain,* 1st ed. London: Smith, Elder and Co., 1876.

9. Ferrier D, Yeo GF. On the effects of brain lesions in monkeys. *Phil Trans R Soc* 1884;ii:494.
10. Glickstein M, May J, Buchbinder S. Visual control of the arm, the wrist and the fingers: pathways through the brain. *Neuropsychologia* 1998;36:981–1001.
11. Haaxma R, Kuypers HGJM. Intrahemispheric cortical connexions and visual guidance of hand and finger movements. *Brain* 1975;98:239–260.
12. Leigh RJ, Zee DS. *The neurology of eye movements.* New York: Oxford University Press, 1999.
13. Myers RE, Sperry RW, McCurdy NM. Neural mechanisms in visual guidance of movement. *Arch Neurol* 1962;7:195–202.
14. Brinkman J, Kuypers HGJM. Cerebral control of contralateral and ipsilateral arm, hand, and finger movements in the split-brain Rhesus monkey. *Brain* 1973;96:653–674.
15. Mercier B, Legg C, Glickstein M. Basal ganglia and cerebellum receive different somatosensory information in rats. *Proc Natl Acad Sci USA* 1990;87:4388–4392.
16. Armstrong-James M, Fox K, Das Gupta A. Flow of excitation within rat barrel cortex in striking a single vibrissa. *J Neurophysiol* 1992;68:1345–1358.
17. Jenkinson EW, Glickstein M. Whiskers, barrels, and cortical efferent pathways in gap-crossing by rats. *J Neurophysiol* 2000;84:1781–1789.
18. Keating EG, Gooley SA. Removing the superior colliculus silences eye movements abnormally evoked from stimulation of the parietal and occipital eye fields. *Brain Res* 1983;13:145–148.
19. Lynch JC, Graybiel AM, Lobeck LJ. The differential projection of two cytoarchitectonic subregions of the inferior parietal lobule of macaque upon the deep layers of the superior colliculus. *J Comp Neurol* 1985;235:241–254.
20. Andersen RA. Visual and eye movement functions of the posterior parietal cortex. *Annu Rev Neurosci* 1989; 12:377–403.
21. Paré M, Wurtz B. Monkey posterior parietal neurons antidromically activated from superior colliculus. *J Neurophysiol* 1997;78:3493–3497.
22. Keizer K, Kuypers HGJM, Ronday HK. Branching cortical neurons in cat which project to the colliculi and to the pons: a retrograde fluorescent double-labelling study. *Exp Brain Res* 1987;67:1–15.
23. Gibson A, Baker J, Mower G, et al. Corticopontine cells in area 18 of the cat. *J Neurophysiol* 1978;41:484–495.
24. Kemp JM, Powell TPS. The corticostriate projection in the monkey. *Brain* 1970;93:525–546.
25. Saint-Cyr J, Ungerleider L, Desimone R. Organization of visual cortical inputs to the striatum and subsequent outputs to the pallido-nigral complex in the monkey. *J Comp Neurol* 1990;298:129–156.
26. Glickstein M, Kralj-Hans I, Legg C, et al. The organisation of fibers within the rat basis pedunculi. *Neurosci Letts* 1991;135:75–79.
27. Dejerine J. *Sémiologie des affections dy système nerveux.* Paris: Masson, 1926.
28. Classen J, Kunesch E, Binkofski F, et al. Subcortical origin of visuomotor apraxia. *Brain* 1995;118:1365–1374.
29. Legg C, Mercier B, Glickstein M. Corticopontine projection in the rat: the distribution of labeled cortical cells after large injections of horseradish peroxidase in the pontine nuclei. *J Comp Neurol* 1989;286:427–436.
30. Baker J, Gibson A, Glickstein M, et al. Visual cells in the pontine nuclei. *J Physiol* 1976;255:414–433.
31. Brodal P. The corticopontine projection from, the visual cortex in the cat II The projection from areas 18 and 19. *Brain Res* 1972;39:319–335.
32. Glickstein M, Mercier B, Whitteridge D. The striate cortex input to the macaque pons: Selectivity and demagnification. *J Physiol* 1983;341:77P.
33. Glickstein M, May J, Mercier B. Corticopontine projections in the macaque: the distribution of labeled cortical cells after large injections of horseradish peroxidase in the pontine nuclei. *J Comp Neurol* 1985; 235:343–359.
34. Glickstein M, May J. Visual control of movement: the circuits that link visual to motor areas of the brain with special reference to the visual input to the pons and cerebellum. In: Neff WD, ed. *Contributions to sensory physiology.* New York: Academic Press, 1982: 103–145.
35. Stein J, Glickstein M. Role of the cerebellum in visual guidance of movement. *Physiol Rev* 1992;72:967–1017.
36. Robinson FR, Cohen JL, May J, et al. Cerebellar targets of visual pontine cells in the cat. *J Comp Neurol* 1984; 223:471–482.
37. Cohen JL, Robinson FR, May J, et al. Corticopontine projections of the lateral suprasylvian cortex: de-emphasis of the central visual field. *Brain Res* 1981;219: 239–248.
38. Rosina A, Provini L. Pontocerebellar system linking two hemispheres by intracerebellar branching. *Brain Res* 1984;296:365–369.
39. Savaki H, Kennedy C, Sokoloff L, et al. Visually guided reaching with the forelimb contralateral to a "blind" hemisphere: a metabolic mapping study in monkeys. *J Neurosci* 1983;13:2772–2789.
40. Savaki H, Kennedy C, Sokoloff, et al. Visually guided reaching with the forelimb contralateral to a "blind" hemisphere in the monkey: contribution of the cerebellum. *Neuroscience* 1996;75:143–159.
41. Sasaki K, Matsuda Y, Kawaguchi S, et al. On the cerebello-thalamo-cerebral pathway for the parietal cortex. *Exp Brain Res* 1972;16:89–103.
42. Clower DM, West RA, Lynch JC, et al. *J Neurosci* 2001;21:6283–6291.
43. Andersen RA. Coordinate transformations and motor planning in posterior parietal cortex. In: Gazzaniga M, ed. *The cognitive neurosciences.* Boston: MIT Press, 1995:519–532.

4

The Organization of Somatosensory Cortex in Anthropoid Primates

Jon H. Kaas and Christine E. Collins

Department of Psychology, Vanderbilt University, Nashville, Tennessee

INTRODUCTION

This chapter is concerned with the organization of parietal cortex in humans and other anthropoid primates. Because it is difficult or impossible to obtain many types of information about brain organization from humans and apes, much of the information relevant to this chapter has been obtained from experimental studies on New World and Old World monkeys. These studies reveal a range of basic similarities in the organization of at least anterior parietal cortex and lateral parietal cortex in these primates that seem compatible with the more limited observations on these regions of parietal cortex in humans. Thus, our emphasis is on features of parietal cortex organization that likely emerged early in the evolution of anthropoid primates, and have been retained in extant monkeys, apes, and humans.

THE AFFERENTS THAT ACTIVATE SOMATOSENSORY CORTEX

Our focus is on parts of the somatosensory system that are devoted to processing tactile information from the skin, and using that information for object identification and guiding motor behavior. This means a concentration on the flow of information from the four major classes of mechanoreceptors and afferents from the skin of the hand, because they provide most of the inputs that allow these important behaviors. The glabrous skin of the hand is known to have two types of slowly adapting afferents (SA-I and SA-II) associated with Merkel and Ruffini receptor types, respectively, and two types of rapidly adapting afferents (RA-I and RA-II), stemming from Meissner and Pacinian receptor types, respectively. These four types of afferents have different sensitivities to tactile stimuli, and different functional roles, although they are likely all active and contributing in most somatosensory perceptions (1,2). However, discriminations of surface features of objects, such as Braille characters, depend on the responses of SA-I and RA-I afferents, because these afferents alone provide an isomorphic image of stimuli scanned over the skin surface (3). In addition, different sensitivities to surface features suggest that the SA-I system is specialized for providing information about the shapes of objects, whereas the RA-I system is more involved in providing information about movement on the skin, especially for grip control. RA-II (Pacinian) afferents are capable of detecting distant, low-amplitude, high-frequency vibrations, whereas SA-II afferents respond to skin stretch, and may have a role in indicating finger movements and positions. All of these four afferent channels activate neurons in the dorsal horn of the spinal cord, while also sending afferent branches to the relay complex in the lower brainstem that consists of the gracile (for the lower body), cuneate (for the arm and hand), and trigeminal (for the face) subnuclei. This medullary somatosensory complex maintains a high degree of segregation of the SA and RA classes of inputs, and relays this information to the contralateral somatosensory thalamus via the medial lemniscus.

Another source of relevant sensory information is from receptors in deep body tissues, especially the muscle spindle receptors, because these inputs provide much of the information about limb position sense (together with SA-II afferents, sensitive to skin stretch as fingers and joints bend). Position sense is critically important in providing information about the locations of skin surfaces because tactile information is used to explore objects. Muscle spindle afferents also course in the dorsal spinal cord to terminate in separate groups of cells (subnuclei) in the medullary relay, where the neurons they activate project via the medial lemniscus to the contralateral somatosensory thalamus.

THE SOMATOSENSORY THALAMUS

Two nuclei of the somatosensory thalamus are involved in receiving and processing SA and RA tactile and muscle spindle inputs, and projecting this information to areas of the somatosensory cortex. The somatosensory thalamus is defined by the terminations of the two major ascending sensory pathways, the dorsal column-medial lemniscus and spinothalamic pathways, and by interconnections with areas of anterior and lateral somatosensory cortex. The medial lemniscus pathway includes a relay of SA and RA cutaneous receptor information, and a relay of deep receptor (muscle spindle) information. The spinothalamic pathway consists of several types of second-order neurons sensitive to temperature, pain, and touch. Some of these neurons respond to light touch and continue to respond at increasing firing rates as stimulus intensity increases into the painful range (wide-dynamic-range neurons). Both pathways relate to a portion of the ventral thalamus that we refer to as the ventroposterior complex, because it contains several nuclei. Although ventroposterior nuclei have been defined traditionally by their distinctiveness in sections stained for cytoarchitecture, a definition based on functional role is more useful. Thus, one major component of the ventroposterior complex is the ventroposterior nucleus (VP), including the traditional ventroposterior medial subnucleus (VPM) and ventroposterior lateral subnucleus (VPL). In addition, a dorsal capping zone of the ventroposterior complex is distinguished here as the ventroposterior superior (VPS) nucleus. In most mammals, VP corresponds to VP as presently defined, but the dorsal border of VP is somewhat indistinct in macaque monkeys and the VPS region is usually included in VP (4). Nevertheless, from the early microelectrode recording experiments, the dorsal capping zone was recognized as different from the larger, more ventral zone, because the cap was activated by stimulating deep receptors, whereas the ventral region was activated by cutaneous receptors (5–8). In addition to these two nuclei, a narrow ventral zone of small, pale-staining cells has long been distinguished (9) as the ventroposterior inferior nucleus (VPI). This nucleus has not been recognized in most mammals, but it is obvious in anthropoid primates. However, VPI is also distinct in raccoons, possibly because they also use the glabrous skin of the hand as an extremely sensitive tactile surface.

In addition to the ventroposterior complex, the spinothalamic afferents terminate in a poorly defined region just posterior to VP. This region includes the posterior part of the ventromedial nucleus (VMpo) that has been recently identified as a nucleus involved in mediating pain and temperature sensibilities (10,11). Other terminations in this region, sometimes called the posterior group (Po), are likely to have other functions. The other somatosensory structure in the thalamus is the anterior pulvinar (PA) of the pulvinar complex. This nucleus, also called the oral pulvinar, receives inputs from somatosensory cortex and projects back to somatosensory cortex (12). Its functions are largely unknown, but PA is generally thought to modulate activity patterns in somatosensory cortex on the basis of activations from the somatosensory cortex. However, a role in the transfer of information from one cortical area to another has been proposed for such "intrinsic" (without ascending inputs) nuclei of the thalamus (13).

VP receives a topological pattern of SA and RA types of inputs from the dorsolateral dorsal column–trigeminal complex. Microelectrode recordings demonstrate a systematic representation of skin mechanoreceptors in VP so that the face is represented in VPM and the rest of the body in VPL (14,15). A narrow cell-poor septum separates most of VPM from VPL, and similar but less obvious septa separate regions of VPM and VPL that are related to distinct body parts. Thus, "hand" and "foot" subnuclei are anatomically isolated in VPL. At a finer level, even regions devoted to individual fingers can be distinguished sometimes. The septa represent regions where terminating afferents from different body parts are separated from each other. The somatotopic pattern in VP proceeds from a representation of the tongue and oral cavity medially (next to the taste nucleus, VPMpc), upper and lower face, hand with digit tips ventral, foot with toe tips ventral, and (for monkeys) tail. The trunk, arm, and leg dorsally cap the hand and foot subnuclei.

Within VP, the SA and RA classes of input terminate on separate clusters of cells, so that VP actually contains two separate representations that are interdigitated in some fashion, but the details of this pattern are unknown. These zones appear to be related to the RA-I and SA-I types of peripheral nerve afferents, and the nature of the distributions of the RA-II (Pacinian) and SA-II (Ruffini-skin stretch) classes are uncertain. Because RA-II afferents poorly localize tactile stimuli, these sparser inputs may not be precisely distributed in the overall somatotopic map. Functionally, SA-II inputs related to skin stretch, especially around joints, would most appropriately terminate in VPS. VPS is largely activated by muscle spindle receptors, although joint receptors may contribute, and responses to touch (SA-II?) occur. The inputs terminate in a pattern to create a somatotopic representation with the face medial and the foot lateral, but the details are not known.

Spinothalamic axons terminate in VPI and in protrusions of VPI into the septal regions of VP (16,17). VPI contains a third representation of the body in a face to foot sequence from medial to lateral. There have been few recordings from the small cells that constitute VPI, but responsiveness to touch at various intensities seems likely (wide-dynamic-range neurons). Spinothalamics also terminate in a series of separate pufflike foci along the caudal border of VP (18), suggesting that a functional subdivision of the nucleus exists in this region. Because the distribution of spinothalamic inputs is highly discontinuous, the existence of a complete somatotopic map of these inputs seems unlikely, and it is difficult to suggest a functional role.

The connections of somatosensory nuclei with subdivisions of somatosensory cortex have been described in New World and Old World monkeys. VP projects densely to layer 4 of area 3b (19), where RA and SA classes of inputs appear to be highly isolated from each other within small, modular regions of cortex (20). Terminal arbors have dense regions, but they may also distribute in several less dense puffs over several millimeters of area 3b, especially in the rostrocaudal direction (21). The terminal arbors do not seem to cross the representations of major body subdivisions in area 3b, such as the lower face and adjoining hand (22), and even terminations related to individual digits of the hand may be highly segregated from each other. VP also projects less densely to area 1 (23), where terminations are largely in layer 3. These inputs in area 1 are unlikely to have the major activating role of the layer 4 terminations in area 3b. Although the conclusion is controversial, there is evidence that about 20% of the VP neurons projecting to area 1 also project to area 3b (24). Based on the response characteristics of area 1 neurons, the VP inputs are most likely from the RA class of neurons. VP also projects rather sparsely and unevenly to area 2, providing what would appear to be a rather weak influence. VPS projects to layer 4 of both area 3a and area 2. Perhaps as many as 20% of VPS neurons project to both areas (24). Thus, some of the activating inputs to areas 3a and 2 are identical.

VPI projects weakly to the superficial layers of areas 3a, 3b, 1, and 2, and more densely to somatosensory areas of the lateral sulcus, especially areas S2 and PV (the parietal ventral area) (25,26). These VPI inputs to cortex are likely to be modulatory rather than activating above threshold, because they terminate above layer 4. In addition, these regions of cortex are unresponsive to tactile stimuli after lesions of the dorsal column afferents (27), showing that dorsal column and not spinothalamic afferents are responsible for activating these regions of somatosensory cortex. Other spinothalamic afferents to the Po region are likely have activating power when related to other areas of lateral somatosensory cortex that are related to pain and temperature. Finally, the PA projects sparsely to anterior parietal cortex and more densely to lateral parietal cortex and the rostral part of posterior parietal cortex (areas 5 and 7) (12,25,28). The connection patterns suggest that a crude somatotopic representation exists in PA (28). Inputs from PA are likely to be modulatory because terminations are in superficial cortical layers.

THE ORGANIZATION OF ANTERIOR PARIETAL CORTEX (AREAS 3A, 3B, 1, AND 2)

Anterior parietal cortex of anthropoid primates contains the territories of four distinctively different architectonic zones, areas 3a, 3b, 1, and 2 (Fig. 4-1). These four fields were known from the early descriptions of Brodmann (1909) (29), who described areas 3, 2, and 1 and included a subdivision of area 3b that was "transitional" in appearance between area 3b proper and area 4 of motor cortex. Vogt and Vogt (1919) (30) subsequently labeled the two subdivisions, areas 3a and 3b. Anterior parietal cortex has long been known to be important in tactile sensibilities as the result of impairments following lesions in humans (31). In addition, some recognition of an orderly representation of the body in this cortex was implied by the order of progressions of sensations that sometimes occurred during epileptic seizures in humans (the "Jacksonian march") as the brain activity spread from a focus to more distant cortex. More details were obtained from humans when portions of anterior parietal cortex were electrically stimulated while exposed during surgery. Because the stimulations produced feelings of numbness or tingling in body parts, an appreciation of an orderly representation of the body from toes to tongue in a mediolateral progression across cortex emerged from results obtained from a large number of patients (32). Similar maps of this mediolateral sequence of cortical representation soon were obtained from anterior parietal cortex of monkeys by recording potentials evoked by taps on the skin with brain surface electrodes (33). Because much of anterior parietal cortex in macaque monkeys and humans is buried on the posterior bank of the central sulcus, these results were largely obtained from area 1 and the adjoining part of area 2, and there was little appreciation of the significance of the four striplike architectonic fields. Instead, they were assumed to be parts of a single larger field, primary somatosensory cortex or S1.

As microelectrode recordings from investigators in several laboratories revealed more details about the organization of anterior parietal cortex in monkeys, it became apparent that the region was not adequately described as a single, systematic representation. Instead, recordings in owl monkeys, which lack a central sulcus so all of anterior parietal cortex is exposed on the cortical surface, demonstrated that each of the four architectonic fields contains a separate representation of the contralateral body (34). The four representations were organized in parallel, so that they all proceeded from tail, leg, trunk, arm, and face in a mediolateral direction, and they approximated mirror reversals of each other in somatotopy. Thus, for example, digit tips were rep-

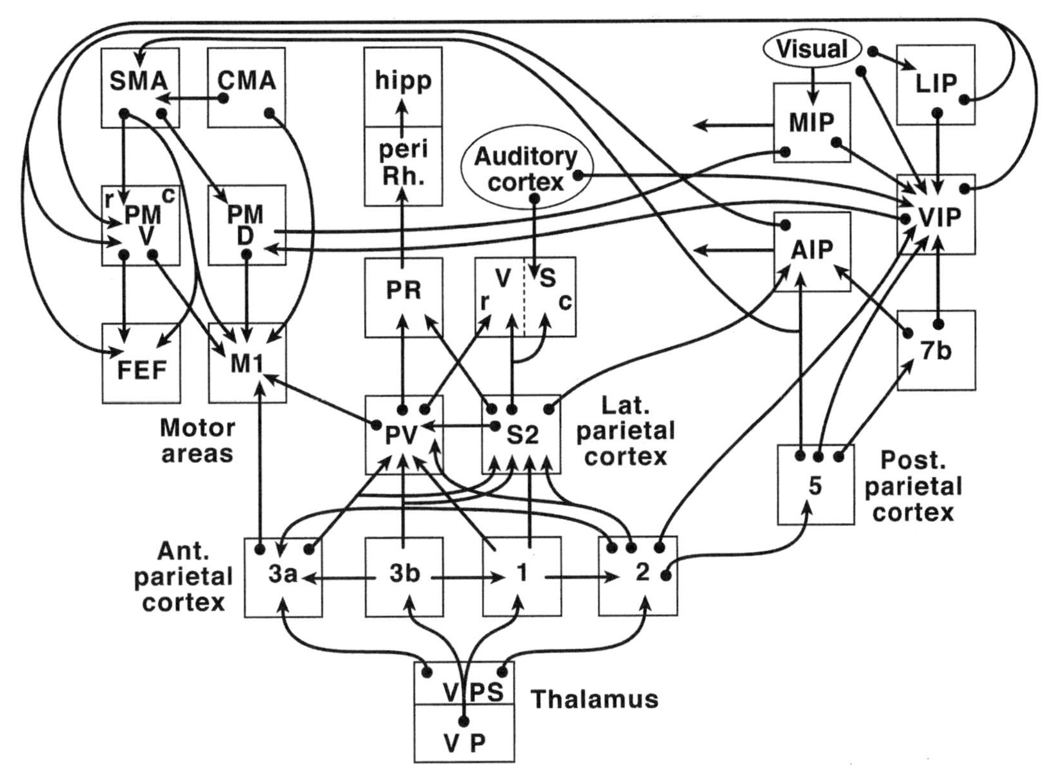

FIG. 4-1. Subdivisions of somatosensory cortex in macaque monkeys. The central, lateral, and intraparietal fissures have been opened to show areas buried in these fissures. The region of flattened cortex comes from a portion of the cerebral hemisphere shown on the upper left. Somatosensory areas include areas 3a, 3b, 1, and 2 of the anterior parietal cortex, the second somatosensory area (S2), the parietal ventral area (PV), the parietal rostral area (PR), and the ventral somatosensory area (VS). The retroinsular region (Ri) and 7b also have somatosensory functions. In posterior parietal cortex, dorsal (5D) and ventral (5V) parts of area 5 are somatosensory. The intraparietal areas are named by location (medial, ventral, lateral, and anterior intraparietal areas, MIP, etc.). Auditory areas A1, rostral (R) and caudomedial (CM) are shown for reference, as is primary motor cortex (M1). Features of somatotopy are indicated for some areas (hand, foot, etc.).

resented in rostral area 3b and caudal area 1, whereas representations of the palm and proximal digits adjoined along the 3b/1 border. Similar features of somatotopic organization were soon found in macaque monkeys (35). Now the evidence for four representations in all anthropoid primates, including humans, is extensive (36). Of course, it seems confusing to refer to four separate representations as S1, although it is still common to do so, because traditions tend to live on. There is considerable comparative evidence that the region termed S1 in rats, cats, and a range of other mammals is a single representation, and that this representation is the homolog (same area) as area 3b of primates (37). Thus, area 3b has been termed "S1 proper" (34) to encourage appropriate comparisons across species. Some of the features of area 3b and the other areas of anterior parietal cortex are described in the following sections.

Area 3b

This has become the most studied area of the somatosensory cortex. The area is distinguished architectonically by having a pronounced "sensory appearance," so that layer 4 is well-differentiated and densely packed with small stellate and granule cells. Area 3b, and the primary auditory and visual cortices, have been called koniocortical areas because they are characterized by these small (konio) neurons in layer 4 (38). Area 3b has other distinguishing architectonic features that are associated with the primary sensory cortex, such as a layer 4 that highly expresses cytochrome oxidase and densely myelinated middle layers of cortex.

The representational divisions in area 3b are proportional to the densities of RA and SA receptors in the skin, so that digit tips, lips, and tongue have large representations, and the trunk has a small representation (in proportion to the size of the skin surface). Thus, cortical space is at least roughly proportional to the magnitude of the receptor inputs from any body location; skin surfaces with densely packed receptors get more cortex. This has been referred to as cortical magnification (39).

Receptive fields are the smallest for skin surfaces that are densely packed with receptors, showing an inverse relation of receptive field size to cortical magnification (39), so that the amount of tissue in area 3b that processes any input is about the same throughout the area. Similar relationships likely hold for other somatosensory areas, but they have not been systematically studied. Compared to other cortical areas, area 3b neurons have the smallest receptive fields, and responses are most resistant to the suppressive effects of anesthesia.

As in the VP nucleus of the thalamus, parts of the body are represented in anatomically distinct divisions of area 3b. Most notably, five myelin-dense bands, separated by myelin-light septa, represent each of the digits of the hand (40), and a series of myelin-dense ovals represent the upper face, lower face, teeth, and tongue (41). Such ovals and bands form in development because they contain cells that are connectionally isolated from those in the adjoining structures. Thus, inputs from the thalamus are from different populations of neurons, and intrinsic connections are densely distributed within, but not across, bands and ovals (22).

Another feature of area 3b is that it is composed of a mosaic of small cell territories or modules that are activated by a relay of either SA or RA afferents (14). The neurons in layer 4 of area 3b closely reflect the properties of the SA and RA afferents, but local circuits alter response properties in other layers to dampen sustained responses. Nevertheless, stimuli that selectively activate SA or RA afferents more effectively activate different modules in area 3b in optical imaging experiments (42).

Area 3b provides dense, feed-forward projections to area 1 and areas S2 and PV of lateral parietal cortex (Fig. 2). Sparser feed-forward projections are to areas 2 and 3a. Area 3b has only limited callosal connections that are largely with area 3b of the other hemisphere. These connections are especially sparse in positions representing the hand and lips (43).

Area 1

This field represents the contralateral body surface a second time, roughly as a mirror image of the representation in area 3b. The representation appears to depend on inputs from area 3b (44), and the neuron responses are of the RA type, possibly as a result of a selective relay from area 3b and VP, and as a result of a reshaping of SA activity that is relayed from area 3b. The feed-back connections from area 1 to area 3b generally decrease the responses of neurons in area 3b to tactile stimulation, suggesting selective terminations on inhibitory neurons, but other modulatory effects have been described as well. As a second level of cortical processing, neurons in area 1 tend to have somewhat larger receptive fields than neurons in area 3b (45), and some neurons are selective for direction of movement on the skin (46). The re-

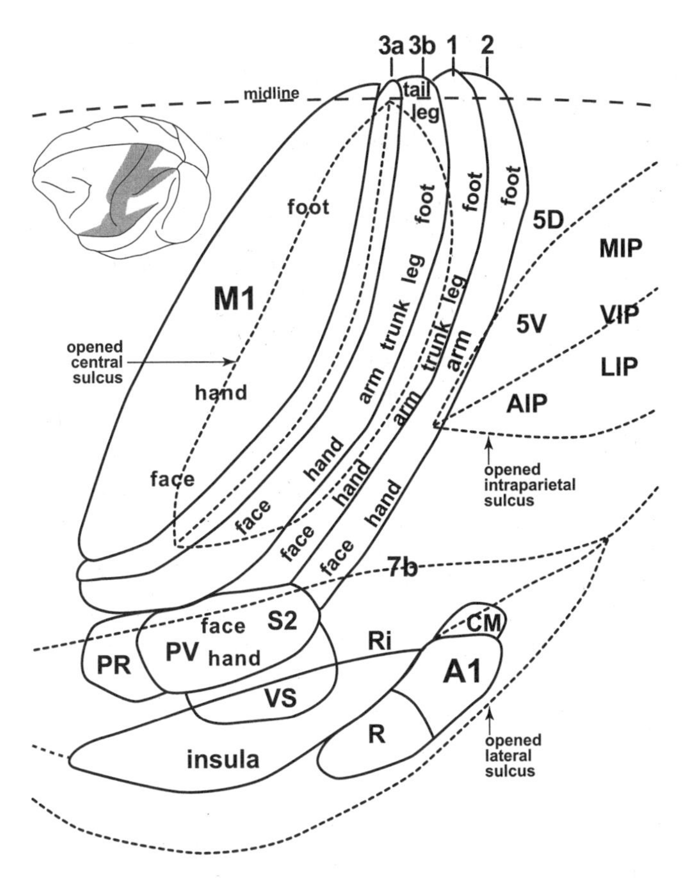

FIG. 4-2. Some of the thalamic and cortical connections of somatosensory areas. See Figure 4.1 for somatosensory areas. The ventral posterior nucleus (VP) and the ventroposterior superior nucleus (VPS) project to somatosensory cortex. Motor areas include M1, dorsal (PMD), and ventral (PMV) premotor areas, the supplementary motor area (SMA), the cingulate motor area (CMA), and the frontal eye field (FEF). Perirhinal cortex projects to the hippocampus.

sponses of area 1 neurons to tactile stimuli are often modified by motor behavior (47). Some neurons have receptive fields that extend across the body midline (48), but callosal connections are sparse in regions representing the hand, foot, and parts of the face (43). Major outputs are to area 2 and S2 and PV of lateral cortex.

Area 2

This caudal field also represents the body in a pattern that approximates a mirror reversal of the area 1 representation (12). However, the representation is more complexly organized in that some skin surfaces, most notably those of the glabrous digits, are represented twice in the field, suggesting a type of modular organization. Architectonically, the posterior border of area 2 with areas 5 and 7 of posterior parietal cortex is not always distinct, and various investigators have estimated this border to be located in slightly different positions. Area 2 is activated by inputs from area 1 and area 3b, and by a relay of muscle spindle receptor information from VPS of the somatosensory thalamus (Fig. 4-2). Major connections are with area 3a, the posterior parietal cortex, S2, and PV. Feed-back connections are to area 1. Callosal connections are denser and more evenly distributed than for areas 3b and 1. Neurons in area 2 are activated by both tactile stimuli and body movement, with high levels of activity occurring during manual exploration of objects. Receptive fields are often, but not always, larger than those in area 1 (12,45). Lesions of area 2 in monkeys impair finger coordination (49).

Area 3a

This area has been studied the least of the four, possibly because of its location in the depths of the central sulcus in macaque monkeys and humans. The area forms a representation of the contralateral body surface that parallels those in areas 3b, 1, and 2. Area 3a receives muscle spindle information from VPS, movement and tactile information from area 2, and a sparse amount of tactile information from area 3b. Major projections are to the primary motor cortex (M1), S2, and PV (50,51). Neurons respond to the movement of body parts, and less strongly and more variably to tactile stimuli. Neurons are influenced by motor plans (47). Area 3a appears to provide the major source of sensory information to area M1.

Neurons in all four fields have subcortical projections, including feedback to the somatosensory thalamus and dorsal column nuclei, the basal ganglia, and even the spinal cord (8). Inputs to the pons are relayed to the cerebellum as part of a motor control circuit that involves motor cortex via cerebellar projections to the anterior ventral thalamus.

THE ORGANIZATION OF THE LATERAL PARIETAL CORTEX

Cortex lateral and posterior to S1 is involved in somatosensory functions in all mammals. Early studies described a second representation (S2) of tactile receptors of the body in cortex on the lateral (ventral) margin of S1 in cats (52) and other mammals (53). Later research indicated that the S2 region actually contains two adjoining representations that are approximately mirror images of each other (54). The term S2 was retained for the more posterior representation, whereas the more anterior representation was called the parietal ventral area, PV. In primates, all or most of S2 and PV are in cortex of the upper bank of the lateral sulcus (Fig. 4-1). Microelectrode recordings and studies of connections indicate that much of cortex around S2 and PV, including more posterior portions of the upper bank of the lateral sulcus and the surface in the depths of the sulcus (the insula), is somatosensory in function. The adjoining cortex in the insula and lower bank of the lateral sulcus is auditory (55). Cortex in the lateral sulcus also includes at least one region where neurons respond to both taste and tactile stimuli, known as the gustatory cortex, and a region devoted to pain and temperature (11). Some of this cortex receives a relay of vestibular information (56). The complete organization of this region of the lateral parietal cortex is unknown, but there has been considerable progress.

S2 and PV

These two fields occupy small ovals of cortex along the lateral border of area 3b that represents the face. Both fields appear to represent the face next to the 3b border, although it is possible that area 1 extends along this border to separate area 3b from S2 and PV. Both areas represent the face, forelimb, and hindlimb in locations progressively more distant from the 3b border, with the digits of the hand in the two representations adjoining along their common border (57–60). Functional imaging studies in humans provide evidence for PV and S2 representations in humans (61). The two areas are similar in that their neurons have larger receptive fields than neurons in the anterior parietal cortex, and some of the neurons can be activated by touch on the ipsilateral as well as the contralateral side of the body. Callosal connections are more evenly distributed in the two areas, and they are thought to have roles in the bilateral coordination of the hands (62). The two areas receive feed-forward inputs from all four areas of anterior parietal cortex (60,63), although much of the activation depends on area 3b (64,65). Although S2 and PV are interconnected, and the neurons in each area are at least similar in their response properties, PV gets feed-forward inputs from S2; thus, PV may represent a higher level of processing (66,67). S2 has been regarded as a first step after anterior parietal cortex, in the processing of tactile and position sense information identifying objects by touch (63). Although both PV and S2 project to other regions of the lateral sulcus (Fig. 4-2), PV has more connections with primary motor cortex (M1), and a more rostral parietal region, the parietal rostral area (PR), which also projects to M1. Thus, PV may have a greater role in guiding motor behavior. PV and S2 are bordered by a ventral somatosensory area, VS, deep in the lateral fissure (57,66,67). The organization of the VS region is complex, with neurons being variously responsive to tactile stimuli, sometimes having bilateral receptive fields, and often responding to both auditory and cutaneous stimuli. Sensory inputs include feed-forward projections from PV and S2, but other connections are not known.

Cortex just caudal to S2 also responds to somatosensory stimuli (59,68), but the organization of this region is not well understood. Generally, cortex of the upper bank of the lateral fissure, just posterior to S2, is considered part of area 7b, whereas cortex just posterior to the insula and VS is called retroinsular (Ri). S2 and PV project to this portion of 7b. A nearby region of the auditory cortex on the lower bank of the lateral sulcus, the caudomedial auditory area (CM), responds to both auditory and somatosensory stimuli (69). The inputs responsible for the responsiveness to somatosensory stimuli are unknown. Finally, as mentioned, S2 and PV project rostrally to the parietal rostral area, PR, but little is known about this field.

Overall, cortex of the lateral fissure is involved in a number of functions, including having roles in the perceptions of taste, pain, and temperature. However, most of the somatosensory areas of the lateral sulcus are thought to be parts of a network involved in identifying surfaces and objects by touch, in a manner analogous to the ventral stream of visual processing for object identification (70). Thus, these lateral areas have a role in tactile memory, and some of them connect with perirhinal cortex and thus the hippocampus (63).

THE ORGANIZATION OF THE POSTERIOR PARIETAL CORTEX

The organization and functions of posterior parietal cortex are reviewed in other chapters in this volume (see Chapters 7, 8, 9, and 10). Somatosensory information is relayed from the anterior and lateral parietal cortex to regions of posterior parietal cortex for the planning and guidance of motor behavior. The direct somatosensory inputs are largely confined to the more rostral parts of posterior parietal cortex that are generally termed areas 5 and 7b (Figs. 4.1 and 4.2). In macaque monkeys, much of posterior parietal cortex is in the intraparietal sulcus, and it is useful to refer to this as the IP region, even in primates without this sulcus. The caudal parts of the region are devoted to areas involved in reaching and looking, and they have major inputs from occipital and temporal visual areas. More anterior regions may be more involved in guiding grasping hand movements and they receive more somatosensory inputs, as well as inputs from more caudal IP areas. Neurons in several regions are bimodal, responding to both visual and somatosensory stimuli, and some neurons can be activated or are influenced by auditory stimuli. Feed-forward projections from the areas are to motor, premotor, and visuomotor areas of the frontal lobe (Fig. 4-2), as well as the dorsolateral prefrontal cortex involved in working memory. Because of considerable interest in the posterior parietal cortex, and intensive study in monkeys and humans, our understanding of the functional organization of portions of posterior parietal cortex is starting to emerge.

REFERENCES

1. Bolanowski SJ, Gescheider GA, Verrillo RT. Hairy skin: psychophysical channels and their physiological substrates. *Somatosens Mot Res* 1994;11:279–290.
2. Bolanowski SJ, Gescheider GA, Verrillo RT, et al. Four channels mediate the mechanical aspects of touch. *J Acoust Soc Am* 1988;84:1680–1694.
3. Johnson KO. The roles and functions of cutaneous mechanoreceptors. *Curr Opin Neurobiol* 2001;11: 455–461.
4. Paxinos G, Huang XF, Toya AW. *The rhesus monkey brain.* San Diego: Academic Press, 2000.
5. Poggio GF, Mountcastle VB. The functional properties of ventrobasal thalamic neurons studied in unanesthetized monkeys. *J Neurophysiol* 1963;26:775–806.
6. Loe PR, Whitsel BL, Dreyer DA, et al. Body representation in ventrobasal thalamus of macaque: a single unit analysis. *J Neurophysiol* 1977;40:1339–1354.
7. Wiesendanger M, Miles TS. Ascending pathway of low-threshold muscle afferents to the cerebral cortex and its possible role in motor control. *Physiol Rev* 1982;62: 1234–1270.
8. Kaas JH, Pons TP. The somatosensory system of primates. In: Steklis HP, Erwin J, eds. *Comparative primate biology.* New York: Liss, 1988;421–468.
9. Jones EG. *The Thalamus.* New York: Plenum, 1985.
10. Craig AD, Bushnell MC, Zhang ET, et al. A thalamic nucleus specific for pain and temperature sensation. *Nature* 1994;372:770–773.
11. Craig AD. How do you feel? Interoception: the sense of the physiological condition of the body. *Nat Rev* 2002; 3:655–666.
12. Pons TP, Kaas JH. Connections of area 2 of somatosensory cortex with the anterior pulvinar and subdivisions of the ventroposterior complex in macaque monkeys. *J Comp Neurol* 1985;240:16–36.
13. Sherman SM, Guillery RW. *Exploring the thalamus.* San Diego: Academic Press, 2001.
14. Kaas JH, Nelson RJ, Sur M, et al. The somatotopic organization of the ventroposterior thalamus of the squirrel monkey, Saimiri sciureus. *J Comp Neurol* 1984;226: 111–140.
15. Rausell E, Jones EG. Histochemical and immunocytochemical compartments of the thalamic VPM nucleus in monkeys and their relationship to the representational map. *J Neurosci* 1991;11:210–225.
16. Apkarian AV, Hodge CV. Primate spinothalamic pathways: III. Thalamic terminations of the dorsolateral and ventral spinothalamic pathways. *J Comp Neurol* 1989; 288:493–511.
17. Apkarian AV, Shi T. Squirrel monkey lateral thalamus. I. Somatic nociresponsive neurons and their relation to spinothalamic terminals. *J Neurosci* 1994;14: 6779–6795.
18. Stepniewska I, Sakai ST, Qi HX, et al. Interrelationship between cerebellothalamic and spinothalamic projections in the macaque monkey: a potential substrate for parkinsonian tremor. *Soc Neurosci Abstr* 2001;27:824.10.
19. Jones EG, Friedman DP. Projection pattern of functional components of thalamic ventrobasal complex on monkey somatosensory cortex. *J Neurophysiol* 1982; 48:521–544.
20. Sur M, Wall JT, Kaas JH. Modular distribution of neurons with slowly adapting and rapidly adapting responses in area 3b of somatosensory cortex in monkeys. *J Neurophysiol* 1984;51:724–744.
21. Garraghty PE, Pons TP, Sur M, et al. The arbors of axons terminating in middle cortical layers of somatosensory area 3b in owl monkeys. *Somatosens Mot Res* 1989;6:401–411.
22. Fang PC, Jain N, Kaas JH. Few intrinsic connections cross the hand-face border of area 3b of New World monkeys. *J Comp Neurol* 2002;454:310–319.
23. Nelson RJ, Kaas JH. Connections of the ventroposterior nucleus of the thalamus with the body surface representations in cortical areas 3b and 1 of the cynomolgus macaque (Macaca fascicularis). *J Comp Neurol* 1981; 199:29–64.
24. Cusick CG, Steindler DA, Kaas JH. Corticocortical and collateral thalamocortical connections of postcentral somatosensory cortical areas in squirrel monkeys: a double-labeling study with radiolabeled wheatgerm agglutinin and wheatgerm agglutinin conjugated to horseradish peroxidase. *Somatosens Res* 1985;3:1–31.
25. Krubitzer LA, Kaas JH. The somatosensory thalamus of monkeys: cortical connections and a redefinition of nuclei in marmosets. *J Comp Neurol* 1992;319:123–140.
26. Stevens RT, London SM, Apkarian AV. Spinothalamocortical projections to secondary somatosensory cortex (SII) in squirrel monkeys. *Brain Res* 1993;631: 241–246.
27. Jain N, Florence SL, Kaas JH. Limits on plasticity in somatosensory cortex of adult rats: hindlimb cortex is not reactivated after dorsal column section. *J Neurophysiol* 1995;73:1537–1546.
28. Cusick CG, Gould HJ. Connections between area 3b of somatosensory cortex and subdivisions of the ventroposterior nuclear complex and the anterior pulvinar nucleus in squirrel monkeys. *J Comp Neurol* 1990;292: 83–102.
29. Brodmann K. *Vergleichende Lokalisationslehre der Grosshirnrinde.* Leipzig: Barth, 1909.
30. Vogt C, Vogt O. Allgemeinere ergebnisse unsere hirnforschung. *J Physiol Neurol* 1919;25:399–462.
31. Head H, Holmes G. Sensory disturbances from cerebral lesions. *Brain* 1911;34:102–254.
32. Penfield W, Boldrey E. Somatic motor and sensory representation in the cerebral cortex of man as studied by electrical stimulation. *Brain* 1937;60:389–443.
33. Marshall W, Woolsey CN, Bard P. Cortical representation of tactile sensibility as indicated by cortical potentials. *Science* 1937;85:388–390.
34. Merzenich MM, Kaas JH, Sur M, et al. Double representation of the body surface within cytoarchitectonic areas 3b and 1 in "SI" in the owl monkey (Aotus trivirgatus). *J Comp Neurol* 1978;181:41–73.
35. Kaas JH, Nelson RJ, Sur M, et al. Multiple representations of the body within primary somatosensory cortex of primates. *Science* 1979;204:521–523.
36. Kaas JH. How sensory cortex is subdivided in mammals: implications for studies of prefrontal cortex. *Prog Brain Res* 1990;85:3–10.
37. Kaas JH. What, if anything, is SI? Organization of first somatosensory area of cortex. *Physiol Rev* 1983;63: 206–231.
38. Sanides F, Krishnamurti A. Cytoarchitectonic subdivisions of sensorimotor and prefrontal regions and of bor-

dering insular and limbic fields in slow loris (Nycticebus coucang coucang). *Journal für Hirnforschung* 1967;9:225–252.
39. Sur M, Merzenich MM, Kaas JH. Magnification, receptive-field area, and "hypercolumn" size in areas 3b and 1 of somatosensory cortex in owl monkeys. *J Neurophysiol* 1980;44:295–311.
40. Jain N, Catania KC, Kaas JH. A histologically visible representation of the fingers and palm in primate area 3b and its immutability following long-term deafferentations. *Cereb Cortex* 1998;8:227–236.
41. Jain N, Qi HX, Catania KC, et al. Anatomic correlates of the face and oral cavity representations in the somatosensory cortical area 3b of monkeys. *J Comp Neurol* 2001;429:455–468.
42. Chen LM, Friedman RM, Ramsden BM, et al. Fine-scale organization of primary somatosensory cortex (area 3b) in the squirrel monkey revealed with intrinsic optical imaging. *J Neurophysiol* 2001;86:3011–3029.
43. Killackey HP, Gould HJ, Cusick CG, et al. The relation of corpus callosum connections to architectonic fields and body surface maps in sensorimotor cortex of new and old world monkeys. *J Comp Neurol* 1983;219: 384–419.
44. Garraghty PE, Florence SL, Kaas JH. Ablations of areas 3a and 3b of monkey somatosensory cortex abolish cutaneous responsivity in area 1. *Brain Res* 1990;528: 165–169.
45. Iwamura Y, Tanaka M, Sakamoto M, et al. Rostrocaudal gradients in the neuronal receptive field complexity in the finger region of the alert monkey's postcentral gyrus. *Exp Brain Res* 1993;92:360–368.
46. Hyvarinen J, Poranen A. Movement-sensitive and direction and orientation-selective cutaneous receptive fields in the hand area of the post-central gyrus in monkeys. *J Physiol* 1978;283:523–537.
47. Nelson RJ. Responsiveness of monkey primary somatosensory cortical neurons to peripheral stimulation depends on 'motor-set'. *Brain Res* 1984;304:143–148.
48. Taoka M, Toda T, Iwamura Y. Representation of the body midline trunk, bilateral arms, and shoulders in the monkey postcentral somatosensory cortex. *Exp Brain Res* 1998;123:315–322.
49. Hikosaka O, Tanaka M, Sokamoto M, et al. Deficits in manipulative behaviors induced by local injections of muscimol in the first somatosensory cortex of the conscious monkey. *Brain Res* 1985;325:375–380.
50. Huerta MF, Pons T.P. Primary motor cortex receives input from area 3a in macaques. *Brain Res* 1990;537: 367–371.
51. Huffman KJ, Krubitzer L. Thalamo-cortical connections of areas 3a and M1 in marmoset monkeys. *J Comp Neurol* 2001;435:291–310.
52. Adrian ED. Double representation of the feet in the sensory cortex of the cat. *J Physiol* 1940;98:16–18.
53. Woolsey CN, Fairman D. Contralateral, ipsilateral, and bilateral representation of cutaneous receptors in somatic Areas I and II of the cerebral cortex of pig. *Surgery* 1946;19:684–702.
54. Krubitzer LA, Sesma MA, Kaas JH. Microelectrode maps, myeloarchitecture, and cortical connections of three somatotopically organized representations of the body surface in parietal cortex of squirrels. *J Comp Neurol* 1986;250:403–430.
55. Kaas JH, Hackett TA. Subdivisions of auditory cortex and processing streams in primates. *Proc Natl Acad Sci USA* 2000;97:11793–11799.
56. Guldin WO, Akbarian S, Grusser OJ. Cortico-cortical connections and cytoarchitectonics of the primate vestibular cortex: a study in squirrel monkeys (Saimiri sciureus). *J Comp Neurol* 1992;326:375–401.
57. Cusick CG, Wall JT, Felleman DJ, et al. Somatotopic organization of the lateral sulcus of owl monkeys: area 3b, S-II, and a ventral somatosensory area. *J Comp Neurol* 1989;282:169–190.
58. Burton H, Fabri M, Alloway K. Cortical areas within the lateral sulcus connected to cutaneous representations in areas 3b and 1: a revised interpretation of the second somatosensory area in macaque monkeys. *J Comp Neurol* 1995;355:539–562.
59. Krubitzer L, Clarey JC, Tweedale R, et al. A redefinition of somatosensory areas in the lateral sulcus of macaque monkeys. *J Neurosci* 1995;15:3821–3839.
60. Krubitzer LA, Kaas JH. The organization and connections of somatosensory cortex in marmosets. *J Neurosci* 1990;10:952–974.
61. Disbrow E, Roberts T, Krubitzer L. Somatotopic organization of cortical fields in the lateral sulcus of Homo sapiens: evidence for SII and PV. *J Comp Neurol* 2000; 418:1–21.
62. Disbrow E, Roberts T, Poeppel D, et al. Evidence for interhemispheric processing of inputs from the hands in human S2 and PV. *J Neurophysiol* 2001;85:2236–2244.
63. Friedman DP, Murray EA, O'Neill JB, et al. Cortical connections of the somatosensory fields of the lateral sulcus of macaques: evidence for a corticolimbic pathway for touch. *J Comp Neurol* 1986;252:323–347.
64. Garraghty PE, Pons TP, Kaas JH. Ablations of areas 3b (SI proper) and 3a of somatosensory cortex in marmosets deactivate the second and parietal ventral somatosensory areas. *Somatosens Mot Res* 1990;7: 125–135.
65. Pons TP, Wall JT, Garraghty PE, et al. Consistent features of the representation of the hand in area 3b of macaque monkeys. *Somatosens Res* 1987;4:309–331.
66. Qi HX, Lyon DC, Kaas JH. Cortical and thalamic connections of the parietal ventral somatosensory area in marmoset monkeys (Callithrix jacchus). *J Comp Neurol* 2002;443:168–182.
67. Disbrow E, Litinas E, Recanzone GH, et al. Cortical connections of the second somatosensory area and the parietal ventral area in macaque monkeys. *Thal Rel Syst* 2002;1:289–302.
68. Robinson CJ, Burton H. Organization of somatosensory receptive fields in cortical areas 7b, retroinsula, postauditory and granular insula of M. fascicularis. *J Comp Neurol* 1980;192:69–92.
69. Schroeder CE, Lindsley RW, Specht C, et al. Somatosensory input to auditory association cortex in the macaque monkey. *J Neurophysiol* 2001;85:1322–1327.
70. Mishkin M. Analogous neural models for tactual and visual learning. *Neuropsychologia* 1979;17:139–151.

5

Is There a Role for the Parietal Lobes in the Perception of Pain?

Gary H. Duncan* and Marie-Claire Albanese†

Departments of Neurology and Neurosurgery and Physiology†, Montreal Neurological Institute, McGill University, Montreal, Quebec, Canada; *Faculté de Médecin Dentaire, Centre de Recherche en Sciences Neurologiques, Université de Montréal, Montréal, Quebec, Canada*

INTRODUCTION

The parietal cortex remains one of the least successfully explored cytoarchitectonic regions of the human cerebral cortex (1), and its exact role in the conscious experience of pain is still an ongoing debate (2–6). The parietal lobe, demarcated anteriorly by the central sulcus and posteriorly by the parieto-occipital fissure (7), is described by Brodmann as comprising the postcentral and parietal regions (8). Within the parietal lobe, three functional subregions have been recognized for their complementary roles in the processing of *innocuous* somatosensory stimuli: the primary somatosensory cortex (SI) within Brodmann areas (BA) 3, 1, and 2 of the postcentral region; the secondary somatosensory cortex (SII), generally considered to occupy BA 43 in the parietal operculum; and the posterior parietal region—especially BA 5 (Chapters 7 and 8). In contrast to the wealth of evidence demonstrating the importance of these regions in the processing of innocuous stimuli, little consensus exists concerning their possible roles in the human perception of pain.

Undoubtedly, the major reason for this difficulty in elucidating cerebral mechanisms of pain—parietal or otherwise—is the complex nature of pain perception, itself. The characterization of "pain" has been a puzzle for generations of scholars and scientists and has ranged from the Aristotelean concept of pain as an emotion, apart from the senses, to the recent emphasis on pain as an interoceptive signal for maintaining homeostasis (9). Throughout much of the 20th century, however, pain has been recognized as a multidimensional phenomenon, rather than a single emotion or simple sensation. Certainly, an emotional component of pain perception is a key factor that facilitates the motivation and learning that helps an organism avoid noxious, tissue-damaging stimuli. Likewise, monitoring the intensity and location of noxious stimuli is an important factor in the sensorimotor integration required to plan and execute avoidance and escape responses. However, although the processing of sensory aspects of nociceptive information by the somatosensory cortices is an intuitively compelling proposition, experimental and clinical data have not always yielded convincing evidence.

In this chapter we present a brief review of the literature that has contributed to our current ideas regarding the role of the parietal lobes in human pain perception. Although much supporting evidence has been derived from anatomic and neurophysiologic studies conducted in nonhuman primates and other animals, the emphasis of this chapter is on human research, from the early studies of brain lesions in patients to recent neuroimaging studies in patients and normal volunteers.

LESIONS AND ELECTRICAL STIMULATION OF THE PARIETAL CORTEX

Observation of sensory and behavioral sequelae in patients with cortical wounds, surgical resections, and epilepsy shaped early concepts concerning the possible role ascribed to the human cerebral cortex in the experience of pain. At the beginning of the 20th century, initial studies of "faradic" (electrical) stimulation of the human cortex established that stimulating the postcentral region elicited tactile sensations, but not pain (10). The conscious experience of pain was thought to occur at the thalamic level (11), with "little if any cortical representation" (12). By the middle of the 20th century, an opposing view began to emerge when J. Marshall (13) presented several cases involving soldiers with superficial wounds involving the parietal cortex, who suffered from either an impairment of pain or temperature sense. Thus, he concluded that the cortex is needed for an appreciation of pain and that small, but not large, lesions seemed to interfere with pain perception. Similarly, Lewin and Phillips (14) found that electrical stimulation of the postcentral gyrus in patients suffering from phantom limb pain, elicited a sensation of spontaneous pain. They concluded, however, that in normal uninjured individuals, a conscious appreciation of painful peripheral stimuli did not necessarily involve the primary somatosensory cortex. Although these early lesion and cortical stimulation studies are seminal in their importance, their results are not easily compared or interpreted. The exact localization of the lesion or resected cortical area is not easily determined. Moreover, clinicians and researchers without a full appreciation of the complex nature of pain may not ask the appropriate questions, and their patients' responses may not reflect a uniform component of the perceptual experience.

More recent case studies have contributed additional evidence regarding the involvement of the parietal lobes in pain perception. Soria and Fine (15) described a patient with hyperpathic pain in the right arm and a right hemisensory syndrome consequent to a lacunar infarct in the left thalamus. This chronic pain condition disappeared following a second cerebrovascular accident, which damaged the subcortical parietal white matter of the left corona radiata, further interrupting thalamoparietal connections. In 1997, Potagas and colleagues (16) described the case of a young woman who developed episodic pain in her right arm caused by a subcortical tumor in the left parietal operculum; the pain resolved following surgical excision of the glioma. These two studies suggest that the interruption of normal communication between thalamus and the parietal lobe—either from the tumor-related compression of thalamoparietal fibers (16) or damage to the thalamus itself (15)—can be a causative factor in the development of chronic pain. These studies also illustrate the complex nature of central pain syndromes—in one case chronic pain is relieved when thalamoparietal communication is restored by removal of the tumor (16), whereas in the other chronic pain disappears on further disruption of the thalamoparietal communication (15).

Lesion studies that incorporate psychophysical methods to assess the patients' responses to noxious stimuli shed additional light on the role of the parietal lobes in normal pain perception. Greenspan and Winfield described a patient with a tumor located near the most posterior portion of the insula and parietal operculum (17). Psychophysical testing before any surgical intervention revealed deficits in both pain and tactile perception similar to those described by Bassetti et al. as a pseudo-thalamic sensory syndrome (18); following surgical removal of the tumor, the patient's perceptual capacities returned to normal, suggesting to the authors that SII and the posterior insular region are essential for normal pain and tactile perception (17). In a more recent report, Ploner and colleagues (19) described a patient who had suffered a selective ischemic lesion involving the right postcentral gyrus and parietal operculum, encompassing the hand area of SI and SII somatosensory cortices. The patient's symptoms included hypesthesia as well as anesthesia of the left hand and arm such that innocuous thermal stimuli did not evoke any sensation. Although unable to describe the quality, location, or intensity of either warm or painful stimuli presented to the anesthetic limb, the patient complained of a "clearly unpleasant" feeling that he wanted to avoid in response to noxious levels of stimulation. This selective impairment of the sensory-discriminative aspects of pain sensation, associated with a restricted lesion within cortical somatosensory areas, suggests a specialization of nociceptive processing within these anterior regions of the parietal lobe. These findings are consistent with the segregation of pain pathways, proposed by Albe-Fessard (20), into a lateral, sensory-discriminative pain system, composed of the lateral thalamic nuclei, SI and SII (21), and a medial motivational-affective pain system, composed of the medial thalamic nuclei and of the anterior cingulate gyrus (22).

Supporting this division of nociceptive processing, Greenspan and colleagues (23) recently assessed six patients suffering from unilateral lesions to the parietal operculum or insula. In these patients, impairments in pain threshold (decreased ability to identify noxious stimuli as painful) were associated only with lesions of the contralateral posterior parietal operculum, whereas disturbances in motivational and affective responses to noxious stimuli (as indicated by a cold pain tolerance test) were associated with lesions of the insula. Although these studies lend support to the notion of a critical role for the parietal lobes in pain perception, their implications are limited by problems inherent in human lesion data; for example, difficulty in determining the precise delineation of the cortical lesions, possible damage to fibers of passage altering the function of distant regions, and the problematic situation of predicting normal cortical function based on abnormal and possibly compensatory behavior observed consequent to cortical damage.

PARIETAL LOBE EPILEPSY

In the 20th century, the study of patients inflicted with epilepsy was one of the more fruitful approaches to investigating human cerebral function (12). Physicians came to understand that the natural progression of an epileptic seizure presented a potential portal through which one could observe the behavioral and sensory consequences of regional cerebral activation. From the patient's realization of the initial aura, to the onset of the seizure itself, a stereotypical pattern of perceptions emerges from the abnormal (but naturally occurring) focus of neuronal hyperactivity—the epileptogenic focus. The precise region of cerebral cortex associated with the patient's particular perceptual symptom could be approximated anatomically by localizing the brain tumor, the most common etiology for parietal lobe epilepsy (see Chapter 21), or by directly recording neuronal activity from the epileptogenic focus. Likewise, direct electrical stimulation, in the awake patient, of the cortex around the epileptogenic region (as a preliminary phase of surgical excision), provided a wealth of information regarding perceptual, behavioral, and emotional consequences of activating (albeit in an artificial fashion) selected regions of cerebral cortex.

In general, the perception of pain is an extremely rare consequence of seizure activity, or of electrical stimulation of the cerebral cortex (12), as cited—suggesting to early researchers that pain perception was likely a subcortical function. However, although less than 3% all epileptic patients suffer from ictal pain (24,25), the proportion is remarkably greater—23.6%—when considering patients specifically diagnosed with parietal epilepsy (26). In addition, painful somatosensory auras, preceding parietal lobe seizures, are manifest contralateral to the epileptic hemisphere in perirolandic parietal epilepsy (27), but not consistently in temporal lobe epilepsy (25). Nair and colleagues suggested that these painful auras more likely originate in SI rather than SII (where receptive fields are generally found to be larger and often bilateral), and concluded that the effectiveness of focal cortical resections in reducing both aural and ictal somatosensory pain argues for a cortical representation of pain (27).

The perception of pain during seizures (ictal pain) has been recognized for over a century (24). The most prevalent type of pain is a burning dysesthesia that is localized to the abdomen or to part or all of the contralateral hemibody. A recent case report described ictal abdominal pain in a patient suffering from simple partial seizures associated with an acute hemorrhage in the right precentral and postcentral gyrus (28). The authors concluded that the ictal pain was likely mediated by SI, because the hemorrhage was far from SII, anterior cingulate cortex (ACC), and insula. Moreover, the patient did not have a "psychic" aura (e.g., fear), usually associated with posterior parietal lobe seizures (28).

Although information gleaned from case histories of epilepsy suggests a critical role for the parietal lobes in the perception of pain, one must remain aware of the limitations of such research. The perceptions described by patients resulting from the aural or ictal phase of a seizure, as well as those resulting from electrical stimulation of the cortex near an epileptic focus, may reflect how the brain adjusts to a pathologic condition rather than how the brain normally functions. The intricate organization of the cerebral cortex in general, and the potential interaction of the many regions of the cortex—especially in processing the multifaceted aspects of nociceptive behavior—suggest that spatial summation of electrical or ictal stimulation is not necessarily the normal manner in which the brain processes nociceptive information.

BRAIN IMAGING STUDIES OF PAIN PERCEPTION

Modern brain imaging has added an important perspective to the field of pain research, allowing an opportunity to investigate possible cerebral mechanisms of pain in awake normal subjects by recording simultaneously across the entire brain changes in function related to the presentation of noxious stimuli. The closing years of the 20th century saw refinements in brain imaging techniques such as positron emission tomography (PET), functional magnetic resonance imaging (fMRI), electroencephalographic (EEG) dipole source analysis, magnetoencephalography (MEG), and single-photon emission computed tomography (SPECT). Each of these techniques has advantages and disadvantages in terms of spatial and temporal resolution, sensitivity, and cost. However, all provide measures that can be used as indirect indices of neuronal activity.

With improvements in spatial and temporal resolution, as well as in statistical sensitivity, most recent imaging studies have described a growing list of cortical and subcortical sites activated by noxious stimuli, although the most commonly observed sites continue to include the ACC, the insular cortex (IC), and the somatosensory cortices (SI and SII) of the parietal lobes (29,30). Interestingly, it has been activation within the parietal lobes that has been the most controversial finding in pain imaging studies. The present chapter concentrates on this research and the growing evidence that parietal lobe function plays in important role in sensory discriminative aspects of human pain perception.

Anterior Parietal Function and Pain Perception

The first three modern brain imaging studies of pain, published in the early 1990s, produced vastly different results in terms of SI cortex. Using PET and repeated 5-second heat stimuli presented to six spots on the arm, our laboratory reported a significant activation focus in SI cortex contralateral to the stimulated arm (31). Using similar heat stimuli, but repetitively presented to a single spot on the dorsal hand, Jones et al. (32) failed to observe significant activation in SI cortex. Finally, using SPECT, Apkarian et al. (33) found that submerging the fingers in hot water for 3 minutes led to a decrease in SI activity.

Jones and colleagues (34,35) postulated that attentional factors, rather than painful stimuli, produced the SI activation observed by Talbot, et al. Although recent studies support the idea that attention can significantly modulate pain-evoked SI activity, little evidence supports the premise that pain is not a major determinant of SI activity during painful stimulation.

Table 5-1 shows the methods and results of a number of human brain imaging studies of pain, using PET, SPECT, fMRI, and MEG. Some studies involving noxious thermal, chemical, or electrical stimulation reveal SI activation, whereas others using similar stimuli do not. Factors that may contribute to these differential results include: (a) influences of cognitive modulation in SI activity; and (b) a possible combination of excitatory and inhibitory effects of nociceptive input to SI.

Cognitive Modulation of Primary Somatosensory Cortex Activity

SI pain-related activation is highly modulated by cognitive factors that alter pain perception, including attention and previous experience. In our laboratory, we have shown that when the subject's attention is directed away from a painful stimulus, the activity of SI cortex is dramatically reduced (6,36). Subjects were presented concurrent sequences of tones and contact heat stimuli and were required to discriminate, during separate PET scans, changes in either thermal intensity or auditory frequency. The subjects' ratings of pain intensity were higher in the thermal than the auditory task ($p = 0.01$), indicating that pain perception was modulated by the attentional demands of the discrimination tasks. Likewise, pain-evoked rCBF was significantly larger in the thermal than in the auditory task ($p < 0.01$). A similar result was recently obtained in another PET study in which attention to a computerized perceptual maze test reduced pain-related activation in contralateral SI, as well as in bilateral SII, ACC, and midinsula (37).

Other data from our laboratory also indicate that attention to sensory aspects of the pain experience can alter SI activity. Using hypnosis, we found that suggestions specifically directed toward changing the perceived intensity of pain evoked by a noxious stimulus significantly modulated pain-related activity in SI (38). In contrast, suggestions directed toward changing the *unpleasantness* of the pain (39) had no effect on pain-related activity in SI, but produced instead a robust modulation of activity in anterior cingulate cortex directly correlated with the subjects' perception of unpleasantness ($p = 0.005$).

In these hypnosis experiments (38,39), we also found evidence that experience with the hypnotic suggestions may have produced long-term changes in the subjects' neural processing of pain. At least a week before participating in a PET scanning session, all subjects received the same hypnotic induction, suggestions, and painful stimuli that were to be used during the scanning experiment. In subsequent PET sessions, control scans performed *before* the subjects underwent hypnotic induction indicated that those previously trained to attend to the *intensity* of the painful stimuli showed substantially greater pain-related activity in SI than did those who had been trained to attend to the *unpleasantness* of those stimuli (38) ($t = 5.05$) (39) ($t = 3.01$).

Attentional modulation within SI cortex is not restricted to pain-related activity. Other investigators have found that rCBF in SI, evoked by tactile stimuli, is reduced when subjects attend to another stimulus modality (40). Similarly, neuronal recordings in SI cortex of trained monkeys reveal low-threshold neurons whose activity is enhanced by attention to the tactile stimulus (41,42). Despite the extensive nature of attentional modulation of SI activity, there is little evidence that attention activates SI neurons without the concurrent presence of sensory-evoked activation.

Inhibitory Effects of Noxious Stimuli in Primary Somatosensory Cortex Activity

Tommerdahl et al. (43) found in monkey SI cortex that the presence of noxious heat reduced the intrinsic optical-imaging signal evoked by low threshold mechanical stimulation of the skin. These data are consistent with the findings of Apkarian et al. (33), which

TABLE 5-1. *Human Brain Imaging Studies of Pain: Anterior Parietal Activation*

Study	Modality	Subjects	*n*	Stimulation device	Stimulation	Stimulated area	SI	SII	Posterior parietal cortex
Talbot et al., 1991 (31)	PET $H_2{}^{15}O$	Healthy	8	Thermode 1 cm^2	42°C, 47°C–48°C	Right volar forearm	Yes, contra	Yes, contra	No
Jones et al., 1991 (32)	PET $^{15}CO_2$	Healthy	6	Thermode 2.5 × 5.0 cm	36.1°C, 41.3°C, and 46.4°C	Dorsum of right hand	No	No	No
Apkarian et al., 1992 (33)	SPECT	Healthy	3	Water bath	Moderate heat pain	Fingers of left hand	Inhib, contra	No	No
Crawford et al., 1993 (76)	PET ^{133}Xn	Healthy	11	Tourniquet and hypnosis	Ischemic pain	Both arms	Yes, contra	No	No
Casey et al., 1994 (77)	PET $H_2{}^{15}O$	Healthy	18	Thermode 254 mm^2	40°C, 50°C	Left volar forearm (six sites)	Yes, contra	Yes, bilat	No
Coghill et al., 1994 (55)	PET $H_2{}^{15}O$	Healthy	9	Thermode 1 cm^2	34°C, 47°C–48°C	Left nondominant forearm	Yes, contra	Yes, contra	No
Derbyshire et al., 1994 (78)	PET $H_2{}^{15}O$	Healthy	6	Thermode	Ramp 25°C–43°C	Dorsum of right hand	No	No	No
		Facial pain	6						
Di Piero et al., 1994 (79)	SPET ^{133}Xn	Healthy	7	Water bath	0°C ± 1°C	Left hand	Yes, contra	No	No
Casey et al., 1996 (85)	PET $H_2{}^{15}O$	Healthy	27	Thermode 254 mm^2	40°C, 50°C	Left nondominant arm	NS, contra	Yes, contra	No
					20°C, 6°C		Yes, contra	No	No
Hsieh et al., 1996 (86)	PET [^{15}O] butanol	Cluster headache	7	Sublingual nitroglycerin	1 mg	Head pain	No	No	No
Andersson et al., 1997 (71)	PET $H_2{}^{15}O$	Healthy	6	Intracutaneous injection of capsaicin	1% capsaicin in a volume of 10 L of vehicle	Dorsum of right hand	Yes, contra	NS	No
						Dorsum of right foot	Yes, contra	NS	No
Antognini et al., 1997 (87)	fMRI	Healthy	5	Electrical stimulator	10–30 mA	Right index finger	Yes, contra	Yes, bilat	No
Aziz et al., 1997 (88)	PET $H_2{}^{15}O$	Healthy	8	2-cm long silicone balloon inflation	Painful stimulation	Lower esophagus	Yes, bilat	No	No
Derbyshire et al., 1997 (89)	PET $H_2{}^{15}O$	Healthy	12	CO_2 laser	Painful stimulation	Dorsum of right hand	Yes, contra	No	Yes, bilat BA 39/40
Di Piero et al., 1997 (90)	SPET ^{133}Xn	Healthy	12	Water bath (cold pressor water test)	Ice water	Immersion of one hand	Yes, contra	No	No
		Cluster headache	7						
Rainville et al., 1997 (39)	PET $H_2{}^{15}O$	Healthy	11	Water bath	35°C, 47°C	Passive immersion of left hand	Yes, contra	Yes, contra	No

Rosen et al., 1994 (80)	PET $H_2{}^{15}O$	Angina pectoris	12	Dobutamine infusion	10 g/kg^{-1}	Chest pain (angina)	No	No	No
Hsieh et al:, 1995 (66)	PET [^{15}O] butanol	Mono-neuropathy	8	None	Spontaneous pain	Pain in lower extremity	No	No	Yes, bilat BA 7/40
Hsieh et al., 1995 (65)	PET [^{15}O] butanol	Healthy	4	Intracutaneous injection	Ethanol (20 L, 70%)	Lateral right upper arm	Yes, bilat	No	Yes, bilat BA 39/40
Davis et al., 1995 (81)	fmRI	Healthy	9	Electrical nerve stimulator	Pain stimulation	Right median nerve	Yes, contra	No	No
Weiller et al., 1995 (82)	PET $H_2{}^{15}O$	Migraine patients	9	None	Spontaneous migraine	Head pain	No	No	No
Howland et al., 1995 (83)	MEG	Healthy	5	Electric intracutaneous stimulation	Voltage adjusted to produce pain 4/10	Digit 5 of nondominant hand	Yes, bilat	Yes, bilat	No
Kitamura et al., 1995 (59)	MEG	Healthy	5	Electric transcutaneous stimulation	2–4 mA: weak	Right digit 2	Yes, contra	No	No
					5–7 mA: moderately pain		Yes, contra	Yes, bilat	No
					10–13 mA: very painful		Yes, contra	Yes, bilat	No
Craig et al., 1996 (84)	PET $H_2{}^{15}O$	Healthy	11	Thermal grill	Alternating bars of 20°C and 40°C	Palmar surface of right hand	Yes, contra	Yes, contra	No
Silverman et al., 1997 (91)	PET $H_2{}^{15}O$	Healthy	12	Latex balloon catheter	20 mm Hg	Rectum	No	No	No
					40 mm Hg				
Svensson et al., 1997 (92)	PET $H_2{}^{15}O$	Healthy	11	Cutaneous CO^2 laser (79 mm^2)	0.5 Hz	Left forearm	No	Yes, contra	No
				Intramuscular electrical stimulator	50-s square-wave pulse at frequency 20 Hz	Brachioradialis muscle of left nondominant forearm	NS, contra	Yes, contra	No
Xu et al., 1997 (93)	PET $H_2{}^{15}O$	Healthy	6	CO^2 laser	23 mJ mm^2	Marked squares of 2.5 cm^2 of dorsum of left hand and foot	Yes, contra at hand area	Yes, bilat	No
Binkofski et al., 1998 (94)	fMRI	Healthy	5	2-cm long silicone balloon	Constant stimulation with a volume of 10 and 20 mL	Distal part of esophagus	Yes, bilat	Yes, bilat	No
Derbyshire and Jones 1998 (95)	PET $H_2{}^{15}O$	Healthy	6	Thermode	Ramp, 25°C, 43°C	Dorsum of left hand	No	No	No
Derbyshire and Jones, 1998 (95)	PET $H_2{}^{15}O$	Healthy	12	Water bath	Continuous heat pain	Right hand	No	No	No

continued

TABLE 5-1. *Continued*

Study	Modality	Subjects	*n*	Stimulation device	Stimulation	Stimulated area	SI	SII	Posterior parietal cortex
Disbrow et al., 1998 (96)	fMRI	Healthy	12	Electrical shocks	20.8 mA, 2 Hz		Yes, contra	Yes, bilat	No
				Thermode 4 cm^2	38°C, 48.5°C	Digit 2 of right hand	No	No	No
				Mechanical	Painful pinch		No	No	No
Iadarola et al., 1998 (70)	PET $H_2{}^{15}O$	Healthy	13	Intradermal injection of capsaicin	250 g in 20 L vehicle	Left volar forearm	Yes, contra	Yes, contra	Yes, bilat
				Light brush (allodynia)			Yes, contra	Yes, ipsi	Yes, ipsi BA 40
May et al., 1998 (97)	PET $H_2{}^{15}O$	Healthy	7	Capsaicin	0.05 mL of 0.1% solution	Subcutaneous right forehead	No	No	No
Oshiro et al., 1998 (98)	fMRI	Healthy	6	Electrical stimulator	8 Hz	Fifth digit of one hand at a time	Yes, contra	Yes, bilat	No
Paulson et al., 1998 (99)	PET $H_2{}^{15}O$	Healthy	20	Thermode 254 mm^2	40°C, 50°C	Left volar forearm	No	No	No
Porro et al., 1998 (100)	fMRI	Healthy	24	Subcutaneous ascorbic acid	0.5 mL, 20%, pH 6.7		Yes, contra	No	No
			16	Subcutaneous saline	0.5 mL	Dorsum of one foot	No	No	No
			16	Innocuous touch with a needle	Touch for 20 s		No	No	No
Baciu et al., 1999 (101)	fMRI	Healthy	6	Latex balloon catheter	Inflation until pain: 80/100	Rectum	Yes, bilat	No	Yes, bilat BA 39/40
Baron et al., 1999 (102)	fMRI	Healthy	9	Intracutaneous capsaicin	20 L of 0.05%	Right volar forearm	No	No	No
Becerra et al., 1999 (103)	fMRI	Healthy	12	Thermode 9 cm^2	41°C, 46°C	Dorsum of left hand	Yes, contra	Yes, contra	No
Coghill et al., 1999 (104)	PET $H_2{}^{15}O$	Healthy	16	Heated probe 1-cm diameter	35°C, 46°C, 48°C	Upper right arm or right-handed subjects (six areas)	Yes, contra	Yes, bilat	No
Derbyshire et al., 1999 (105)	PET $C^{15}O_2$	Post-molar extraction surgery	6	Thermode 2.5 × 1 cm	41.1°C–44.8°C 46.2°C–47.6°C	Dorsum of right hand	No	No	Yes, contra BA 40
Petrovic et al., 1999 (69)	PET [^{15}O] butanol	Mono-neuropathic	5	Soft camel hair brush (0.5-cm diameter)	1 stroke/s	Painful skin area and homologous contralateral nonpainful area	Yes, bilat	Yes, bilat	Yes, contra BA 7

Peyron et al., 1999 (75)	PET	Healthy	12	Thermode 9 cm^2 and attention to stimuli	39°C, 46.6°C	Dorsum of one hand	Inhibit SI ipsi	Yes, bilat	Yes, right BA 40
Rainville et al., 1999 (106)	PET $H_2{}^{15}O$	Healthy	8	Water bath and hypnosis	35°C, 47°C	Left hand	No	No	Inhibition of contra BA 40 with hypnosis
Tölle et al., 1999 (107)	PET $H_2{}^{15}O$	Healthy	12	Thermode (1.6 × 3.6 cm)	1°C below pain threshold	Right volar forearm	No	No	No
					1°C above pain threshold		No	No	No
Casey et al., 2000 (108)	PET $H_2{}^{15}O$	Healthy	20	Hand-held vibrator	130 Hz (nonpainful)	Left volar forearm	Yes, contra	No	No
				Water bath	1°C	Left hand	NS, contra	Yes, contra	No
Creac'h et al., 2000 (109)	fMRI	Healthy	11	Tensiometer	15–18 mm Hg	Dorsal left first metacarpophalangeal joint	Yes, contra, bilat	Yes, contra, bilat	No
Grachev et al., 2000 (110)	^{1}H-MRS	Chronic back pain	9	None	Spontaneous back pain	Back	No	No	No
Ploghaus et al., 2000 (111)	fMRI	Healthy	12	Thermode 9 cm^2 and expectation of painful stimulation	Painful stimulation	Dorsum of left hand	No	No	Yes, bilat BA 7
Ploner et al., 2000 (112)	MEG	Healthy	6	Electrical nerve stimulator	40–60 V (tactile)	Dorsum of each hand	Yes, contra	Yes, bilat	No
				CO_2 laser stimulation	600–700 mJ (pain)		Yes, contra	Yes, bilat	No
Sawamoto et al., 2000 (113)	fMRI	Healthy	10	CO_2 laser and expectation of pain	60 ms duration	Dorsum of right hand	No	Yes, bilat	No
Tracey et al., 2000 (114)	fMRI	Healthy	6	Thermode	5°C, 46°C	Dorsum of left hand	Yes, contra	Yes, contra	Yes, bilat BA 5,7, 40
Apkarian et al., 2001 (115)	fMRI	Healthy	1	None	Straight leg rising to exacerbate pain	Back	No	No	No
		Chronic back pain	1						
Bentley et al., 2001 (116)	LEPs	Healthy	1	CO_2 laser heat stimuli	15.3 mJ mm^{-2}	Right forearm	No	No	Yes, contra
Coghill et al., 2001 (67)	PET $H_2{}^{15}O$	Healthy	9	Thermode 1-cm diameter	35°C, 49°C	Six spots on both arms of right-handed subjects	Yes, contra	Yes, contra	Yes, right BA 40

continued

TABLE 5-1. *Continued*

Study	Modality	Subjects	*n*	Stimulation device	Stimulation	Stimulated area	SI	SII	Posterior parietal cortex
Chang et al., 2001 (117)	EEG	Healthy	15	Intramuscular injection capsaicin vehicle	50 g/0.5 mL 0.5 mL	Nondominant left brachioradialis muscle	No	No	Decrease in alpha-1 and alpha-2 activity, bilat
De Pascalis et al., 2001 (118)	ERPs	Healthy	29	Electrical stimulator + hypnosis	0.5 mA above pain threshold	Ventral right wrist	No	No	Inhibition with hypnosis
Hofbauer et al., 2001 (38)	PET $H_2{}^{15}O$	Healthy	10	Water bath + hypnosis	35°C, 46°C–47.5°C	Left hand	Yes, contra	Yes, contra	No
Lotze et al., 2001 (119)	fMRI	Upper limb amputee	14	Hand and lip movement		Hand/lip	Yes, contra for all	No	No
		Healthy	7	Imagined movement of phantom limb or left hand			Yes, contra for PTs	No	No
Olausson et al., 2001 (57)	fMRI	Hemispherectomized	4	Soft brush (7 cm wide)	20 cm/s, distance 10 cm	Medial leg, 15 cm inferior to the patella (of paretic leg for patients)	Yes, contra for healthy; ipsi for patients	Yes, bilat for healthy; no for patients	No
		Healthy	4	Thermode 9 cm²	34°C–36°C, 44°C–47°C		Yes, contra for healthy; NS ipsi for patients	Yes, contra for healthy; NS ipsi for patients	No
Witting et al., 2001 (68)	$H_2{}^{15}O$ PET	Healthy	8	Intradermal capsaicin	10 g/20 L	Left non-dominant volar forearm	No	No	No
				Brush-evoked allodynia	Soft brush, frequency of 0.25 Hz	Brush over a 2-cm² area	No	No	Yes, contra BA 5/7
Chen et al., 2002 (56)	fMRI	Healthy	4	Soft brush (2-cm wide)	2 Hz, 10-cm region	Left inner calf	Yes, contra	Yes, contra	No
				Thermode 9cm²	35-36°C, 45-46°C	Left inner calf	Yes, contra	Yes, contra	No
Davis et al., 2002 (120)	fMRI	Healthy	7	Thermode 5cm²	32°C, 3°C	Thenar eminance, right hand	No	Yes, contra No	Yes, bilat BA 7/40 No

DaSilva et al., (2002) (121)	fMRI	Healthy	9	Thermode 2.6cm^2	32°C, 46°C	ophtalmic, maxillary, mandibular divisions right side of the face and palmar right thumb	Yes, contra	Yes, bilat	No
Inui et al., (2002) (122)	MEG	Healthy	13	intraepidermal, square wave pulse	0.16±0.09mA	Dorsum of left hand	Yes, contra in 5 subjects	Yes, bilat	No
					0.23±0.09mA	Left lateral elbow joint			
Laureys et al., (2002) (123)	H_2 ^{15}O PET	Healthy	15	Electrical square wave pulses	7.4±5.9mA	Right median nerve at the wrist	Yes, contra	Yes, contra	Yes, bilat BA 40
		Patients (PVS)	15	Electrical square wave pulses	14.2±8.7mA		Yes, contra	No	No
Maihöfner et al., (2002) (124)	MEG	Healthy	7	Thermode 3.5cm diameter	32°C, -3±5°C	Dorsum of right hand	No	Yes, contra in 4 and bilat in 3 subjects	No
Nakamura et al., (2002) (125)	MEG	Healthy	6	Cutaneous YAG laser stimulator	600-650mJ	Dorsum of left hand	No	Yes, contra	No
Peyron et al., (2002) (126)	H_2 ^{15}O PET	Healthy	12	Thermode 9cm^2	39, 46.6°C	Dorsum of one hand	No	Yes, bilat	No
	fMRI	Healthy	8	Thermode 9cm^2	37, 46.9°C	Dorsum of one hand	No	Yes, contra and bilat in 6 subjects	No
	Intracerebral LEPs	Epilepsy patients	13	CO_2 laser stimulator 38.5mm^2	1.3 x pain threshold	Dorsum of hand contra to inplantation of electrode	No	Yes, contra and bilat in 2 patients	No
	Scalp LEP	Healthy	31	CO_2 laser stimulator 38.5mm^2	10 W	Superficial right radial nerve	No	Yes, bilat	No
Porro et al., (2002) (127)	fMRI	Healthy	26	23 gauge needle	s.c. ascorbic acid (20%)	Dorsum of one foot	Yes, contra	No	No
Torquati et al., (2002) (128)	MEG	Healthy	10	electric rectangular pulses	59±18.8mA	Right median nerve at wrist	Yes, contra	Yes, bilat	No
Tran et al., (2002) (129)	MEG	Healthy	9	CO_2 laser stimulator 2mm diameter	5-12 W	Dorsum of left hand	Yes, contra in 4 subjects	Yes, bilat	No
Wise et al., (2002) (130)	fMRI	Healthy	9	Thermal resistor 3cm^2	56.4±1.4°C	Dorsum of left hand	No	Yes, bilat	No

BA, Brodmann area; bilat, bilateral activation; contra, contralateral to the stimulated side; EEG, electroencephalography; ERP, evoked-related potential; fMRI, functional magnetic resonance imaging; ^{1}H-MRS, single-voxel proton magnetic resonance spectroscopy; ipsi, ipsilateral to the stimulated side; LEP, laser-evoked potential; MEG, magnetoencephalography; NS, not significant; PET, positron emission tomography; SPECT, single-photon emission computed tomography; SPET, single-photon emission tomography.

showed a decrease in blood flow to SI cortex in human subjects during the presentation of a tonic heat stimulus. Consonant with the idea that noxious stimulation produces inhibition of tactile sensitivity in SI cortex are psychophysical data showing that the presence of pain reduces tactile perception, a phenomenon described by Apkarian et al. as a "touch gate" (44).

Parietal Operculum and Pain Perception

Pain-related activation during brain imaging studies has been observed in various regions within the parietal operculum; however, the nature and function of this activation is still open to conjecture. Whereas the functional organization of the postcentral gyrus has been extensively detailed and defined as SI with its exquisitely arranged somatotopic representation of cutaneous receptors, no consensus yet exists concerning the exact role of activity observed within the parietal operculum. This region, located on the dorsal bank of the sylvian fissure just lateral to SI, includes SII—the secondary somatosensory cortex; but even the borders of SII are in dispute (9,30) and its functional relationship to SI appears to vary from one species to another (45–47).

Early electrophysiologic studies in monkey revealed responses in SII evoked by both noxious and innocuous stimulation. Whitsel et al. (48) described a well-localized region at the posterior margin of SII with cells responsive to nociceptive stimuli and a separate anterior part of SII in which most neurons responded to gentle tactile stimuli. Some evidence from human studies also suggests separate representations for pain and touch within the posterior parietal cortex; however, differences between monkey and human cortical anatomy and inconsistencies in the designation of SII proper complicate a direct comparison of these results. Treede and colleagues argue for the separate processing of innocuous stimuli in human SII and posterior insula, and reserve the deeper areas of the parietal operculum (medial to SII) and anterior insula for the processing of nociceptive stimuli (30). More recent data, from stereotactically placed electrodes in patients, are not entirely consistent with this view (49) and underscore the continuing controversy surrounding this issue.

The majority of electrophysiologic studies in the monkey have underscored, as was the case for SI, that most neurons in SII and surrounding areas respond to innocuous somatic stimulation, whereas only a few display nociceptive properties (50,51). However, unlike the detailed somatotopic organization of the contralateral body surface found in SI, neurons in SII demonstrate receptive fields that may be of variable size and responsive to bilateral as well as contralateral stimulation of the body. Therefore, based on the animal literature, this parietal region is unlikely to contribute toward a fine spatial discrimination of cutaneous stimuli—at either innocuous or noxious intensities.

Anatomic evidence from the monkey is consistent with at least a presence of nociceptive neurons in the SII region, as demonstrated by direct projections to the parietal operculum from thalamic nuclei (52) that convey nociceptive information from the spinal cord (53). Furthermore, corticocortical projections from SI may contribute to nociceptive activity observed in SII (47,54).

In human subjects, evidence gained from many different experimental approaches clearly demonstrates activation of SII (and/or surrounding regions of the parietal operculum) by noxious as wells as innocuous stimuli (see Table 5-1). Studies comparing directly different modalities of noxious and innocuous somatosensory stimulation have generally observed significant SII activation by both. In our own laboratory, both innocuous vibrotactile and noxious heat stimulation of the forearm evoked significant increases in SII rCBF, as detected by PET (55); likewise, significant increases in fMRI-detected activation were observed in individual subjects within SII during the presentation to the leg of either innocuous brush of noxious thermal stimuli (56,57). In other laboratories, similar results have been obtained for noxious heat and innocuous vibrotactile stimuli, using fMRI

(58), and for innocuous electrical and noxious CO_2 laser stimuli, using evoked potential analysis (49).

Although the region of SII may be activated during the presentation of either innocuous or noxious stimulation, this by no means implies that SII activity cannot contribute to a subject's ability to distinguish painful and nonpainful sensations. Studies using a single stimulus modality have demonstrated intensity dependent pain-related responses in SII contralateral to the site of stimulation (59–62), indicating a functional capacity to discriminate between noxious stimuli of different intensities, as well as between innocuous and noxious levels of stimulation. These recent studies (59–62) also demonstrate a similar degree of activation of SII, ipsilateral to the stimulation site. This bilateral activation of the parietal operculum—rarely, if ever, observed in SI—may account in part for the ability of hemispherectomized patients to perceive the intensity as well as the affective components of heat stimuli presented to the paretic leg (57).

Posterior Parietal Function and Pain Perception

Evidence in humans for a direct involvement of posterior regions of the parietal cortex in pain processing is considerably less convincing than that documented for other areas of the parietal lobes. On the other hand, a few studies in the nonhuman primate have, indeed, suggested a role for this region in nociception. Electrophysiologic data from the monkey revealed a small number of neurons responsive to noxious stimuli in the vicinity of area 7b, near the posterior border of SII (48,63). Neurons within this area demonstrated intensity-dependent responses correlated with the monkey's tendency to escape noxious stimulation, and a unilateral lesion of the region, in one monkey, altered thermal pain tolerance, although intensity discrimination remained intact (64). The specific relevance of these data to posterior parietal function is difficult to assess, however, owing to the small number of neurons recorded and the extension of the experimental lesion in this case into the adjoining parietal operculum.

In humans, a few studies have suggested that the posterior parietal cortex may process attentional aspects associated with sensory integration and body orientation to potentially damaging (painful) stimuli. Hsieh et al. (65) reported that acute experimental pain, evoked by an intracutaneous injection of ethanol in the right upper arm, was associated with a significant increase in parietal rCBF within contralateral SI and bilateral posterior parietal cortices. Such a scenario is consistent with a potential divergence of roles for SI and posterior parietal regions—in that activation within the contralateral SI is appropriate for localization of the stimulus to the upper arm, whereas activation in both contralateral and ipsilateral posterior parietal cortices would be expected for orientation toward the stimulus and an integration of motor responses involving a reaction of the opposite hand to care for the painful upper arm. Additional evidence for a role of posterior parietal regions in attentional aspects of pain was observed in a PET study of patients suffering from chronic pain elicited by mononeuropathy (66). An increase in rCBF was observed during periods of chronic pain in bilateral posterior parietal cortex—among other areas—but not in SI or SII. The preponderance of activation in the posterior regions of the parietal lobe suggested to the authors that processing of chronic pain, as opposed to phasic experimental pain, may reflect the increased attention and vigilance that patients devote to their clinical condition.

Additional evidence for the involvement of the posterior parietal lobe in attention and spatial orientation toward noxious stimuli is suggested by studies of moving noxious stimuli. For example, Coghill et al. suggested that stimulating different areas of the skin may solicit the subject's attention and be more likely to implicate the right posterior parietal lobe, a reflection of "spatial attentional/awareness components of somatosensory information" (67). A similar rationale may be applied to brush-evoked allodynia; that is, pain evoked by innocuous brushing stimuli applied to sensitized skin. Witting and colleagues (68) recently studied brush-evoked allodynia in healthy volunteers by injecting intradermal capsaicin in their nondominant forearm. Using PET, these authors found that brush-evoked allodynia, but not capsaicin pain alone, activated BA 5 and 7 in the contralateral posterior parietal lobe. Likewise, Petrovic and colleagues (69) reported that brush-induced allodynia in mononeuropathic patients was associated with significant activations of the contralateral posterior parietal cortex (along with bilateral activation of SI, SII, and thalamus). Similarly, Iadarola et al. (70) had described bilateral activation in the posterior parietal lobe (e.g., BA 40) and adjacent regions of SII during brush-evoked allodynia; however, in this study capsaicin pain was also associated with increased rCBF in bilateral posterior parietal cortex. Thus, whereas possible attention toward a moving pain presents a compelling argument for posterior parietal involvement in spatial orientation toward noxious stimulation, results across studies are not entirely consistent and indicate that movement of the stimulus is neither sufficient nor necessary for stimulus-related activation of this region.

SUMMARY

Converging lines of evidence confirm a role for the anterior parietal cortex in pain processing and extend the traditional view of SI to include discriminative aspects of somatic stimulation that is potentially tissue-damaging (e.g., painful). Recent studies more specifically implicate SI in the sensory aspect of pain perception by demonstrating that SI activation is modulated by cognitive manipulations that alter perceived pain intensity, but not by manipulations that alter unpleasantness, independent of pain intensity. Nevertheless, despite the probable role of SI in the encoding of the various sensory features of pain, considerable evidence suggests that nociceptive input to SI may also serve to modulate tactile perception. Thus, SI cortex may be involved in both the perception and modulation of both painful and nonpainful somatosensory sensations.

Defining a role in pain processing for the parietal operculum is somewhat more problematic. The absence of a fine somatotopic organization of cutaneous (or visceral) receptors virtually eliminates a substantial role for this region in localizing noxious stimuli. Several studies suggest separate representations for pain and touch within the posterior parietal cortex and SII, respectively; however, interspecies differences in cortical anatomy and inconsistencies in the designation of SII proper preclude a clear reconciliation of the data. Likewise, suggestions that SII activation is predominantly related to processing the nociceptive quality of the stimulus (60,61) are inconsistent with many studies in both human and nonhuman subjects, which show a strong functional relationship between SII activity and innocuous (especially, vibrotactile) stimulation. Nevertheless, the numerous studies indicating pain-related activation within the parietal operculum (and/or SII) underscore the potential importance of this region in the perception of pain and the need for continued research.

Finally, a possible role of posterior parietal cortex (BA 5/7, 39/40) in orientation and attention toward painful sensory stimuli is consistent with existing literature describing this region as a poly modal association area concerned with intrapersonal and extrapersonal space; however, results from studies that actually manipulate the subjects' level of attention relative to painful stimuli have not uniformly supported this hypothesis (75). Future studies assessing both attentional demand and direct manipulation or motor interactions involving noxious stimuli may help to resolve this issue.

In spite of some discrepant results concerning specific details of the nociceptive process, the weight of human pain research now firmly establishes a role for the parietal lobes in the conscious appreciation of the sensation of pain.

ACKNOWLEDGMENTS

The authors would like to express their appreciation to the staff and faculty of the Brain Imaging Center at the Montreal Neurological Institute. This work was supported by grants from the Medical Research Council of Canada; M.-C. Albanese is supported by a student scholarship awarded by the Natural Sciences & Engineering Research Council of Canada (NSERC) and by le Fonds pour la Formation de Chercheurs et l'Aide à la Recherche (FCAR).

REFERENCES

1. Zilles K, Palomero-Gallagher N. Cyto-, myelo-, and receptor architectonics of the human parietal cortex. *Neuroimage* 2001;14:S8–20.
2. Roland P. Cortical representation of pain. *Trends Neurosci* 1992;15:3–5.
3. Roland PE. Pain and activation in the thalamus. Reply. *Trends Neurosci* 1992;15:252–253.
4. Duncan GH, Bushnell MC, Talbot JD, et al. Pain and activation in the thalamus. *Trends Neurosci* 1992;15: 252–253.
5. Stea RA, Apkarian AV. Pain and somatosensory activation. *Trends Neurosci* 1992;15:250–251.
6. Bushnell MC, Duncan GH, Hofbauer RK, et al. Pain perception: is there a role for primary somatosensory cortex? *Proc Natl Acad Sci USA* 1999;96:7705–7709.
7. Duvernoy HM. *The human brain: surface, three-dimensional sectional anatomy with MRI, and blood supply,* 2nd ed. Berlin: Springer-Verlag, 1999.
8. Brodmann K. *Vergleichende Lokalisationslehre der Grossßhirnrinde in Ihren Prinzipien dargestellt auf Grund des Zellenbaues.* Barth Leipzig, 1909.
9. Craig AD. How do you feel? Interoception: the sense of temperature, pain, itch, and "the material me." *Soc Neurosci Abstr* 2001;341.
10. Cushing H. A note upon the faradic stimulation of the postcentral gyrus in conscious patients. *Brain* 1909; 32:44–53.
11. Head H, Holmes G. Sensory disturbances from cerebral lesions. *Brain* 1911;34:102–254.
12. Penfield W, Boldrey E. Somatic motor and sensory representation in the cerebral cortex of man as studied by electrical stimulation. *Brain* 1937;60:389–443.
13. Marshall J. Sensory disturbances in cortical wounds with special reference to pain. *J Neurol Neurosurg Psychiatr* 1951;14:187–204.
14. Lewin W, Phillips CG. Observations on partial removal of the post-central gyrus for pain. *J Neurol Neurosurg Psychiatr* 1952;15:143-147.
15. Soria ED, Fine EJ. Disappearance of thalamic pain after parietal subcortical stroke. *Pain* 1991;44:285–288.
16. Potagas C, Avdelidis D, Singounas E, et al. Episodic pain associated with a tumor in the parietal operculum: a case report and literature review. *Pain* 1997;72: 201–208.
17. Greenspan JD, Winfield JA. Somatosensory alterations associated with a tumor compressing the retroinsular cerebral cortex. *IBRO World Congr Neurosci* 1991;3:188.
18. Bassetti C, Bogousslavsky J, Regli F. Sensory syndromes in parietal stroke. *Neurology* 1993;43: 1942–1949.
19. Ploner M, Freund H-J, Schnitzler A. Pain affect without pain sensation in a patient with a postcentral lesion. *Pain* 1999;81:211–214.
20. Albe-Fessard D, Berkley KJ, Kruger L, et al. Diencephalic mechanisms of pain sensation. *Brain Res Rev* 1985;9:217–296.
21. Kenshalo DR, Willis WD. The role of cerebral cortex in pain sensation. In: Peters A, Jones EG, eds. *Cerebral cortex, normal and altered states of function.* New York: Plenum Press, 1991:153–212.
22. Vogt BA, Sikes RW, Vogt LJ. Anterior cingulate cortex and the medial pain system. In: Vogt BA, Gabriel M, eds. *Neurobiology of cingulate cortex and limbic thalamus: a comprehensive handbook.* Boston: Birkhauser, 1993:313–344.
23. Greenspan JD, Lee RR, Lenz FA. Pain sensitivity alterations as a function of lesion location in the parasylvian cortex. *Pain* 1999;81:273–282.
24. Siegel AM, Williamson PD. Roberts DW, et al. Localized pain associated with seizure origin in the parietal lobe. *Epilepsia* 1999;40:845–855.
25. Young GB, Blume WT. Painful epileptic seizures. *Brain* 1983;106:537–554.
26. Mauguière F, Courjon J. Somatosensory epilepsy. A review of 127 cases. *Brain* 1978;101:307–332.
27. Nair DR, Najm I, Bulacio J, et al. Painful auras in focal epilepsy. *Neurology* 2001;57:700–702.
28. Phan TG, Cascino GD, Fulgham J. Ictal abdominal pain heralding parietal lobe haemorrhage. *Seizure* 2001;10:56–59.
29. Bushnell MC, Duncan GH, Ha B, et al. Non-invasive brain imaging during experimental and clinical pain. In: Devor M, Rowbotham M, Wisenfeld-Hallin Z, eds. *Proceedings of the XVth World Congress on Pain.* Seattle: IASP Press, 2000:485–496.
30. Treede R, Apkarian AV, Bromm B, et al. Cortical representation of pain: functional characterization of nociceptive areas near the lateral sulcus. *Pain* 2000;87: 113–119.
31. Talbot JD, Marrett S, Evans AC, et al. Multiple representations of pain in human cerebral cortex. *Science* 1991;251:1355–1358.
32. Jones AKP, Brown WD, Friston KJ, et al. Cortical and subcortical localization of response to pain in man using positron emission tomography. *Proc R Soc Lond Biol* 1991;244:39–44.
33. Apkarian AV, Stea RA, Manglos SH, et al. Persistent pain inhibits contralateral somatosensory cortical activity in humans. *Neurosci Lett* 1992;140:141–147.
34. Jones AK, Friston K, Frackowiak RS. Localization of responses to pain in human cerebral cortex. *Science* 1992;255:215–216.
35. Jones AKP, Derbyshire SWG. Cerebral mechanisms operating in the presence and absence of inflammatory pain. *Ann Rheum Dis* 1996;55:411–420.
36. Carrier B, Rainville P, Paus T, et al. Attentional modulation of pain-related activity in human cerebral cortex. *Soc Neurosci Abstr* 1998;24.

37. Petrovic P, Petersson KM, Ghatan PH, et al. Pain-related cerebral activation is altered by a distracting cognitive task. *Pain* 2000;85:19–30.
38. Hofbauer RK, Rainville P, Duncan GH, et al. Cortical representation of the sensory dimension of pain. *J Neurophysiol* 2001;86:402–411.
39. Rainville P, Duncan GH, Price DD, et al. Pain affect encoded in human anterior cingulate but not somatosensory cortex. *Science* 1997;227:968–971.
40. Meyer E, Ferguson SG, Zatorre RJ, et al. Attention modulates somatosensory cerebral blood flow response to vibrotactile stimulation as measured by positron emission tomography. *Ann Neurol* 1991;29:440–443.
41. Hyvärinen J, Poranen A, Jokinen Y. Influence of attentive behavior on neural responses to vibration in primary somatosensory cortex of the monkey. *J Neurophysiol* 1980;43:870–882.
42. Poranen A, Hyvärinen J. Effects of attention on multiunit responses to vibration in the somatosensory regions of the monkey's brain. *Electroencephalograph Clin Neurophysiol* 1982;53:535–537.
43. Tommerdahl M, Delemos KA, Vierck CJ Jr, et al. Anterior parietal cortical response to tactile and skin-heating stimuli applied to the same skin site. *J Neurophysiol* 1996;75:2662–2670.
44. Apkarian AV, Stea RA, Bolanowski SJ. Heat-induced pain diminishes vibrotactile perception: a touch gate. *Somatosens Motil Res* 1994;11:259–267.
45. Disbrow E, Roberts T, Krubitzer L. Somatotopic organization of cortical fields in the lateral sulcus of Homo sapiens: evidence for SII and PV. *J Comp Neurol* 2000; 418:1–21.
46. Murray GM, Zhang HQ, Kaye AN, et al. Parallel processing in rabbit first (SI) and second (SII) somatosensory cortical areas: effects of reversible inactivation by cooling of SI on responses in SII. *J Neurophysiol* 1992;68:703–710.
47. Pons TP, Garraghty PE, Mishkin M. Serial and parallel processing of tactual information in somatosensory cortex of rhesus monkeys. *J Neurophysiol* 1992;68: 518–527.
48. Whitsel BL, Petrucelli LM, Werner G. Symmetry and connectivity in the map of the body surface in somatosensory area II of primates. *J Neurophysiol* 1969;32: 170–183.
49. Frot M, Garcia-Larrea L, Guenot M, et al. Responses of the supra-sylvian (SII) cortex in humans to painful and innocuous stimuli. A study using intra-cerebral recordings. *Pain* 2001;94:65–73.
50. Robinson CJ, Burton H. Somatotopographic organization in the second somatosensory area of M. fascicularis. *J Comp Neurol* 1980;192:43–67.
51. Dong WK, Salonen LD, Kawakami Y, et al. Nociceptive responses of trigeminal neurons in SII-7b cortex of awake monkeys. *Brain Res* 1989;484:314–324.
52. Friedman DP, Murray EA. Thalamic connectivity of the second somatosensory area and neighboring somatosensory fields of the lateral sulcus of the macaque. *J Comp Neurol* 1986;252:348–374.
53. Stevens RT, London SM, Apkarian AV. Spinothalamocortical projections to the secondary somatosensory cortex (SII) in squirrel monkey. *Brain Res* 1993;631: 241–246.
54. Pons TP, Kaas JH. Corticocortical connections of area 2 of somatosensory cortex in macaque monkeys: a correlative anatomic and electrophysiologic study. *J Comp Neurol* 1986;248:313–335.
55. Coghill RC, Talbot JD, Evans AC, et al. Distributed processing of pain and vibration by the human brain. *J Neurosci* 1994;14:4095–4108.
56. Chen J-I, Ha B, Bushnell MC, et al. Differentiating noxious- and innocuous-related activation of human somatosensory cortices using temporal analysis of fMRI. *J Neurophysiol* 2002;88:464–474.
57. Olausson H, Ha B, Duncan GH, et al. Cortical activation by tactile and painful stimuli in hemispherectomized patients. *Brain* 2001;124:916–927.
58. Gelnar PA, Krauss BR, Sheehe PR, et al. A comparative fMRI study of cortical representations for thermal painful, vibrotactile, and motor performance tasks. *NeuroImage* 1999;10:460–482.
59. Kitamura Y, Kakigi R, Hoshiyama M, et al. Pain-related somatosensory evoked magnetic fields. *Electroencephalograph Clin Neurophysiol* 1995;95:463–474.
60. Valeriani M, Le Pera D, Niddam D, et al. Dipolar source modeling of somatosensory evoked potentials to painful and nonpainful median nerve stimulation. *Muscle Nerve* 2000;23:1194–1203.
61. Timmermann L, Ploner M, Haucke K, et al. Differential coding of pain intensity in the human primary and secondary somatosensory cortex. *J Neurophysiol* 2001;86:1499–1503.
62. Opsommer E, Weiss T, Plaghki L, et al. Dipole analysis of ultralate (C-fibres) evoked potentials after laser stimulation of tiny cutaneous surface areas in humans. *Neurosci Lett* 2001;298:41–44.
63. Dong WK, Chudler EH, Sugiyama K, et al. Somatosensory, multisensory, and task-related neurons in cortical area 7b (PF) of unanesthetized monkeys. *J Neurophysiol* 1994;72:542–564.
64. Dong WK, Hayashi T, Roberts VJ, et al. Behavioral outcome of posterior parietal cortex injury in the monkey. *Pain* 1996;64:579–587.
65. Hsieh JC, Ståhle-Bäckdahl M, Hägermarko, et al. Traumatic nociceptive *Pain* activates the hypothalamus and the periaqueductal gray: a positron emission tomography study. *Pain* 1996;64:303–314.
66. Hsieh JC, Belfrage M, Stone-Elander S, et al. Central representation of chronic ongoing neuropathic pain studied positron emission tomography. *Pain* 1995;63: 225–236.
67. Coghill RC, Gilron I, Iadarola MJ. Hemispheric lateralization of somatosensory processing. *J Neurophysiol* 2001;85:2602–2612.
68. Witting N, Kupers RC, Svensson P, et al. Experimental brush-evoked allodynia activates posterior parietal cortex. *Neurology* 2001;57:1817–1824.
69. Petrovic P, Ingvar M, Stone-Elander S, et al. A PET activation study of dynamic mechanical allodynia in patients with mononeuropathy. *Pain* 1999;83:459–470.
70. Iadarola MJ, Berman KF, Zeffiro TA, et al. Neural activation during acute capsaicin-evoked pain and allodynia assessed with positron emission tomography. *Brain* 1998;121:931–947.
71. Andersson, JLR, Lilja A, Hartvig P, et al. Somatotopic organization along the central sulcus, for pain localization in humans, as revealed by positron emission tomography. *Exp Brain Res* 1997;117:192–199.
72. Zhang HQ, Murray GM, Coleman GT, et al. Functional characteristics of the parallel SI- and SII-pro-

jecting neurons of the thalamic ventral posterior nucleus in the marmoset. *J Neurophysiol* 2001;85: 1805–1822.
73. Sinclair RJ, Burton H. Neuronal activity in the second somatosensory cortex of monkeys (Macaca mulatta) during active touch of gratings. *J Neurophysiol* 1993; 70:331–350.
74. Burton H, Sinclair RJ. Second somatosensory cortical area in macaque monkeys: 2. Neuronal responses to punctate vibrotactile stimulation of glabrous skin on the hand. *Brain Res* 1991;538:127–135.
75. Peyron R, Garcia-Larrea L, Gregoire MC, et al. Haemodynamic brain responses to acute pain in humans: sensory and attentional networks. *Brain* 1999; 122:1765–1780.
76. Crawford HJ, Gur RC, Skolnick B, et al. Effects of hypnosis on regional cerebral blood flow during ischemic pain with and without suggested hypnotic analgesia. *Int J Psychophysiol* 1993;15:181–195.
77. Casey KL, Minoshima S, Berger KL, et al. Positron emission tomographic analysis of cerebral structures activated specifically by repetitive noxious heat stimuli. *J Neurophysiol* 1994;71:802–807.
78. Derbyshire SWG, Jones AKP, Devani P, et al. Cerebral responses to pain in patients with atypical facial pain measured by positron emission tomography. *J Neurol Neurosurg Psychiatr* 1994;57:1166–1172.
79. Di Piero V, Ferracuti S, Sabatini U, et al. A cerebral blood flow study on tonic pain activation in man. *Pain* 1994;56:167–173.
80. Rosen SD, Paulesu E, Frith CD, et al. Central nervous pathways mediating angina pectoris. *Lancet* 1994;344: 147–150.
81. Davis KD, Wood ML, Crawley AP, et al. fMRI of human somatosensory and cingulate cortex during painful electrical nerve stimulation. *NeuroReport* 1995;7:321–325.
82. Weiller C, May A, Limmroth V, et al. Brainstem activation in spontaneous human migraine attacks. *Nat Med* 1995;1:658–660.
83. Howland EW, Wakai RT, Mjaanes BA, et al. Whole head mapping of magnetic fields following painful electric finger shock. *Cogn Brain Res* 1995;2:165–172.
84. Craig AD, Reiman EM, Evans AC, et al. Functional imaging of an illusion of pain. *Nature* 1996;384: 258–260.
85. Casey KL, Minoshima S, Morrow TJ, et al. Comparison of human cerebral activation patterns during cutaneous warmth, heat pain, and deep cold pain. *J Neurophysiol* 1996;76:571–581.
86. Hsieh JC, Hannerz J, Ingvar M. Right-lateralised central processing for pain of nitroglycerin- induced cluster headache. *Pain* 1996;67:59–68.
87. Antognini JF, Buonocore MH, Disbrow EA, et al. Isoflurane anesthesia blunts cerebral responses to noxious and innocuous stimuli: a fMRI study. *Life Sci* 1997;61:349–354.
88. Aziz Q, Andersson JLR, Valind S, et al. Identification of human brain loci processing esophageal sensation using positron emission tomography. *Gastroenterology* 1997;113:50–59.
89. Derbyshire SW, Jones AK, Gyulai F, et al. Pain processing during three levels of noxious stimulation produces differential patterns of central activity. *Pain* 1997;73:431–445.
90. Di Piero V, Fiacco F, Tombari D, et al. Tonic pain: a SPET study in normal subjects and cluster headache patients. *Pain* 1997;70:185–191.
91. Silverman DHS, Munakata JA, Ennes H, et al. Regional cerebral activity in normal and pathologic perception of visceral pain. *Gastroenterology* 1997;112:64–72.
92. Svensson P, Minoshima S, Beydoun A, et al. Cerebral processing of acute skin and muscle pain in humans. *J Neurophysiol* 1997;78:450–460.
93. Xu XP, Fukuyama H, Yazawa S, et al. Functional localization of pain perception in the human brain studied by PET. *NeuroReport* 1997;8:555–559.
94. Binkofski F, Schnitzler A, Enck P, et al. Somatic and limbic cortex activation in esophageal distention: a functional magnetic resonance imaging study. *Ann Neurol* 1998;44:811–815.
95. Derbyshire SW, Jones AK. Cerebral responses to a continual tonic pain stimulus measured using positron emission tomography. *Pain* 1998;76:127–135.
96. Disbrow E, Buonocore M, Antognini J, et al. Somatosensory cortex: a comparison of the response to noxious thermal, mechanical, and electrical stimuli using functional magnetic resonance imaging. *Hum Brain Map* 1998;6:150–159.
97. May A, Kaube H, Büchel C, et al. Experimental cranial pain elicited by capsaicin: a PET study. *Pain* 1998;74:61–66.
98. Oshiro Y, Fuiji N, Tanaka H, et al. Functional mapping of pain-related activation with echo-planar MRI: significance of the SII-insular region. *NeuroReport* 1998; 9:2285–2289.
99. Paulson PE, Minoshima S, Morrow TJ, et al. Gender differences in pain perception and patterns of cerebral activation during noxious heat stimulation in humans. *Pain* 1998;76:223–229.
100. Porro CA, Cettolo V, Francescato MP, et al. Temporal and intensity coding of pain in human cortex. *J Neurophysiol* 1998;80:3312–3320.
101. Baciu MV, Bonaz BL, Papillon E, et al. Central processing of rectal pain: a functional MR imaging study. *AJNR Am J Neuroradiol* 1999;20:1920–1924.
102. Baron R, Baron Y, Disbrow EA, et al. Brain processing of capsaicin-induced secondary hyperalgesia: a functional MRI study. *Neurology* 1999;53:548–557.
103. Becerra LR, Breiter HC, Stojanovic M, et al. Human brain activation under controlled thermal stimulation and habituation to noxious heat: an fMRI study. *Magn Reson Med* 1999;41:1044–1057.
104. Coghill RC, Sang CN, Maisog JM, et al. Pain intensity processing within the human brain: a bilateral, distributed mechanism. *J Neurophysiol* 1999;82:1934–1943.
105. Derbyshire SWG, Jones AKP, Collins M, et al. Cerebral responses to pain in patients suffering acute postdental extraction pain measured by positron emission tomography (PET). *Eur J Pain* 1999;3:103–113.
106. Rainville P, Hofbauer RK, Paus T, et al. Cerebral mechanisms of hypnotic induction and suggestion. *J Cogn Neurosci* 1999;11:110–125.
107. Tölle TR, Kaufmann T, Siessmeier T, et al. Region-specific encoding of sensory and affective components of pain in the human brain: a positron emission tomography correlation analysis. *Ann Neurol* 1999;45:40–47.
108. Casey KL, Svensson P, Morrow TJ, et al. Selective opiate modulation of nociceptive processing in the human brain. *J Neurophysiol* 2000;84:525–533.

109. Creac'h C, Henry P, Caille JM, et al. Functional MR imaging analysis of pain-related brain activation after acute mechanical stimulation. *AJNR Am J Neuroradiol* 2000;21:1402–1406.
110. Grachev ID, Fredrickson BE, Apkarian AV. Abnormal brain chemistry in chronic back pain: an in vivo proton magnetic resonance spectroscopy study. *Pain* 2000;89: 7–18.
111. Ploghaus A, Tracey I, Clare S, et al. Learning about pain: the neural substrate of the prediction error for aversive events. *Proc Natl Acad Sci USA* 2000;97: 9281–9286.
112. Ploner M, Schmitz F, Freund H-J, et al. Differential organization of touch and pain in human primary somatosensory cortex. *J Neurophysiol* 2000;83:1770–1776.
113. Sawamoto N, Honda M, Okada T, et al. Expectation of pain enhances responses to nonpainful somatosensory stimulation in the anterior cingulate cortex and parietal operculum/posterior insula: an event-related functional magnetic resonance imaging study. *J Neurosci* 2000;20:7438–7445.
114. Tracey I, Becerra L, Chang I, et al. Noxious hot and cold stimulation produce common patterns of brain activation in humans: a functional magnetic resonance imaging study. *Neurosci Lett* 2000;288:159–162.
115. Apkarian AV, Krauss BR, Fredrickson BE, et al. Imaging the pain of low back pain: functional magnetic resonance imaging in combination with monitoring subjective pain perception allows the study of clinical pain states. *Neurosci Lett* 2001;299:57–60.
116. Bentley DE, Youell PD, Crossman AR, et al. Source localisation of 62-electrode human laser pain evoked potential data using a realistic head model. *Int J Psychophysiol* 2001;41:187–193.
117. Chang PF, Arendt-Nielsen L, et al. Topographic effects of tonic cutaneous nociceptive stimulation on human electroencephalograph. *Neurosci Lett* 2001;305:49–52.
118. De Pascalis V, Magurano MR, Bellusci A, et al. Somatosensory event-related potential and autonomic activity to varying pain reduction cognitive strategies in hypnosis. *Clin Neurophysiol* 2001;112:1475–1485.
119. Lotze M, Flor H, Grodd W, et al. Phantom movements and pain. An fMRI study in upper limb amputees. *Brain* 2001;124:2268–2277.
120. Davis KD, Pope GE, Crawley AP, and Mikulis DJ. Neural correlates of prickle sensation: a percept-related fMRI study. *Nat Neurosci* 2002;5(11): 1121–1122.
121. DaSilva AFM, Becerra L, Makris N, et al. Somatotopic activation in the human trigeminal pain pathway. *J Neurosci* 2002;22(18):8183–8192.
122. Inui K, Tran TD, Qiu Y, Wang X, Hoshiyama M, and Kakigi R. Pain-related magnetic fields evoked by intra-epidermal electrical stimulation in humans. *Clin Neurophys* 2002;113:298–304.
123. Laureys S, Faymonville ME, Peigneux P, et al. Cortical processing of noxious somatosensory stimuli in the persistent vegetative state. *Neuroimage* 2002;17: 732–741.
124. Maihöfner C, Kaltenhäuser M, Neundörfer B, and Lang E. Temporo-spatial analysis of cortical activation by phasic innocuous and noxious cold stimuli—a magnetoencephalographic study. *Pain* 2002;100:281–290.
125. Nakamura Y, Paur R, Zimmermann R, and Bromm B. Attentional modulation of human pain processing in the secondary somatosensory cortex: a magnetoencephalographic study. *Neurosci Lett* 2002;328:29–32.
126. Peyron R, Frot M, Schneider F, et al. Role of operculoinsular cortices in human pain processing: converging eveidence from PET, fMRI, dipole modeling, and intracerebral recordings of evoked potentials. *Neuroimage* 2002;17:1336–1346.
127. Porro CA, Baraldi P, Pagnoni G, et al. Does anticipation of pain affect cortical nociceptive systems? *J Neurosci* 2002;22(8):3206–3214.
128. Torquati K, Pizzella V, Della Penna S, et al. Comparison between SI and SII responses as a function of stimulus intensity. *NeuroReport* 2002;13(6):813–819.
129. Tran TD, Inui K, Hoshiyama M, Lam K, Qiu Y, and Kakigi, R. Cerebral activation by the signals ascending through unmyelinated c-fibers in humans: a magnetoencephalographic study. *Neurosci* 2002;113(2): 375–386.
130. Wise RG, Rogers R, Painter D, et al. Combining fMRI with a pharmacokinetic model to determine which brain areas activated by painful stimulation are specifically modulated by remifentanil. *Neuroimage* 2002; 16:999–1014.

6

Anatomic and Functional Reorganization of Somatosensory Cortex in Mature Primates After Peripheral Nerve and Spinal Cord Injury

Jon H. Kaas and Christine E. Collins

Department of Psychology, Vanderbilt University, Nashville, Tennessee

INTRODUCTION

Somatosensory cortex is characterized by orderly representations of the body (see Chapter 4). In anterior parietal cortex of anthropoid primates (monkeys, apes, and humans), four histologically distinct strips of cortex, areas 3a, 3b, 1, and 2, individually represent the contralateral body surface and muscles (1,2). The representations all proceed from medial representations of the foot to successively more lateral representations of the trunk, forelimb, face, and tongue. The representation of cutaneous receptors in area 3b is the most precise. Once the organization of these representations became known in detail, it became possible to determine the consequences of sensory loss on the somatotopy of these maps.

Early studies of the consequences of sensory loss reported the effects of deactivating the median nerve to the hand or the nerves to a single finger on the organization of the hand representation in area 3b of monkeys (3–5). Initially, neurons in the deprived region of the cortex failed to respond to tactile stimuli; however, after a recovery period of weeks to months, the deprived parts of the hand representation were no longer unresponsive to somatosensory stimuli. Instead, neurons responded vigorously to light touch on parts of the hand that adjoin the denervated region. The basic result is portrayed schematically in Fig. 6-1, where the receptor sheet (in this case, the skin of the hand) is subserved by a grid of afferents that project to the brainstem (in this case, the cuneate nucleus of the medullary trigeminal–cuneate–gracilis complex), which projects in turn to the hand subnucleus of the contralateral ventroposterior nucleus (VP), which relays to the hand representation in lateral area 3b. At each level, a topographic representation of the arrangement of peripheral afferents in the skin is preserved, although distorted by variations in receptor densities, splits, and discontinuities in the central map. The loss of sensory afferents from the skin of region 1 deprives cortical neurons within the sector of area 3b that is normally activated by touching skin region 1. These neurons, however, do not remain unresponsive to touch, but become responsive to other nearby regions of skin over a period of time. Thus, neurons in cortical territory 1 become activated by stimulating skin sites in 2 and 3. The consequence is that the cortical territories of skin sites 2 and 3 expand so that these skin sites have a larger than normal representation. This basic result has been repeatedly confirmed (6), and it raises a number of questions. First, what changes in the brain mediate this reactivation? Does the reactivation involve or depend on subcortical

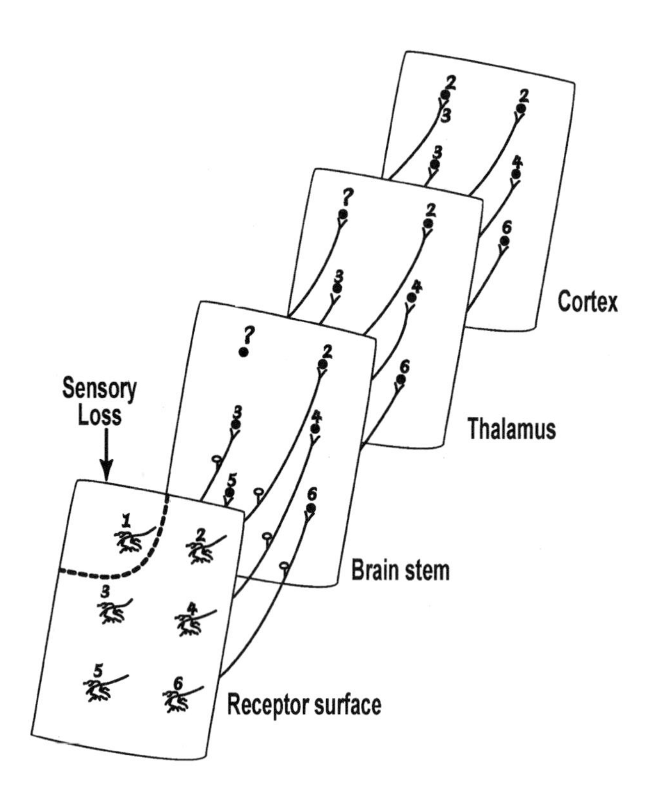

FIG. 6-1. A schematic of the topographic relay of projections from a receptor surface to the cortex in the somatosensory system. A loss of inputs from part of the receptor surface can be followed by a reorganization of cortex, so that the deprived zone of cortex becomes responsive to intact inputs (region 1 becomes responsive to inputs 2 and 3). Comparable changes may occur in the brainstem and thalamus and be relayed to cortex.

changes? As the loss of afferents from skin location 1 (Fig. 6-1) deprives parts of the representations in the brainstem and thalamus, as well as cortex, do similar reorganizations occur at these other levels? As a further question, does the extent of the sensory loss matter so that reactivations occur more fully or more rapidly for a restricted than extensive losses of afferents? In addition, what are the functional consequences of cortical reorganization? Is sensory performance recovered, improved, or impaired in some way? Finally, can treatments alter the outcomes so that beneficial reorganizations are promoted, while detrimental changes are prevented?

As a start toward answering such questions, it is useful to consider the complexity of the somatosensory system. The diagram in Fig. 6-1 is obviously a greatly simplified schematic of the early stages of processing in the somatosensory system. Fig. 6-2 includes more of the relevant complexity. First, terminal arbors of sensory afferents and the axon arbors of projection neurons at each relay are divergent so that they overlap the territories of the arbors of other neurons. Thus, the potential spread of information across representations at each stage and across stages is significant. As a result, the electrical stimulation of peripheral afferents activate regions of cortex that are many times larger than those activated by natural stimulation (7). However, this lateral spread of activation is dampened by inhibitory neurons at each stage, so that the receptive fields of neurons in area 3b are not much larger than those of neurons in the brainstem or the thalamus. When the inhibitory neurons are locally suppressed at any level, the receptive fields of neurons increase in size (8). As another potential source of plasticity, the cortex sends feedback projections to the thalamus and brainstem. Altered activity patterns in cortex would perturb this feedback, some of which would involve the inhibitory neurons in the reticular nucleus of the thalamus and their inhibitory influence on VP neurons. Thus, blocking activity in area 3b of somatosensory cortex results in a striking enlargement of the receptive fields of neurons in the ventroposterior nucleus (9). Another

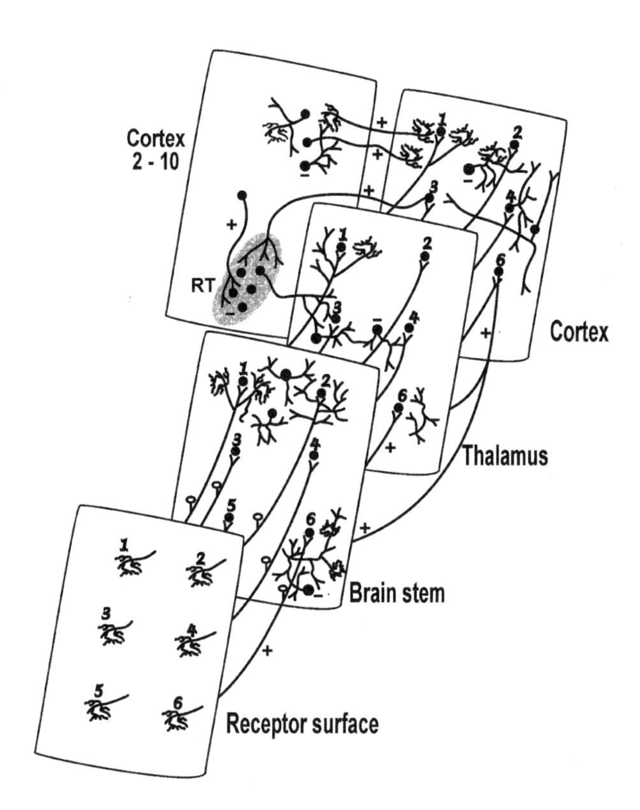

FIG. 6-2. The connectional framework for processing in the somatosensory system is more complex than indicated in Fig. 6-1, allowing for more opportunities for modifying cortical activation patterns and reorganizing somatotopic representations. Excitatory neurons (+); inhibitory neurons (–).

relevant complication is that somatosensory cortex includes a number of interconnected areas that influence each other. Area 3b, as the first stage of cortical processing, sends feedforward, activating inputs to other areas, while receiving modulatory inputs back from those areas. As feedback connections are more divergent than feedforward connections, deprived regions of area 3b, for example, would receive feedback from normally activated portions of other areas, as well as from deprived portions. Finally, cortical areas have intrinsic connections that spread over portions of representations to contribute to modulatory influences on receptive fields. The normally weak activation caused by these intrinsic connections could be potentiated in deprived regions of cortex.

The dynamics of this richly interconnected somatosensory system allow a number of opportunities for an immediate rebalancing after restricted sensory loss. To this, we can add more slowly emerging consequences of altered gene expression and other cellular functions when sensory loss reduces the evoked activity in some central neurons (6). We can expect changes in the expression of neurotransmitters and modulators, and membrane receptors. In addition, changes in activity patterns may induce axon and dendrite growth or retraction. Synapses may be potentiated or weakened by activity-dependent mechanisms. From this thick brew of possibilities, what mechanisms are evoked to mediate cortical reactivation and reorganization after sensory loss?

IMMEDIATE REORGANIZATIONS REFLECT A DYNAMIC REBALANCING

Recording in somatosensory cortex immediately after the loss of afferents from a digit, sometimes reveals neurons along the border of the deprived zone of cortex that immediately acquire new receptive fields that are slightly displaced onto adjacent skin from original locations on the inactivated skin (10). In some cases, such apparently new receptive fields could reflect responses that were there all the time, but not noticed in the presence of more powerful activations from the receptive field center. This remaining activation has been called the "residue of prelesion responses" (11). More often, the new receptive field is likely to be the result of reduced inhibition so that previously suppressed parts of the receptive field emerge. The deafferentated skin no longer excites central neurons, including those that inhibit activation by intact afferents from adjoining skin (Fig. 6-3). This emergence of previously ineffective sources of activation has been called the "unmasking" of silent synapses (12), or disinhibition (13), or sometimes the iceberg effect, as a change in the level of inhibition allows previously submerged margins of receptive fields to appear. Receptive fields may also enlarge simply because nearby regions are not stimulated, or reduce because they are stimulated (14), again reflecting reductions or increases in afferent drive that alter the dynamic balance of the system. We classify these residue responses and immediate alterations as examples of pseudoplasticity, as they are expressions of the normal functions of neural circuits. Nevertheless, such instant and reversible changes in cortical activity patterns may persist if they potentiate newly expressed synapses (15). Thus, a recovery of lost afferents may not restore the original balance. However, after even long periods (months) of sensory loss produced by a nerve crush, normal cortical organization returns after nerve regeneration (16). This suggests that rebalanced cortical circuits usually return to their original state if lost sensory inputs are restored.

A FEW SURVIVING AFFERENTS CAN ACTIVATE THE CORTICAL TERRITORY OF THE ENTIRE AFFERENT GROUP

After the loss of some but not all, of the afferents subserving a skin region, surviving af-

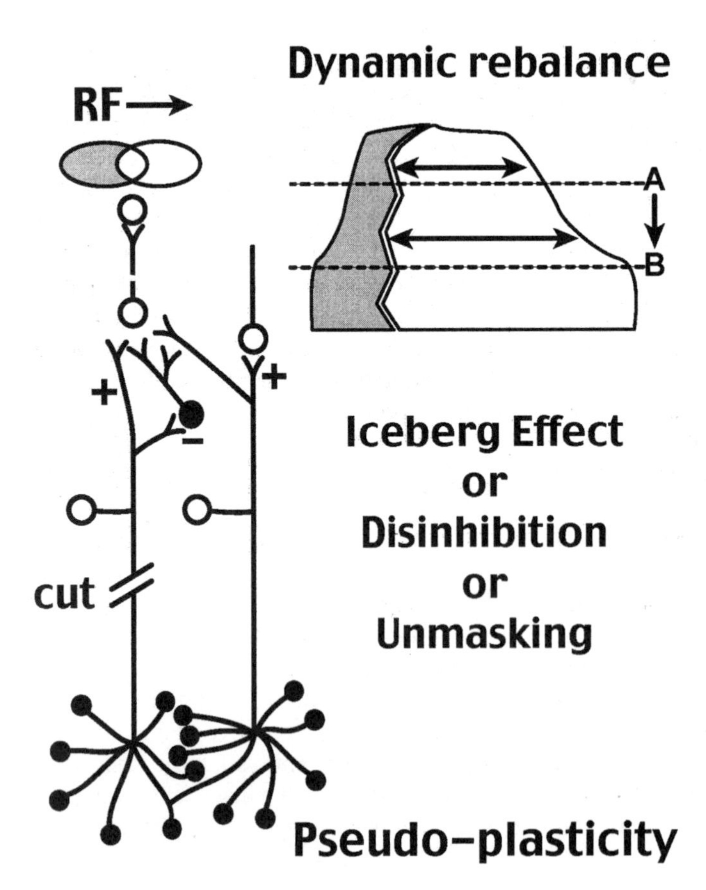

FIG. 6-3. Some cortical reorganization is simply the result of a rebalancing of excitatory and inhibitory influences after a sensory loss. As a result of reduced inhibition (A to B), previously suppressed regions of a receptive field may be expressed and produce a shift in the location of a receptive field. The shaded part of the receptive field is not expressed after the sensory loss because its cut afferents were the source of activation.

ferents clearly become more effective in activating cortical neurons. This was most effectively demonstrated in experiments where a number of dorsal roots in the spinal cord were cut so that most, but not all, of the afferents from a skin region were lost (17) (Fig. 6-4). Such a partial loss can occur because afferents from a given skin region enter the spinal cord over several spinal cord roots. The result of a nearly complete loss of afferents from a region of skin was that a deprived region of somatosensory cortex was initially unresponsive to tactile stimuli, but responses to the partially deafferentated skin recovered over a period of months. Apparently, the few remaining afferents from the deafferentated skin were initially insufficient to activate their normal cortical territory, but they gained potency over time. Although a number of mechanisms could be responsible for such a recovery, the initial reduction in cortical activity, would downregulate the expression of the inhibitory neurotransmitter, GABA, and the receptor sites for GABA on cortical neurons (18,19). This would allow remaining afferents to be expressed and gain potency with use. An even smaller loss of activating connections produced by thalamic lesions, estimated as removing 30% of the neurons representing a skin surface, had no obvious effect on the cortical representation of that skin surface (20). Similarly, small lesions of the cuneate nucleus had little or no effect on the responses of neurons in the ventroposterior nucleus (21). These results indicate that only a portion of the afferents from a skin region, or a portion

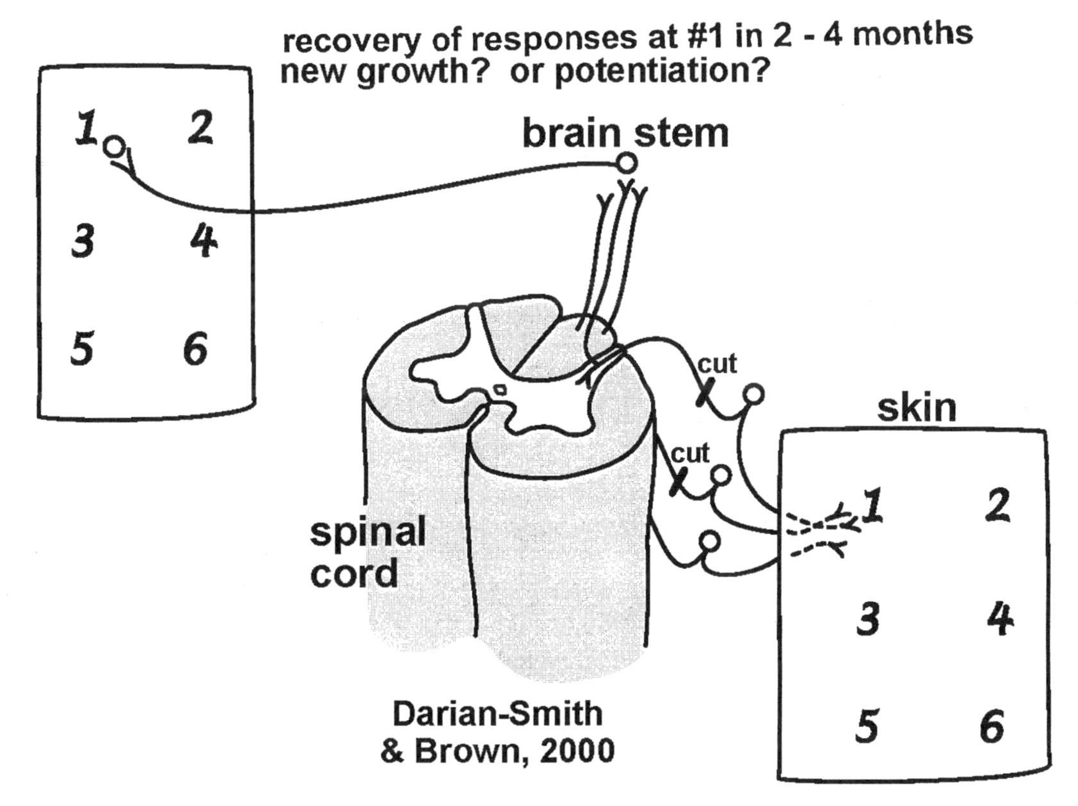

FIG. 6-4. Only a few remaining inputs from a skin area are capable of fully activating regions of cortex, although a recovery period may be needed in which synaptic strengths are potentiated, and possibly new connections formed. In the study of Darian-Smith and Brown (2000) (17), most of the afferents from a given region of skin were eliminated by cutting afferents at the level of the dorsal roots of peripheral nerves. Neurons in the deprived region of cortex were initially unresponsive to touch on the partially denervated skin, but responsiveness recovered in 2 to 8 months.

of brainstem neurons relaying information from a skin region, are needed to activate the cortical territory of that skin region.

WHEN ONLY A FEW AFFERENTS FROM A LARGE SKIN REGION SURVIVE, THEY CAN GREATLY EXPAND THEIR CORTICAL TERRITORY

Under some conditions, a few intact afferents become capable of activating not only their original cortical territory, but the territories of nearby totally deafferentated parts of the body as well. In one series of relevant experiments, afferents from the forearm and lower body were sectioned as they coursed rostrally in the dorsal columns of the spinal cord (22). With sections at a high cervical level, dorsal column afferents from all of the forelimb but a small portion of the anterior arm can be cut. However, functionally important afferent terminations remain in the spinal cord to mediate local reflexes and activate spinothalamic pathways. In spite of the remaining spinothalamic connections, a complete section of the dorsal columns totally deactivated the hand representation in area 3b. Thus, spinothalamic afferents do not appear to be capable of independently activating this region of somatosensory cortex. However, when the section of dorsal column afferents was incomplete and a few afferents from the hand remained to reach the cuneate nucleus, these afferents remained capable of activating their normal territory in the hand representation. In addition, over weeks of recovery, this activated territory greatly expanded to encompass most of the deprived hand representation. Preserved inputs from a single digit or pad of the hand, for example, not only activated cortex normally responsive to that input, but most or all of the entire hand representation.

The mechanisms mediating this extensive reactivation are uncertain, but it is likely that they involved the previously existing network of intrinsic connections within area 3b. We know from studies of the anatomy of such connections that they interconnect representations of digits with each other, and digits with pads of the palm, but interconnections between the adjoining hand and face representations are extremely limited (23). As the few remaining hand afferents are soon capable of activating much of the hand representation, but the intact face inputs are not, a role for the intrinsic cortical connections is suggested. The intrinsic connections normally have weak, modulatory effects, but they could become potentiated to activate nearby neurons above threshold. In addition, it is likely that intrinsic connections in deactivated cortex grow to become more effective and encounter more neurons (24,25). Of course, axon arbors at brainstem and thalamic levels may grow to contact more neurons than normal to contribute to the cortical reactivation.

SUBCORTICAL CHANGES ARE RELAYED TO CONTRIBUTE TO CORTICAL REORGANIZATION

One of the first convincing demonstrations of altered somatotopy in cortical representations after sensory loss involved cutting the median nerve to the hand of monkeys (3,4). This nerve subserves the thumb half of the glabrous hand. Over a period of days to weeks of recovery, the cortex formerly activated by the glabrous hand became responsive to somatotopically matching parts of the dorsal hand, which normally activates little cortex (26–29). These findings are consistent with the electrophysiologic evidence that there is a subthreshold, latent representation of the dorsal hand in area 3b of monkeys that is topographically matched with the manifest representation of the glabrous hand (30). A superimposition of maps also appears to occur at subcortical levels. Thus, when glabrous hand afferents are removed, dorsal hand afferents rapidly substitute at the brainstem, thalamic, and cortical levels (Fig. 6-5). Part of the reactivation occurs in the cuneate nucleus of the medullary relay (31), and this partial recovery is relayed to the ventroposterior nucleus where more extensive recovery is ob-

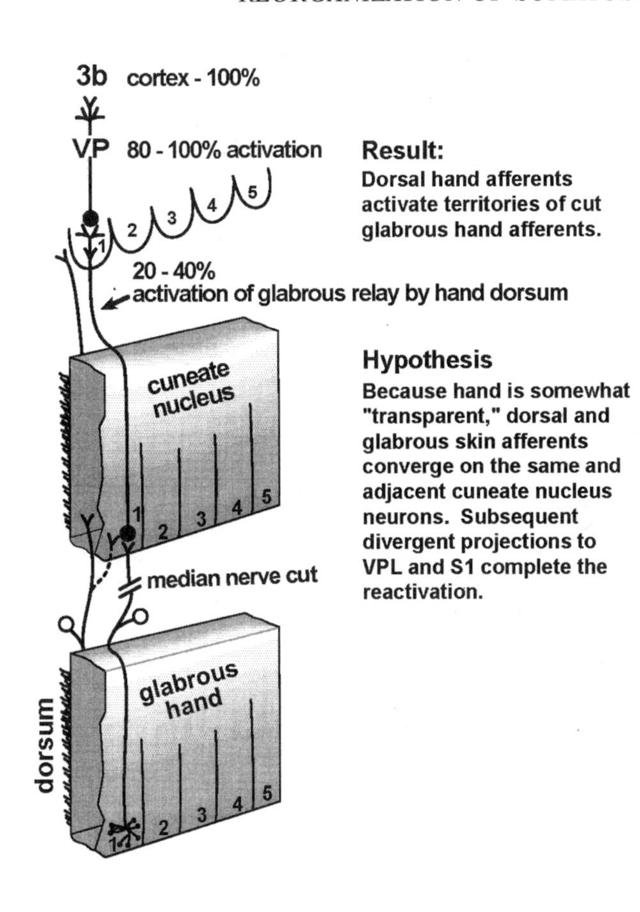

FIG. 6-5. A schematic of afferents from the glabrous skin and the dorsal surface of the hand *(digits 1 to 5 are indicated),* and their relay to a corresponding representation of the hand in the cuneate nucleus. Subsequent relays include the ventroposterior nucleus of the thalamus and area 3b of somatosensory cortex. After afferents from the glabrous hand have been cut, afferents from the hand dorsum become capable of activating the deprived neurons in the cuneate nucleus, either by forming new connections or potentiating previous connections *(dashed axon).* A limited reactivation at the level of the brainstem is amplified at each subsequent level.

served (26). The reactivations at each level could be mediated by reductions in inhibition, the potentiation of weak synapses, and the local growth of new connections. Changes at each level would be relayed to the next, to accumulate over levels.

EXTENSIVE CORTICAL REACTIVATIONS DEPEND IN PART ON THE GROWTH OF NEW SUBCORTICAL CONNECTIONS

In monkeys with a complete section of afferents in the dorsal columns at a high cervical level, the hand region of cortex remains unresponsive to tactile stimuli for months. However, after 6 to 8 months of recovery, hand cortex is activated throughout by light touch on the face (22). This recovery is accompanied by the sprouting and growth of afferents of the face from their normal territories in the trigeminal nucleus to the deprived cuneate nucleus (32,33). These new connections apparently activate some relay neurons in the cuneate nucleus that in turn activate many neurons in the hand subnucleus of the ventroposterior nucleus of the thalamus (34). A relay of this thalamic pattern to cortex would allow much of the hand representation in area 3b to become responsive to the face. Intrinsic connections in area 3b, either potentiated normal distributions of such connections or those enlarged by new growth, could amplify this reactivation. The growth of new connections at both brainstem and cortical levels seems to occur after major sensory losses, such as after therapeutic forelimb amputations or deafferentations in monkeys (33, 35,36) and humans (37,38). In such cases, both the deprived hand regions of the ventroposterior nucleus and somatosensory cortex become activated by inputs from the stump or the face. In addition, there is evidence from monkeys for the growth of face afferents or shoulder afferents into the hand region of the cuneate nucleus (32,33,39).

FUNCTIONAL CONSEQUENCES OF CORTICAL REORGANIZATION

The experimental results described in the preceding indicate that cortical neurons, when deprived of their normal sources of activation, can acquire new sources of activation. It is less clear how these changes in activation patterns affect perception and performance. One possibility is that the larger pool of cortical neurons that is activated by the remaining peripheral inputs from a skin surface result in an improved tactile acuity or some other perceptual ability. In this case, the reactivated neurons would be reassigned to functional circuits mediating perceptions from the innervated skin, and the greater number of neurons in the enlarged network would lead to improved performance. In some studies when training or sensory experience led to an increase in the size of cortical representations, improved behavioral performances were also noted (40–42). There is also evidence that sensory discriminations are improved on the skin of the stump of an amputated limb (43). Another possibility is that the reactivated neurons do not become integrated into functionally relevant circuits. Instead, the new sources of activation are not distinguished from the original sources, and misperceptions occur and persist. Thus humans with a missing limb typically feel that the missing limb is still present, and touch on the stump of the limb or on the face also may be felt on the missing limb (44). The reasonable interpretation of such phantom sensations and misperceptions is that cortical neurons are abnormally activated by inputs from intact parts of the body, but they continue to contribute to cortical circuits that signal the presence of stimuli on the missing parts of the body (38). Although most of these misperceptions are easily tolerated, those that involve the perception of pain in missing limbs are not. Thus, it is important to investigate the possibilities of promoting or preventing cortical reorganization in order to obtain the most desirable clinical outcomes (45).

REFERENCES

1. Merzenich MM, Kaas JH, Sur M, et al. Double representation of the body surface within cytoarchitectonic areas 3b and 1 in "SI" in the owl monkey (Aotus trivirgatus). *J Comp Neurol* 1978;181:41–73.
2. Kaas JH, Nelson RJ, Sur M, et al. Multiple representations of the body within the primary somatosensory cortex of primates. *Science* 1979;204:521–523.
3. Merzenich MM, Kaas JH, Wall J, et al. Topographic reorganization of somatosensory cortical areas 3b and 1 in adult monkeys following restricted deafferentation. *Neuroscience* 1983a;8:33–55.
4. Merzenich MM, Kaas JH, Wall JT, et al. Progression of change following median nerve section in the cortical representation of the hand in areas 3b and 1 in adult owl and squirrel monkeys. *Neuroscience* 1983b;10: 639–665.
5. Merzenich MM, Nelson RJ, Stryker MP, et al. Somatosensory cortical map changes following digit amputation in adult monkeys. *J Comp Neurol* 1984;224: 591–605.
6. Kaas JH, Florence SL. Reorganization of sensory and motor systems in adult mammals after injury. In: Kaas JH, ed. *The mutable brain.* Amsterdam: Harwood, 2001: 165–241.
7. Bradley WE, Farrell DF, Ojemann GA. Human cerebrocortical potentials evoked by stimulation of the dorsal nerve of the penis. *Somatosens Mot Res* 1998;15: 118–127.
8. Dykes RW, Landry P, Metherate R, et al. Functional role of GABA in cat somatosensory cortex: shaping receptive fields of cortical neurons. *J Neurophysiol* 1984;52: 1066–1093.
9. Ergenzinger ER, Glasier MM, Hahm JO, et al. Cortically induced thalamic plasticity in the primate somatosensory system. *Nat Neurosci* 1998;1:226–229.
10. Calford MB, Tweedale R. Immediate expansion of receptive fields of neurons in area 3b of macaque monkeys after digit denervation. *Somatosens Mot Res* 1991; 8:249–260.
11. Rajan R, Irvine DR, Wise LZ, et al. Effect of unilateral partial cochlear lesions in adult cats on the representation of lesioned and unlesioned cochleas in primary auditory cortex. *J Comp Neurol* 1993;338:17–49.
12. Wall PD. The presence of ineffective synapses and the circumstances which unmask them. *Phil Trans R Soc Lond B* 1977;278:361–372.
13. Calford MB, Tweedale R. Immediate and chronic changes in responses of somatosensory cortex in adult flying fox after digit amputation. *Nature* 1988;332: 446–448.
14. Gilbert CD, Das A, Ilo M, et al. Spatial integration and

cortical dynamics. *Proc Natl Acad Sci USA* 1996;93: 615–622.
15. Rauschecker JP. Mechanisms of visual plasticity: Hebb synapses, NMDA receptors, and beyond. *Physiol Rev* 1991;71:587–615.
16. Wall JT, Felleman DJ, Kaas JH. Recovery of normal topography in the somatosensory cortex of monkeys after nerve crush and regeneration. *Science* 1983;221: 771–773.
17. Darian-Smith C, Brown S. Functional changes at periphery and cortex following dorsal root lesions in adult monkeys. *Nat Neurosci* 2000;3:476–481.
18. Jones EG. GABAergic neurons and their role in cortical plasticity in primates. *Cereb Cortex* 1993;3:361–372.
19. Garraghty PE, LaChica EA, Kaas JH. Injury-induced reorganization of somatosensory cortex is accompanied by reductions in GABA staining. *Somatosens Mot Res* 1991;8:347–354.
20. Jones EG, Manger PR, Woods TM. Maintenance of a somatotopic cortical map in the face of diminishing thalamocortical inputs. *Proc Natl Acad Sci USA* 1997; 94:11003–11007.
21. Alloway KD, Aaron GB. Adaptive changes in the somatotopic properties of individual thalamic neurons immediately following microlesions in connected regions of the nucleus cuneatus. *Synapse* 1996;22:1–14.
22. Jain N, Florence SL, Kaas JH. Limits on plasticity in somatosensory cortex of adult rats: hindlimb cortex is not reactivated after dorsal column section. *J Neurophysiol* 1995;73:1537–1546.
23. Fang PC, Jain N, Kaas JH. Few intrinsic connections cross the hand-face border of area 3b of New World monkeys. *J Comp Neurol* 2002;454:310–319
24. Florence SL, Taub HB, Kaas JH. Large-scale sprouting of cortical connections after peripheral injury in adult macaque monkeys. *Science* 1998;282:1117–1121.
25. Darian-Smith C, Gilbert CD. Axonal sprouting accompanies functional reorganization in adult cat striate cortex. *Nature* 1994;368:737–740.
26. Garraghty PE, Kaas JH. Large-scale functional reorganization in adult monkey cortex after peripheral nerve injury. *Proc Natl Acad Sci USA* 1991;88:6976–6780.
27. Wall JT, Huerta MF, Kaas JH. Changes in the cortical map of the hand following postnatal median nerve injury in monkeys: modification of somatotopic aggregates. *J Neurosci* 1992;12:3445–3455.
28. Kolarik RC, Rasey SK, Wall JT. The consistency, extent, and locations of early-onset changes in cortical nerve dominance aggregates following injury of nerves to primate hands. *J Neurosci* 1994;14:4269–4288.
29. Silva AC, Rasey SK, Wu X, et al. Initial cortical reactions to injury of the median and radial nerves to the hands of adult primates. *J Comp Neurol* 1996;366: 700–716.
30. Schroeder CE, Seto S, Arezzo JC, et al. Electrophysiologic evidence for overlapping dominant and latent inputs to somatosensory cortex in squirrel monkeys. *J Neurophysiol* 1995;74:722–732.
31. Xu J, Wall JT. Rapid changes in brainstem maps of adult primates after peripheral injury. *Brain Res* 1997;774: 211–215.
32. Jain N, Florence SL, Qi HX, et al. Growth of new brainstem connections in adult monkeys with massive sensory loss. *Proc Natl Acad Sci USA* 2000;97:5546–5550.
33. Florence SL, Kaas JH. Large-scale reorganization at multiple levels of the somatosensory pathway follows therapeutic amputation of the hand in monkeys. *J Neurosci* 1995;15:8083–8095.
34. Jain N, Qi H-X, Catania KC, et al. Large-scale reorganization in the somatosensory cortex and thalamus after dorsal column lesions in macaque monkeys. *Soc Neurosci Abstr* 2000;26:146.2.
35. Pons TP, Garraghty PE, Ommaya AK, et al. Massive cortical reorganization after sensory deafferentation in adult macaques. *Science* 1991;252:1857–1860.
36. Florence SL, Hackett TA, Strata F. Thalamic and cortical contributions to neural plasticity after limb amputation. *J Neurophysiol* 2000;83:3154–3159.
37. Flor H, Elbert T, Muhlnickel W, et al. Cortical reorganization and phantom phenomena in congenital and traumatic upper-extremity amputees. *Exp Brain Res* 1998; 119:205–212.
38. Davis KD, Kiss ZH, Luo L, et al. Phantom sensations generated by thalamic microstimulation. *Nature* 1998; 391:385–387.
39. Wu CWH, Kaas JH. The effects of long-standing limb loss on anatomical reorganizaiton of somatosensory afferents in the brainstem and spinal cord. *Somatosens Motil Res* 2002;19:153–163.
40. Recanzone GH, Schreiner CE, Merzenich MM. Topographic reorganization of the hand representation in cortical area 3b of owl monkeys trained in a frequency-discrimination task. *J Neurophysiol* 1992;67: 1031–1056.
41. Xerri C, Stern JM, Merzenich MM. Alterations of the cortical representation of the rat ventrum induced by nursing behavior. *J Neurosci* 1994;14:1710–1721.
42. Zohary E, Celebrini S, Britten KH, et al. Neuronal plasticity that underlies improvement in perceptual performance. *Science* 1994;263:1289–1292.
43. Wall JT, Kaas JH. Long-term cortical consequences of reinnervation errors after nerve regeneration in monkeys. *Brain Res* 1986;372:400–404.
44. Ramachandran VS, Hirstein W. The perception of phantom limbs. *Brain* 1998;121:1603–1630.
45. Kilgard MP, Merzenich MM. Cortical map reorganization enabled by nucleus basalis activity. *Science* 1998; 279:1714–1718.

7

Mechanisms of Selection and Guidance of Reaching Movements in the Parietal Lobe

John F. Kalaska, Paul Cisek*, and Nadia Gosselin-Kessiby

*Département de Physiologie, Pavillon Paul-G-Desmarais, Université de Montréal, Québec, Canada; *Section of Neurophysiology, Laboratory of Systems Neuroscience, National Institute of Mental Health, Bethesda, Maryland*

INTRODUCTION

For many years, the parietal lobe was regarded as a functionally homogeneous cortical associative area that integrates multimodal sensory information to generate a single unified and veridical representation of the external world and of the body in a common reference framework. This general-purpose spatial representation was presumably used by sensory-perceptual systems to generate an introspective perceptual image of the spatial structure of the world. Information extracted from it about spatial relationships was presumably also relayed to neural circuits controlling different motor effectors such as the eyes, head, arm, and hand, to guide motor actions and motor interactions with the external world. However, it is now increasingly recognized that different parts of the posterior parietal cortex (PPC) generate different sensory representations in different coordinate frameworks appropriate to different classes of actions, such as directing gaze, orienting, reaching, and grasping (1–6). The parietal cortex does not contain one common unified representation of external space used by all effector systems, but rather a collection of parallel representations each specialized for the unique demands imposed by the sensorimotor control of specific effectors (1,3,7).

We discuss a series of recent studies on the parietal cortex in terms of several key theoretical issues relevant to reaching movements, including the selection and planning of a reaching action as well as its overt performance. This chapter builds on two earlier reviews of parietal function in arm movement control (8,9), and focuses mainly on research done in the past 10 years. A more comprehensive discussion of earlier work may be found in those articles. In this chapter, we focus on the control of arm reaching movements made to visual targets, and emphasize structures in the superior parietal lobule (SPL) medial to the intraparietal sulcus (IPS). Although more lateral cortical regions in the inferior parietal lobule (IPL) have been more intensely studied over the years, a number of recent developments have provided new insights and perspectives on the functional role of the SPL in visuomotor behavior and sensorimotor guidance. We develop two major themes. The first relates to the role of medial parietal regions in the overt performance of reaching movements. The second concerns the role the parietal cortex may play in the processes that use salient sensory inputs to select a particular movement from among the multiple options that are usually available during natural behavior. Where relevant, we also briefly discuss findings on the inferior parietal lobule, frontal cortex, and cerebellum. We begin with a review of theoretical and conceptual issues related to the performance of visually guided reaching that provides a context for subsequent sections.

PERFORMANCE OF VISUALLY GUIDED REACHING

Many motor control theories assume that the neural mechanisms that implement visually guided reaching movements must exhibit several features. First, they must have up-to-date information on the current configuration of the limbs and body posture. Second, they must have access to up-to-date information specifying the target or goal of the movement. Third, they must transform information about the current position of the body, arm, and target into motor commands to drive the hand to the target.

All three of these functions introduce critical issues related to reference frames and sensorimotor coordinate transformations. For instance, to estimate arm posture, neural mechanisms must integrate multimodal signals from many different classes of somatic mechanoreceptors with visual input about arm position and efference copies of outgoing motor commands. To initiate a reach, neural circuits must convert retinal input about target locations into motor commands about causal muscle activity patterns. This planning process prior to movement onset is usually described in terms of a sequence of sensorimotor transformations of reach-related signals in different reference frames, progressing from retinal and eye-centered coordinates in early stages ("extrinsic" coordinates, to signify frameworks that do not require signals about the current status of the arm) to increasingly limb- or even muscle-centered coordinates in later stages ("intrinsic" coordinates, to signify frameworks that do require signals about arm posture and movements or muscle biomechanics). However, the nature of the coordinate transformations intervening between retinal input and motor output is still under investigation (1,6,10,11).

What is the nature and parameter space in which the movement "plan" is expressed? Does this plan include all the moment-to-moment details of movement production from start to end, prior to movement onset (12), or is the a priori plan fairly minimal and most of the temporal details arise dynamically during movement execution? According to the first view, the motor system preplans the movement in full detail in one coordinate space or another in advance of its initiation. According to the second view, all that is strictly necessary before initiation is a signal specifying the direction of a movement, a "motor error" or "difference vector" between the current (initial) and desired (final) hand position. Consistent with this second general scheme, there is extensive behavioral and modeling evidence supporting the hypothesis that an early stage in the planning of reaching movements involves a motor error signal in hand-centered coordinates (13–19). Before generating the motor commands, many schemes propose that at least one more intervening transformation is necessary to convert the intended action from extrinsic spatial coordinates of hand motion to an intrinsic reference frame of limb motion (20–22). For instance, Soechting and Flanders (21,22) have proposed that before the motor command can be calculated, the final target location is redefined in terms of the desired final arm posture after reaching by several separate psychophysical channels that each signal one of the spatial orientation angles of the upper arm or forearm segments when the hand is at the target location.

Even more fundamental than the issue of sensorimotor transformations is the question of the overall computational structure of the neural systems involved in visuomotor control. Engineering theory distinguishes two basic approaches to control. The first is closed-loop or feedback control, in which a signal about the current state of the system, provided by an external sensory feedback loop from the periphery, is continuously compared to the desired goal state and this difference ("motor error") is used to generate and update the motor command in real time throughout the movement (Fig. 7-1A). This simple scheme at once brings the controlled effector to the desired goal state with minimal a priori planning and compensates for any perturbations that might occur during the movement, if feedback loop gains are sufficiently large. How-

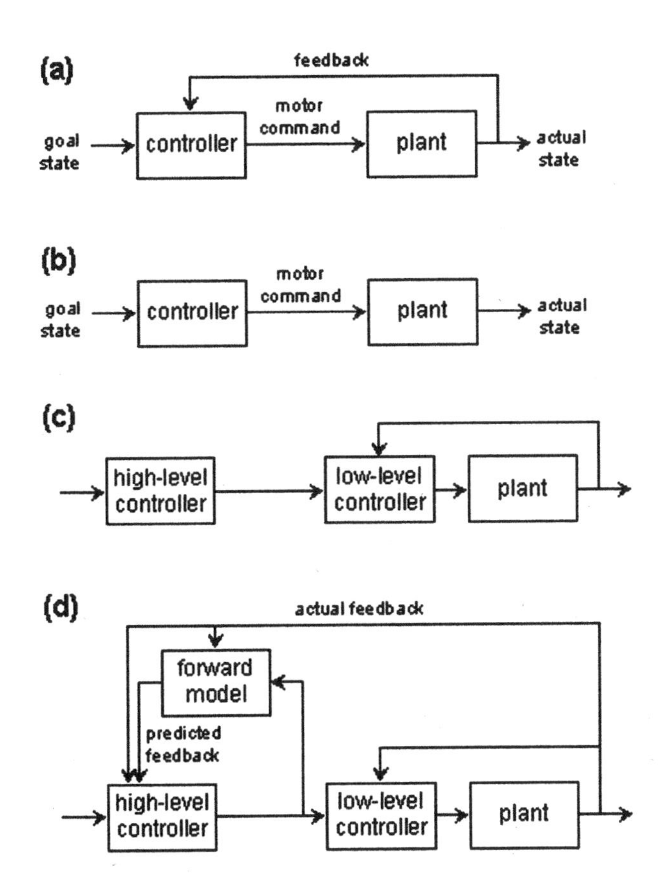

FIG. 7-1. Four different kinds of control architectures. **A:** Feedback control. **B:** Feed-forward control. **C:** Feed-forward control with assistance from feedback correction. **D:** feedback control with assistance from a predictive forward model.

ever, closed-loop control can be unstable if the conduction delays in the feedback loop are too long, because the revised motor command is no longer appropriate by the time it arrives at the plant. Human reaction times to even the simplest visual stimuli are on the order of hundreds of milliseconds. If this delay is representative of the time it takes visual input to influence motor output, then pure feedback control is not a viable foundation for guiding complex movements. For this reason, traditional motor control theory has been primarily based on a second type of control, called open-loop or feed-forward control. In pure open-loop control, the entire time course of the motor command is programmed autonomously and executed without using feedback information (Fig. 7-1B). This kind of scheme is more demanding because it has to compute the command precisely, and must anticipate and compensate for perturbations. However, provided the motor command is sufficiently accurate, an open-loop controller is not subject to instabilities owing to long conduction delays.

Many control systems used in engineering take advantage of both approaches. For example, robotic controllers often specify the main components of their motor commands in an open-loop feed-forward fashion but then include a fast closed-loop feedback system to fine tune the command to take care of residual errors (Fig. 7-1C). This scheme also implies a strict sequential separation between planning and execution of movements. Inspired by such systems, theories of biological motor control have proposed that the descending command from cortical motor centers is generated predominantly in a feed-forward fashion, as a precomputed "motor program" (12), which is fine-tuned during execution by fast spinal feedback reflexes. Of course, unlike the robotic schema in Fig. 7-1C, feedback circuits also project to the cortex, but it was presumed that because of conduction delays, they could only have a significant effect on the descending motor command toward the end of a movement (23). From this perspective, the basic organization of the motor system is feed-forward, with feedback playing a supportive role.

However, there is mounting evidence that the opposite description of the motor system might be more accurate, such that the basic organization is a *feedback* circuit in which *feed-forward* processes play a supportive role (Fig. 7-1D). According to this view, a representation of the current state of the system is compared with a representation of its desired goal state, and this difference is used to compute the motor command (14,24). Because of conduction delays, however, the representation of the current state cannot be based purely on sensory feedback from the periphery. Instead, during well-practiced movements an *estimate* of the current state may be provided by an internal feedback system that has learned to anticipate the likely outcome (i.e., prediction of likely future sensory feedback) resulting from particular motor commands within a given task context. A internal feedback system that provides predictions of likely outcomes is called a "forward model" (24–26). In its simplest form, the forward model could simply use an efference copy of motor commands for prediction, but in more complex versions known as "observers" it also uses external feedback signals (Fig. 7-1D). The internal estimate of current state provided by the forward model minimizes stability problems imposed by conduction delays in the external feedback loop and permits the system to function in a fundamentally feedback mode. A second major computational advantage of such an internal feedback loop is that by providing the system with predicted future states with minimal delays, it can theoretically permit online correction of the outgoing motor command to compensate for anticipated errors in motor output before they actually occur. In this computational architecture, the motor system is predominantly a series of nested internal and external feedback loops coupled to feed-forward signals, that from a control perspective has more in common with the closed-loop scheme of Fig. 7-1D than with the open-loop schemes of Fig. 7-1B,C.

In the sections that follow, we review recent studies of the SPL in the context of such issues as current state estimation, sensorimotor coordinate transformations, multiple reference frames, online control, and forward models.

The Superior Parietal Lobule

In monkeys, the superior parietal cortex just posterior to the primary somatosensory cortex (SI) was originally defined as a single architectonic region, designated as area 5 or area PE in different cytoarchitectonic schemes. Anatomic and neurophysiologic evidence has now compartmentalized the SPL into four to five subregions with differing nomenclatures (Fig. 7-2A). A rostrodorsal region (area 5d, PE) extends from the crown of the IPS to the posterior border of SI proper. A more caudal medial region (area 5m, PEc, V6A) is located just medial to the junction of the IPS with the lunate sulcus. The cortex in the medial rostral bank of the IPS has been subdivided into separate medial intraparietal (MIP) and medial rostral intraparietal cortex (area 5c, PEip). The cortex occupying the fundus of the IPS (ventral intraparietal cortex, VIP) is considered transitional with the cortex of the IPL. The cortex in the lateral bank has been subdivided into anterior intraparietal (AIP) and lateral intraparietal areas (LIP).

Each of these subdivisions possesses specific combinations of corticocortical inputs and outputs, implying functional specialization in each area (3–6,27,28). The connections with the precentral gyrus have received particular attention because of the presumed importance of parietofrontal circuits in visuomotor control. The consensus is that there is an overall mirror-inverted rostrocaudal orga-

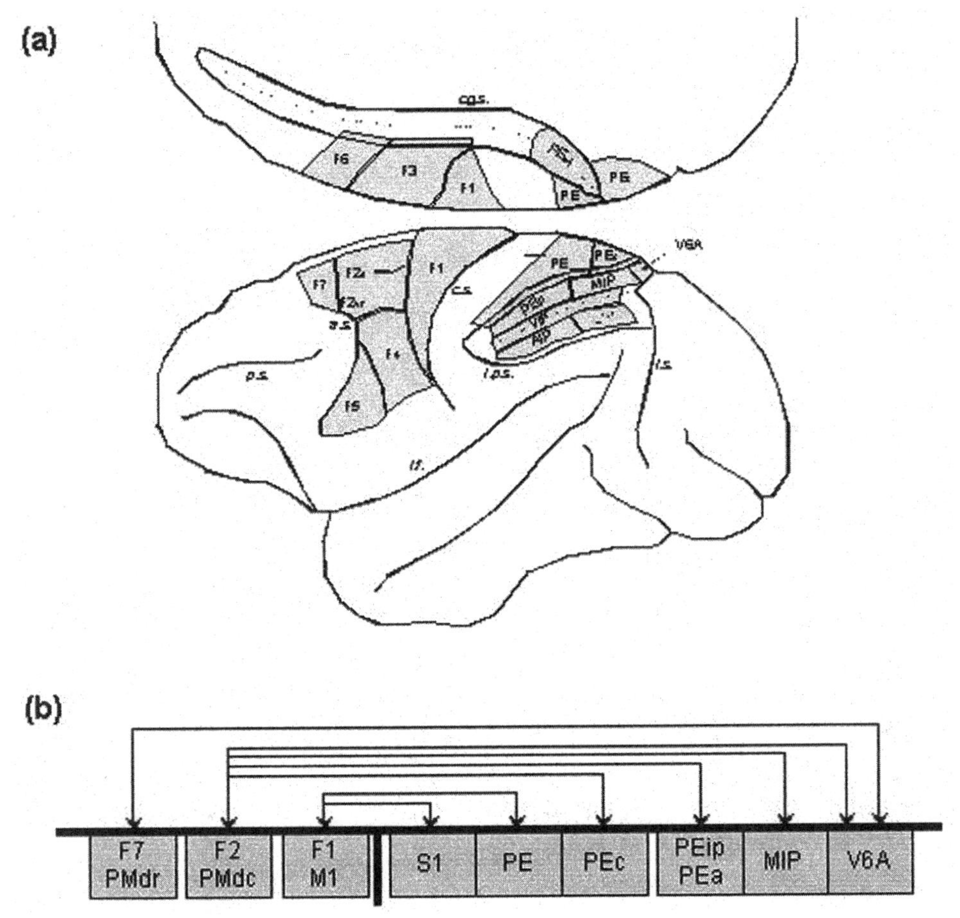

FIG. 7-2. A: Mesial and lateral views of the macaque cerebral cortex. The cingulate and intraparietal sulci have been unfolded to reveal regions within their interior, with the fundus indicated by a dashed line. AIP, anterior intraparietal area; as, arcuate sulcus; cs, central sulcus; cgs, cingulate sulcus; F1, primary motor cortex (or M1); F2d, dorsal premotor cortex (PMd); F2vr, ventral/rostral part of PMd; F3, supplementary motor area (SMA); F4,F5, ventral premotor cortex (PMv); F6, presupplementary motor area (pre-SMA); F7, rostral PMd, or pre-PMd; ips, intraparietal sulcus; lf, lateral fissure; LIP, lateral intraparietal area; ls, lunate sulcus; MIP, medial intraparietal area; PE, anterior part of area 5 on the postcentral gyrus; PEc, caudal part of anterior area 5; PEci, cingulate part of area 5; PEip, intraparietal/posterior area 5; ps, principal sulcus; V6A, visual area 6A; VIP, ventral intraparietal area. (Adapted from: Rizzolatti G, Luppino G. The cortical motor system. *Neuron* 2001;31:889–901.) **B:** Highly schematic and simplified diagram of the major corticocortical connections between frontal areas and the superior parietal lobule.

nization spanning the central sulcus (28–30). The most rostral parietal regions (PE) project mainly to the most caudal precentral cortex (primary motor cortex, MI or F1). Progressively more caudal and medial parietal regions (PEip, PEc, MIP) project to progressively more rostral precentral sites (PMd, PMdr, or F2 and F7). A similar gradient is seen for the parietal regions projecting to the ventral premotor cortex lateral to the genu of the arcuate sulcus (F4, F5) (5). These parietofrontal connections are reciprocal, permitting recurrent re-entrant processing of signals between the parietal and motor structures (Fig. 7-2B) reminiscent of the nested internal feedback loops in Fig. 7-1D (27,31). Whereas some evidence supports a continuous gradient of overlapping reciprocal connections across areas (28,31), other work favors a more selective compartmentalization of connections between specific areas (4,5).

Along with the somatosensory input into the SPL, it is also now well established that there are extensive corticocortical visual inputs from medial components of the dorsal visual stream into medial parts of the SPL (31,32). Because these same parts of the SPL project to PMd and PMdr (Fig. 7-2), they may form the critical corticocortical route by which salient arm movement-related visual information projects into premotor regions (28,31–33).

Finally, Andersen, Snyder and colleagues (11,34) have defined on functional grounds a predominantly arm movement-related region in the medial SPL and immediately adjacent IPL that they have named the parietal reach region (PRR). The PRR has not been localized using rigorous histological criteria. However, magnetic resonance imaging (35) suggests that it overlaps parts of PEc, MIP and V6A in the SPL and the adjacent cortex on the lateral bank of the IPS.

Monitoring Body Posture and Movement: Somatic Inputs, Visual Inputs, and Efference Copies

Because the SPL in monkeys receives a massive corticocortical input from SI, it was traditionally regarded to be a higher-order somatic sensory association area (8,9). Many area 5 cells, especially in PE, PEip, and MIP, respond to complex combinations of multijoint motions, limb postures, and tactile inputs. This is consistent with extensive evidence that the parietal lobe plays a critical role in generating the so-called "body schema," the neural mechanisms underlying the perception and introspective awareness of body form, posture, and movement (8,9,36, 37). However, the body schema is a highly dynamic construct dependent not only on inputs from somatic receptors, but also on a complex integration of visual inputs about the body and even efference copies of motor commands (26,38–40). A similar combination of signals is required to permit monitoring and prediction of current state by a forward model (Fig. 7-1D).

In that light, it is an important finding that many area 5 cells, especially in PEip, MIP, and PRR, discharge as a function of the intended direction of arm movements during the delay period of instructed-delay tasks (9, 35,41–45). The cells in PE and PEip also typically begin to discharge prior to movement onset in reaction-time tasks, yet many of these same cells are recruited into activity 50 to 60 ms later than cells in MI or PMd (9,46,47). Furthermore, the reach-related discharge of area 5 cells is more strongly coupled to the position of the arm and to the directionality and spatiotemporal form of active arm movements, and less to the causal forces and muscle activity underlying motor output than are cells in primary motor cortex (48–51). This combination of properties is consistent with the hypothesis that parts of area 5 receive convergent peripheral inputs about unfolding movement via SI and internally generated signals about motor intentions from precentral motor areas to generate a representation of body form, posture, and movements that could be used primarily to monitor motor performance (9,14,36,50). However, Burbaud et al. (52) reported that cells in the posterior portion of area 5 deep within the IPS discharged *before* cells in primary motor cortex in reaction-time tasks, whereas the discharge of more anterior area 5 cells lagged behind motor cortex, which is consistent with different functional roles for different parts of the SPL.

This representation of limb posture and movement in SPL is not based only on somatosensory and somatomotor information. Many cells in MIP and more lateral parts of area 5 respond to combinations of both somatic and visual inputs (53,54). One potential function of these neurons could be to generate a cross-modal representation of the spatial relations between body parts and nearby visual stimuli that may contribute to the organization of movements directed toward nearby objects in immediate interpersonal space. It may also play a critical role in monitoring limb position and guiding reaching movements themselves. Metabolic studies have shown that anterior and lateral parts of area 5 are equally active during reaching movements to targets in the light and dark (55). In contrast, caudal and medial parts of area 5 are less active when visual feedback about the arm is eliminated during reaching movements in the dark. Imaging studies likewise show an activation of parietal cortex by vision of the arm during pointing movements (56).

Graziano et al. (36) recently reported important new evidence that visual input contributes to the internal neural representation of static body form and posture. First, they confirmed that many area 5 cells (in PE, PEip, and MIP) responded in a graded fashion to somatic signals about different static arm positions when the contralateral arm was extended out in front of the body on either the right or left side of the midline, while it was out of sight behind an opaque barrier. When they placed a stuffed monkey cadaver arm onto the opaque barrier while blocking the monkey's vision and then revealed it to the monkey, some area 5 cells changed their discharge. Typically, the discharge was maximal when the stuffed arm was aligned with the actual position of the monkey's own arm immediately below the barrier, and was reduced when the stuffed arm was placed on the opposite side of the body to the monkey's arm. Although the effect was often modest on a single-cell basis, it produced a statistically significant reduction of total population activity (about 10%) when the actual arm was in the preferred posture for each cell and the stuffed arm was on the opposite side. These cells that integrated somatic and visual signals about arm spatial location were found in both PE and PEip/MIP with about equal frequency, but were rarely seen in SI proper. This response was not caused simply by attention being directed to a novel or foreign object, nor was it a nonspecific response any visual stimulus. The visual modulation was remarkably specific, in that the object had to be recognizably limblike and in the proper orientation. Turning the stuffed arm around so that the proximal stump was where the hand should be, or placing a stuffed left arm where the right arm should be, substantially reduced or eliminated the effect.

These results show that a subpopulation of SPL cells receives an input signal about the spatial location of a complex limblike visual stimulus and integrates this information with somatic sensory feedback to synthesize a multimodal signal about limb spatial location. These results are consistent with psychophysical evidence that occlusion of vision of the arm results in an immediate shift in the perceived position of the arm (57) and in pointing errors to targets (58,59), which can be improved by intermittent glimpses of the arm. This integration of visual and somatic input about static arm position is accomplished in a statistically efficient manner, in that the variance of errors of estimates of arm position when both sources of information are available is significantly less than that expected from the variances seen when only one or the other is used in isolation (60,61). Similarly, estimates of hand location based on somatic sensory feedback and efference copies of motor commands are combined in a statistically optimal manner (26). Consistent with these findings in neurologically intact subjects, Wolpert et al. (37) have described a patient with a lesion of the left SPL who, like many other parietal-lesioned patients, showed deficits in awareness of the posture and movement of the contralateral arm. However, unlike most such cases, this effect was not permanent. This patient lost awareness of her arm gradually over several tens of seconds after acquiring a static posture at the end of a movement. She could never recover awareness of her arm by any conscious introspective mental effort, but could immediately restore awareness by merely looking at it. Wolpert et al. (37) concluded that the SPL contains a mechanism to integrate and store an internal state estimate of body form and movement based on somatic, visual, and motor signals that are used to update the state estimate. Whereas most patients with asomatognosia have deficits in both the updating and storage mechanisms of the state estimate, this particular patient uniquely had a deficit specifically in the storage mechanism.

There is other recent evidence of the experience-dependent lability of the body schema and the role of visual input into the SPL in that process. For instance, tools held in the hand often become a psychophysical extension of the arm itself (62). Iriki et al. (63,64) trained monkeys to use a rakelike tool to retrieve food pellets located out of reach of the arm. Single-unit studies in medial SPL regions found that many cells had bimodal somatic and visual receptive fields and that the visual component extended out beyond the arm to include the tool (63). This showed that in the process of acquiring the ability to use the tool, polysensory neurons in medial SPL began to respond to visual inputs in spatial locations congruent with the tip of the tool that would be needed to guide its action. Imaging studies showed a strong activation of the medial IPL related specifically to use of the tool independent of the arm movement itself (64). Tool use-specific activations were also seen in a number of other motor-related structures, including the pre-SMA, premotor cortex, and cerebellum, indicating that the altered body schema in the SPL may be an integral component of a distributed system that performs the visuomotor transformations underlying the visual guidance of skilled movements.

Consistent with this hypothesis, Sekiyama et al. (65) asked subjects to wear reversing prisms continuously for more than a month. Reaching accuracy to visual targets and the accuracy of perceived locations of visual stimuli improved gradually over the first 2 to 3 weeks. This was paralleled by the gradual emergence of an altered body image based on visual inputs, as assessed by the ability of the subjects to correctly identify a drawing of a hand in different spatial orientations as being either a right or left hand. Imaging with fMRI during the latter perceptual body-image task likewise revealed activity in several neural structures, including the SPL, premotor cortex, and cerebellum, which are commonly associated with visuospatial processing and motor planning. They concluded that the modified body image acquired during prism adaptation resolves not only the prism-reversed mapping between visual and somatic inputs about body form, but also the reversed visuomotor transformation between visual inputs and motor outputs. These findings extend earlier evidence that the parietal cortex plays a critical role in prism adaptation during reaching (66), and further indicate that spatial representations in the parietal cortex serve not only perceptual functions but also or even primarily permit efficient motor interactions with the environment.

Coordinate Systems for the Planning and Execution of Reaching Movements

Cells in all subdivisions of the SPL discharge during arm movements. This discharge is characterized by broad symmetrical tuning functions for different directions of movement, and systematic graded modulations of tonic activity with different arm postures (8,9). A more specific question is the nature of the coordinate framework for reach-related activity. Do SPL neurons encode arm movements predominantly in terms of motor error signals in hand-centered, body-centered, gaze-centered, or other alternative reference frames? To what degree do they signal current or future states? Is there one coordinate framework in the SPL, or many?

Lacquaniti et al. (51) performed the first rigorous neurophysiological study of this question. They trained monkeys to make reaching movements from the center to the eight corners of three cubes in three-dimensional space. The cubes were positioned in three adjacent workspace locations, one centered on the midline and the other two to the left and right of the central cube. They found that area 5 cells (mainly in PE and PEip) typically showed broad tuning for movement direction in one or more of the three cubes. However, the preferred movement direction axis of the directional tuning function was usually not the same in each cube, suggesting that movement was not encoded explicitly in a framework of intended hand-centered direction of displacement. Instead, the tuning functions tended to rotate about the vertical axis by an amount that on average was similar to the change in shoulder angle while holding the arm at the central starting location in each cube. No systematic trend was observed for rotations about the other two axes. This suggested that the cells were coding arm posture and movement in a limb- or body-centered coordinate system.

To test this further, they subjected the static postural and dynamic movement-related activity of the cells across all three workspace regions to three different regression models that encoded arm postures and movement directions in terms of static hand location (posture) and intended final hand location (movement) in a single body-centered coordinate framework. Each model defined hand location in a different body-centered parameter space. A spherical coordinate model defined hand location in terms of its elevation relative to a horizontal plane, azimuth angle relative to the sagittal plane and radial distance, all centered on the shoulder. A spherangular model used the same elevation and azimuth angles but defined distance from the shoulder in terms of elbow extension angle. Finally, an orientation-angle model defined hand location in terms of the elevation and yaw angles of the spatial orientation of both the upper arm and forearm segments (22). All three models accounted about equally well for the tonic discharge of area 5 cells during static hold at the target locations and their dynamic activity during movements toward the targets. For most cells, the regression models that expressed movement-related activity in terms of the spatial locations of the intended target locations in different body-centered coordinate frameworks accounted for more of their reach-related activity than did a fourth regression model based only on the direction of displacement of the hand from the central target locations in each workspace (i.e., hand-centered coordinates that did not capture any changes in spatial locations or arm geometry). These analyses suggested that area 5 cells were signaling reaching movements predominantly in terms of the actual and intended spatial position of the arm, and less about the hand movement trajectory between actual and intended positions (51).

Another striking finding of their analyses was that each cell tended to show a correlation predominantly with only one of the parameters that defined hand location in each of the regression models. In other words, the cells did not code movement in terms of motor error vectors distributed uniformly in the different parameter spaces of each regression model. Instead, they tended to cluster along the axes of each coordinate system. This was consistent with Soechting and Flander's (22) hypothesis for separate psychophysical channels for each parameter defining final hand location.

Lacquaniti et al. (51) proposed a body-centered reference frame for reaching movements, but their analyses could not distinguish between purely extrinsic spatial parameter spaces (cartesian, spherical) and intrinsic parameter spaces that incorporated information about specific limb geometries. The reason was that in their highly overtrained animals, each target location was associated with a highly stereotypical arm posture, thereby confounding extrinsic and intrinsic parameter spaces. Scott et al. (67) trained monkeys to make reaching movements along the same spatial hand paths to target locations in a horizontal plane at shoulder height while holding the arm either up in the horizontal plane at shoulder level or suspended below the plane of movement in a more natural parasagittal orientation. This manipulation dissociated the trajectories of the reaching movements in intrinsic parametric coordinates, while they remained identical in extrinsic space. They found that many area 5 cells (mainly in PE and PEip) changed their reach-related activity (overall activity level, directional tuning, or both) between arm orientations. This confirmed that the parameter space in which reaching movements are expressed in that part of area 5 incorporates information about intrinsic arm posture and is not limited only to terms related to body-centered spatial locations and hand-centered errors.

In contrast, Batista et al. (68) found evidence that cells in PRR, more medial than the region explored by Lacquaniti et al. (51) and Scott et al. (67), may code reaching movements in gaze-centered coordinates. Monkeys made reaching movements between targets in a vertical plane on a rectangular grid positioned in front of them. Throughout the reaching movements, the monkeys were required to fixate constantly one of the targets on the panel, independent of start and end target locations. PRR cells were broadly tuned for movement direction or target location relative to a given starting location, as has been found in many other studies of reach-related activity in the SPL. The critical manipulation in this study came when the monkeys were required to change the target that they fixated. This shift in gaze direction often caused a more pronounced shift in the reach-related tuning of the cells than did shifts in starting hand location. Correlation analysis of the tuning functions observed for different starting hand locations and gaze directions indicated that most PRR reach-related activity was better explained in a reference framework in which target locations were defined relative to the current direction of gaze, rather than in absolute spatial or body-, limb-, or hand-centered coordinates.

This finding initially might seem to be at odds with the extensive behavioral evidence that reaching movements are first planned in terms of a motor error signal in hand-centered coordinates. However, the overwhelming majority of those studies did not control for or otherwise account for the effect of gaze direction. Furthermore, Batista et al. (68) recorded in a more medial part of the SPL than in the earlier reaching studies by Lacquaniti et al. (51) and Scott et al. (67).

The latter factor was clearly significant. The study by Batista et al. (68) was followed up by recordings using similar tasks in more lateral area 5, encompassing parts of MIP, PE, and PEip (41) that likely overlapped with the recording sites in earlier studies. Monkeys again made reaching movements between targets on a vertical grid that involved different combinations of starting and final target locations and gaze directions. In sharp contrast to the findings in PRR, reach-related activity in lateral area 5 during both an instructed-delay period and during movement itself was *least* well explained in a pure gaze direction-centered reference frame. Hand-, body-, or hand-and-body-centered coordinates provided a substantial improvement in explanatory power. Nevertheless, the best fit came when cell activity was represented in a coordinate framework in which targets were always in the same position relative to both the starting hand location and the direction of gaze simultaneously, independent of their absolute spatial locations. This finding is consistent with the hypothesis that cells in area 5 represent potential movements as a vector difference between an estimate of the current position of the hand and the intended target location (14), and suggests that this vector is represented in eye-centered coordinates.

Despite differences in their conclusions, the results of Lacquaniti et al. (51), Scott et al. (67), and Buneo et al. (41) concur that the reach-related reference frame in more lateral parts of the IPS incorporates information about starting arm position. Recent results in our own laboratory suggest that the trend toward limb-centered coordinates continues in PMd and MI (69). Cells were studied during reaching to targets in an instructed-delay task without oculomotor control. Cell activity was analyzed as a function of gaze direction during the spontaneous fixation episodes. Cells often showed planar modulations in activity as a function of momentary gaze direction, as described previously in PMd in a different task (70). However, this gaze modulation was generally modest, accounting for substantially less than 20% of total task-related discharge variance in the large majority of cells. Regression of discharge in a reference frame of intended movement direction relative to the starting hand location systematically accounted for far more of the variance. Moreover, specific tests failed to find evidence consistent with gaze-centered coordinates in virtually all cells tested. In contrast, preliminary results using the same task in a parietal region overlapping the recording sites in Buneo et al. (41) and Batista et al. (68) showed far stronger effects of gaze direction and evidence for a gaze-centered reference framework (69).

The most medial parts of the SPL, including PEc, MIP, V6A, and 7m, have also been the subject of a number of recent studies. Oculomotor signals are particularly prominent in these medial areas such as V6A (71, 72). Each of these regions has also been studied in a series of tasks that required reaching to or looking at targets in a variety of different conditions (2,31,73–75). One of the striking features of the cells in these areas is that they are typically modulated systematically as a function of many different factors during reaching movement tasks. Different cells respond to different combinations of inputs, including the direction of intended or executed arm movements, static arm postures, retinal inputs about target stimuli, motion of the target, and visual feedback about movements of the arm itself, as well as the direction of gaze and eye movements. An even more striking feature of these multiple convergent inputs is that the major axis of the tuning function of each of the factors that modulate the activity of a given cell all tend to cluster in a similar region of direction space called the cell's "global tuning field." Different cells possess different global tuning fields for their inputs. This multimodal convergence of different inputs with a common directional bias was hypothesized to be the single-neuron substrate for a process of successive matching of different sources of information needed to organize an arm movement directed toward a spatial target and to perform some of the early intervening sensorimotor transformations necessary to convert target location into limb motor commands (27,73). Furthermore, this multimodal convergence may permit the dynamic context-dependent construction of different coordinate frameworks as a function of task requirements and the needs of different effectors (1,2,6,31,33,74,76).

The results of these various studies indicate a progressive trend in the response properties of cells along the IPS that is not consistent with a single homogeneous coordinate framework for reaching in the SPL. The properties of cells in the most medial parts of the SPL (MIP, PEc, V6A, PRR) show a strong influence of visual and oculomotor signals, and reach-related activity is apparently coded in gaze-centered coordinates. These properties are consistent with a role in the putative early stages of sensorimotor transformations (33,68). In more lateral regions (PE, PEip), there is a much stronger representation of static arm posture and movement based on both somatic and visual inputs and a shift in coordinate frameworks toward a more body- or limb-centered representation. Even here, however, gaze direction makes a significant contribution to reach-related activity. The properties of cells in this part of the SPL are more consistent with a role in the generation of a neural representation of body form, posture, and movement for perceptual purposes and for the monitoring of motor performance, than with a predominant causal role in motor planning itself (9). This putative functional dichotomy between more medial and lateral parts of the SPL requires further experimental validation (36).

This progressive shift in reference frames is consistent with the sensorimotor coordinate transformation hypothesis for the planning and control of reaching movements. Nevertheless, it is noteworthy that one does not see abrupt and complete transitions between retina-, gaze-, body-, or limb-centered frameworks predicted by standard sensorimotor transformation models. Instead, cells throughout the SPL show convergence of a greater range of different types of inputs than expected at any putative sequential coordinate transformation stage. For instance, gaze-related signals are seen to different degrees throughout the SPL, and extend into the premotor cortex. Likewise, somatosensory and somatomotor signals related to current limb position and intended arm movements modulate cell activity even in medial parts of the SPL that are presumably implicated in some of the earliest stages of visuomotor processing. Clearly, parietal neural mechanisms do not respect the arbitrary mathematical and geometrical formalisms of the standard coordinate transformation hypothesis.

Furthermore, there may well be computational and strategic advantages for this extensive degree of convergence of multimodal reach-related signals. In one computational scheme (6,27,33,74), the direction of gaze and current arm position are the two unifying frames of reference on which the visuomotor control system learns the stable relationships between multiple convergent sources of information required to perform the visuomotor transformations from retinal input to motor output by a process of progressive matching of combinations of inputs. Alternatively, the parietofrontal visuomotor circuits may be implementing a neuronal approximation of the complex multidimensional nonlinear transformation underlying visually guided reaching via linear combinations of radial basis function representations (1,76–78). Although approaching the problem using different formalisms and vocabularies, both computational schemes possess many features in common. Both show that simultaneous coding of multiple inputs in different reference frames by single neurons can permit extraction of reach-related information in a specific appropriate reference frame in a context-dependent manner. Both resolve the combinatorial explosion problem inherent in trying to integrate simultaneously many different sources of input signals by decomposing the global computational problem into a series of smaller computational stages that integrate subsets of inputs expressed in different reference frames in a disciplined fashion (27,76). In that sense, the sequential coordinate transformation hypothesis retains general heuristic value for describing the functional role of the global shifts in response properties from more visual, gaze-centered, and extrinsic spatial representations to more limb-centered and intrinsic parametric representations within the distributed parietofrontal circuitry for reaching movements. Where these computational schemes differ from the most simplistic literal neural implementation of the coordinate transformation hypothesis is that they imply that cortical neurons contribute to visuomotor control by attempting to capture the relationships between different combinations of input and output signals in different contexts, rather than attempting to represent movement explicitly in any particular parameter space or reference frame (27,76). Therefore, they make no explicit distinction between the representation of movement in a given coordinate framework and the causal transformation mechanisms between different putative coordinate frameworks.

Online Control

The role of the parietal cortex in visuomotor control is most often discussed in terms of the putative sensorimotor transformations required to generate a motor command prior to movement onset. However, this represents only one aspect of the visual guidance of action. Another important aspect concerns the real-time mechanisms that ensure that the hand arrives successfully at the target. There is mounting evidence that the parietal cortex plays a critical role in online control throughout movement (25). Online control itself involves a number of functions including the real-time generation of the time-varying motor command and processes that monitor performance and correct for errors in motor commands. The errors can arise both from imprecision in the processes that generate the motor command itself and from unexpected changes in the external world, such as shifts in target location. We focus mainly on the issue of real-time error correction.

There is extensive evidence that the motor system possesses efficient mechanisms to correct for certain types of performance errors at latencies that are substantially shorter than standard visual reaction time delays. For instance, a number of studies have inferred that the initial motor command for reaching movements is only an approximation of the correct motor command, and that errors in the initial command are fine-tuned in real time as the movement unfolds (16,17,20,79). Visual feedback received early in the movement contributes to this rapid online correction process (80). This online process can also correct for initial output errors evoked by an optical illusion, even when vision of the arm during movement is not permitted (81), implying that the correction mechanism also uses somatic feedback or efference copies of motor commands. The latter finding emphasizes the distinction between planning and online control, and suggests that context, such as visual inputs that evoke a perceptual error, has more influence on planning than on online control (81).

Another major line of evidence supporting rapid online control and correction of errors comes from target-shift experiments, in which the target of a reach is displaced just prior to or during the arm movement (82,83). In such conditions, arm movements show a rapid correction of the trajectory in-flight at latencies that are often much shorter than standard visual reaction times (83,84). These findings suggest that the activation levels for the appropriate muscles are continuously adjusted to move the hand in the direction of the current internal representation of target location.

These fast corrections appear to be mediated by an unconscious "automatic" process (85). Their occurrence does not require conscious detection of target displacement (82, 85–88). They even occur when there is no visual feedback of the moving hand (86,87). The corrections also appear to be relatively resistant to voluntary modifications such as stopping instead of correcting or moving in the opposite direction to the target shift (84, 89,90).

A number of lines of evidence suggest that rapid automatic corrections during ongoing movements involve the PPC. For example, automatic corrections to unperceived target shifts were abolished in 4 of 5 normal subjects when transcranial magnetic stimulation (TMS) was applied over the PPC contralateral to the performing arm (86). An imaging study found a specific activation within the PPC associated with corrections of reaching movements in response to target shifts, independent of other confounding factors (91). Finally, Pisella et al. (90,92) reported that a patient with a bilateral lesion of the PPC was able to point accurately to and grasp stationary targets but was unable to correct such movements rapidly when the target was suddenly displaced. Normal subjects performed such corrections reliably, even when instructed not to do so.

Some of the most compelling evidence for the role of the PPC in visuomotor control comes from the deficits incurred after parietal lobe injuries. One of the most striking of those deficits is the optic ataxia that can follow primarily from lesions of the SPL in humans. It is characterized in part by gross spatial and directional errors during visually guided arm movements. There is no primary motor deficit, as evidenced by the observation that reaching under proprioceptive guidance, such as pointing to different parts of the body, is usually intact. This indicates that the cortical system that controls arm movements that are not dependent on visual guidance remains functionally intact. Optic ataxia often has been interpreted to result from the disruption of a spatial representation of visual information within the parietal cortex, or as a disconnection syndrome within a serial process that projects spatial information from the parietal cortex to frontal motor areas. However, its origin is being reinterpreted in the light of recent insights into SPL function. Battaglia-Mayer et al. (33,93) have proposed that optic ataxia results primarily from a disruption of the dynamic integration of visual, eye position, and arm position information with recurrent internal feedback signals about motor intentions at the single-cell level in medial SPL, thereby perturbing the organization of appropriately directed visually guided arm movements. The neural substrates of that integration are the multimodal global tuning fields of medial SPL neurons and recurrent re-entrant loops of connections between parietal and frontal motor regions. Desmurget et al. (85,86,90) argue that optic ataxia does not reflect a deficit in initial motor planning per se prior to movement onset or a general visuomotor dysfunction, but rather a specific deficit in rapid automatic online control of arm movements using visual inputs during the movements. In their view, the PPC is implicated in many visuospatial and visuomotor functions, including online control, but the latter is particularly dependent on the functional integrity of the SPL. These two hypotheses are not necessarily mutually exclusive, because the global tuning functions and re-entrant processing described by Battaglia et al. (33,93) could subserve both a priori planning processes and online control.

The deficits observed with parietal damage also suggest that the PPC is primarily involved in movements guided online by *currently available* visual information, as opposed to movements guided from memory. For example, Milner et al. (94,95) studied the performance of patients (A.T. and I.G.) with severe bilateral damage to the parietal lobes, resulting in optic ataxia, who exhibit large errors during normal pointing. Interestingly, however, the pointing accuracy of parietal subject A.T. *improved* when she was required to wait 5 seconds before pointing, unlike normal subjects whose accuracy was always worse for delayed movements (94). A similar result was reported for patient I.G., who was first unable to scale the grip size when a simple immediate grasp was required, even when the target object had been previewed (5 second delay). However, a good grip scaling was observed in pantomimed grasping to remembered target objects (95). After practice, I.G. became able to scale the grip when grasping a real target object that had been previewed earlier. Her scaling was guided by *memorized* visual information, because she did not recalibrate grip when a different object was covertly substituted for the original target during a delay period. In contrast with normal subjects who invariably ignored the previewed information, I.G. depended on the previewed information to program a grasp without regard to the current visual information.

It is possible that patients with parietal damage make use of an alternative route for visuomotor guidance, normally used when guiding movements from memory, which involves the ventral visual stream (95). The presence of this alternative route is supported by the observation that a patient with damage to the ventral stream (D.F.) showed normal grip scaling when grasping real objects (96), but could not pantomime the same movements to remembered targets (97).

Paradoxically, the behavioral evidence that the parietal cortex predominantly plays a role in movements guided by *current* sensory information (94,95,97) appears to be at odds with the many single-unit studies that have shown sustained activity in various regions of the PPC during a memorized instructed-delay period, when targets' signals are no longer visible (3,43,44,98). However, the two sets of findings are not necessarily contradictory. For instance, it is possible that when intact subjects perform movements to memorized targets, spatial information is sustained in PPC via alternative pathways involving the ventral visual system and prefrontal cortex. In intact monkeys, the activity of LIP output neurons is significantly reduced when visual stimuli are not present (99), and it remains to be seen whether such memory-period activity is present at all when ventral processing has been disrupted. Future research with both human subjects and with monkeys is needed to resolve these questions more fully.

A Neural Circuit for Online Control

The view of visuomotor control as a feedback circuit assisted with a forward model (Fig. 7-1D) raises the question of where in the brain the various putative computations take place. There is strong evidence that the PPC, and the entire parieto-frontal network, is part of a neural mechanism for visually guided reaching (2,5,6,10), and neural correlates of directional motor commands are found throughout the SPL (9,35,41–45). The computation of these directional signals requires two kinds of information: a representation of target position, and a representation of the current hand position. This raises additional questions. First, there is the issue of how and where information on target locations undergoes transformations from a visual coordinate system to an intrinsic effector-related representation, that is, the internal model of *inverse kinematics*. Imaging studies strongly implicate the PPC in this process (66). Furthermore, we have already reviewed results from single-unit recordings that suggest that medial parts of PPC (PEc, MIP, V6A) represent movements in eye-centered coordinates (41, 68,74), whereas movement-related activity in more lateral parts of the SPL (PE, PEip) appears to be more strongly influenced by limb posture (51,67). Second, the PPC also appears to be involved in the integration of multimodal information on body posture and hand position (36,54–56,63). This integration can use current proprioceptive and visual information to estimate limb position, and it can also use a predictive estimate of position based on the efference copy of motor commands, that is, an internal model of forward kinematics, or simply, a forward model. Although some authors have suggested that the forward model may reside in the PPC itself (86), it is also possible that the PPC simply receives input from a forward model located elsewhere, such as the cerebellum (24).

The results reviewed in the preceding sections may be used to sketch out the general neural circuit for visuomotor guidance. This simple model has the basic structure of the control scheme in Fig. 7-D: a feedback system assisted with a predictive forward model. In this view, parietal cortex compares visual information on the location of a reach target with an estimate of the current limb position. In novel movements, this estimate is based on overt visual and proprioceptive feedback, but in well-learned movements, a forward model predicts the current limb position using the efference copy of recent motor commands. Thus, the parietofrontal circuit acts as an "automatic pilot" for the hand (90), using current visual information and an estimate of current position to generate the command for moving the hand to the target.

This simple circuit may be used to interpret a large variety of the studies described in the preceding sections, including evidence for representations of target location, body configuration, and reach directions in parietal cortex (9,35,36,41–45,51,54–56,63,67, 68,74), online guidance of movement (84, 86,88,90), and immediate versus delayed reaches by intact and neurologically damaged subjects (24,94,97,100). For example, during normal movements with full visual feedback, the parietal circuit smoothly guides the motor command to its target, automatically correcting for any perturbations such as target shifts. When damage or inactivation of the parietal cortex prevents the operation of this automatic pilot, an alternative ventral route for visuomotor guidance comes into play. Together with a forward model that predicts current hand position, this remembered target representation can be used to guide movements, albeit with reduced accuracy. Patients with parietal damage (90,92) and normal subjects whose PPC has been temporarily inactivated (86) use this alternate route to guide their pointing movements, but they do not correct for a target jump because they are using stored target location information. This alternate route is also used by normal subjects when they reach to a remembered target. However, when full visual feedback is available, the online parietal system is engaged instead. This may be the reason why patients such as A.T. and I.G., with PPC lesions, perform poorly during online movements (because their damaged parietal circuit is in control), but perform fairly well when reaching or scaling grip to a remembered target (94,95).

To summarize, there is growing evidence that the superior parietal lobule forms part of a large circuit involved in the online visual guidance of reaching movements. During the course of reaching, this circuit continuously computes the difference between the location of the hand and the location of the target (expressed within an eye-centered coordinate framework, at least in some parts of the SPL), and uses this difference to shape the ongoing motor command. For this reason, it automatically corrects for any perturbations such as target motion. The circuit does not operate in a simple closed-loop fashion, however. It uses a forward model (located either in the cerebellum or in the parietal cortex itself), that estimates the current position of the effector using an efference copy of recent motor commands and information about the context in which the movement is performed. Alternative routes for visuomotor guidance appear to exist, but the fast parietal system appears to be critical to most of our visually guided interaction with the everchanging environment.

Selecting an Action

To this point, we have discussed control issues related to the performance of a reaching movement to a single visually defined target. However, the world around us normally presents many opportunities for action at any given moment, including many potential reach targets such as your morning cup of coffee, glass of juice, and bagel. It has often been assumed that the brain decides on an action, such as choosing to reach toward one of many alternative targets, via cognitive processes that are separate from and antecedent to the motor mechanisms involved in overt execution of the selected action. These decision-making processes supposedly operate on perceptual representations of the world. The posterior parietal cortex has long been implicated in the construction of such perceptual representations for sensory guidance of behavior.

However, the sensory representation of the world in the parietal cortex is often highly impoverished when the subject is not actively engaged in a behavioral task, highly context-dependent when so engaged, and highly selective for different stimulus attributes in different parietal regions (1,3,54,98). In particular, neurons in many parietal regions seem to be strongly modulated by the salience of sensory stimuli. This property is best documented in LIP. Many LIP neurons discharge in relation to visually guided saccades. However, it has been argued that LIP activity does not represent the saccade plan per se, but instead represents a "salience map" of visual space prior to saccadic target selection (3,101). The visual response of LIP neurons is strongest to stimuli that draw attention by standing out from the background, suddenly appearing, or briefly moving, whether or not the stimulus serves as a target for saccades (101). LIP neurons also respond during saccades, but this response is strongest when the saccade is guided by a visual stimulus than when it is made from memory (99,102). Finally, LIP cells respond to visual "distractor" stimuli even when a saccade is already planned toward a different location (103).

In contrast, other evidence supports a direct role for LIP neural activity in saccade planning (45,98). For example, LIP neurons were preferentially activated when a salient sensory stimulus serves as the target for a saccade rather than a reach (34). Conversely, when the same saccadic movement is guided by different sensory cues (visual versus auditory), LIP neurons show similar activity patterns (104). These results have been interpreted as evidence that the parietal cortex represents the intention to make a specific movement regardless of the sensory information used to specify that movement (98).

Despite much effort, it has proven impossible to resolve whether the parietal cortex is primarily involved with representing perceptual space or with representing potential actions (105,106). This difficulty is itself informative. Debates among salience, attention, and intention in LIP (3,45,98), or between the sensory versus motor interpretations of area 5 activity (9) presuppose mutually exclusive distinctions among perceptual, cognitive, and motor executive processes, but these distinctions may be too rigidly defined (106–108).

A growing body of research on the neural substrates of the decision-making processes involved in saccade selection suggests that these decisions are made within the same structures that also plan and execute the selected movement, distributed throughout parietal and frontal cortex and even subcortical systems (109–118). In this view, potential saccadic movements are specified concurrently (106,119–121) and engage in a competition for release into overt execution. This competition is influenced by various factors, such as the probability of a given saccade or the size of the associated reward (115), or the sensory evidence favoring a particular choice (113,118). LIP appears to be directly involved in this competitive decision-making process (115,118) together with the frontal eye fields (111,116,117) and dorsolateral prefrontal cortex (113). Even cells as close to the motor periphery as those in the superior colliculus respond to multiple saccade targets and reflect the certainty of selection of a particular target over others (119,122). All of these results suggest that the decision to perform a saccade does not occur within a localized central "cognitive" system prior to engaging motor executive circuits. It appears instead to involve a distributed network of cortical and subcortical regions, some of which are directly involved in saccade execution. From this perspective, cell activity in LIP is neither uniquely related to a spatial representation nor to a specific motor intention, but also reflects the simultaneous processing of decision-making variables that influence the selection of one saccadic command from among multiple potential saccadic movement choices.

The powerful effect of attention on LIP neuron activity also can be viewed within the context of competitive selection from among multiple potential saccade targets. Directing attention to different spatial locations is tightly coupled to overt orienting behavior, of which saccades are a part, and the ability to direct attention covertly to a location independent of eye and/or head movement may have evolved recently within the context of overt orientation behavior (123,124). In support of this functional association between attention and intention, saccade-related cells in the superior colliculus are engaged even during covert attentional shifts (123), and cortical systems for shifting spatial attention and for making overt saccades overlap extensively (125).

Furthermore, it has been suggested that attention itself is an integral part of the larger process of action selection (126–130). Rushworth and colleagues have proposed that the parietal lobe contains partially distinct attention mechanisms for selecting gaze and reach targets. Functional imaging suggests that these mechanisms involve, respectively, the lateral and medial banks of the intraparietal sulcus (131), and lesion studies particularly implicate the left parietal lobe in the selection of hand actions (132). From this perspective, this form of "motor attention" can be regarded as a centrally generated bias favoring one action over another that contributes to the competitive selection process in much the same way as other decision-making variables such as salience, response probability, and reward expectancy. It would be the motor equivalent of the more familiar sensory attention processes that serve to selectively enhance or suppress sensory signals on the basis of an a priori central bias toward certain sensory properties such as modality or spatial location.

The concept of "motor attention" as a predisposing decisional factor in response selection is supported by recent studies by Snyder et al. (34,35,45). They trained monkeys to prepare either saccades or reaching movements to a target presented in an instructed-delay task (45). The animals were subsequently informed of the effector choice (eyes, arm) by a second arbitrary stimulus. Many cells in both PRR and LIP discharged during the initial delay period when the monkeys knew the target location but before they knew which response to make. Subsequent changes in activity of these cells after the choice cue was presented indicated that many of the PRR cells were predominantly related to arm movements, whereas others were coupled to eye movements. Their coactivation prior to the choice cue indicated that the monkey was attending to a spatial location and preparing simultaneously to perform two different actions toward it, and the choice between them was eventually made using an arbitrary stimulus-response rule. When a stimulus presented at the intended target location informed the monkey to change the intended response from a saccade to a reach or vice versa, activity changes related to the effector change were seen in PRR and LIP (34). Because the effector-change cues were presented at the intended target location, the activity changes could not be explained by a shift in the location of spatial attention. Finally, Calton et al. (35) studied PRR in a task in which the effector choice cue was presented before the target location information. Many reach-related cells changed their activity as soon as the reach cue was presented, even before the monkeys could know in which direction to move. These results revealed the existence of a centrally generated bias to perform one class of actions (reach) over others that could be independent of any information about a spatial locus of attention or the specific metrics of the response. This bias may be a neuronal correlate of motor attention (131,132) or motor "set" that contributes to the competitive selection of actions.

Further evidence of the simultaneous preparation of multiple potential actions comes from studies in our laboratory. In one study, monkeys were trained to move toward a target, away from it, or not to move at all, in response to the colors of cues presented in different target locations in an instructed-delay task (9,44,133). Cells in both medial area 5 and PMd systematically reflected the directionality of reaching movements when the monkeys prepared to move toward or away from the instructional cue. However, when the NO-GO instruction was presented, a clear dichotomy arose between area 5 and PMd. Whereas PMd cell discharge reflected the decision not to move, cells in area 5 continued to signal the directionality of the potential but unexecuted movement toward the target location (9,44).

In a second study, monkeys were presented briefly with two potential target locations for reaching movements, and later were presented with a nonspatial cue signaling to which of the two memorized target locations to move (134). As soon as the two possible targets were presented, directional signals related to both potential movements appeared in dorsal premotor cortex (PMd). Once the decision between these options was made according to the second nonspatial choice cue, the signal associated with the rejected target was suppressed and the remaining directional activity indicated the monkey's choice. This activity was not simply a representation of the sensory information by which the monkey made its choice, because the PMd activity during the delay period of trials in which the monkey incorrectly moved to the rejected target signaled their erroneous choice and not the sensory evidence leading to that decision. In a preliminary investigation, we have found that 16 of 34 task-related cells recorded in MIP/PRR in the two-target task likewise responded when either of the two targets was presented near their preferred direction (135). Once the decision was made, 15 of these 16 cells reflected the selection of one target over the other. These observations suggest that the primate brain can begin to perform in parallel the sensorimotor transformations to specify two mutually exclusive reaching movements before finally selecting one of them for overt execution.

These various studies have a number of important implications. They show that the brain can begin to formulate a limited number of response options simultaneously before sufficient information is available to settle on one action (108,109,113,136). The selection process is competitive and can be biased by many factors. This competition occurs not only between possible targets for a saccade or reach, but also between choices of effector. The decision-making process involves distributed parieto-frontal circuits and overlaps the circuits implicated in the overt execution of the final selected choice. Finally, the processes involved in the selection, planning, and execution of a voluntary movement are not organized in a strictly serial manner.

This hypothesis makes a number of empirically testable predictions. First, if the parietal subregions specialized for different kinds of movements (reaching, grasping, orienting, etc.) are all capable of simultaneously representing different potential actions, then cells in many parts of the PPC should be modulated by the kinds of biasing factors that influence the competition for action selection. Such influences have been documented in LIP in saccade tasks (115,118), but they should be evident in other parts of the posterior parietal cortex also. For instance, if a monkey is presented with two possible targets for reaching movements, then neural signals in MIP corresponding to the two potential actions should be modulated by the decision variables involved in selecting between them. Many different kinds of biasing factors may be reflected in parietal activity, including influences of salience, predicted reward size, selection probability, and so on. For example, if two potential movements are presented simultaneously in a free-choice paradigm but one requires more effort (i.e., larger force production), then cells in MIP may discharge less for the harder movement. Sensitivity to force output levels is generally low in the parietal cortex in situations in which a single response choice is available and force levels are only pertinent for specifying the requisite motor command for response execution (50,137). In contrast, when force output level becomes a decision variable, the prediction is that the parietal cortex cell activity will become more strongly modulated by force. Furthermore, activity may be reduced for movements requiring larger forces not because it represents the force output per se, but because it reflects the monkey's process of deciding between two options and its preference for choosing the one that yields the best return for effort expended.

CONCLUSIONS

This chapter has touched on a wide range of themes, including representations of body schema, online guidance of movements, forward models, and selection of actions. This array of concepts seems appropriate when reviewing a structure with such a diversity of functions as the superior parietal cortex. We believe that this diversity is not without order, however. The apparently multiplexed sensory and motor representations in parietal regions reflect its role in sensorimotor integration and movement guidance. As discussed, a neural mechanism for motor guidance must have access to information about body configuration and spatial layout of targets and obstacles, and must act during the course of movement to transform this information into motor commands. The posterior parietal cortex appears to meet all of these criteria.

Furthermore, a biological system for sensorimotor guidance need not be rigidly separated from the parts of the brain involved in selecting movements. The world is full of things to look at, reach out to, and touch, and parallel specialized subregions of the PPC may be constantly interpreting visual information to crudely specify potential saccade, reach, and grasp actions, among others. Even during an ongoing movement, information about the location of objects relative to body segments may be useful for modifying the motor command or for switching to a different action altogether.

In summary, the functional roles of the PPC defy the rigid functional decomposition of many models of voluntary behavior into perceptual, cognitive, and motor representations. Instead, the PPC may act both as a blackboard for deciding among currently available potential actions, and as a system for executing the selected action through feed-forward mechanisms to specify the initial motor command and internal and external feedback loops to ensure rapid and accurate sensorimotor control.

REFERENCES

1. Andersen RA, Snyder LH, Bradley DC, et al. Multimodal representation of space in the posterior parietal cortex and its use in planning movements. *Annu Rev Neurosci* 1997;20:303–330.
2. Caminiti R, Ferraina S, Battaglia-Mayer A. Visuomotor transformations: early cortical mechanisms of reaching. *Curr Opin Neurobiol* 1998;8:753–761.
3. Colby CL, Goldberg ME. Space and attention in parietal cortex. *Annu Rev Neurosci* 1999;22:319–349.
4. Matelli M, Luppino G. Parietofrontal circuits for action and space perception in the macaque monkey. *NeuroImage* 2001;14:S27–S32.
5. Rizzolatti G, Luppino G. The cortical motor system. *Neuron* 2001;31:889–901.
6. Wise SP, Boussaoud D, Johnson PB, et al. Premotor and parietal cortex: corticocortical connectivity and combinatorial computations. *Annu Rev Neurosci* 1997; 20:25–42.
7. Stein JF. The representation of egocentric space in the posterior parietal cortex. *Behav Brain Sci* 1992;15: 691–700.
8. Kalaska JF. The representation of arm movements in postcentral and parietal cortex. *Can J Physiol Pharmacol* 1988;66:455–463.
9. Kalaska JF. Parietal cortex area 5 and visuomotor behavior. *Can J Physiol Pharmacol* 1996;74:483–498.
10. Kalaska JF, Scott SH, Cisek P, et al. Cortical control of reaching movements. *Curr Opin Neurobiol* 1997;7: 849–859.
11. Snyder LH, Batista AP, Andersen RA. Intention-related activity in the posterior parietal cortex: a review. *Vision Res* 2000;40:1433–1441.
12. Keele SW. Movement control in skilled motor performance. *Psychol Bull* 1968;70:387–403.
13. Bhat RB, Sanes JN. Cognitive channels computing action distance and direction. *J Neurosci* 1998;18: 7566–7580.
14. Bullock D, Cisek P, Grossberg S. Cortical networks for control of voluntary arm movements under variable force conditions. *Cereb Cortex* 1998;8:48–62.
15. Bullock D, Grossberg S. Neural dynamics of planned arm movements: emergent invariants and speed-accuracy properties during trajectory formation. *Psychol Rev* 1988;95:49–90.
16. Gordon J, Ghilardi MF, Cooper SE, et al. Accuracy of planar reaching movements: II. Systematic extent errors resulting from inertial anisotropy. *Exp Brain Res* 1994;99:112–130.
17. Gordon J, Ghilardi MF, Ghez C. Accuracy of planar reaching movements. I. Independence of direction and extent variability. *Exp Brain Res* 1994;99: 97–111.
18. Messier J, Kalaska JF. Differential effect of task conditions on errors of direction and extent of reaching movements. *Exp Brain Res* 1997;115:469–478.
19. Morasso PG. Spatial control of arm movements. *Exp Brain Res* 1981;42:223–227.
20. Karst GM, Hasan Z. Initiation rules for planar, two-joint arm movements: Agonist selection for movements throughout the work space. *J Neurophysiol* 1991;66:1579–1593.
21. Soechting JF, Flanders M. Errors in pointing are owing to approximations in sensorimotor transformations. *J Neurophysiol* 1989a;62:595–608.
22. Soechting JF, Flanders M. Sensorimotor representations for pointing to targets in three-dimensional space. *J Neurophysiol* 1989b;62:582–594.
23. Keele SW, Posner MI. Processing of visual feedback in rapid movements. *J Exp Psychol* 1968;77:155–158.
24. Miall RC, Wolpert DM. Forward models for physiological motor control. *Neural Networks* 1996;9: 1265–1279.
25. Sabes PN. The planning and control of reaching movements. *Curr Opin Neurobiol* 2000;10:740–746.
26. Wolpert DM, Ghahramani Z, Jordan MI. An internal model for sensorimotor integration. *Science* 1995;269: 1880–1882.
27. Burnod Y, Baraduc P, Battaglia-Mayer A, et al. Parieto-frontal coding of reaching: an integrated framework. *Exp Brain Res* 1999;129:325–346.
28. Johnson PB, Ferraina S, Bianchi L, et al. Cortical networks for visual reaching: physiological and anatomic organization of frontal and parietal arm regions. *Cereb Cortex* 1996;6:102–119.
29. Jones EG, Coulter JD, Hendry HC. Intracortical connectivity of architectonic fields in the somatic sensory, motor and parietal cortex of monkeys. *J Comp Neurol* 1978;181:291–348.
30. Pandya DN, Kuypers HGJM. Cortico-cortical connections in the rhesus monkey. *Brain Res* 1969;13:13–36.
31. Marconi B, Genovesio A, Battaglia-Mayer A, et al. Eye-hand coordination during reaching. I. Anatomic relationships between parietal and frontal cortex. *Cereb Cortex* 2001;11:513–527.
32. Caminiti R, Ferraina S, Johnson PB. The sources of visual information to the primate frontal lobe: a novel role for the superior parietal lobule. *Cereb Cortex* 1996;6:319–328.
33. Battaglia-Mayer A, Caminiti R. Optic ataxia as a result of the breakdown of the global tuning fields of parietal neurones. *Brain* 2002;125:225–237.
34. Snyder LH, Batista AP, Andersen RA. Change in motor plan, without a change in the spatial locus of attention, modulates activity in posterior parietal cortex. *J Neurophysiol* 1998;79:2814–2819.
35. Calton JL, Dickinson AR, Snyder LH. Non-spatial, motor-specific activation in posterior parietal cortex. *Nat Neurosci* 2002;5:580–588.
36. Graziano MSA, Cooke DF, Taylor CSR. Coding the location of the arm by sight. *Science* 2000;290: 1782–1786.
37. Wolpert DM, Goodbody SJ, Husain M. Maintaining internal representations: the role of the human superior parietal lobe. *Nat Neurosci* 1998;1:529–533.
38. Goodbody SJ, Wolpert DM. The effect of visuomotor displacements on arm movement paths. *Exp Brain Res* 1999;127:213–223.
39. Lackner JR. Some proprioceptive influences on the perceptual representation of body shape and orientation. *Brain* 1988;111:281–297.
40. Ramachandran VS, Hirstein W. The perception of phantom limbs. The D. O. Hebb lecture. *Brain* 1998; 121:1603–1630.
41. Buneo CA, Jarvis MR, Batista AP, et al. Direct visuomotor transformations for reaching. *Nature* 2002;416: 632–636.

42. Crammond DJ, Kalaska JF. Neuronal activity in primate parietal cortex area 5 varies with intended movement direction during an instructed-delay period. *Exp Brain Res* 1989;76:458–462.
43. Ferraina S, Bianchi L. Posterior parietal cortex: functional properties of neurons in area 5 during an instructed-delay reaching task within different parts of space. *Exp Brain Res* 1994;99:175–178.
44. Kalaska JF, Crammond DJ. Deciding not to GO: neuronal correlates of response selection in a GO/NOGO task in primate premotor and parietal cortex. *Cereb Cortex* 1995;5:410–428.
45. Snyder LH, Batista AP, Andersen RA. Coding of intention in the posterior parietal cortex. *Nature* 1997; 386:167–170.
46. Chapman CE, Spidalieri G, Lamarre Y. Discharge properties of area 5 neurones during arm movements triggered by sensory stimuli in the monkey. *Brain Res* 1984;309:63–77.
47. Kalaska JF, Crammond DJ. Cerebral cortical mechanisms of reaching movements. *Science* 1992;255: 1517–1523.
48. Georgopoulos AP, Caminiti R, Kalaska JF. Static spatial effects in motor cortex and area 5: quantitative relations in a two-dimensional space. *Exp Brain Res* 1984;54:446–454.
49. Kalaska JF, Cohen DAD, Hyde ML, et al. A comparison of movement direction-related versus load direction-related activity in primate motor cortex, using a two-dimensional reaching task. *J Neurosci* 1989;9: 2080–2102.
50. Kalaska JF, Cohen DAD, Prud'homme MJ, et al. Parietal area 5 neuronal activity encodes movement kinematics, not movement dynamics. *Exp Brain Res* 1990;80:351–364.
51. Lacquaniti F, Guigon E, Bianchi L, et al. Representing spatial information for limb movement: role of area 5 in the monkey. *Cereb Cortex* 1995;5:391–409.
52. Burbaud P, Doegle C, Gross CG, et al. A quantitative study of neuronal discharge in areas 5, 2, and 4 of the monkey during fast arm movements. *J Neurophysiol* 1991;66:429–443.
53. Colby CL, Duhamel JR. Spatial representations for action in parietal cortex. Brain *Res Cogn Brain Res* 1996;5:105–115.
54. Colby CL. Action-oriented spatial reference frames in cortex. *Neuron* 1998;20:15–24.
55. Gregoriou GG, Savaki HE. The intraparietal cortex: subregions involved in fixation, saccades, and in the visual and somatosensory guidance of reaching. *J Cereb Blood Flow Metab* 2001;21:671–682.
56. Inoue K, Kawashima R, Satoh K, et al. PET study of pointing with visual feedback of moving hands. *J Neurophysiol* 1998;79:117–125.
57. Wann JP, Ibrahim SF. Does limb proprioception drift? *Exp Brain Res* 1992;91:162–166.
58. Prablanc C, Echallier JE, Jeannerod M, et al. Optimal response of eye and hand motor systems in pointing at a visual target. II. Static and dynamic visual cues in the control of hand movement. *Biol Cybernet* 1979;35: 183–187.
59. Vindras P, Desmurget M, Prablanc C, et al. Pointing errors reflect biases in the perception of the initial hand position. *J Neurophysiol* 1998;79:3290–3294.
60. van Beers RJ, Sittig AC, Denier van der Gon JJ. Integration of proprioceptive and visual position-information: an experimentally supported model. *J Neurophysiol* 1999a;81:1355–1364.
61. van Beers RJ, Sittig AC, Denier van der Gon JJ. Localization of a seen finger is based exclusively on proprioception and on vision of the finger. *Exp Brain Res* 1999b;125:43–49.
62. Yamamoto S, Kitazawa S. Sensation at the tips of invisible tools. *Nat Neurosci* 2001;4:979–980.
63. Iriki A, Tanaka M, Iwamura Y. Coding of modified body schema during tool use by macaque postcentral neurones. *Neuroreport* 1996;7:2325–2330.
64. Obayashi S, Suhara T, Kawabe K, et al. Functional brain mapping of monkey tool use. *NeuroImage* 2001; 14:853–861.
65. Sekiyama K, Miyauchi S, Imaruoka T, et al. Body image as a visuomotor transformation device revealed in adaptation to reversed vision. *Nature* 2000;407: 374–377.
66. Clower DM, Hoffman JM, Votaw JR, et al. Role of the posterior parietal cortex in the recalibration of visually guided reaching. *Nature* 1996;383:618–621.
67. Scott SH, Sergio LE, Kalaska JF. Reaching movements with similar hand paths but different arm orientations. II. Activity of individual cells in dorsal premotor cortex and parietal area 5. *J Neurophysiol* 1997;78: 2413–2426.
68. Batista AP, Buneo CA, Snyder LH, et al. Reach plans in eye-centered coordinates. *Science* 1999;285: 257–260.
69. Cisek P, Kalaska JF. Modest gaze-related discharge modulation in monkey dorsal premotor cortex during a reaching task performed with free fixation. *J Neurophysiol* 2002a;88:1064–1071.
70. Boussaoud D, Jouffrais C, Bremmer F. Eye position effects on the neuronal activity of dorsal premotor cortex in the macaque monkey. *J Neurophysiol* 1998;80: 1132–1150.
71. Galletti C, Battaglini PP, Fattori P. Eye position influence on the parieto-occipital area PO (V6) of the macaque monkey. *Eur J Neurosci* 1995;7:2486–2501.
72. Nakamura K, Chung HH, Graziano MS, et al. Dynamic representation of eye position in the parieto-occipital sulcus. *J Neurophysiol* 1999;81:2374–2385.
73. Battaglia-Mayer A, Ferraina S, Mitsuda T, et al. Early coding of reaching in the parietooccipital cortex. *J Neurophysiol* 2000;83:2374–2391.
74. Battaglia-Mayer A, Ferraina S, Genovesio A, et al. Eye-hand coordination during reaching. II. An analysis of the relationships between visuomanual signals in parietal cortex and parieto-frontal association projections. *Cereb Cortex* 2001;11:528–544.
75. Ferraina S, Johnson PB, Garasto MR, et al. Combination of hand and gaze signals during reaching: activity in parietal area 7m of the monkey. *J Neurophysiol* 1997;77:1034–1038.
76. Pouget A, Sejnowski TJ. Spatial transformations in the parietal cortex using basis functions. *J Cogn Neurosci* 1997;9:222–237.
77. Pouget A, Snyder LH. Computational approaches to sensorimotor transformations. *Nat Neurosci* 2000;3: 1192–1198.
78. Snyder LH. Coordinate transformations for eye and

arm movements in the brain. *Curr Opin Neurobiol* 2000;10:747–754.
79. Messier J, Kalaska JF. Comparison of variability of initial kinematics and endpoints of reaching movements. *Exp Brain Res* 1999;125:139–152.
80. Sheth BR, Shimojo S. How the lack of visuomotor feedback affects even the early stages of goal-directed pointing movements. *Exp Brain Res* 2002;143: 181–190.
81. Glover S, Dixon P. Motor adaptation to an optical illusion. *Exp Brain Res* 2001;137:254–258.
82. Pelisson D, Goodale MA, Jeannerod M. Visual control of reaching movements without vision of the limb. II. Evidence of fast unconscious processes correcting the trajectory of the hand to the final position of a double-step stimulus. *Exp Brain Res* 1986;62:303–311.
83. van Sonderen JF, Gielen CC, Denier van der Gon JJ. Motor programmes for goal-directed movements are continuously adjusted according to changes in target location. *Exp Brain Res* 1989;78:139–146.
84. Day BL, Lyon IN. Voluntary modification of automatic arm movements evoked by motion of a visual target. *Exp Brain Res* 2000;130:159–168.
85. Desmurget M, Grafton ST. Forward modeling allows feedback control for fast reaching movements. *Trends Cog Sci* 2000;4:423–431.
86. Desmurget M, Epstein CM, Turner RS, et al. Role of the posterior parietal cortex in updating reaching movements to a visual target. *Nat Neurosci* 1999;2: 563–567.
87. Goodale MA, Pelisson D, Prablanc C. Large adjustments in visually guided reaching do not depend on vision of the hand or perception of target displacement. *Nature* 1986;320:748–750.
88. Prablanc C, Martin O. Automatic control during hand reaching at undetected two-dimensional target displacements. *J Neurophysiol* 1992;67:455–469.
89. Pisella L, Arzi M, Rossetti Y. The timing of color and location processing in the motor context. *Exp Brain Res* 1998;121:270–276.
90. Pisella L, Grea H, Tilikete C, et al. An 'automatic pilot' for the hand in human posterior parietal cortex: toward reinterpreting optic ataxia. *Nat Neurosci* 2000;3: 729–736.
91. Desmurget M, Grea H, Grethe JS, et al. Functional anatomy of nonvisual feedback loops during reaching: a positron emission tomography study. *J Neurosci* 2001;21:2919–2928.
92. Grea H, Pisella L, Rossetti Y, et al. A lesion of the posterior parietal cortex disrupts on-line adjustments during aiming movements. *Neuropsychologia* 2002;40: 2471–2480.
93. Battaglia-Mayer A, Ferraina S, Marconi B, et al. Early motor influences on visuomotor transformations for reaching: a positive image of optic ataxia. *Exp Brain Res* 1998;123:172–189.
94. Milner AD, Paulignan Y, Dijkerman HC, et al. A paradoxical improvement of misreaching in optic ataxia: new evidence for two separate neural systems for visual localization. *Proc R Soc Lond B Biol Sci* 1999; 266:2225–2229.
95. Milner AD, Dijkerman HC, Pisella L, et al. Grasping the past. delay can improve visuomotor performance. *Curr Biol* 2001;11:1896–1901.
96. Goodale MA, Milner AD, Carey DP. A neurological dissociation between perceiving objects and grasping them. *Nature* 1991;349:154–156.
97. Goodale MA, Jakobson LS, Keillor JM. Differences in the visual control of pantomimed and natural grasping movements. *Neuropsychologia* 1994;32:1159–1178.
98. Andersen RA. Encoding of intention and spatial location in the posterior parietal cortex. *Cereb Cortex* 1995;5:457–469.
99. Paré M, Wurtz RH. Progression in neuronal processing for saccadic eye movements from parietal cortex area lip to superior colliculus. *J Neurophysiol* 2001;85: 2545–2562.
100. Kawato M, Gomi H. The cerebellum and VOR/OKR learning models. *Trends Neurosci* 1992;15:445–453.
101. Kusunoki M, Gottlieb J, Goldberg ME. The lateral intraparietal area as a salience map: the representation of abrupt onset, stimulus motion, and task relevance. *Vision Res* 2000;40:1459–1468.
102. Colby CL, Duhamel J-R, Goldberg ME. Visual, presaccadic, and cognitive activation of single neurons in monkey lateral intraparietal area. *J Neurophysiol* 1996;76:2841–2852.
103. Powell KD, Goldberg ME. Response of neurons in the lateral intraparietal area to a distractor flashed during the delay period of a memory-guided saccade. *J Neurophysiol* 2000;84:301–310.
104. Mazzoni P, Bracewell RM, Barash S, et al. Spatially tuned auditory responses in area LIP of macaques performing delayed memory saccades to acoustic targets. *J Neurophysiol* 1996;75:1233–1241.
105. Gottlieb J. Parietal mechanisms of target representation. *Curr Opin Neurobiol* 2002;12:134–140.
106. Platt ML, Glimcher PW. Responses of intraparietal neurons to saccadic targets and visual distractors. *J Neurophysiol* 1997;78:1574–1589.
107. Andersen RA, Essick GK, Siegel RM. Neurons of area 7 activated by both visual stimuli and oculomotor behavior. *Exp Brain Res* 1987;67:316–322.
108. Kalaska JF, Sergio LE, Cisek P. Cortical control of whole-arm motor tasks. In: Glickstein M, ed. *Sensory guidance of movement.* Chichester, UK: Wiley, 1998: 176–201.
109. Glimcher PW. Making choices: the neurophysiology of visual-saccadic decision making. *Trends Neurosci* 2001;24:654–659.
110. Glimcher PW, Sparks DL. Movement selection in advance of action in the superior colliculus. *Nature* 1992; 355:542–545.
111. Gold JI, Shadlen MN. Representation of a perceptual decision in developing oculomotor commands. *Nature* 2000;404:390–394.
112. Gold JI, Shadlen MN. Neural computations that underlie decisions about sensory stimuli. *Trends Cog Sci* 2001;5:10–16.
113. Kim J-N, Shadlen MN. Neural correlates of a decision in the dorsolateral prefrontal cortex of the macaque. *Nat Neurosci* 1999;2:176–185.
114. Platt ML. Neural correlates of decisions. *Curr Opin Neurobiol* 2002;12:141–148.
115. Platt ML, Glimcher PW. Neural correlates of decision variables in parietal cortex. *Nature* 1999;400:233–238.
116. Schall JD, Bichot NP. Neural correlates of visual and motor decision processes. *Curr Opin Neurobiol* 1998; 8:211–217.
117. Schall JD, Thompson KG. Neural selection and con-

trol of visually guided eye movements. *Annu Rev Neurosci* 1999;22:241–259.
118. Shadlen MN, Newsome WT. Neural basis of a perceptual decision in the parietal cortex (area lip) of the rhesus monkey. *J Neurophysiol* 2001;86:1916–1936.
119. Basso MA, Wurtz RH. Modulation of neuronal activity in superior colliculus by changes in target probability. *J Neurosci* 1998;18:7519–7534.
120. McPeek RM, Keller EL. Superior colliculus activity related to concurrent processing of saccade goals in a visual search task. *J Neurophysiol* 2002;87: 1805–1815.
121. McPeek RM, Skavenski AA, Nakayama K. Concurrent processing of saccades in visual search. *Vision Res* 2000;40:2499–2516.
122. Munoz DP, Wurtz RH. Saccade-related activity in monkey superior colliculus. I. Characteristics of burst and buildup cells. *J Neurophysiol* 1995;73:2313–2333.
123. Kustov AA, Robinson DL. Shared neural control of attentional shifts and eye movements. *Nature* 1996;384: 74–77.
124. Rizzolatti G, Riggio L, Dascola I, et al. Reorienting attention across the horizontal and vertical meridians: evidence in favor of a premotor theory of attention. *Neuropsychologia* 1987;25:31–40.
125. Corbetta M, Akbudak E, Conturo TE, et al. A common network of functional areas for attention and eye movements. *Neuron* 1998;21:761–773.
126. Allport DA. Selection for action: some behavioral and neurophysiologic considerations of attention and action. In: Heuer H, Sanders AF, eds. *Perspectives on perception and action.* Hillsdale, NJ: Erlbaum, 1987:395–419.
127. Castiello U. Mechanisms of selection for the control of hand action. *Trends Cogn Sci* 1999;3:264–271.
128. Neumann O. Visual attention and action. In: Neumann O, Prinz W, eds. *Relationships between perception and action: current approaches.* Berlin: Springer-Verlag, 1990:227–267.
129. Tipper SP, Howard LA, Houghton G. Action-based mechanisms of attention. *Phil Trans R Soc Lond B* 1998;353:1385–1393.
130. Tipper SP, Lortie C, Baylis GC. Selective reaching: evidence for action-centered attention. *J Exp Psychol Hum Percep Perform* 1992;18:891–905.
131. Rushworth MF, Paus T, Sipila PK. Attention systems and the organization of the human parietal cortex. *J Neurosci* 2001;21:5262–5271.
132. Rushworth MF, Nixon PD, Renowden S, et al. The left parietal cortex and motor attention. *Neuropsychologia* 1997;35:1261–1273.
133. Crammond DJ, Kalaska JF. Modulation of preparatory neuronal activity in dorsal premotor cortex owing to stimulus-response compatibility. *J Neurophysiol* 1994; 71:1281–1284.
134. Cisek P, Kalaska JF. Simultaneous encoding of multiple potential reach directions in dorsal premotor cortex. *J Neurophysiol* 2002b;87:1149–1154.
135. Cisek P, Kalaska JF, unpublished data.
136. Cisek P. Embodiment is all in the head. *Behav Brain Sci* 2001;24:36–38.
137. Kalaska JF, Hyde ML. Area 4 and area 5: differences between the load direction- dependent discharge variability of cells during active postural fixation. *Exp Brain Res* 1985;59:197–202.

8

The Role of the Parietal Cortex in Grasping

Hideo Sakata

Departments of Anatomy and Physiology, Seitoku Junior College of Nutrition, Katsushika-ku, Tokyo, Japan

INTRODUCTION

The primate hand is well adapted for grasping. The difference is clear if one compares the hand of the macaque monkey with the forepaw of the cat. In monkeys, palmar side of the hand is covered with glabrous skin, whereas in cats only fingers and palmar pads are covered with glabrous skin (Fig. 8-1) (1). Feline pads are mainly for stepping, and the claws are more important for grasping, albeit they are somewhat destructive. In monkeys, the nails are flat and they are used for digging and scratching rather than grasping. Moreover, thanks to its flexibility, the monkey's hand can be shaped according to the shape of the object being grasped. The prototype of the prehensile hand emerged in the tarsier, together with the large frontal eyes for full-fledged stereopsis (2). Emergence of these two major functional features of the primates coincided with the development of the parietal association cortex. When Hill (1955) (3) compared the cortical areas of tarsier with those of tupias, the closest infraprimate mammals, he found an intermediate area between visual and somatosensory areas in tarsiers and designated it as parietal association cortex. Evolution of the hand culminated in the hominid, with the remarkable development of opposable thumbs and the advanced precision grip, key elements for tool using

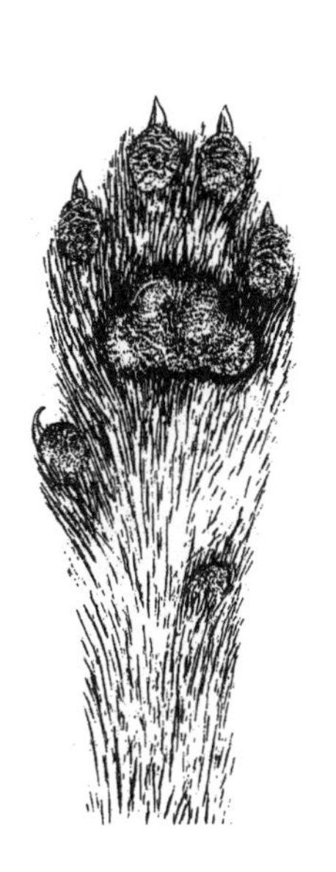

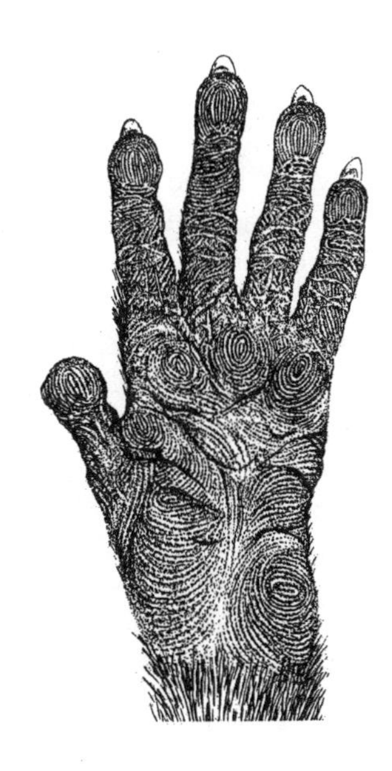

FIG. 8-1. Comparison of the monkey's hand and cat's forepaw. Note glabrous ridged skin covers the palmar side of the hand of the monkey, but in cat, hairy skin is mixed with the glabrous skin of finger and palmar pads. (From Ref. 1, with permission.)

and tool making (4). Again, the evolution of the hand coincided with the further development of the parietal association cortex in early hominids.

Balint (1909) first described the relation of the parietal cortex to visually guided hand movement (5). The patient with bilateral parietal lesions had difficulty grasping an object he could see. This symptom was designated optic ataxia, and was found together with the deficit in shifting gaze to a visual target (so-called psychic paralysis of eye movement) (6) and visual inattention. Balint's syndrome is the complex of these three symptoms. At about the same time, Liepman (1905) (7) described a disturbance of hand manipulation as the loss of limb kinetic representation *(Der Verlust der gliedkinetischen Vorstellung)*. The patient showed marked abnormality in the action of his right hand when grasping a pencil or coin, buttoning his shirt, and putting on socks. The lesion was centered in the left postcentral and supramarginal gyri (area 40 of Brodmann). Liepman (1920) (8) designated a similar symptom as limb kinetic apraxia. This is a deficit of tactile guidance of hand movement in this author's view.

However, little is known about the neural mechanisms of tactile and visual guidance of hand movement, although the parietal cortex is known to be concerned with space vision, in contrast to the inferotemporal cortex, which is concerned with form or object vision (9). In this chapter, I review our single unit analysis of the parietal cortical neurons in alert behaving monkeys, together with studies of Iwamura's group in the somatosensory cortex to illustrate the neural processing of visual and tactile signals in the parietal cortex underlying visual and tactile guidance of hand movement.

HIERARCHICAL PROCESSING OF SOMATOSENSORY INFORMATION IN THE FIRST SOMATOSENSORY CORTEX

For the purpose of tactile guidance of hand movement, it is necessary to represent functionally unified surface extending beyond the anatomic borders between individual phalanges, fingers, or palm and fingers. Convergence of multiple joints is necessary to represent finger posture as a whole. Also, the convergence of cutaneous and deep submodality is necessary in order to represent the pattern of contact of the object to be grasped. Hierarchical processing is inevitable in order to attain the integration of receptive fields and submodalities as well as to extract features such as the direction of movement and the orientation of edge (10).

The presence of hierarchical processing in the cortical somatosensory areas was first suggested by Duffy and Burchfiel (1971) (11) and Sakata et al. (1973) (12) on the basis of a single unit recording in area 5 of awake monkeys. Sakata et al. (1973) reported that many neurons in area 5 responded maximally to specific combinations of multiple joints or to the combination of joint and skin stimulation. These neurons were likely to represent specific limb postures or animal poses. The first somatosensory cortex (SI) of the monkey is composed of areas 3a, 3b, 1, and 2. These areas are serially connected with corticocortical fibers (13), except area 3b, which receives cutaneous afferents, and area 3a, which receives proprioceptive afferents of joints and muscles are in parallel (14). Later, Hyvaerinen and Poranen (1978) (15) showed that the increase in the size and complexity of cutaneous receptive fields (RFs) does in fact start in area 1; the complexity of the functional properties of the neurons increases toward the posterior part of SI; and the convergence of the cutaneous and deep submodalities occurs mainly in area 2, the most posterior part of SI.

Iwamura et al. (1983) (16) made a more extensive study of the receptive field properties of the neurons of areas 1 and 2 of the hand region of SI and found that many neurons have multifinger-type RFs in these areas. Fig. 8-2 illustrates the RFs of neurons recorded in a penetration made along the posterior bank of

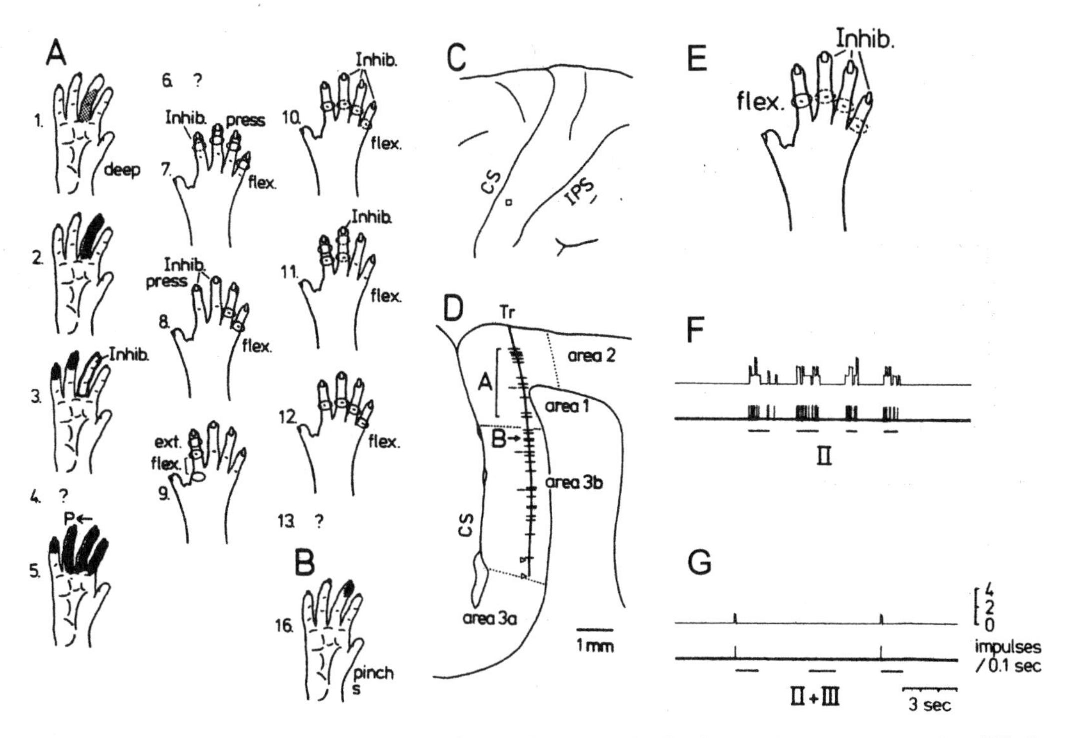

FIG. 8-2. Cutaneous and deep receptive fields of neurons in the first somatosensory cortex (SI). **A:** Neuron 1 responded to the deep stimulation in the middle finger. Neuron 2,3,5 had cutaneous receptive fields shown in black with inhibitory field shown with outer contour, P with an arrow indicates preferred direction. Seven- to 12-joint neurons, receptive fields shown with circles. Inhib press, inhibition by pressure on finger pad; ?, unclassified neurons. **B:** Cutaneous receptive field of an area 3b neuron localized in the distal phalanx of the middle finger. **C,D:** The site of penetration and the sites of unit. **E,F,G:** Receptive field and response of neuron 10. **F:** Excitation by flexion of index finger *(II)*. **G:** Response was suppressed by combining flexion of the middle finger *(II + III)*. Upper trace is the impulse histogram. Lower trace is impulse displayed with small spikes. (From Ref. 16, with permission.)

the central sulcus traversing areas 1 and 3b. Many neurons responded to the flexion of the interphalangeal (IP) joints of fingers 2, 3, 4, and 5. For the neuron shown on the right, flexion of index finger IP joint was excitatory (E), whereas the flexion of the third, fourth, and fifth fingers was inhibitory. Thus, the best stimulus for the neuron was likely to specify a specific finger posture (i.e., flexion of index finger alone while the rest of the fingers were extended). It should be noted that most of the neurons recorded in area 3b had small localized RF in the distal phalanx of the third finger. The combination of excitatory and inhibitory RF also may be useful to distinguish different types of grip, as illustrated in Fig. 8-3. This neuron was recorded in area 2 and had a small excitatory RF on the ulnar side of the thumb and a large inhibitory RF on the palmar side of the second to fifth fingers, extending to palm whirls. The cell was briskly activated when the monkey picked up a small pellet of food with the side opposition of thumb and index finger, whereas its background activity was suppressed when the monkey grasped a piece of orange with four fingers.

Directionally selective neurons for the movement of stimulus over the cutaneous RF were reported first by Whitsel et al. (1972)

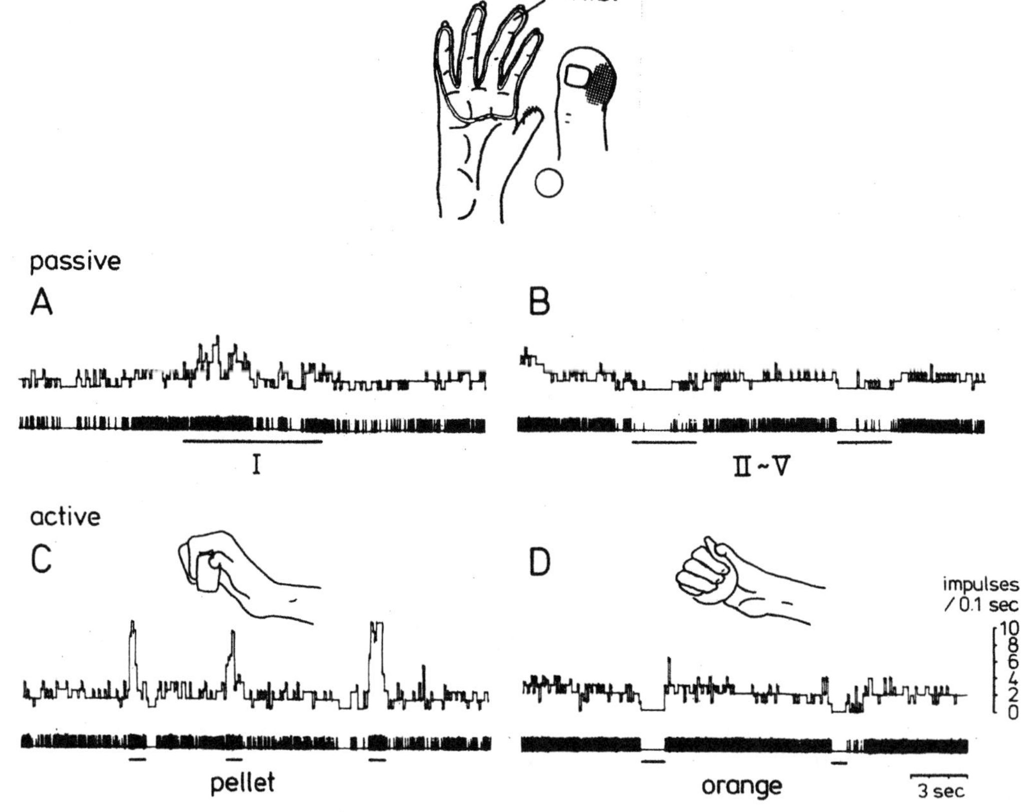

FIG. 8-3. Combination of excitatory and inhibitory receptive field of an area 1 neuron *(upper insert figure)* and its relation to the pattern of grip. **A:** Passive response to the cutaneous stimulus in the receptive field of the thumb *(I)*. **B:** Suppression of the spontaneous activity by the stimulus in the inhibitory receptive field extending from second to fifth fingers *(II–V)*. **C:** Activation during side grip of food pellet between the thumb and index finger. **D:** Suppression during whole hand grip of an orange. (From Ref. 22, with permission.)

(17) and studied in detail by Costanzo and Gardner (1980) (18). They were quite common among neurons in area 1. Some of them had a large receptive field and were maximally activated when the tactile object was grasped in a specific way. The dependence of directionally selective neurons on the type of grip is illustrated in Fig. 8-4. The neuron shown in Fig. 8-4A,B was activated vigorously when the monkey's hand, grasping the edge of a table, was moved in the ulnar direction, whereas the stationary contact on various objects suppressed the background activity of the cell (16). The neurons shown in Fig. 8-4C were recorded in area 2 (Fig. 8-4D) (19). Three of five neurons recorded in a nearly vertical penetration were directionally selective and activated maximally when the hand grasped different types of objects, such as a sphere, cylinder, and barrel. The responses of such directionally selective neurons may be useful as a feed-back signal to prevent slipping of the object grasped in the hand.

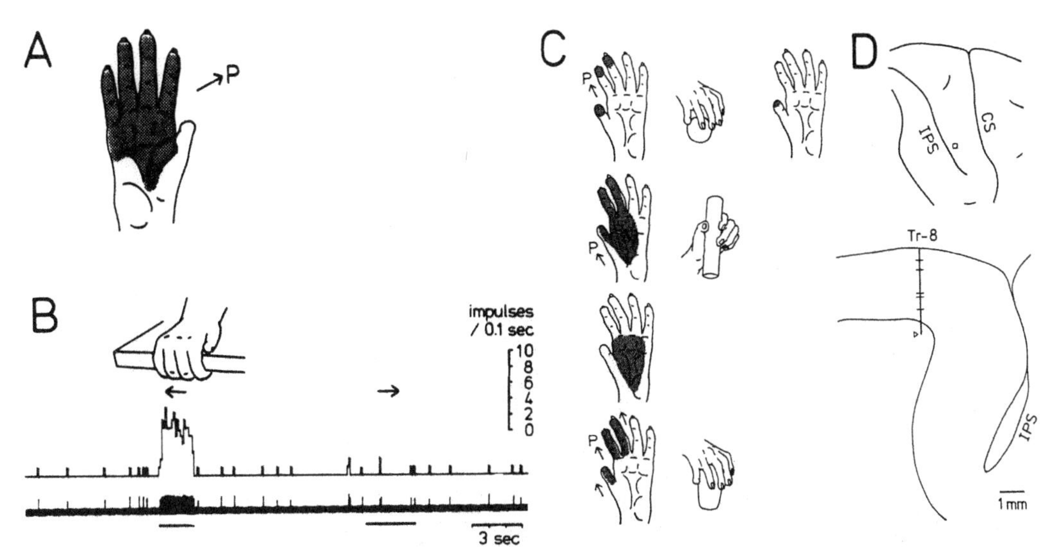

FIG. 8-4. A: Receptive field of an area 1 neuron covering second to fifth fingers, palm whirls, and the tip of thenar eminence. **B:** Activation of the neuron when the monkey's hand, grasping the plate of the chair, was moved to the right side. The cell was suppressed during the movement in opposite direction (to the left). **C:** Receptive fields of area 2 neurons activated by slipping movements of objects of different shapes; sphere, cylinder, and barrel. **D:** Recording sites of neurons shown in Fig. 8-4C. (From Refs. 16 and 19, with permission.)

FUNCTIONAL BLOCKING WITH MUSCIMOL MICROINJECTION

In 1985, Hikosaka et al. (20) reported interesting results concerning the functional blocking of the hand region of SI by a local injection of muscimol (a GABA agonist). They made a single unit recording in area 2 before muscimol injection. In an illustrative penetration shown in Fig. 8-5A,B, neuron 4 was activated by flexion of the distal and proximal IP joints of the third, fourth, and fifth fingers, and was activated maximally while the monkey attempted to pick up a small piece of apple from a small hole on a wooden block (Fig. 8-5C). A local injection of muscimol to the recording site of this neuron disrupted the finger grasping. The third to fifth fingers of the right hand (contralateral to the injection) lost the stable flexed configuration and repeated clumsy flexion and extension. As a result, it took much longer for the monkey to pick up the apple piece. Thus, the disruption of monkey's performance had a good correlation with the characteristics of neurons close to the injection site. The lack of finger coordination was clearly revealed when the monkey was required to pick up a piece of apple from a funnel. While the monkey reached out her left hand, its shape was already adapted to that of the funnel. The right (contralateral) hand was severely disorganized after the injection. As soon as the fingers touched the inner wall of the funnel, the fourth and fifth fingers were extended and protruded outside the funnel.

These disorders of hand movement in monkeys remind us of the hemianesthesia of a patient with a parietal lobe lesion (including the postcentral gyrus and the superior parietal lobule) reported by Jeannerod and Michel (1984) (21). This patient showed a complete loss of shaping of the hand on the contralesional side before reaching the target object

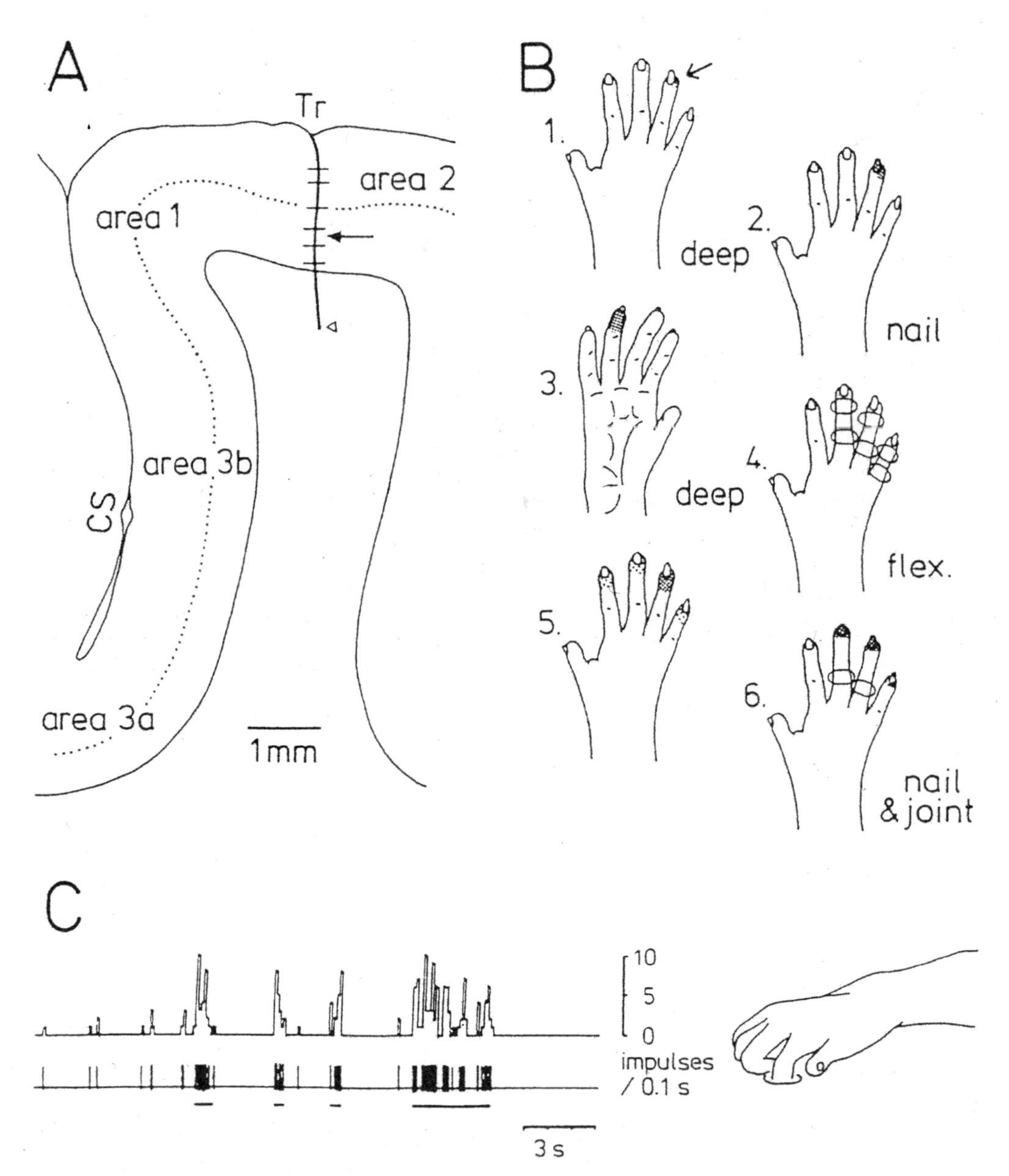

FIG. 8-5. A: A cluster of neurons activated during active grasping of objects; appropriate shapes of grasping are shown with small pictures. Neurons marked with a black star are those activated during active touch. (From Ref. 20, with permission.)

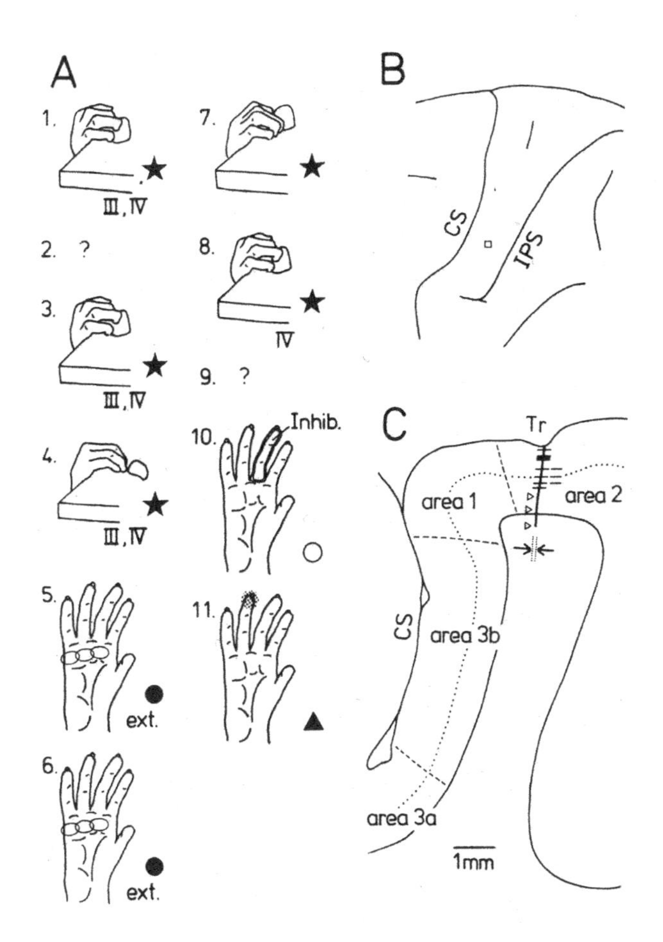

FIG. 8-6. A: Injection site of muscimol *(arrow)*. **B:** Receptive fields of neurons recorded along the penetration of the needle before injection. **C:** Response of neuron 4, strongly activated when the monkey picked up a piece of apple from a small hole with index finger, while the other fingers (third to fifth) were flexed. (From Ref. 22, with permission.)

without visual feed-back. It is most likely that the monkey lost the tactile perception of the object and the grasping hand.

NEURONAL ARRAYS SIGNALING ACTIVE TOUCH

Iwamura et al. (1985) (22) found neuronal arrays signaling active contact of fingers to the object in the hand region of the postcentral gyrus. Fig. 8-6 illustrates that many neurons did not have receptive fields as plotted by cutaneous stimulation or passive movement of joints, but their activity was more directly related to the active contact of fingers to the object. Although response to passive stimulation was found in some neurons, all but one of these neurons were activated more vigorously in relation to the monkey's active finger movements. Most of the neurons along this penetration (6/11) were activated when the animal conducted a particular hand action in order to contact and grasp a piece of food with its ulnar fingers. Hyvaerinen and Poranen (1978) (15) described a similar type of neurons in the posterior part of the postcentral gyrus (area 2). They were only moderately activated by passive stimulation of the skin but strongly activated when the monkey herself actively pressed the receptive field against an obstacle. Iwamura et al. (1985) (22) observed 22 such cells out of total sample of 632, and found that the neurons with large and complex receptive fields were activated most effectively when the animal's active hand movements brought the receptive field into contact with an object. Iwamura and Tanaka (1996) (23) also recorded SI neurons that were activated only during active grasping and were selective for the pattern or shape of grasping. The source of activation caused by active touch is most likely an efference copy of the command signals of the manipulative movements of the hand, as suggested by Soso and Felz (1980) (24) in a study of postcentral neurons comparing their response during active and passive joint movement. Therefore, it may be useful to monitor finger movements and adjust finger posture to the shape, size, and orientation of the object, although it is also likely that some area 2 neurons related to active touch may contribute to the form discrimination of tactile objects (25).

DEFECT OF VISUALLY GUIDED GRASPING CAUSED BY PARIETAL LESION

In early neuropsychologic studies, the major symptom of optic ataxia owing to the parietooccipital lesion was described as misreaching (6,26). Patients groped for an object with overextended fingers, and the abnormal finger posture was attributed to a strategy for compensating for the reaching deficit, thus increasing the probability of hand contact with the object. This idea was supported by lesion studies in monkeys that showed deficits in visual reaching after bilateral ablation of the posterior parietal cortex or the inferior parietal lobule (IPL) (27,28). Jeannerod (1986) (29) found a remarkable disturbance of the formation of finger grip before reaching the target objects, which he called "preshaping," in patients with parietooccipital lesions. Perenin and Vighetto (1988) (30) found a disturbance in adjustment of hand orientation to match that of the target in patients with optic ataxia. More recently, Jeannerod et al. (1994) (31) found a patient with bilateral deficit in grasping without having a deficit in reaching. The patient had both abnormal preshaping with overextended fingers and an inability to adjust hand orientation. The dissociation of the deficits in reaching and grasping suggested that separate regions may exist for the control of reaching and grasping.

The disturbance of preshaping (32) or improper orientation of the contralateral hand or fingers (33) was demonstrated in monkeys by the lesion of the inferior parietal lobule. We induced deficits of preshaping by means of a functional block of the anterior intraparietal (AIP) area on the posterolateral bank of the rostral intraparietal sulcus (34). For example, the preshaping of the hand for grasping a small plate buried in a groove with tip opposition of the index finger and the thumb was lost after muscimol injection, and the animal extended all of the fingers and failed to insert the index finger smoothly into the groove.

HAND MANIPULATION–RELATED NEURONS IN THE ANTERIOR INTRAPARIETAL AREA

When the cortical neurons related to grasping or manipulation of objects were first recorded in the IPL in the alert behaving macaque by Mountcastle et al. (1975) (35) and Hyvaerinen and Poranen (1974) (36), they distinguished them from the neurons related to reaching or arm projection, and designated them hand manipulation neurons. Further studies of reaching neurons in the parietal cortex revealed that the superior parietal lobule (SPL, area 5) plays an important role in reaching (37), although reach-related neurons were also recorded in area 7a of the IPL (38). The parietal reach region (PRR) was found to be localized in both the medial and lateral bank of the middle IPS (39). The neurons that involved in hand movement were found to be concentrated in a small zone within the rostral part of the posterolateral bank of IPS, designated as the anterior intraparietal (AIP) area (40,41). This area is strongly interconnected with area F5 of the ventral premotor cortex, in which Rizzolatti et al. (1988) (42) recorded "grasping-with-the-hand" neurons. We recorded the activity of neurons in this area in monkeys that had been trained to manipulate various types of switches: pushbutton, pull lever, pull knob, and pull knob in a groove (40,43). Many of these hand manipulation task–related neurons (we now call them "hand manipulation–related" neurons) were highly selective and preferentially activated during the manipulation of one of four routinely used objects.

Separation of Visual and Motor Components

In order to determine the contribution of visual signals to the activation of these neurons, we let the monkeys perform the same task in the dark, guided only by a small spot of light on the object (43). We then classified the hand manipulation–related neurons into three groups according to the difference between the levels of activity during manipulation of objects in the light and in the dark. "Motor-dominant" neurons (Fig. 8-7A) did not show any significant difference in the level of activity between the two conditions; "visual-and-motor" neurons (Fig. 8.7B) were less active during manipulation in the dark than during that in the light; and "visual-dominant" neurons (Fig. 8.7C) were not activated during manipulation in the dark. Many of the latter two visually responsive neurons were activated by the sight of objects during fixation without grasping (object type, Fig. 8-

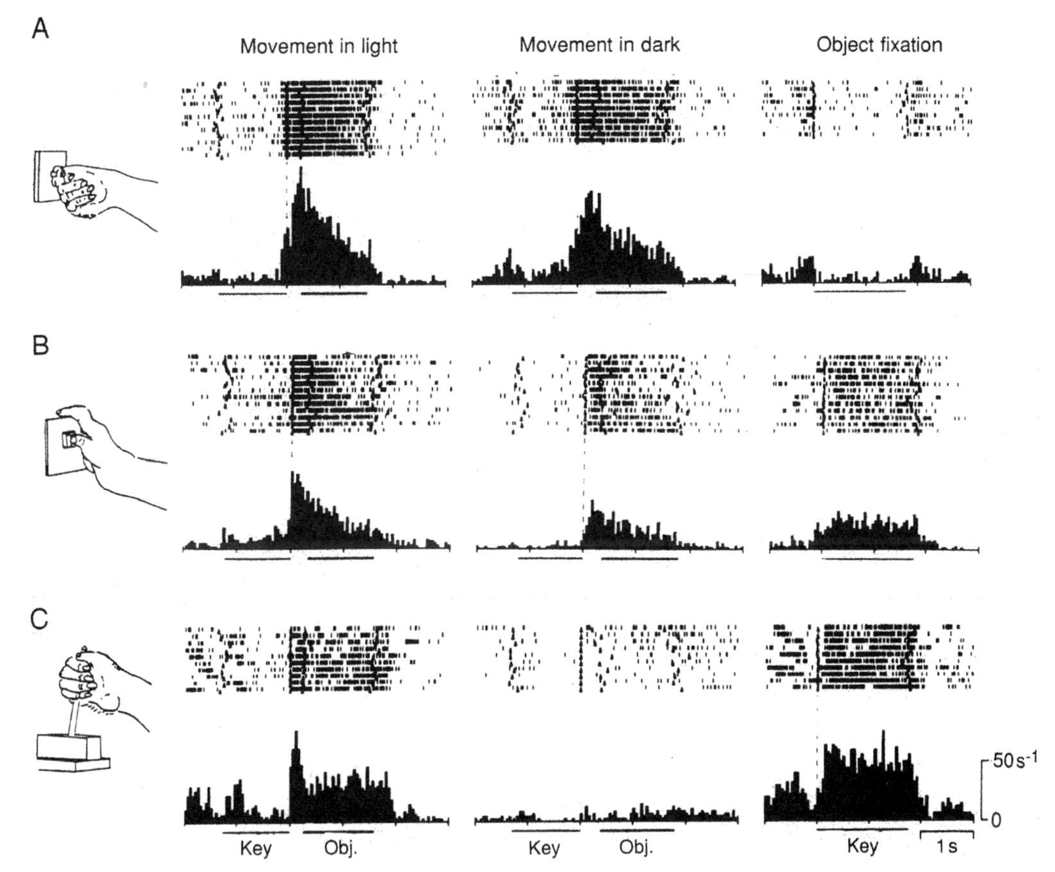

FIG. 8-7. Three types of hand manipulation–related neurons. **A:** Motor-dominant neurons that were activated during the grasping of a pull knob; equally well in the lighted room and in the dark room. **B:** A visual-and-motor neuron that was activated during the pushing of a button. The response was maximum in the lighted room, but reduced in the dark room. This neuron was activated also during fixation of object without grasping. **C:** Visual dominant neurons that were activated in the lighted room during the grasping of a pull lever. The cell was not activated in the dark room. It was fully activated during fixation of object without grasping. It was also selective for orientation of the lever, preferring upward orientation. (From Ref. 56, with permission.)

7B,C). The other visually responsive neurons were not activated during object fixation (non–object type), but seemed to require other visual stimuli for activation, such as the sight of the moving hand.

Almost all of the highly selective visual-and-motor neurons preferred the same object for manipulation and fixation, showing precise correspondence between the patterns of hand action and the object. This suggests that visual-and-motor neurons in AIP play an important role in matching the pattern of hand movement to the spatial characteristics of the manipulated object (40). These results correspond closely to the clinical evidence that parietal lobe lesions cause disturbances in "preshaping" the hand according to the shape and orientation of the object to be grasped (29,30).

Selectivity for the Three-Dimensional Shape of Objects

In more recent experiments on hand manipulation–related neurons we used six different objects of simple geometric shape: sphere, cone, cylinder, cube, ring, and square plate (41). These objects connected to microswitches that were set in six sectors of a turntable and presented to the monkey in random order one at a time. The monkey was required to grasp and pull the object to turn on the microswitch or fixate the spot reflected on a half-mirror and superimposed on the object (Fig. 8-8A). More than one fourth of the manipulation-related neurons were highly selective for one particular object. The activity profile of one highly selective object-type visual-dominant neuron during object fixation is shown in Fig. 8-8B. The cell responded preferentially to the view of the ring among the six objects. We found highly selective visual-dominant neurons for every one of the six objects. The square plate and circular ring were the most commonly preferred objects, and many of the cells that preferred these two objects showed selectivity to the orientation of the plane. We also found moderately selective neurons that responded to two or more objects equally strongly. Some of these moderately selective neurons showed a preference for a certain category of geometric shapes, such as round (sphere, cone, and cylinder), angular (cube and square plate), or flat objects (plate and ring). These results suggest that the visually responsive neurons in AIP represent spatial characteristics of objects for manipulation, and that at least some of these neurons represent the three-dimensional (3D) shapes of the objects, categorized into a limited number of simple geometric shapes, as suggested by Biederman (1987) (44) in his theory of recognition by components. Where these neurons receive signals of 3D shape from was not clear at the beginning. The neurons of the inferotemporal cortex that play a major role in object vision did not show much difference in intensity between responses to actual 3D objects and responses to two-dimensional images of the same objects (45). Therefore, the inferotemporal cortex cannot provide adequate signals of 3D shape for object manipulation. It is more likely that 3D shape may be represented within the parietal cortex, because we found a higher order area of stereopsis in the caudal part of the parietal cortex, as described in the following.

SELECTIVITY TO THREE-DIMENSIONAL ORIENTATION OF VISUAL STIMULUS IN THE NEURONS OF THE CAUDAL INTRAPARIETAL AREA

Discrimination of object orientation in a viewer-centered coordinate system is important for hand manipulation of objects. Thus, we found orientation-selective neurons during our study of manipulation-related neurons in the natural course of events. This led us to find a group of visual neurons specifically selective to the axis- and surface-orientation in space.

Axis Orientation–Selective Neurons

We found that most of the cells that preferred the pull lever were selective for the axis orientation of the lever. During a further investigation, we found a group of visual neu-

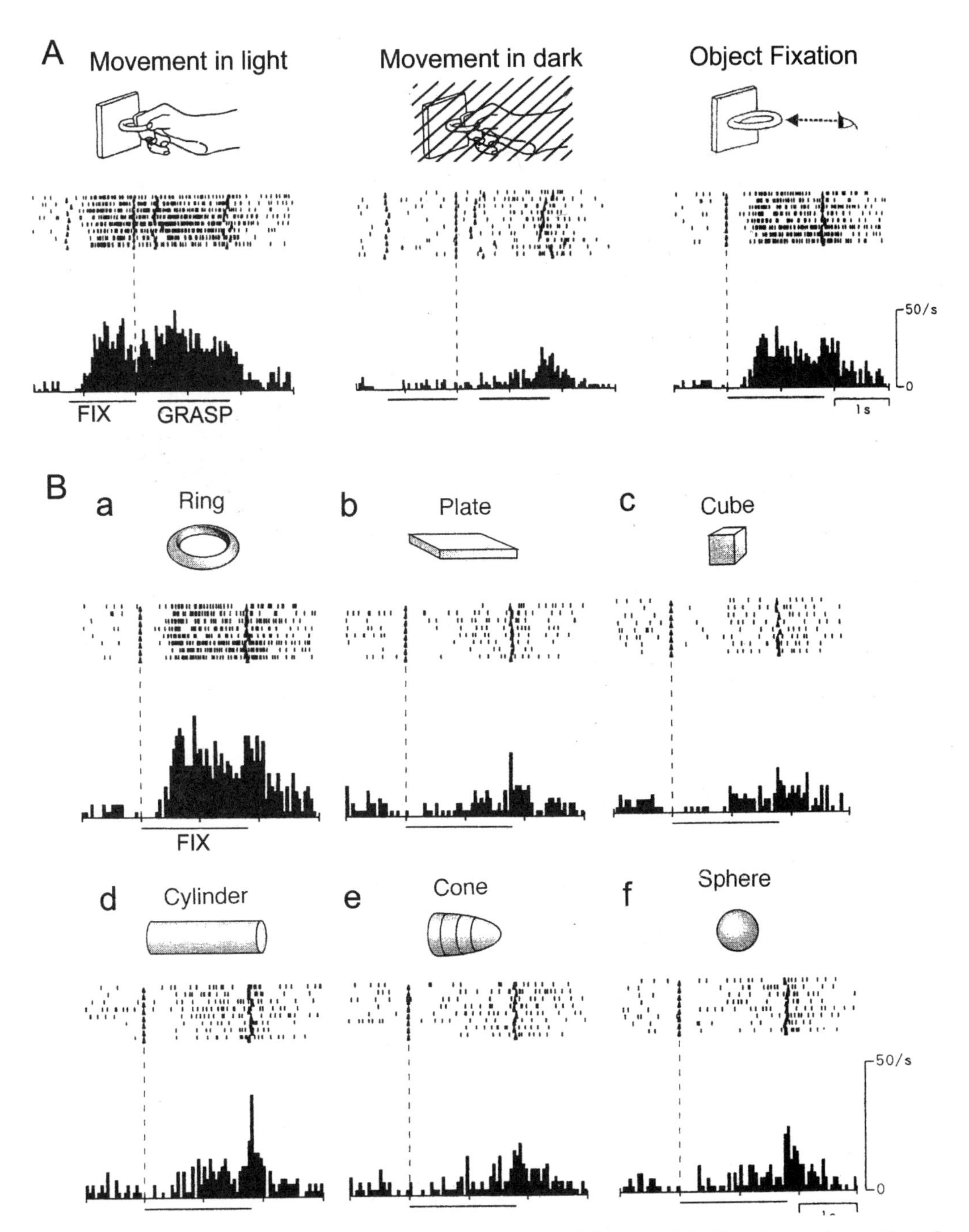

FIG. 8-8. Selectivity of a hand-manipulation neuron (visual-dominant) for the shape of an object. **A:** Response of the neuron during fixation and grasping in the light *(left).* Note that there is no response of the same neuron during fixation and grasping in the dark *(middle).* There is a strong response during object fixation in the light *(right).* **B:** Response of the neuron during fixation on the objects of six different shapes. The cell was highly selective for the ring. (From Ref. 41, with permission.)

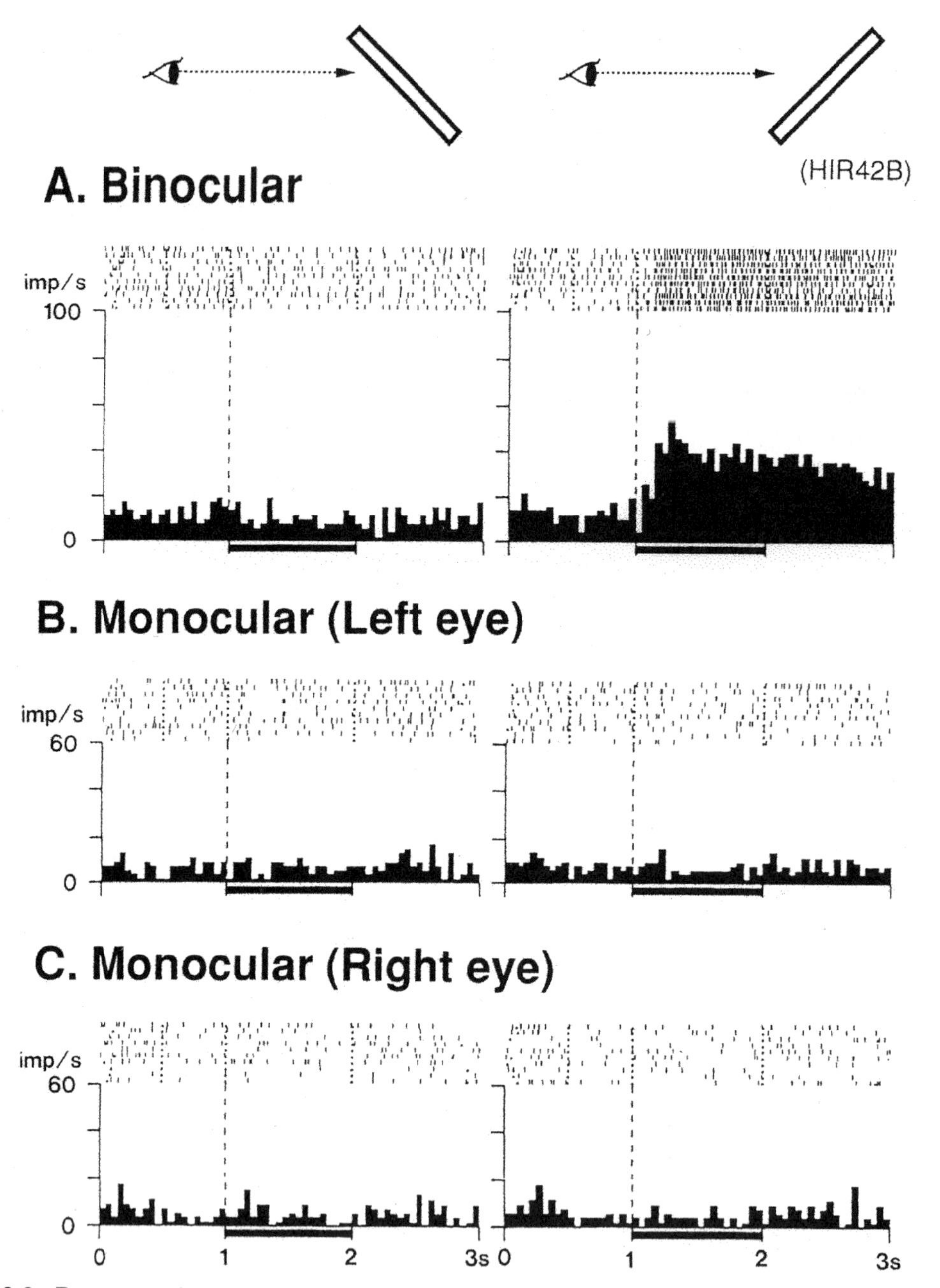

FIG. 8-9. Response of axis orientation selective (AOS) neurons to an inclined luminous bar. **A:** Response of AOS neuron to a luminous bar inclined forward *(left)* and backward *(right)* in binocular viewing condition. **B,C:** Response to the same stimulus in a monocular viewing condition (with left eye and right eye). The bar under each histogram indicates the stimulus period, which is controlled electronically by a light-emitting diode. (From Ref. 46, with permission.)

rons in the lateral bank of the caudal part of the intraparietal sulcus (CIP) that showed selectivity to the orientation of a luminous bar in depth. We designated these neurons as axis orientation selective (AOS) neurons. Fig. 8-9 shows an example of AOS neurons that preferentially responded to a bar tilted 45 degrees backward (46); no response was obtained when the bar was tilted 45 degrees forward (Fig. 8-9A), although the silhouettes of these two stimuli were nearly the same. No response was obtained when the right or left eye was closed, and consequently a monocular viewing condition was created (Fig. 8-9B,C). We recorded AOS neurons that preferred vertical, horizontal, or sagittal bars, or bars tilted or slanted 45 degrees in the frontal, horizontal, or sagittal plane. It was clear that these neurons exhibited orientation selectivity in 3D space in a viewer-centered coordinate system. Additionally, most of these AOS neurons were binocular visual neurons, which responded much less strongly under monocular than binocular viewing conditions. The discharge rate of most of the cells increased monotonically with increasing length of the stimulus; most of the cells preferred a thinner stimulus to a thicker one. However, the AOS neurons typically showed the same orientation preference across a wide range of change in stimulus thickness and length, as well as across wide receptive fields. Therefore, in our more recent experiments, we used 3D computer graphics and stereoscopic display to present stimuli with binocular disparity and in different orientation, sizes, and positions in space (47). These results suggest that AOS neurons represent the orientation of the longitudinal axes of objects in 3D space.

Bender and Jung (1948) (48) reported that in human parietal patients vertical and horizontal axes are tilted considerably to the contralesional side. McFie et al. (1950) (49) described a similar finding in several cases of patients with a right occipitoparietal lesion. Similar deficits in the perception of line orientation owing to parietal lobe lesion were reported more recently by von Cramon and Kerkhoff (1993) (50). Impairment of the judgment of orientation of a rod in space was reported by De Renzi et al. (1971) (51) in patients with posterior region (including parietal and occipital cortex) lesions in the right hemisphere. Therefore, the discrimination of axis orientation in space is one of the prominent functions of the parietal cortex in the domain of space perception.

Surface Orientation–Selective Neurons

According to Marr's theory of vision (52), the main purpose of vision is object-centered representation of the 3D shape and spatial arrangement of an object. The main stepping stone toward this goal is representation of the geometry of the visible surface. Therefore, if 3D shape is to be represented somewhere in the cerebral cortex, then there should be some area in the visual cortical pathways to represent surface orientation and curvature. In our experiments on hand manipulation–related neurons, we found that some visual-dominant neurons that preferred the square plate showed selectivity to the orientation of the plate. This suggests that some parietal visual neurons discriminate surface orientation. Because a disparity gradient is the most important cue for perceiving the orientation of a surface in depth, we used a stereoscopic display of 3D computer graphics and performed an extensive study of binocular neurons in the caudal part of the lateral bank of the IPS that were sensitive to changes in surface orientation (53).

We first compared the responses of the cells to a flat stimulus with those to an elongated stimulus, because we found parietal visual neurons that responded to a square plate or luminous checkerboard in the same region as the axis orientation selective neurons. Most of the cells that preferred the flat to the elongated stimulus presented on a stereoscopic display screen were selective for the orientation of the flat surface and defined as surface orientation–selective (SOS) neurons. Almost all SOS neurons responded more strongly to a binocular than monocular stimulus. The intensity of their responses increased with the width of the stimulus. Most SOS neurons showed no change in response intensity with a change in stimulus shape (e.g., disk versus square plate) or even thickness, suggesting that the coding of surface orientation in space is independent of shape.

In more recent experiments, we used surface (square) of different orientations in random dot stereogram (RDS), as well as solid figure stereogram (SFS) as stimulus, and found that many SOS neurons responded to the square in RDS with surface orientation tuning (54). This suggests that at least a part of SOS neurons represent surface orientation by computing the disparity gradient. Moreover, in the most recent experiments, we found that some SOS neurons showed tuning to the surface orientation either on the basis of disparity cues or the monocular cues of depth, as shown in Fig. 8-10 (55). The neuron shown in Fig. 8-10 responded maximally to a square

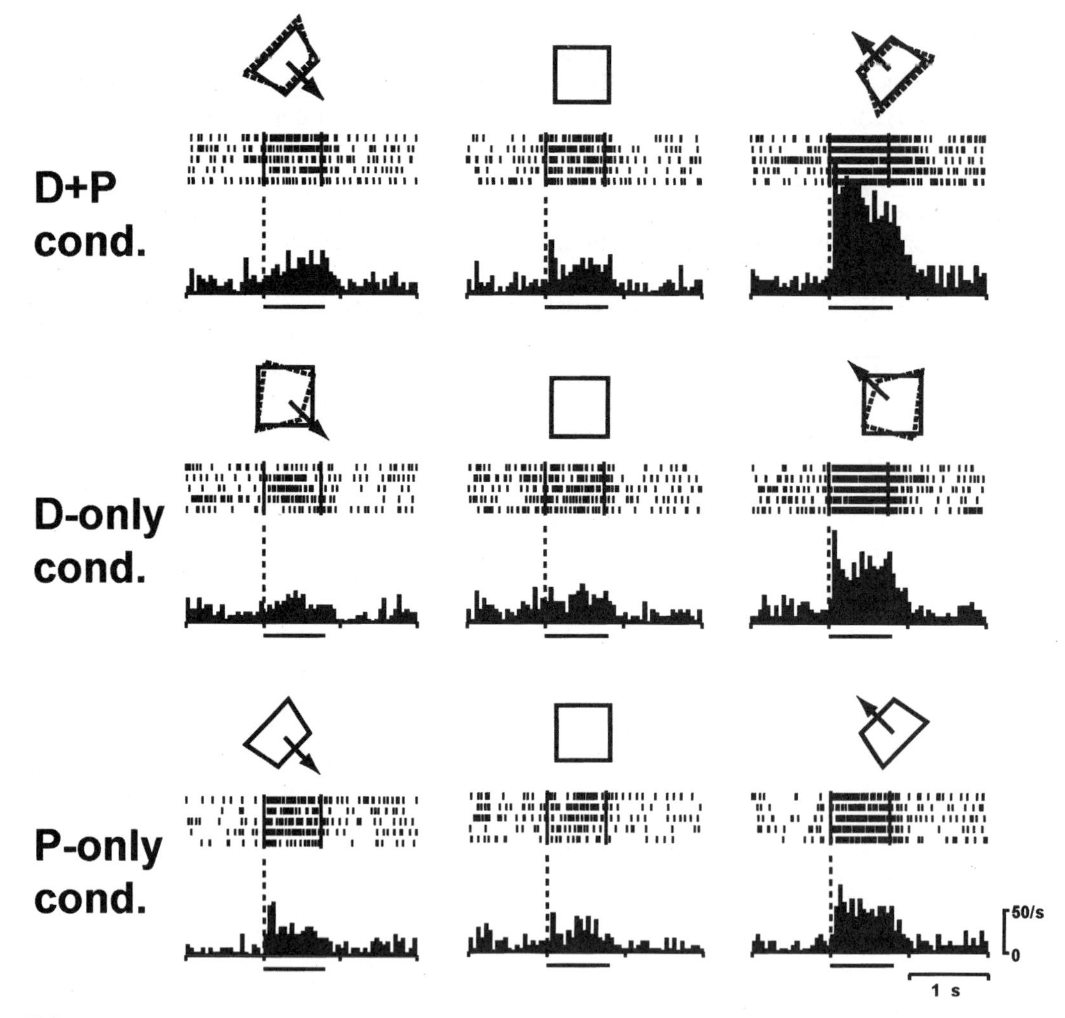

FIG. 8-10. Response of surface orientation selective (SOS) neurons to solid figure stereogram in three different orientations from left to right; inclined forward and slanted rightward, in frontoparallel plane, and inclined backward and slanted leftward. **Upper row:** The stimulus contained both disparity (D) cues and perspective (P) cues. **Middle row:** The stimulus contained only disparity cues, orientation, and width disparity of the contour of the square. **Lower row:** The stimulus contained the monocular cue of linear perspective (P) (trapezoidal contour) alone. Note the maximal response to the stimulus with D and P cues together. (From: Tsutsui et al., 2001, with permission.)

plate in diagonal orientation, inclined backward 45 degrees and slanted 45 degrees toward the right side. The response was very strong when the binocular disparity cues of the edges and the cue of linear perspective (i.e., the trapezoidal silhouette of a slanted square) were combined. The response to the stimulus in the equivalent orientation, either with the perspective cue alone or with the disparity cue alone, was much smaller than that to the stimulus with the combined cues of perspective and disparity together. It looked as if the effects of perspective and disparity cues were summatated to elicit maximal response in the SOS neuron. This is also true in the case of texture gradient, which is a powerful monocular cue of depth. These findings regarding SOS neurons in the CIP area suggest that this area adjacent to area V3A is a higher center of stereopsis for coding of the 3D orientation and probably the 3D shape of objects. Moreover, this area seems to integrate binocular and monocular cues of depth to represent surface geometry in 3D space.

TWO SUBSYSTEMS OF THE DORSAL VISUAL PATHWAY

The current view of the dorsal visual pathway is that area V5/MT and V5A/MST are the main gateways to convey visual information of motion and depth to the parietal association cortex. However, our finding of area CIP as the higher center of stereopsis and area AIP to receive signals of 3D shape and orientation from CIP and adjacent areas of the most posterior part of IPL changed this traditional view of two visual pathways.

Thus, the function of the parietal cortex in the dorsal cortical visual pathway is not limited to the perception of position and movement, but includes the perception of 3D features of objects mediated by stereopsis (56). Binocular disparity signals for stereopsis is mediated mainly by the magnocellular system of the dorsal visual pathway through thick stripes of the V2 (57) and V3–V3A complex in which many disparity-sensitive neurons have been recorded (58), although the contribution of projections from the areas TE and TEO of the ventral visual pathway to the lateral bank of the IPS (59) may be implicated as well.

The dorsal cortical visual pathway projecting to the parietal cortex may be subdivided into two subsystems, as shown in the diagram in Fig. 8-11A of the entire cortical visual pathway, modified from the diagrams constructed by Zeki and Shipp (1988) (60) and Felleman and Van Essen (1991) (61). One subsystem contains a relay at area V5/MT and terminates in the areas V5A /MST and VIP, subserving motion vision in extrapersonal and peripersonal space. The other subsystem contains a relay at the V3–V3A complex and terminates in the areas around the parietooccipital and intraparietal sulci (areas V6, CIP, LIP, and AIP), subserving the perception of spatial position and the 3D features of objects. The areas V6, LIP, and 7a are concerned with the representation of position in egocentric space (62).

ROLE OF THE PARIETAL ASSOCIATION CORTEX IN GRASPING

Murata et al. (1999) (63) studied the functional properties of neurons of the ventral premotor cortex (area F5 of Rizzolatti et al.) (42) using the same apparatus and tasks we used to study AIP neurons. They found visual-and-motor and motor-dominant neurons in area F5 whose responses were similar to that of the AIP neurons. Some of visual-and-motor neurons were object-type; they responded partially to the view of object during fixation and were highly selective for shape. However, they never recorded visual-dominant neurons in area F5 as in area AIP. These results led to a conceptual model of cortical circuit of visual guidance of grasping, as shown in Fig. 8-12. The flow of information within this system is as follows. (a) The visual signals related to the 3D shape and orientation of object is processed in area CIP and fed into visual-dominant neurons of AIP. (b) This information of the target for grasping is conveyed

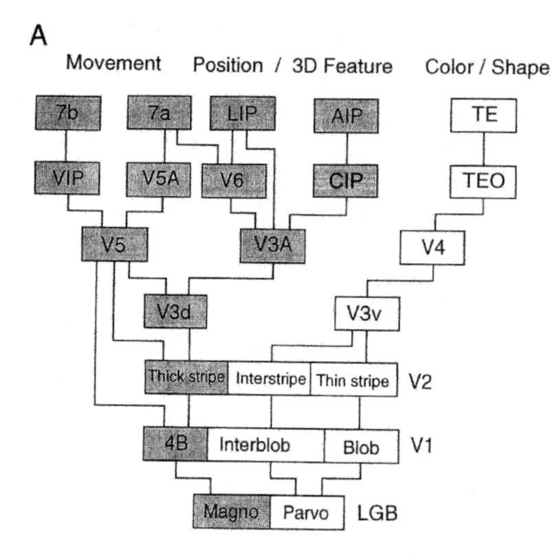

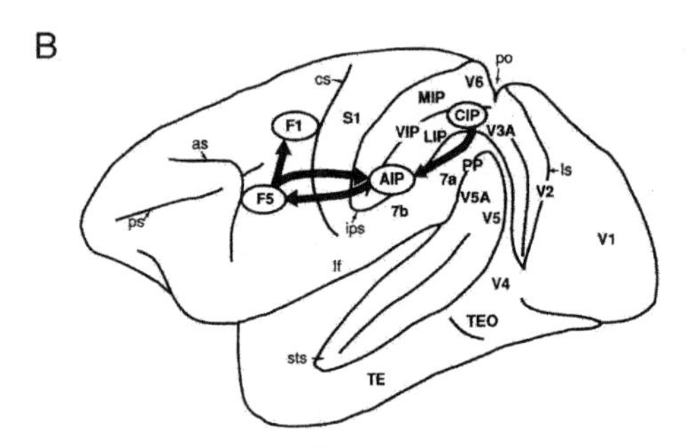

FIG. 8-11. Hierarchical organization of the cortical visual areas of the monkey. **A:** Dorsal stream *(shaded)* start from α ganglion cells of retina, relayed in magnocellular layer of the lateral geniculate body (LGB), layer 4B of the primary visual cortex V1, the thick stripe of V2, V3, and then divided into two subsystems; one is related to motion vision through V5/MT to V5A/MST and VIP, another to spatial position through V3A to V6, and to stereopsis through V3A to CIP. Ventral stream *(open)* starting from β ganglion cells, relayed in the parvocellular layer of LGB, blob, and interblob of V1, thin and interstripe of V2, V3, V4, and project to TEO and TE in the temporal lobe. **B:** A diagram of the monkey cerebral cortex (outer view of the left hemisphere), showing the location of areas included in the diagram A. Thick arrows indicate corticocortical connections related to the visual guidance of grasping. (From Ref. 56, 1997, with permission.)

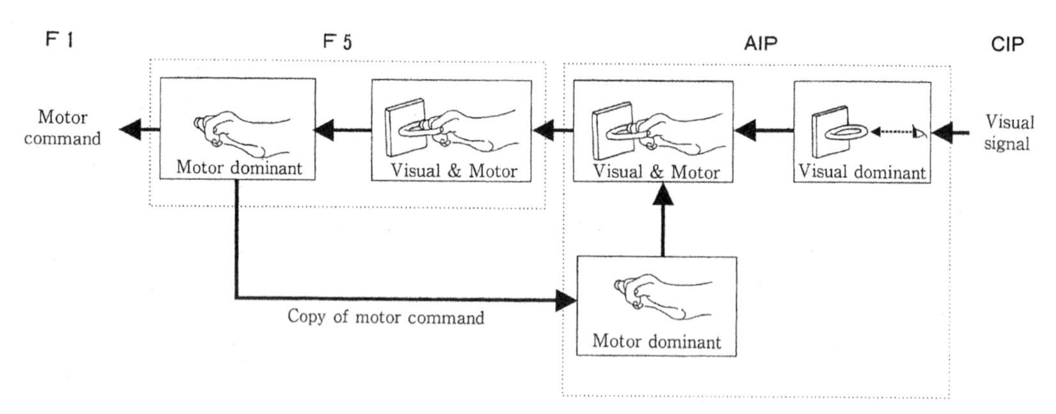

FIG. 8-12. A diagram of the model of the cortical network for visually guided grasping. (From Ref. 63, with permission.)

to the visual-and-motor (object-type) neurons of AIP. (c) The combined visual-and-motor signal is sent through the corticocortical pathway to area F5 to activate visual-and-motor neurons there. (d) The activity of visual-and-motor neurons triggers motor-dominant neurons in F5 to send appropriate command signals to F1, the primary motor cortex. (e) Finally, the collateral of F5 motor-dominant neurons send the efference copy of this command to motor-dominant neurons in AIP, and the latter are connected to visual-and-motor neurons of AIP. This closed circuit may constitute a positive feed-back circuit to maintain a proper motor command signal sent from F5 to F1.

The control of grasping is divided into two stages. The first stage is the preshaping of the hand before it reaches the object. The control of this stage is visually guided by the corticocortical circuit described in the preceding. The second stage is the adjustment of grip after touching the object. The control of the second stage is tactually guided by the corticocortical circuit between the premotor cortex and areas 2 and 5 of the postcentral gyrus. Complex somatosensory responses with the combination of joint and skin receptive fields were recorded in the premotor cortex neurons by Rizzolatti et al. (1981) (64). Hierarchical processing of somatosensory information in SI and area 5 (10) is important for the tactile object recognition necessary for tactile motor guidance. In addition, bimodal visual-somatosensory neurons in area VIP (65) and adjacent area 5 may be important to combine these two stages of visual and tactile motor guidance. In this respect, Iriki et al. (1996) (66) found interesting visual-somatosensory neurons in area 5 that changed their visual receptive fields after training the animal to use a small rake as a tool to retrieve food morsels. Their visual receptive fields extended to the end of the tool after the monkey acquired the skill to use the rake.

Thus, the parietal cortex plays an important role in the control of grasping by providing visual and somatosensory perceptual information about the object to the premotor cortex to trigger the proper motor command and then monitor the active motor signal to ensure matching of the motor signal with the perceptual signal. Further studies are necessary to elucidate the process of fine adjustment of the hand and finger movements by monitoring the perceptual image of the hand and fingers. In conclusion, the role of the parietal cortex in grasping is to bring the hand movements under conscious perceptual control.

REFERENCES

1. Sakata H. Contribution of primate research to sensory physiology. In: Bourne GH, ed. *Nonhuman primates and medical research.* New York: Academic Press, 1973:381–405.
2. Polyak S. *The vertebrate visual system.* Chicago: University of Chicago Press, 1957.
3. Hill WCO. *Primates, comparative anatomy and taxonomy.* Vol. 2. *Tarsioidea.* Edinburgh: Edinburgh University Press, 1955.
4. Susman RL. Fossil evidence for early hominid tool use. *Science* 1994;265:1570–1573.
5. Balint R. Seelenlaehmug des "Schauens," optische Ataxie, räumliche Stoerung der Aufmerksamkeit. *Monatsschr Psychiat Neurol* 1909;25:51–81.
6. Hécaen H, de Ajuriaguerra J. Balint's syndrome (psychic paralysis of visual fixation) and its minor forms. *Brain* 1954;77:373–400.
7. Liepman H. *Der Verlust der gliedkinetischen Vorstellungen, Ueber Stoerungen des Handelns bei Gehirnkranken.* Berlin: Karger, 1905:146.
8. Liepman H. Apraxia. *Ergbn Ges Med* 1920;1:516–543.
9. Ungerleider LG, Mishkin M. Two cortical visual systems. In: Ingle DJ, Goodale MA, Mansfield RJW, eds. *Analysis of visual behavior.* Cambridge, MA: MIT Press, 1982:549–586.
10. Iwamura Y. Hierarchical somatosensory processing. *Curr Opin Neurobiol* 1998;8:522–528.
11. Duffy FH, Burchfiel JL. Somatosensory system: organizational hierarchy from single unit in monkey area 5. *Science* 1971;172:273–275.
12. Sakata H, Takaoka Y, Kawarasaki A, et al. Somatosensory properties of neurons in the superior parietal cortex (area 5) of the rhesus monkey. *Brain Res* 1973;64: 85–106.
13. Jones EG, Coulter JD, Hendry SHC. Intracortical connectivity of architectonic fields in the somatic sensory, motor and parietal cortex of monkeys. *J Comp Neurol* 1978;181:291–348.
14. Jones EG, Wise SP, Coulter JD. Differential thalamic relationships of sensory, motor and parietal cortical fields in monkeys. *J Comp Neurol* 1979;183:833–882.
15. Hyvaerinen J, Poranen A. Receptive field integration and submodality convergence in the hand area of the post-central gyrus of the alert monkey. *J Physiol* 1978; 283:539–556.
16. Iwamura Y, Tanaka M, Sakamoto M, et al. Converging patterns of finger representation and complex response properties of neurons in area 1 of the first somatosen-

sory cortex of the conscious monkey. *Exp Brain Res* 1983;51:327–337.
17. Whitsel BL, Roppolo JR, Werner G. Cortical information processing of stimulus motion on primate skin. *J Neurophysiol* 1972;35:691–717.
18. Costanzo RM, Gardner EP. A quantitative analysis of responses of direction-sensitive neurons in somatosensory cortex of awake monkeys. *J Neurophysiol* 1980;43: 1319–1341.
19. Iwamura Y, Tanaka M, Hikosaka O. Cortical neuronal mechanisms of tactile perception studied in the conscious monkey. In: Katsuki Y, Norgren R, Sato M, eds. *Brain mechanisms of sensation.* New York: Wiley, 1981: 61–70.
20. Hikosaka O, Tanaka M, Sakamoto M, et al. Deficits in manipulative behavior induced by local injection of muscimol in the first somatosensory cortex of the conscious monkey. *Brain Res* 1985;325:375–380.
21. Jeannerod M, Michel F, Prablanc C. The control of hand movements in a case of hemianaesthesia following a parietal lesion. *Brain* 1984;107:899–920.
22. Iwamura Y, Tanaka M, Sakamoto M, et al. Vertical neuronal arrays in the postcentral gyrus signaling active touch: a receptive field study in the conscious monkey. *Exp Brain Res* 1985;58:412–420.
23. Iwamura Y, Tanaka M. Representation of reaching and grasping in the monkey postcentral gyrus. *Neurosci Lett* 1996;214:147–150.
24. Soso MJ, Fetz EE. Responses of identified cells in postcentral cortex of awake monkeys during comparable active and passive joint movements. *J Neurophysiol* 1980; 43:1009–1110.
25. Iwamura Y, Tanaka M. Postcentral neurons in hand region of area 2: their possible role in the form discrimination of tactile objects. *Brain Res* 1978;150:662–666.
26. Damasio AR, Benton AL. Impairment of hand movements under visual guidance. *Neurology* 1979;29: 170–174.
27. Bates JAV, Ettlinger G. Posterior biparietal ablations in the monkey. *Arch Neurol* 1960;3:177–192.
28. Ettlinger G, Kalsbeck JE. Change in tactile discrimination and in visual reaching after successive and simultaneous bilateral posterior parietal ablations in the monkey. *J Neurol Neurosurg Psychiat* 1962;25:256–268.
29. Jeannerod M. The formation of finger grip during prehension. A cortically mediated visuomotor pattern. *Bebav Brain Res* 1986;19:99–116.
30. Perenin MT, Vighetto A. Optic ataxia: a specific disruption in visuomotor mechanisms. I. Different aspects of the deficit in reaching for objects. *Brain* 1988;111: 643–674.
31. Jeannerod M, Decety J, Michel F. Impairment of grasping movements following a bilateral parietal lesion. *Neuropsychologia* 1994;32:369–380.
32. Faugier-Grimaud S, Frenois C, Stein DG. Effects of posterior parietal lesions on visually guided behavior in monkeys. *Neuropsychologia* 1978;16:151–168.
33. Haaxma R, Kuypers HGJM. Intrahemispheric cortical connections and visual guidance of hand and finger movements in the rhesus monkey. *Brain* 1975;98: 239–260.
34. Gallese V, Murata A, Kaseda M, et al. Deficit of hand preshaping after muscimol injection in monkey parietal cortex. *NeuroReport* 1994;5:1525–1529.
35. Mountcastle VB, Lynch JC, Georgopoulos A, et al. Posterior parietal association cortex of the monkey: command functions for operations within extrapersonal space. *J Neurophysiol* 1975;38:871–908.
36. Hyvaerinen J, Poranen A. Function of the parietal associative area 7 as revealed from cellular discharges in alert monkeys. *Brain* 1974;97:673–692.
37. Kalasaka JF, Caminiti R, Georgopoulos AP. Cortical mechanisms related to the direction of two dimensional arm movements, relations in parietal area 5 and comparison with motor cortex. *Exp Brain Res* 1983;51: 247–260.
38. MacKay WA. Properties of reach-related neuronal activity in cortical area 7a. *J Neurophysiol* 1992;67: 1335–1345.
39. Snyder LH, Batista AP, Andersen RA. Coding of intention in the posterior parietal cortex. *Nature* 1997;386: 167–170.
40. Sakata H, Taira M, Murata A, et al. Neural mechanisms of visual guidance of hand action in the parietal cortex of the monkey. *Cereb Cortex* 1995;5:429–438.
41. Murata A, Gallese V, Luppino G, et al. Selectivity for the shape, size, and orientation of objects for grasping in neurons of monkey parietal area AIP. *J Neurophysiol* 2000;83:2580–2601.
42. Rizzolatti G, Camarda R, Fogassi L, et al. Functional organization of inferior area 6 in the macaque monkey. Area F5 and the control of distal movements. *Exp Brain Res* 1988;71:491–507.
43. Taira M, Mine S, Georgopoulos AP, et al. Parietal cortex neurons of the monkey related to the visual guidance of hand movement. *Exp Brain Res* 1990;83:29–36.
44. Biederman I. Recognition-by-component: a theory of human image understanding. *Psychol Rev* 1987;94: 115–147.
45. Tanaka K, Saito H, Fukada Y, et al. Coding visual images of objects in the inferotemporal cortex of the macaque monkey. *J Neurophysiol* 1993;66:170–189.
46. Ohtsuka H, Tanaka Y, Kusunoki M, et al. Neurons in monkey parietal association cortex sensitive to axis orientation. *Nippon Ganka Gakkai Zasshi* 1995;99:59–67.
47. Sakata H, Taira M, Kusunoki M, et al. Neural coding of 3D features of objects for hand action in the parietal cortex of the monkey. *Philos Trans R Soc Lond B Biol Sci* 1998;353:1363–1373.
48. Bender MB, Jung R. Abweichungen der subjectiven optischen Vertikalen und Horizontalen bei Gesunder und Hirnverletzten. *Arch Psychiatr Nervenkr* 1948;181: 193–212.
49. McFie J, Piercy NF, Zangwill OL. Visual spatial agnosia associated with lesion of the right cerebral hemisphere. *Brain* 1950;73:167–190.
50. Cramon DY von, Kerkhoff G. On the cerebral organization of elementary visuo-spatial perception. In: Gulyas B, Ottoson D, Roland PE, eds. *Functional organization of the human visual cortex.* Oxford: Pergamon, 1993: 211–231.
51. De Renzi E, Faglioni P, Scotti G. Judgment of spatial orientation in patients with focal brain damage. *J Neurol Neurosurg Psychiat* 1971;34:489–495.
52. Marr D. *Vision.* New York: Freeman, 1982.
53. Shikata E, Tanaka Y, Nakamura H, et al. Selectivity of the parietal visual neurons in 3D orientation of surface of stereoscopic stimuli. *NeuroReport* 1996;7:2389–2394.

54. Taira M, Tsutsui K, Jiang M, et al. Parietal neurons represent surface orientation from the gradient of binocular disparity. *J Neurophysiol* 2000;83:3140–3146.
55. Tsutsui K, Jiang M, Yara K, et al. Integration of perspective and disparity cues in surface-orientation-selective neurons of area CIP. *J Neurophysiol* 2001;86: 2856–2867.
56. Sakata H, Taira M, Kusunoki M, et al. The TINS Lecture. The parietal association cortex in depth perception and visual control of hand action. *Trends Neurosci* 1997;20:350–357.
57. Hubel DH, Livingstone MS. Segregation of form, color and stereopsis in primate area 18. *J Neurosci* 1987;7: 3378–3415.
58. Adams D, Zeki S. Functional organization of macaque V3 for stereoscopic depth. *J Neurophysiol* 2001;86: 2195–2203.
59. Webster MJ, Bachevalier J, Ungerleider LG. Connections of inferior temporal area TEO and TE with parietal and frontal cortex in macaque monkeys. *Cereb Cortex* 1994;4:470–483.
60. Zeki S, Shipp S. The functional logic of cortical connections. *Nature* 1988;335:311–317.
61. Felleman DJ, Van Essen DC. Distributed hierarchical processing in the primate cerebral cortex. *Cereb Cortex* 1991;1:1–47.
62. Galletti C, Battaglini PP, Fattori P. Parietal neurons encoding spatial locations in craniotopic coordinates. *Exp Brain Res* 1993;96:221–229.
63. Murata A, Fadiga L, Fogassi L, et al. Object representation in the ventral premotor cortex(area F5) of the monkey. *J Neurophysiol* 1997;78:2226–2230.
64. Rizzolatti G, Scandolara C, Matelli M, et al. Afferent properties of periarcuate neurons in macaque monkeys. I. Somatosensory responses. *Behav Brain Res* 1981;2: 125–146.
65. Duhamel JR, Colby CL, Goldberg ME. Ventral intraparietal area of the macaque: congruent visual and somatic response properties. *J Neurophysiol* 1998;79:126–136.
66. Iriki A, Tanaka M, Iwamura Y. Coding of modified body schema during tool use by macaque postcentral neurons. *NeuroReport* 1996;7:2325–2330.

9

The Role of the Parietal Cortex in the Neural Processing of Saccadic Eye Movements

James W. Bisley and Michael E. Goldberg

David Mahoney Center of Brain and Behavior, Center for Neurobiology and Behavior, Columbia University College of Physicians and Surgeons, and the New York State Psychiatric Institute, New York, New York

INTRODUCTION

The activity of neurons in parietal cortex is intimately intertwined with the neural processing of saccadic eye movements. Saccades are rapid eye movements that change the point of regard from one spatial location to another, allowing us to scan the world for objects of interest. Patients with right parietal lesions have dramatic deficits: their saccades into the contralesional field are hypometric and have longer reaction times than those into the ipsilesional field; and they have difficulty compensating for saccades into the contralesional field (1,2). To study the mechanisms underlying these deficits, research has been done on an equivalent area in the monkey brain: the lateral intraparietal area (LIP). First described by Andersen, Asanuma, and Cowan (3), LIP is particularly suited for this with strong anatomical connections to areas known to play important roles in saccade planning.

The first physiological evidence that LIP was involved in processing saccades came from its activity in the memory-guided delayed saccade task (4), a task which entails visual, motor and memory components that are separated in time (5). In this task (Fig. 9-1A), the subject fixates a spot during which time another spot, the target, is flashed briefly in the periphery. The subject has to remember the location of the target and when the fixation point is extinguished, the subject must make a saccade to that location. Thus, the task can be broken into three phases: the visual phase, during which the target is flashed; the memory phase, during which the subject must remember where the target was flashed and prepare to make a saccade to that location; and the motor phase, the time around which the subject actually makes the eye movement. Neurons in LIP often respond to several components of the task (Fig. 9-1A), bursting during the visual response, maintaining activity during the memory phase and firing just before and around the time of the eye movement (4,6,7).

THE RELATIONSHIP OF NEURAL ACTIVITY IN LIP TO SACCADES

Understanding the role of LIP in visuomotor processing hinges on the interpretation of this saccade related activity. At one level, the reason for this activity may appear to be simple: the subject is planning a saccade, so this correlation between neural activity and the monkey's behavior may betoken a causal relationship between the cellular discharge and the saccade. It is certainly clear that LIP has a strong signal describing a monkey's saccade plan. Not only does LIP discharge through the period during which a monkey plans a memory-guided saccade, but it also follows the

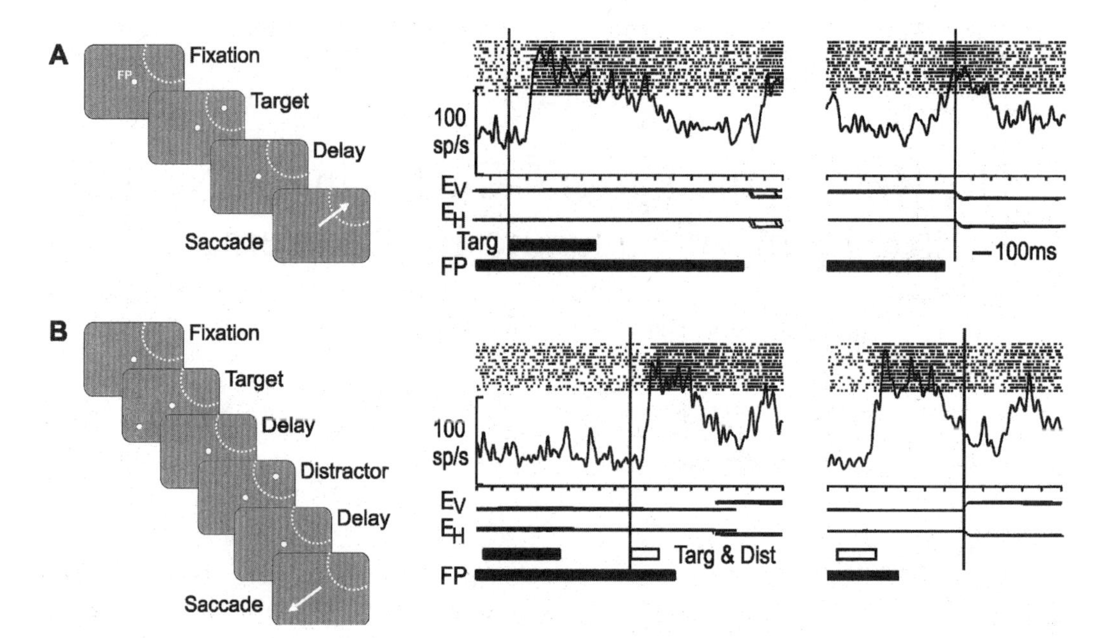

FIG. 9-1. Response of an LIP neuron during the memory-guided saccade task. **A:** The task is shown on the left. The monkey initiated a trial by fixating a small spot (FP). A second small spot (the target) was flashed briefly after which the monkey had to maintain fixation until the central spot was extinguished. To get a reward, the monkey had to make a saccade to the remembered location of the target. In this example the target was flashed in the receptive field (dashed circle). The response of a single neuron during this task is shown on the right. In the raster, each row is a single trial, and each spot is a single action potential. The solid trace is a spike density function calculated every millisecond. The center panel shows the data aligned on the onset of the target and the right panel shows the same data aligned on the onset of the saccade. This neuron shows visual, delay and peri-saccadic activity. **B:** The task illustrated on the left is the same as in A, except that a third small spot (a task irrelevant distractor) was flashed during the delay. The panels on the right show data from trials in which the target was flashed outside of the receptive field and the distractor was flashed in the receptive field. Adapted from Ref. 12.

motor plan when a monkey is told to change his planned saccade to a new target, with elevated activity in regions corresponding to the new goal of the upcoming saccade (8). Other support for the activity in LIP being related to saccades comes from tasks in which monkeys can be instructed to make an arm or eye movement to a target. When the animal has to reach to a target, either independently or at the same time as making a saccade elsewhere, then LIP neurons fire more in saccade trials than in reach trials (9). Furthermore, if in the middle of a trial in which the monkey is planning a saccade, the animal is instructed to change the movement to a reach, then activity is reduced in response to the target than it is to an identical target instructing a change from a reach to a saccade (10). These physiological data, when coupled with the anatomical connections of LIP to the frontal eye fields (3,11) and the superior colliculus (12), seemingly support a role for LIP in the generation of saccades.

Other experiments are less consistent with the idea that "the posterior parietal cortex plays an important role in the planning of action, with one area specialized for eye movements (LIP)" (13). A major problem with this theory is that LIP neurons, even those with robust delay-and presaccadic activity in the memory-guided saccade task often respond to stimuli that are unlikely to be considered saccade targets. For example, when a monkey plans a memory-guided saccade to the receptive field of a neuron, the neuron discharges throughout the delay (Fig. 9-1A). However, when the monkey plans a saccade away from the receptive field and a task-irrelevant distractor appears in the receptive field, the neuron responds as strongly to this behaviorally irrelevant stimulus as it did to the onset of the saccade target (Fig. 9-1B). Even immediately before the saccade the irrelevant stimulus can evoke as great a response as the saccade target, but the distractor has no effect on the performance of the saccade. Saccade velocity, latency, accuracy, and early trajectory are all unaffected by the distractor, despite the vigorous response in LIP (14). LIP neurons can also be driven by a number of features–shape (15), the sudden onset of a stimulus (16), the behavioral relevance of a stimulus (7,17–19), and sudden motion (20). Thus salient targets evoke significant activity in neurons in LIP even when they describe objects and spatial locations to which saccades are both expressively forbidden by the task, and unlikely from the behavior of the monkey. LIP does describe the target of a purposive saccade, but this description is a subset of the activity of LIP. Unlike neurons in the frontal eye field (21) or superior colliculus (22) LIP neurons do not signal where, when, or if a monkey will actually make a saccade.

LIP AS A SALIENCE MAP

Since the development of the fixation task by Wurtz (23), the standard method for determining a visual response of a neuron has been to measure the response of the neuron to a stimulus that appears suddenly in its receptive field. This definition has a problem, however. Abruptly appearing stimuli are not only associated with photons exciting the retina; they are also automatic attractors of attention (24–26). Although a general assumption in neurophysiology has been that the abrupt onset of a stimulus in the receptive field is equivalent to the appearance in the receptive field of a stable object brought in by a saccade, this assumption has not been tested until recently (16,27) The question then arises as to how much of the 'visual responses' of parietal neurons are visual, i.e. responding to photons on the retina, like a retinal ganglion cell, and how much are related to the attentional capture by the stimulus onset.

To answer this question Gottlieb et al. (16) devised a number of tasks in which the stimulus, rather than appearing *de novo* in the receptive field, entered the receptive field by virtue of a saccadic eye movement. This enabled them to stimulate the receptive field using stimuli that did not have the attentional tag of abrupt onset. In these "stable array" tasks, the monkeys were presented with eight stimuli arranged uniformly in a circular array (Fig. 9-2). These stimuli did not appear or disappear from trial to trial. Instead, they were constant for a block of trials. They were positioned so that when the monkey fixated the center of the array at least one stimulus appeared in the receptive field of the neuron under study. In the simplest of these tasks the monkey fixated at a position outside the array so that no stimulus was in the receptive field of the neuron being studied, and then, when the red fixation point jumped, made a saccade to the center of the array (Fig. 9-2). This saccade brought one of the stable stimuli into the receptive field

The typical neuron had a brisk response to the sudden appearance of a stimulus in its receptive field during a fixation task (Fig. 9-3A), and a much smaller response when the same stimulus, as a member of the stable array, was brought into the receptive field by a saccade (Fig. 9-3B). The decrement of response could have been related to the behavioral irrelevance of the stable target, or, it could have been due to a series of other confounds. For example, the movement of the stimulus into the receptive field by the saccade is not exactly the same as its appearance from the flash; the other members of the array might exert some purely visual local inhibition that suppresses the response. To test if these other factors could be responsible for the diminished response to the stable target, the authors had one of the stimuli in the array

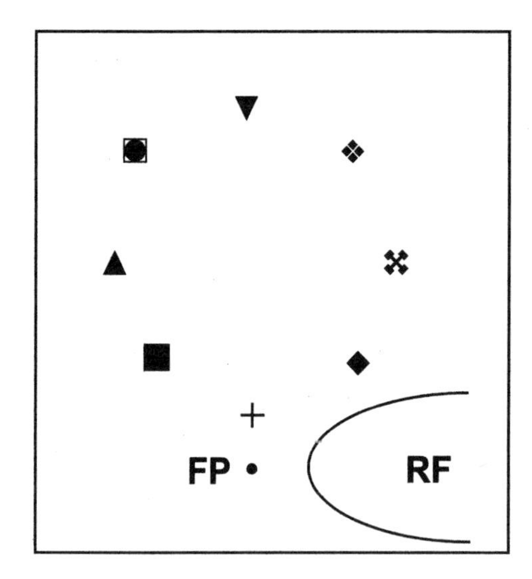

Fixate so all symbols are outside RF

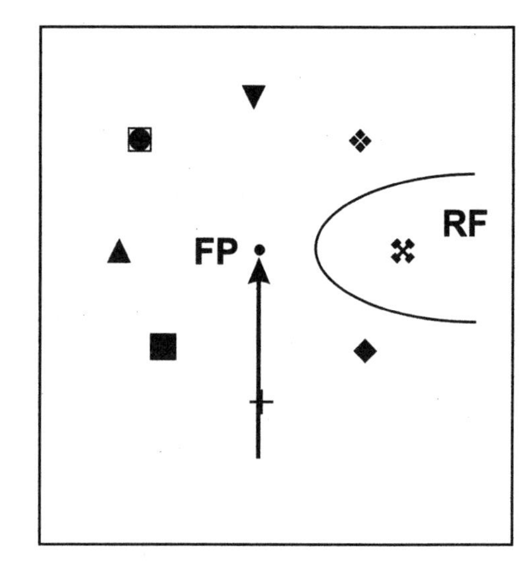

Saccade brings stable stimulus into RF

FIG. 9-2. The stable array task. In this task the eight objects remained on the screen for the duration of the experiment. The monkey initiated the trial by fixating a small spot (FP) that was positioned such that none of the stable objects appeared in the receptive field (RF). After a short delay the fixation point jumped to the center of the array, bringing one of the objects into the receptive field. Adapted from Ref. 20.

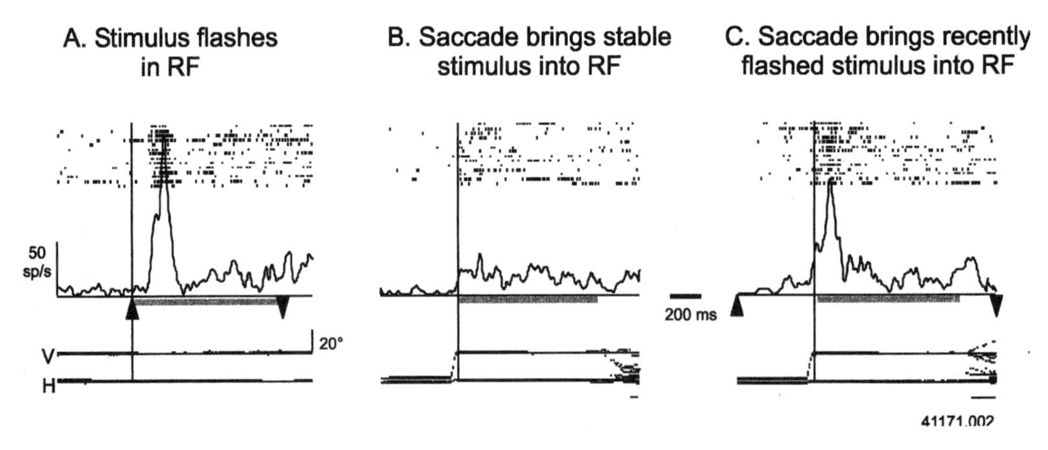

FIG. 9-3. Effect of a recent flash on the response of an LIP neuron. Data are presented in the format seen in Figure 9-1. The gray bar beneath the spike density function shows when, during the trial, the stimulus was in the receptive field of the neuron. The up arrows represent the onset of a flashed stimulus, the down arrows represent its disappearance. **A:** The response to a single stimulus flashing in the receptive field during a fixation task. No other stimuli were present on the screen. Activity is aligned on the stimulus onset. **B:** The response to the same stimulus, as part of the stable array, moving into the receptive field by a saccade. Activity is aligned on saccade end. **C:** The monkey makes a saccade that brings a recently flashed stimulus into the receptive field. The stimulus appears approximately 500 ms before the saccade. Activity is aligned on saccade end. Adapted from Ref. 20.

flash on while the monkey was fixating (up arrow, Fig. 9-3C). The monkey then made a saccade to the center of the array, which brought this recently appeared stimulus into the receptive field. In this case, the neuron responded almost as briskly as it did to the abrupt appearance of the stimulus in the receptive field (Fig. 9-3C, compare with Fig. 9-3A). Therefore the difference between the fixation case and the stable target case was not due to the visual or oculomotor differences between the tasks, but to the lack of salience of a stable component of the visual environment (16). Note that the neuron began to respond at or before the end of the saccade. This was a much shorter latency than when the stimulus appeared in the receptive field abruptly (compare latency in Fig. 9-3A with Fig. 9-3C). Presumably this occurred because of the predictive response, a second influence of eye movements in LIP, which we will describe in detail below.

Salience not only arises from intrinsic properties of the stimulus, but can also arise by virtue of their relevance to current behavior. Under these circumstances a member of a stable array can evoke a response from a neuron in LIP. This can be shown using the stable target task—a more complicated version of the stable array task (Fig. 9-4). In this task, the monkey fixated so that the stimulus was not in the receptive field, and a cue appeared during this fixation. This cue matched one of the symbols in the stable array. The fixation point

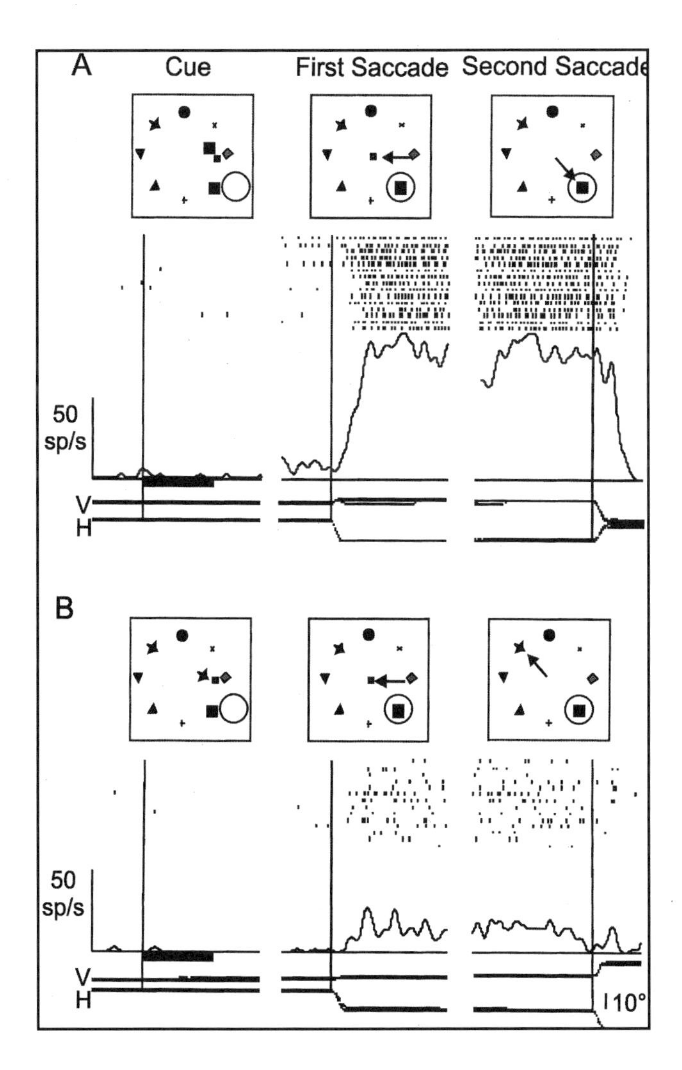

FIG. 9-4. Response of an LIP neuron in a stable array task requiring a saccade to a cued object. While the monkey was fixating outside of the array, a cue was flashed. The fixation point then jumped into the center of the array bringing an object into the receptive field. The animal then waited until the fixation point was extinguished and made a saccade to the cued object. In (**A**) the cued object was within the receptive field after the first saccade, and in (**B**) the same object, this time not the target of the saccade, was brought into the receptive field by the first saccade. Each trio of raster plots shows the response of the neuron in the same trials synchronized on the cue (left), first saccade beginning (middle) and second saccade beginning (right). Adapted from Ref. 20.

then jumped to the center of the array and the monkey tracked it with a saccade. Finally, when the fixation point disappeared, the monkey made a saccade to the member of the array that matched the cue. Neurons responded strongly to stable stimuli brought into their receptive field if these were designated as the target of the next saccade (Fig. 9-4A). The neuron discharged from the first saccade to the second. In contrast, if the identical stable stimulus entered the receptive field but was not designated as the saccade target (i.e. the monkey was instructed to make a saccade to a different stimulus), neurons responded minimally (Fig. 9-4B).

It is possible that LIP neurons respond in the stable target task because the monkey is planning a purposive saccade, and the activity is less related to the salience of the stable target than it is to the processes underlying saccade planning. In a study designed to analyze how activity in LIP can be allocated to the planning and generation of a saccade itself, monkeys performed a task in which they had to make a saccade to a spatial location that had no visual stimulus (the "black hole" task). This task was identical to the task in Figure 9-4, except one member of the array was removed for a block, such that saccades to that object were made into a region with no visual stimulus—the "black hole". LIP neurons active in the stable target task frequently failed to respond when a saccade was made into the receptive field that contained the "black hole", and no neurons responded more in this task than in the stable target task when the cue, and not the target was in the receptive field (16)

In another series of experiments, Gottlieb and Goldberg (28) used an anti-saccade task to show that the pre-saccadic activity in LIP neurons rarely gives a clear indication about the direction in which the impending saccade will go. An anti-saccade is an eye movement made to the location opposite a target. These type of tasks have been used to study the saccadic system and as a neurological tool (for review see Ref. 29). They require that a subject first suppress the reflex of making a saccade to the target and then plan and make a saccade to the opposite location.

Figure 9-5 shows the analysis of the data collected from LIP neurons when pro-and anti-saccades were interleaved within a session. This analysis examined the information transmitted about the two possible stimulus locations and saccade directions, producing a value measured in bits, with a maximum of one bit for each variable (stimulus location and saccade direction). In the upper panel, the information about stimulus location, as measured during an epoch 40-160 ms after the stimulus onset, is compared to the information about saccade direction, as measured during the 60 ms epoch prior to the saccade. For the majority of LIP neurons there is more information about the stimulus location during the period following the stimulus than about the saccade direction during the period preceding the saccade. As the task progressed, information about the stimulus location declined. However, even at the time of saccadic generation there was more information about the stimulus location than the direction of the saccade (lower panel, Fig. 9-5). In fact, only 13/105 LIP neurons were able to reliably indicate the direction of the saccade independent of the target location, whereas almost all the neurons were able to reliably indicate the location of the stimulus soon after it was illuminated. In another series of experiments, (30) showed that in a delayed antisaccade task

FIG. 9-5. Information transmitted about cue location and saccade direction across 105 neurons in an anti-saccade task. Information transmitted by each neuron during the presaccadic epoch (abscissa) is compared with the information transmitted about the cue location during the cue epoch (**A**) and during the presaccadic epoch (**B**). All values represent instantaneous information averaged across the interval of interest. Histograms show the distribution of values along each axis. Adapted from Ref. 28.

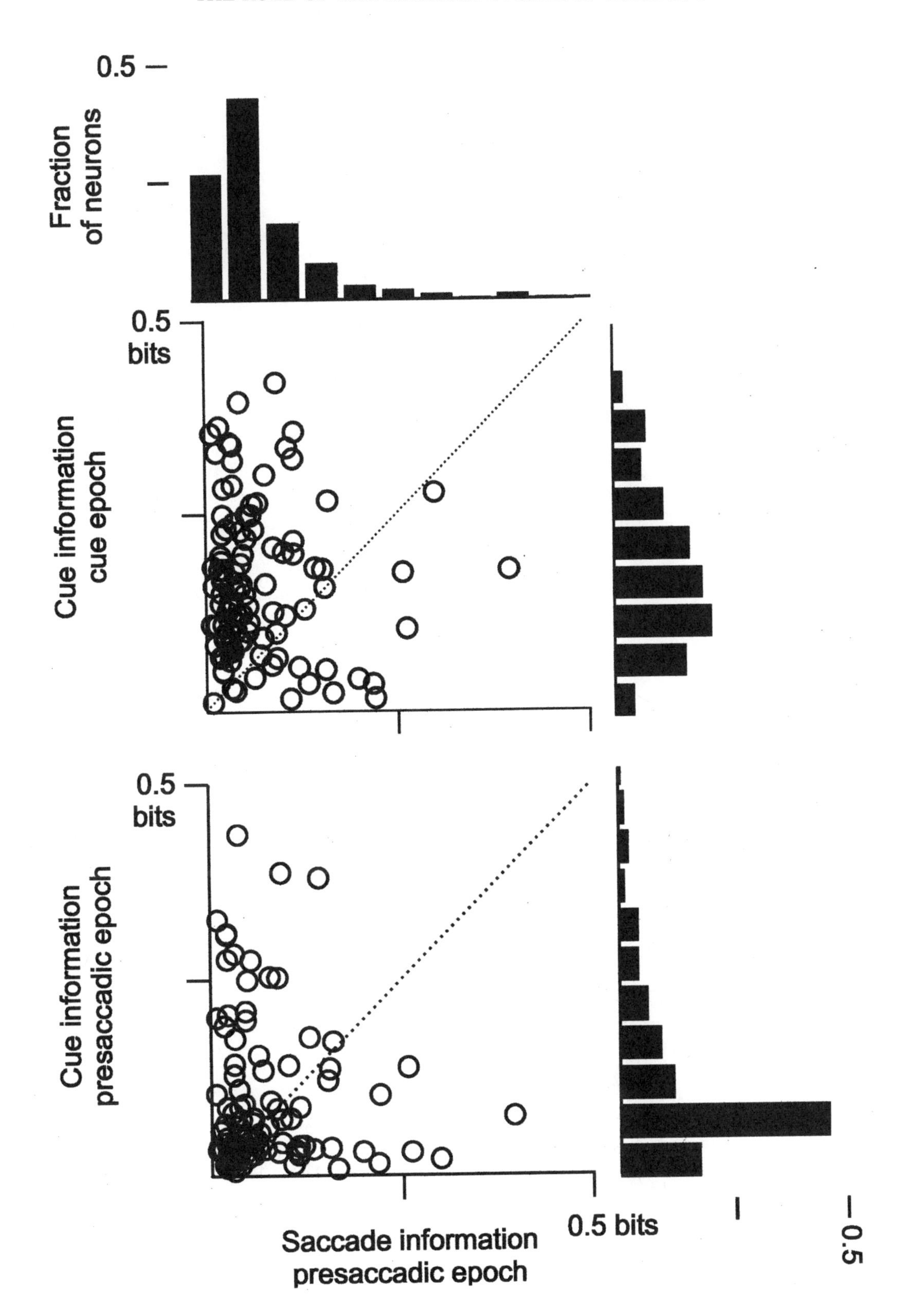
0.5
Fraction of neurons
0.5 bits
Cue information cue epoch
0.5 bits
Cue information presaccadic epoch
0.5 bits
Saccade information presaccadic epoch
0.5

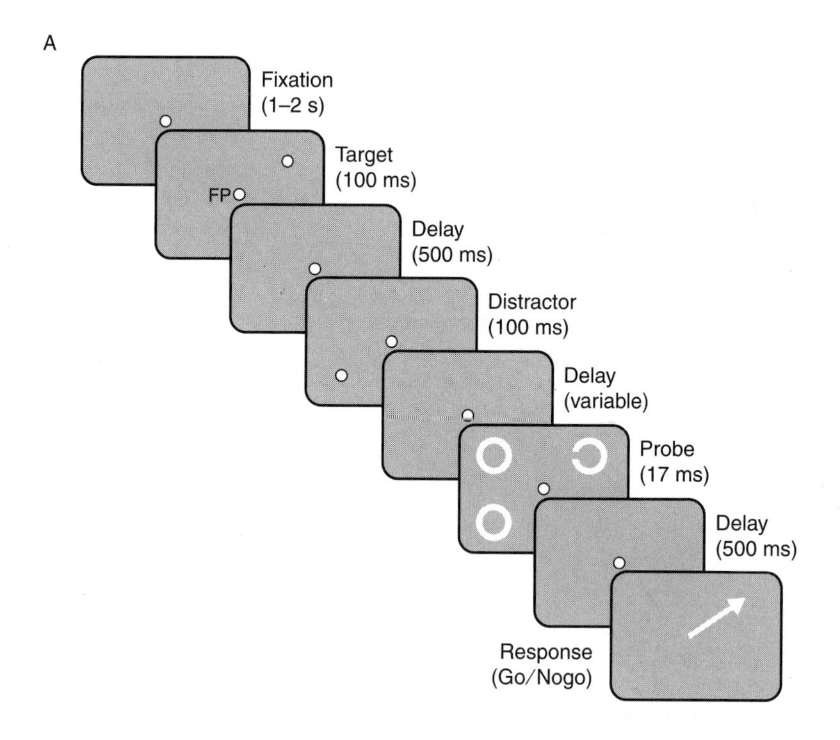
A
Fixation
(1–2 s)
Target
(100 ms)
FP
Delay
(500 ms)
Distractor
(100 ms)
Delay
(variable)
Probe
(17 ms)
Delay
(500 ms)
Response
(Go/Nogo)

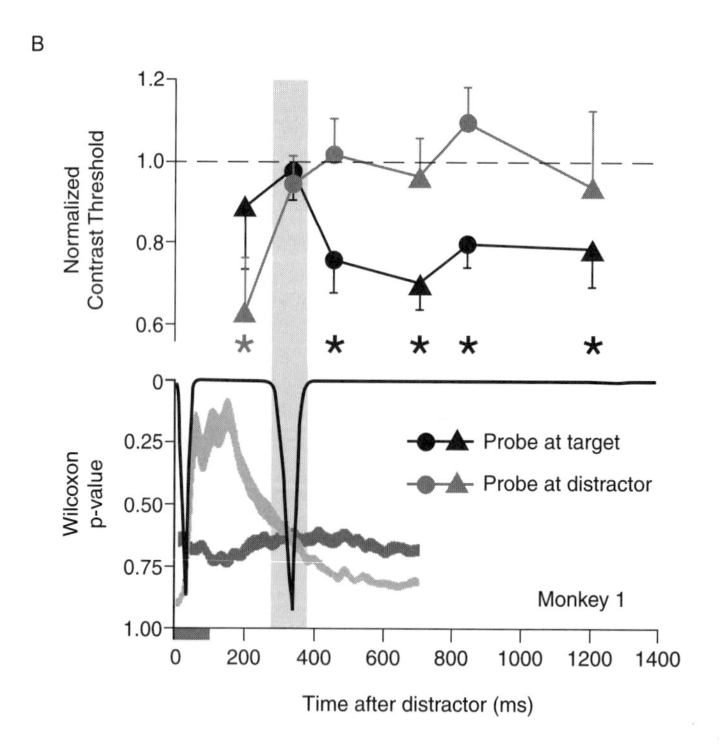
B
Normalized
Contrast Threshold
1.2
1.0
0.8
0.6
Wilcoxon
p-value
0
0.25
0.50
0.75
1.00
Probe at target
Probe at distractor
Monkey 1
0
200
400
600
800
1000
1200
1400
Time after distractor (ms)

some neurons with predominantly visual responses in the delayed prosaccade task develop a late signal that describes the antisaccade goal, switching their response fields from the cue location to the goal location.

The strongest argument that LIP acts as a salience map comes from a study that shows that activity in LIP accurately describes the locus of attention (31). In this study, monkeys were trained on a variation of the memory-guided saccade task. Unlike the standard task (shown in Fig. 9-1A), the monkey not only had to remember the target location, but also had to ignore a distractor and perform a discrimination (Fig. 9-6A) that instructed him to either make the planned saccade or not. The task took advantage of the technique of measuring the locus of attention by determining the spatial locus with the highest sensitivity to contrast (32), and it also could be used to map the responses to the stimuli across LIP. Figure 9-6B shows the activity of a population of neurons in LIP (the thick traces in the lower plot) and the locations of attention measured psychophysically at different times within the trial (the thin traces in the upper plot). Initially only the data represented by the triangles was recorded behaviorally, and it was hypothesized that attention was at the location in LIP with the highest activity. This hypothesis made a prediction: when the activity in the 2 traces was indistinguishable, then there should be no attentional advantage at either site (33). When the authors behaviorally measured attention at this time (340 ms), they found that there was no locus of attention at either location (circles in gray column). When the same experiment was performed on a second monkey, the activity in LIP again matched the behavioral data, showing a period of no attention at the time predicted by the activity in LIP. Moreover, these times differed by 100 ms in the 2 animals (31), suggesting that the activity in LIP describes the temporal and spatial dynamics of attention on a monkey by monkey basis.

IS LIP INVOLVED IN PLANNING SACCADES?

The question then remains –even if LIP is a salience map, is the saccade related activity used by the motor system or is it reflective of the motor plan? In the results discussed thus far, we have suggested that LIP activity is not required by the oculomotor system to make saccades since the activity cannot be used to predict where, when or even if a saccade will occur. This conclusion is further supported by microstimulation and reversible inactivation, two methods commonly used to show causality. Microstimulation of LIP can produce saccades (34,35), however the current needed to elicit them is much higher than that needed in the

FIG. 9-6. The task and data collected to compare the locus of attention with activity in LIP. **A:** The task was based on the memory-guided saccade task, with the task irrelevant distractor (see Fig. 9-1). In this task, before the fixation point (FP) was extinguished four rings appeared. One of the rings had a gap either on the left or the right (the probe). The monkey had to identify the side of ring that the gap was on and indicate it either by making the planned memory guided saccade when the fixation point was extinguished (GO) or by canceling the saccade and maintaining fixation until the end of the trial (NOGO). **B:** Behavioral and physiological data from this task. The thin traces in the top panel show the animal's behavioral performance plotted as normalized threshold. Points that are significantly beneath the black dashed line indicate an attentional advantage (*). The thick traces in the lower panel show the spike density function of a single neuron (the width of the trace shows the SEM). The darker traces show data from trials in which the probe was placed at the target site, and the distractor had flashed elsewhere. The lighter traces show data from trials in which the probe was placed at the distractor site and the target had flashed elsewhere. The thin trace in the lower plot shows the result of a running statistical test showing when the two thick traces were indistinguishable (gray block). Adapted from Ref. 31.

frontal eye fields (FEF) or the superior colliculus (SC) (36–38), 2 areas known to play a part in saccade planning. In fact, the effects are inconsistent even with high currents (34), again in contrast with the FEF or SC. Similar saccadic eye movements can be evoked by high-current stimulation of striate cortex as well (39), although the argument that striate cortex is a center for saccade planning is difficult to sustain. Reversible inactivation of LIP has been shown to be entirely ineffective on visually guided and memory saccades (40), or weakly effective (41) on visually and memory-guided saccades. In the study that showed some effect on memory-guided saccades, the inactivation never actually eradicated them. However, in both studies the inactivation did appear to produce a monkey equivalent of extinction; when given a choice of 2 simultaneous targets the animals made saccades to the target in the intact hemifield, ignoring the stimulus in the inactivated hemifield. Thus, LIP may contribute to the selection of targets, but does not seem to be necessary for the programming and execution of saccadic eye movements (40).

The Importance of Saccade Related Activity in LIP

If delay activity in LIP is not involved in the saccade plan, then why is it present? The answer to this may lie within the role of LIP as a salience map. A number of psychophysical experiments have shown that attention is allocated to the spatial location of an upcoming saccade (31,42–45). Thus, it is likely that LIP uses the information about the saccadic plan to create a representation of the saccadic end point in its salience map. Often, the delay activity that reflects the saccadic plan is weaker than the visual response to a suddenly appearing stimulus.

While such a salient stimulus does not affect saccade metrics (14), it briefly attracts attention, following the same time course as the activity in LIP (31). This supports both the idea of LIP as a salience map and the use that saccade related activity in LIP may have.

These studies ultimately suggest a facilitatory role for LIP in the actual guidance of eye movements. The salience map can be used to drive visual attention, and select targets for saccades; when there is a congruence between the salience map and the goal of a saccade LIP can reinforce that congruence, presumably through its connections to the superior colliculus and frontal eye field. When there is a dissonance, the oculomotor system must ignore the LIP signal. The purpose of the robust saccadic signal, dramatically illustrated in the memory-guided saccade task, cannot be to drive the saccade; instead it must be to relay to other parts of the brain the fact that a saccade goal chosen by the oculomotor system is worthy of visual attention. A deficit in attention, and in choosing saccade targets from the salience map could explain the hypometric, longer latency saccades in patients with right parietal deficits.

SACCADE-INDUCED REMAPPING OF RECEPTIVE FIELDS IN LIP

The brain not only must generate saccades, it must also be able to compensate for their visual effects. Eye movements cause two perceptual problems. The first is that whenever the eye moves, the position of stimuli stable in the world moves on the retina. Because of the visual latency of cortical areas (minimally 60 ms in LIP), at the end of a saccade a purely retinotopic map provides inaccurate information for the visual localization of objects in the world until a new map can be established from retinal information gathered after the saccade but delayed by the visual latency. A second problem induced by eye movements is that of the spatial location of flashed stimuli that appear and disappear before an intervening saccade. This can be shown by laboratory experiment known as the 'double step task' (Fig. 9-7). In this task the first saccade is simple: the retinal vector is identical to the saccade vector.

However, after the first saccade the retinal vector of the second target is no longer the same as the vector of the saccade necessary to acquire it. In order to acquire the target the brain must use the dimensions of the prior saccade to update the vector associated with the spatial location of the target that vanished before the recent saccade. Yet humans (46) and monkeys (47) have little difficulty in performing this task.

The posterior parietal cortex has a mechanism that enables the solution of both of the latency problem and the post-saccade memory problem. Neurons in LIP (4,48) discharge appropriately in the double step task, describing the proper saccade vector even when there is a dissonance between the second saccade and the retinal location at which its target appeared. This mechanism is equivalent to remapping the retinal receptive field of the neuron into the coordinates of the post-saccadic fixation. The remapping mechanism is not, however, limited to the laboratory curiosity of the double step task (49). Whenever a monkey makes a saccade that brings a stimulus into the receptive field, even though that stimulus will be the target of a subsequent saccade, some neurons actually discharge before the saccade begins, as if they predicted that the stimulus would be brought into the receptive field by the saccade (Fig. 9-8). About a third of neurons in LIP show this predictive effect, discharging with a latency shorter than their latency to a stimulus flashed in their receptive field. The neurons do not discharge when the stimulus appears alone in the future receptive field (Fig. 9-8C), nor do they discharge when the monkey makes the saccade alone (Fig. 9-8D). Instead they require the combination of a stimulus that will be brought into the receptive field by the saccade and the saccade itself. In this way the brain can maintain a spatially accurate representation of the visual world without having to wait for the reestablishment of a retinotopic map. The latency advance occurs not only for recently flashed stimuli as in Figure 9-8, but also for stable stimuli that become behaviorally relevant (see Figs. 9-3A and C).

This predictive effect is also seen with memory activity. If a stimulus flashes briefly outside the receptive field of a neuron and the monkey subsequently makes a saccade that brings the spatial location of the vanished stimulus into the receptive field, over

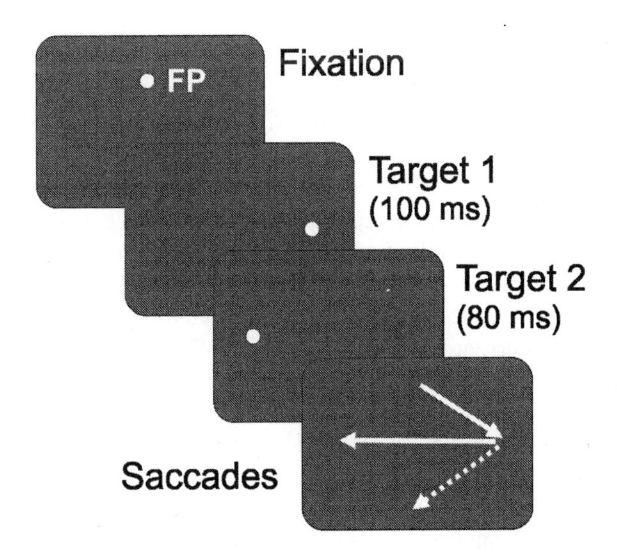

FIG. 9-7. The double step task. In this task the subject fixates a small spot (FP). The point jumps to the first target location (Target 1) for 100 ms and then to the second target location (Target 2) for 80 ms, then disappears. The subject must make successive saccades to Target 1 and Target 2. Because the targets have all disappeared before the first saccade, there is a dissonance between the vector of the second saccade (from Target 1 to Target 2) and the retinal location of Target 2. If the saccadic system were only to use retinotopic data the subject would make an incorrect second saccade (white dotted line).

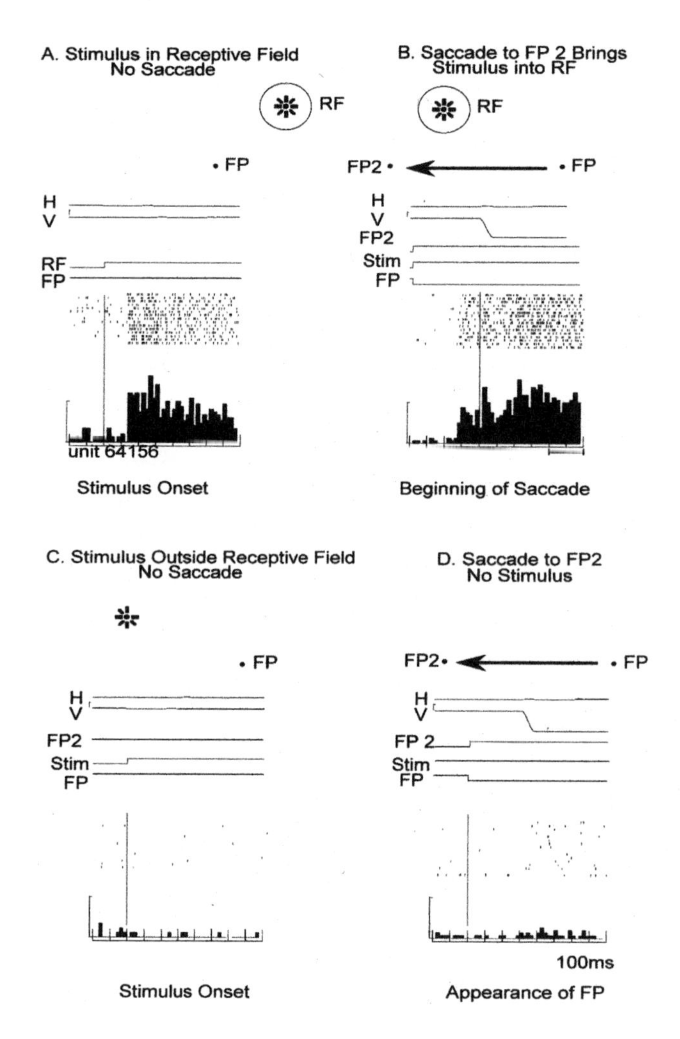

FIG. 9-8. Predictive visual response in a neuron in LIP. **A:** The response of the neuron to the stimulus flashed in its receptive field. Activity is synchronized on the stimulus onset. **B:** The response of the neuron when the monkey makes a saccade that brings the stimulus into its receptive field. Activity is synchronized on the beginning of the saccade. Note that the activity begins before the saccade. **C:** No response of the neuron when the stimulus appears in the future receptive field but there is no saccade. **D:** No response of the neuron when the monkey makes a saccade from FP to FP2 but there is no stimulus in the future receptive field. Adapted from Ref. 49.

80% of neurons in LIP will respond to the stimulus even though it never appeared in their receptive fields as defined in a fixation task (Fig. 9-9).

This remapping mechanism is not instantaneous. Around the time of the saccade there are two important spatial locations: the spatial location in the receptive field before the saccade (the current receptive field) and the spatial location of the receptive field after the saccade (the future receptive field). Excitability decreases in the future receptive field and increases in the current receptive field simultaneously before the saccade. Figure 9-10 shows the neural response of a population of LIP neurons measured 0-300 ms after the saccade, to stimuli flashed for 100 ms before, during, and after the saccade (50). Note that stimuli flashed well before the saccade evoke no signal above background after the saccade if they appear in the current receptive field, but they do evoke a signal in the future receptive field. Stimuli that appear in the interval 150-50 ms before the saccade, however, evoke activity both in the current and future receptive fields, although the activity induced in the current receptive field is less than it would be if there were no impending saccade. This effect is not merely non-specific saccadic suppression: stimuli moved from one location in the receptive field to another are frequently not associated with suppression. The response in the current recep-

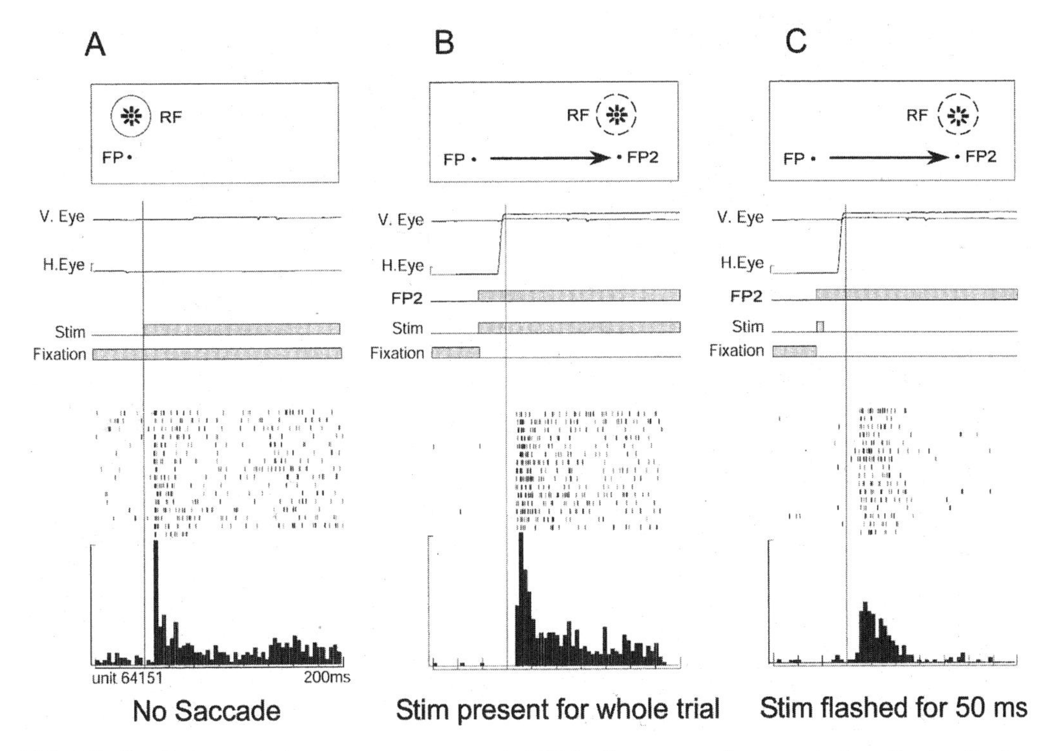

FIG. 9-9. Spatially accurate memory response in LIP. **A:** Response of neuron to the stimulus flashed in its receptive field, synchronized on stimulus onset. **B:** Response of neuron to when saccade brings stimulus into the receptive field. Note that there is no latency advance. **C:** Response of the neuron after a saccade that brings the spatial location of briefly flashed stimulus into the receptive field. The stimulus only appeared in the future receptive field before the saccade. Adapted from Ref. 49.

tive field occurs because the salience of abrupt onset overrides the suppression evoked by the impending saccade. This means that a single stimulus flashed immediately before a saccade will evoke activity in two sets of neurons, those subtending the current receptive field and those subtending the future receptive field. This, in turn, means that such stimuli should be poorly localized, a result that has been found both perceptually (51) and as an inaccuracy in eye movements (52,53).

These remapping mechanisms provide a spatially accurate signal in LIP. However, the signal does not describe the spatial location of the target explicitly: instead it describes a vector from the current or impending center of gaze to the spatial location of the salient object. This output vector is immutable. When there has been no recent saccade the LIP representation is equivalent to a retinotopic representation. If an intervening saccade occurs, a new spatial location is described by the output vector of the cell. The remapping mechanism enables this new spatial location to drive the cell, even though objects in this location never appeared in the retinal location that defines the receptive field in a fixation task. Thus the receptive field of an LIP neuron cannot be defined in the same way as for a purely sensory neuron—rather than describing the part of the retina that drives the cell, the receptive field describes the part of the visual field that currently is associated with the output vector of the cell. Although the retinal location may change according to the recent oculomotor history of the animal, the output meaning of the cell never changes.

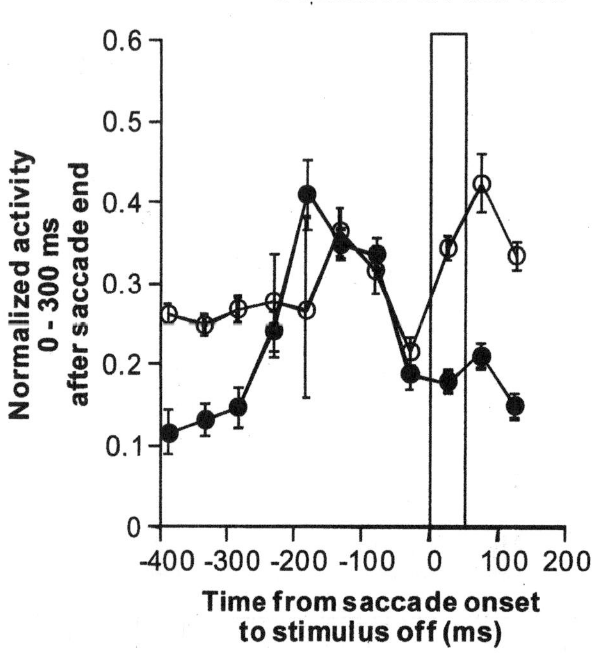

FIG. 9-10. Time course of perisaccadic excitability change. Ordinate: normalized activity of 36 neuron sample of 2 monkeys in LIP Abscissa: time from saccade to the disappearance of the stimulus, which was flashed for 100 ms in each trial. A negative value means that the stimulus appeared only before the saccade. The saccade occurred from roughly 0 to 50 ms on the scale (column). Error bars are standard errors of the means. Adapted from Ref. 50.

Humans with right parietal lesions exhibit a deficit predicted by the remapping mechanism: they cannot compensate for a contralateral saccade in the double step task. Figure 9-11 shows the eye movements of a patient with a right frontoparietal lesion who could make accurate saccades in both directions when each saccade was retinotopic (Fig. 9-11 top panels). When the patient was presented with stimuli that flashed rapidly in the right (ipsilesional) field and then the left (contralesional) field she did fairly well (Fig. 9-11, bottom left). However, when presented with stimuli that flashed first in the contralesional field and then the ipsilesional field, she had great difficult subsequently acquiring the second target, a target that appeared in the intact field, and could be acquired with a saccade in the intact direction (1). This inability to compensate for contralesional saccades is frequently found in patients with right posterior parietal deficits (54). Presumably they are unable to compensate for the contralesionally directed saccade. This finding has not yet been duplicated in monkeys. In one study, Li and Andersen (55) found that the orbital position of the first saccade was more important in determining latency and task error than the direction of the first saccade.

SUMMARY

The lateral intraparietal area has a signal that describes a saccade target, maintains the memory of a saccade plan during a delay, and describes the saccade itself. It is unlikely, however, that this signal generates a plan for the saccade, because most neurons with this delayed saccade activity also respond, sometimes more strongly, to salient stimuli that are unlikely to be saccade targets. Instead, it is more likely that this saccadic signal performs two functions unrelated to saccade planning itself. The first function is to contribute to a salience map: it is well known that attention is located at the goal of a saccadic eye movement, and recent experiments detailed here show that the attentional advantage of the saccade goal is

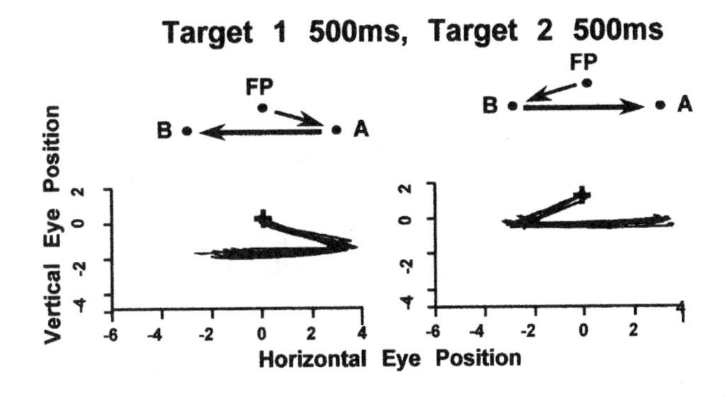

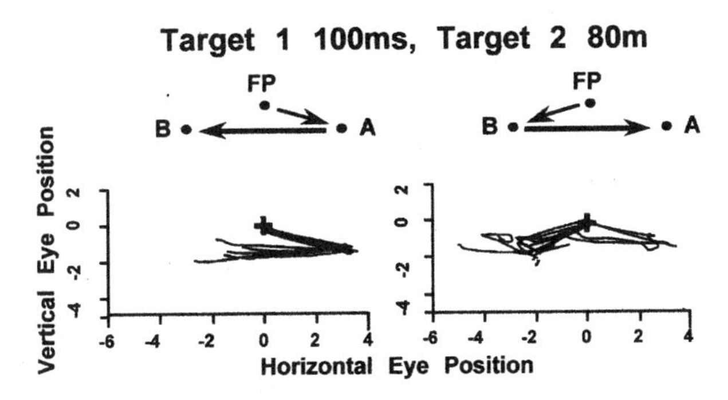

FIG. 9-11. Performance of a patient with a right fronto-parietal lesion in the double step task. In each diagram there is a cartoon of the spatial location of stimuli and saccades. The eye movements for that set of trials are superimposed. Ordinate: the vertical component of the eye movements, abscissa: horizontal component of eye movements. Eight trials are superimposed. Top: eye movements when the patient makes 2 successive simple saccades, with the stimuli flashed for 500 ms, durations long enough so that the first saccade is completed before the second stimulus appears. Left: saccades into the ipsilesional then contralesional field. Right: saccades into the contralesional then ipsilesional field. Bottom: Saccades to the same spots when both targets appear and disappear before the first saccade occurs. Left: saccades into the ipsilesional then contralesional field. Note that the quality of the contralesional saccades are degraded relative to the case when the subject does not have to rely on memory, but they are roughly accurate. Right: saccades into the contralesional then ipsilesional field. The subject usually does not make the second saccade, although, because of neglect, sometimes goes directly to the contralesional stimulus, ignoring the ipsilesional one. Adapted from Ref. 1.

maintained throughout the delay period of a memory-guided saccade. The saccade signal, presumably driven by the frontal eye fields or other prefrontal cortical areas, informs the salience map of a saccade plan, and therefore renders the goal of the saccade a salient location for attentional processes and, possibly, to provide targets for future saccades. The second function is to use the saccade signal to provide information by which the parietal cortex can update the visual representation to compensate for an eye movement, thus maintaining a spatially accurate vector map of the visual world despite a moving eye.

REFERENCES

1. Duhamel JR, Goldberg ME, Fitzgibbon EJ, et al. Saccadic dysmetria in a patient with a right frontoparietal lesion. The importance of corollary discharge for accurate spatial behaviour. *Brain* 1992;115:1387–1402.
2. Heide W, Kompf D. Combined deficits of saccades and

visuo-spatial orientation after cortical lesions. *Exp Brain Res* 1998;123:164–171.
3. Andersen RA, Asanuma C, Cowan WM. Callosal and prefrontal associational projecting cell populations in area 7A of the macaque monkey: a study using retrogradely transported fluorescent dyes. *J Comp Neurol* 1985;232:443–455.
4. Gnadt JW, Andersen RA. Memory related motor planning activity in posterior parietal cortex of macaque. *Exp Brain Res* 1988;70:216–220.
5. Hikosaka O, Wurtz RH. Visual and oculomotor functions of monkey substantia nigra pars reticulata. III. Memory-contingent visual and saccade responses. *J Neurophysiol* 1983;49:1268–1284.
6. Barash S, Bracewell RM, Fogassi L, et al. Saccade-related activity in the lateral intraparietal area. II. Spatial properties. *J Neurophysiol* 1991;66:1109–1124.
7. Barash S, Bracewell RM, Fogassi L, et al. Saccade-related activity in the lateral intraparietal area. I. Temporal properties; comparison with area 7a. *J Neurophysiol* 1991;66:1095–1108.
8. Bracewell RM, Mazzoni P, Barash S, et al. Motor intention activity in the macaque's lateral intraparietal area. II. Changes of motor plan. *J Neurophysiol* 1996;76:1457–1464.
9. Snyder LH, Batista AP, Andersen RA. Coding of intention in the posterior parietal cortex. *Nature* 1997;386:167–170.
10. Snyder LH, Batista AP, Andersen RA. Change in motor plan, without a change in the spatial locus of attention, modulates activity in posterior parietal cortex. *J Neurophysiol* 1998;79:2814–2819.
11. Barbas H, Mesulam MM. Organization of afferent input to subdivisions of area 8 in the rhesus monkey. *J Comp Neurol* 1981;200:407–431.
12. Lynch JC, Graybiel AM, Lobeck LJ. The differential projection of two cytoarchitectonic subregions of the inferior parietal lobule of macaque upon the deep layers of the superior colliculus. *J Comp Neurol* 1985;235:241–254.
13. Snyder LH, Batista AP, Andersen RA. Intention-related activity in the posterior parietal cortex: a review. *Vision Res* 2000;40:1433–1441.
14. Powell KD, Goldberg ME. Response of neurons in the lateral intraparietal area to a distractor flashed during the delay period of a memory-guided saccade. *J Neurophysiol* 2000;84:301–310.
15. Sereno AB, Maunsell JH. Shape selectivity in primate lateral intraparietal cortex. *Nature* 1998;395:500–503.
16. Gottlieb JP, Kusunoki M, Goldberg ME. The representation of visual salience in monkey parietal cortex. *Nature* 1998;391:481–484.
17. Robinson DL, Bowman EM, Kertzman C. Covert orienting of attention in macaques. II. Contributions of parietal cortex. *J Neurophysiol* 1995;74:698–712.
18. Colby CL, Duhamel JR, Goldberg ME. Visual, presaccadic, and cognitive activation of single neurons in monkey lateral intraparietal area. *J Neurophysiol* 1996;76:2841–2852.
19. Platt ML, Glimcher PW. Neural correlates of decision variables in parietal cortex. *Nature* 1999;400:233–238.
20. Kusunoki M, Gottlieb J, Goldberg ME. The lateral intraparietal area as a salience map: the representation of abrupt onset, stimulus motion, and task relevance. *Vision Res* 2000;40:1459–1468.
21. Bruce CJ, Goldberg ME. Primate frontal eye fields: I. Single neurons discharging before saccades. *J Neurophysiol* 1985;53:603–635.
22. Waitzman DM, Ma TP, Optican LM, et al. Superior colliculus neurons provide the saccadic motor error signal. *Exp Brain Res* 1988;72:649–652.
23. Wurtz RH. Visual receptive fields of striate cortex neurons in awake monkeys. *J Neurophysiol* 1969;32:727–742.
24. Yantis S, Jonides J. Abrupt visual onsets and selective attention: evidence from visual search. *J Exp Psychol Hum Percept Perform* 1984;10:601–621.
25. Yantis S, Jonides J. Attentional capture by abrupt onsets: new perceptual objects or visual masking? *J Exp Psychol Hum Percept Perform* 1996;22:1505–1513.
26. Egeth HE, Yantis S. Visual attention: control, representation, and time course. *Annu Rev Psychol* 1997;48:269–297.
27. Gawne TJ, Martin JM. Responses of primate visual cortical neurons to stimuli presented by flash, saccade, blink, and external darkening. *J Neurophysiol* 2002;88:2178–2186.
28. Gottlieb J, Goldberg ME. Activity of neurons in the lateral intraparietal area of the monkey during an antisaccade task. *Nat Neurosci* 1999;2:906–912.
29. Everling S, Fischer B. The antisaccade: a review of basic research and clinical studies. *Neuropsychologia* 1998;36:885–899.
30. Zhang M, Barash S. Neuronal switching of sensorimotor transformations for antisaccades. *Nature* 2000;408:971–975.
31. Bisley JW, Goldberg ME. Neuronal activity in the lateral intraparietal area and spatial attention. *Science* 2003;299:81–86.
32. Bashinski HS, Bacharach VR. Enhancement of perceptual sensitivity as the result of selectively attending to spatial locations. *Percept Psychophysiol* 1980;28:241–248.
33. Sperling G, Weichselgartner E. Episodic theory of the dynamics of spatial attention. *Psychol Rev* 1995;102:503–532.
34. Shibutani H, Sakata H, Hyvarinen J. Saccade and blinking evoked by microstimulation of the posterior parietal association cortex of the monkey. *Exp Brain Res* 1984;55:1–8.
35. Thier P, Andersen RA. Electrical microstimulation distinguishes distinct saccade-related areas in the posterior parietal cortex. *J Neurophysiol* 1998;80:1713–1735.
36. Robinson DA, Fuchs AF. Eye movements evoked by stimulation of frontal eye fields. *J Neurophysiol* 1969;32:637–648.
37. Schiller PH, Stryker M. Single-unit recording and stimulation in superior colliculus of the alert rhesus monkey. *J Neurophysiol* 1972;35:915–924.
38. Bruce CJ, Goldberg ME, Bushnell MC, et al. Primate frontal eye fields. II. Physiologic and anatomic correlates of electrically evoked eye movements. *J Neurophysiol* 1985;54:714–734.
39. Tehovnik EJ, Slocum WM, Schiller PH. Differential effects of laminar stimulation of V1 cortex on target selection by macaque monkeys. *Eur J Neurosci* 2002;16:751–760.
40. Wardek C, Olivier E, Duhamel JR. Saccadic target selection deficits after lateral intraparietal area inactiavtion in monkeys. *J Neurosci* 2002;22:9877–9884.
41. Li CS, Mazzoni P, Andersen RA. Effect of reversible in-

activation of macaque lateral intraparietal area on visual and memory saccades. *J Neurophysiol* 1999;81: 1827–1838.
42. Shepherd M, Findlay JM, Hockey RJ. The relationship between eye movements and spatial attention. *Q J Exp Psychol* 1986;38:475–491.
43. Hoffman JE, Subramaniam B. The role of visual attention in saccadic eye movements. *Percept Psychophysiol* 1995;57:787–795.
44. Kowler E, Anderson E, Dosher B, et al. The role of attention in the programming of saccades. *Vision Res* 1995;35:1897–1916.
45. Deubel H, Schneider WX. Saccade target selection and object recognition: evidence for a common attentional mechanism. *Vision Res* 1996;36:1827–1837.
46. Hallett PE, Lightstone AD. Saccadic eye movements to flashed targets. *Vision Res* 1976;16:107–114.
47. Mays LE, Sparks DL. Dissociation of visual and saccade-related responses in superior colliculus neurons. *J Neurophysiol* 1980;43:207–232.
48. Goldberg ME, Colby CL, Duhamel J-R. The representation of visuomotor space in the parietal lobe of the monkey. *Cold Spring Harbor Symp Quant Biol* 1990; 55:729–739.
49. Duhamel JR, Colby CL, Goldberg ME. The updating of the representation of visual space in parietal cortex by intended eye movements. *Science* 1992;255:90–92.
50. Kusunoki M, Goldberg ME. The time course of perisaccadic receptive field shifts in the lateral intraparietal area of the monkey. *J Neurophysiol* 2003; in press.
51. Morrone MC, Ross J, Burr DC. Apparent position of visual targets during real and simulated saccadic eye movements. *J Neurosci* 1997;17:7941–7953.
52. Honda H. The time courses of visual mislocalization and of extraretinal eye position signals at the time of vertical saccades. *Vision Res* 1991;31:1915–1921.
53. Dassonville P, Schlag J, Schlag-Rey M. Oculomotor localization relies on a damped representation of saccadic eye displacement in human and nonhuman primates. *Vis Neurosci* 1992;9:261–269.
54. Heide W, Blankenburg M, Zimmermann E, Kompf D. Cortical control of double-step saccades: implications for spatial orientation. *Ann Neurol* 1995;38:739–748.
55. Li CS, Andersen RA. Inactivation of macaque lateral intraparietal area delays initiation of the second saccade predominantly from contralesional eye positions in a double-saccade task. *Exp Brain Res* 2001;137: 45–57.

10

Sensorimotor Integration in Posterior Parietal Cortex

Richard A. Andersen and Christopher A. Buneo

Division of Biology, California Institute of Technology, Pasadena, California

INTRODUCTION

The view of the functioning of the posterior parietal cortex (PPC) has evolved over time. In the 19th century this area of cortex was believed to be an "association area," responsible for associating different sensory modalities. In this conceptual framework the PPC was considered to have a purely sensory role. Further refinement of the sensory role of PPC occurred in the last century based on observations of deficits following lesions in humans and nonhuman primates. This and other research led to the influential proposal of Ungerleider and Mishkin (1) that there are two functional pathways in visual cortex, a dorsal pathway that includes the PPC and is involved in spatial perception, and a ventral pathway involved in object perception.

Parallel neurophysiologic studies of behaving monkeys by Mountcastle and colleagues (2), identified correlations between single cell activity and the behaviors of the animals, including reaching, fixation, saccades, and smooth pursuit eye movements. They proposed that PPC was involved in programming motor behaviors. This view was challenged at the time by Goldberg and colleagues (3,4), who found similar results to Mountcastle and associates, but interpreted the neural activity concomitant with movement as resulting from sensory or attentional processes, rather than motor processes.

Examination of cell activity during control experiments that separated sensory- from behavior-related activity determined that both are found in PPC (5). This finding led to the proposal that PPC is neither strictly sensory nor motor in function, but rather is important for sensorimotor integration. More recent clinical data from humans also support this view. For example, Goodale, Milner, and colleagues (6) have shown that shape is represented in the PPC specifically for action planning. They refined the dichotomy of Ungerleider and Mishkin, proposing that the dorsal pathway is involved in sensorimotor processing, with spatial processing being one aspect of action planning.

Thus, the concept that PPC is important for sensorimotor integration is generally accepted. The main points of contention now relate to where the area sits along this sensorimotor continuum (7,8). The goal of the current chapter is to examine four partially overlapping views of the PPC and its role in sensorimotor integration. One view posits that there exists a high degree of functional specificity within the PPC, and that this specificity is in the form of separate cortical fields or subregions. A second proposes that, in regard to the issue of position along the sensorimotor continuum, PPC occupies a largely intermediate position. The third proposes that the different functional subdivisions within PPC are nodes in specific, distributed networks, and share features in common with nodes located in the PPC and frontal lobe. The final view holds that PPC exhibits functions that are represented throughout the cerebral cortex, including attention and learning. However, this view also holds that a role in these functions is only evident during processes specific to the particular function of individual cortical areas. Thus, attention and learning in PPC are related to sensorimotor processes, whereas these same functions in the ventral pathway are related to object recognition.

This chapter examines these four views by concentrating largely on two subregions of the PPC that have been the focus of many studies, the lateral intraparietal area (LIP) and the parietal reach region (PRR), although studies from most parts of PPC are considered as well. At the end of the chapter we attempt to synthesize these four views of PPC function using LIP and PRR as examples. All four views are insightful and complementary avenues for understanding how the PPC performs sensorimotor transformations.

SPECIFICITY

The posterior parietal cortex was originally recognized, based on cytoarchitectural criteria, to contain three different cortical areas, Brodmann's areas 5, 7a, and 7b (9), or areas PE, PG, and PF of von Economo and Koskinas (10), and von Bonin and Bailey (11). In recent years these areas have been further parceled based on anatomic, clinical, and physiologic criteria (12–19). An emerging view is that PPC follows the same rules as sensory and motor cortices, containing a large number of specialized regions (Fig. 10-1). We review evidence for functional specialization within PPC in the following.

Functional Subdivisions of the Posterior Parietal Cortex in Macaque Monkeys

A most remarkable observation is the specificity of connections of different parts of the PPC. Small regions of cortex, often adjacent to one another and no bigger than one fourth of the area of one's small fingernail, can have very different patterns of connectivity (12). In recent years it has become apparent that cells within several of the anatomically defined subregions of the PPC also have very different response properties. The combination of these findings has led to the recognition of a number of cortical areas within the PPC, several of which are described herein.

Perhaps the most studied subregion of the PPC is area LIP, which is specialized for saccadic eye movements (20–27). This area was initially identified anatomically as an eye movement area based on its strong connections to other saccade regions, including the frontal eye fields (FEF) and the superior colliculus (SC) (12,28–30). LIP has been divided into ventral and dorsal subdivisions on myeloarchitectural and connectional grounds (15,29). Cells in both subdivisions have saccade related responses that precede eye movements (29). In instructed-delay tasks LIP neurons typically have stronger activity when monkeys are planning saccades rather than reaches, indicating that a large component of this delay activity is related to saccade planning (26). A similar area has been identified in a number of functional magnetic resonance imaging (fMRI) experiments in humans (31,32). Reversible inactivation of this area produces eye movement deficits in memory saccade tasks (23), and biases decisions toward the healthy field in eye movement tasks (33). Finally, electrical stimulation of this area generates saccades (27).

The cells in LIP represent primarily the contralateral visual field in retinal coordinates (21). These response fields are gain modulated by other body position signals, including eye position signals and proprioceptively derived head position signals (34,35). Thus, this area can conceivably represent other coordinate frames in a distributed fashion (see Gain Fields). Neurons in LIP are also sensitive to auditory stimuli, but generally only when the auditory stimuli are targets for saccades (36,37). The activity of LIP neurons also code decision variables related to eye movements (38).

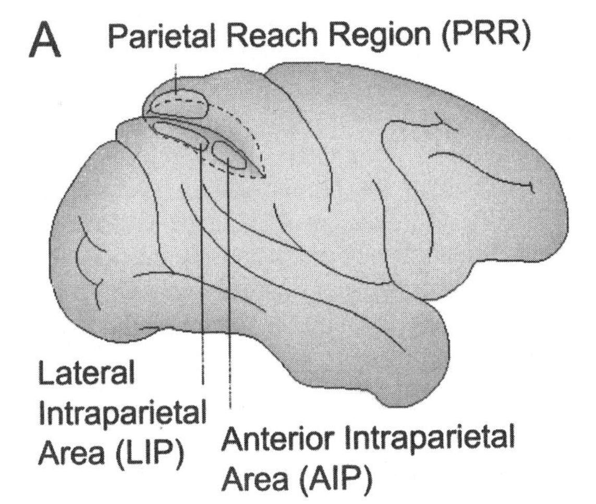

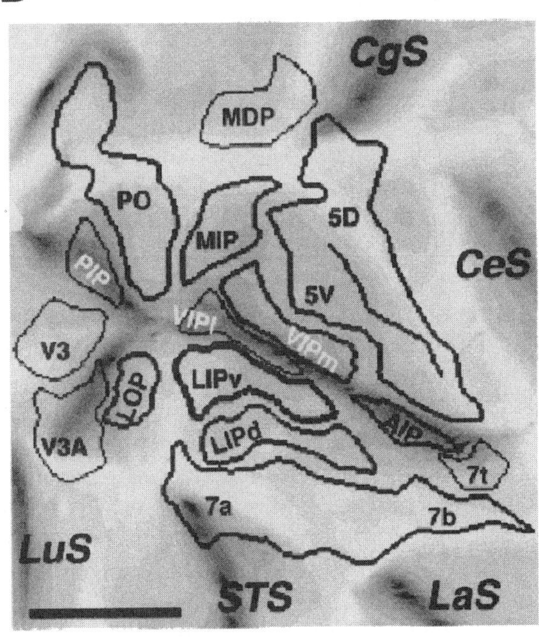

FIG. 10-1. A: Diagram of the cerebral cortex of a macaque monkey, with the intraparietal sulcus (IPS) opened up to reveal the approximate locations of the lateral intraparietal area (LIP), the parietal reach region (PRR), and the anterior intraparietal area (AIP). (Adapted from: Cohen YE, Andersen RA. A common reference frame for movement plans in the posterior parietal cortex. *Nat Rev Neurosci* 2002;3:553–562, with permission.) **B:** Flat map representation of the cortex in and near the IPS of a single macaque monkey. Seventeen architectonically defined subregions are shown, including the dorsal and ventral subdivisions of LIP (LIPd and LIPv, respectively), the medial and lateral subdivisions of the ventral intraparietal area (VIPm and VIPl, respectively), and AIP. The PRR consists of the subdivision labeled MIP (medial intraparietal area) and also possibly parts of those labeled PO (parieto-occipital area) and 5v (ventral subdivision of area 5). (Adapted from Lewis JW, Van Essen DC. Mapping of architectonic subdivisions in the macaque monkey, with emphasis on parieto-occipital cortex. *J Comp Neurol* 2000; 428:79–111, with permission.)

A second eye field, the medial parietal area (MP) has also been identified in PPC (39). This region is located on the medial surface of the hemisphere, above the cingulate sulcus, and has been much less studied than LIP. It appears to be coextensive with anatomically defined area 7ip (40) (also called PGm) (16). The cells in this region show delay and saccade-related responses similar to LIP neurons, and electrical stimulation of this region also provokes eye movements (39).

Several subregions of the PPC have been shown to be active during arm movements including dorsal area 5 (PE), PEc, V6A, 7a, and 7m (41–45). The PRR is a newly described arm movement area of the PPC that includes area MIP and parts of V6A and area 5 residing within the anterior bank of the intraparietal sulcus. Area MIP receives input from areas V6A and 7m, which have direct connections to cortical visual areas, and projects to areas of the prefrontal and frontal cortex that process limb movements (46). Cells in this region have delay activity related to plans for limb movements, but not eye movements (26). A striking feature of this area is that the coding of these limb movements is in eye, not limb, coordinates (47,48).

Area VIP lies in the floor of the intraparietal sulcus. This region appears to be highly multimodal, even for sensory stimuli that are not task relevant (49,50). In particular, many neurons have somatosensory fields on the head, and respond when visual stimuli approach those head locations (50). The visual responses of cells in VIP are found in head and retinal coordinates, with a substantial population being intermediate between the two (51).

Studies by Sakata and colleagues (52,53) point to the anterior intraparietal area (AIP) being specialized for grasping. Cells in this area respond to the shapes of objects and the configuration of the hand for grasping the objects. Reversible inactivations of AIP produce deficits in shaping the hand prior to grasping in monkeys. This deficit is reminiscent of problems in shaping the hands found in humans with parietal lobe damage (54).

The medial superior temporal area (MST) appears to play a specialized role in smooth pursuit eye movements. Cells in this area are active for pursuit, even during brief periods when the pursuit target is extinguished (55). Inactivations of this area produce pursuit errors that are not a result of sensory deficits (56). The cells in this cortical area are sensitive to optic flow stimuli (57), and compensate for visual motion during smooth pursuit eye movements and head movements to maintain invariant representations of heading direction (58,59). These studies suggest that MST plays not only a role in pursuit eye movements, but also a role in navigation using visual motion cues.

Human fMRI experiments suggest the functional anatomy of the human and monkey PPC is similar. Rushworth and colleagues (32) reported that peripheral attention tasks activate the lateral bank of the intraparietal sulcus, whereas planning manual movements activates the medial bank. They concluded that their results have strong similarities to those found in monkey studies, with the medial bank of the human intraparietal sulcus specialized for manual movements and the lateral bank for attention and eye movements. An area specialized for grasping has also been identified in the anterior aspect of the intraparietal sulcus in humans (31,60), which may be homologous to monkey AIP.

INTERMEDIATE PROPERTIES

A second, and very natural view of the PPC is that, given its intermediate anatomic position between sensory and motor cortices, it occupies an intermediate stage in the sensorimotor transformation process. It also appears to be the location where different sensory modalities are combined, again implying an intermediate stage in the sensorimotor pathway.

An issue that has proven tractable for study in sensorimotor cortex is the transformation between coordinate frames. Sensory stimuli are gathered and represented in early parts of the nervous system in coordinate frames that are different from the eventual motor reference frames required to execute movements. For instance, visual stimuli are initially represented in retinal coordinates, but to reach to a visual target requires a transformation to muscle coordinates to move the limb. Studies of this issue in PPC suggest that this area occupies an intermediate processing stage during sensorimotor transformations. We use the topic of coordinate transformations to illustrate the intermediate nature of PPC in the next section. However, other sensorimotor processes also appear intermediate in nature in PPC. For instance, movement plans appear more cognitive and abstract in PPC than at subsequent stages in motor cortex.

Gain Fields

Gain fields represent a form of intermediate representation used to compute transformations between reference frames. For instance, in many of the areas within the PPC cells have eye-centered response fields, but are also gain modulated by body position signals (Fig. 10-2B). These "gain field" effects are found throughout the PPC and include modulation of retinotopic fields by eye, head, body, and limb position signals (34,35, 61–63). Although referred to as gain fields, the modulation effects can be additive as well as multiplicative (64). Theoretical studies suggest that gain fields are a computational mechanism for transforming information between coordinate frames (65,66). In fact, small populations of neurons with retinal response fields, modulated by various body part position signals, can be read out in multiple frames of reference (67,68) as would be needed to direct movements of the eyes, head, or hands. These results suggest that the PPC represents space in a distributed fashion, with groups of cells potentially representing multiple reference frames.

Although originally identified in areas of the PPC, gain effects have been subsequently identified throughout the brain, including the dorsal premotor cortex, V1, V4, and the SC (69–72). These findings suggest that multiplicative and additive interactions between different inputs to neurons may reflect a general method of neural computation. Although the role of gain fields in coordinate transformations has been highlighted in this chapter, gain fields appear to play a role in many other functions, including attention, navigation, decision making, and object recognition. These other functions have been reviewed by Salinas and Thier (73).

Multisensory Integration

Several areas of PPC are considered to be primarily visual in terms of sensory inputs (e.g., areas MST, 7a, PO, and LIP), whereas others are considered primarily somatosensory (e.g., dorsal area 5), or somatosensory and visual (e.g., 7b and VIP). Interestingly, area PRR has been shown recently to code locations in eye-centered coordinates, which is surprising given the area's major role in limb movements (47). However, recent experiments have demonstrated that many neurons in area 5 and VIP have response fields that are partially shifted between eye and limb-centered representations (for area 5) (74) or eye- and head-centered representations (for VIP) (51). These types of responses can be interpreted as reflecting an encoding of spatial location in both reference frames (74,75), or as reflecting the existence of a single but ambiguously defined "intermediate" reference frame (Fig. 10-2D). Regardless, these results suggest that vision exerts a strong influence over spatial behaviors, which is perhaps not surprising given its superior spatial acuity.

Both areas LIP and PRR respond to auditory stimuli if the animal is planning a saccade (in LIP) (36,37,76) or a reach (in PRR) (77) to the auditory target. Interestingly, in both areas many cells encode sound locations in eye-centered coordinates, whereas others encode them in head-centered coordinates, or in coordinates intermediate between these two reference frames (77). Area LIP appears to respond poorly, if at all, to auditory targets when the targets do not have behavioral significance to the animal (36,76). Similarly, LIP neurons only appear to show color selectivity if the color is relevant to eye movement planning (78). These results suggest the areas are visual by default, but other sensory modalities can be gated into them depending on the requirements of the task.

Neural network models that perform multisensory integration typically have an intermediate step in which units have receptive fields that are gain modulated by eye and other body position signals (68,75,79). Often these receptive fields show intermediate coding, for instance, with the integration of auditory and visual signals some intermediate units demonstrate response fields midway between eye-centered and head-centered coordinates. The finding of gain fields and intermediate coordinate frames in PPC is suggestive that multisensory integration and coordinate transformations are taking place in this cortical area. In particular, this seems to be the

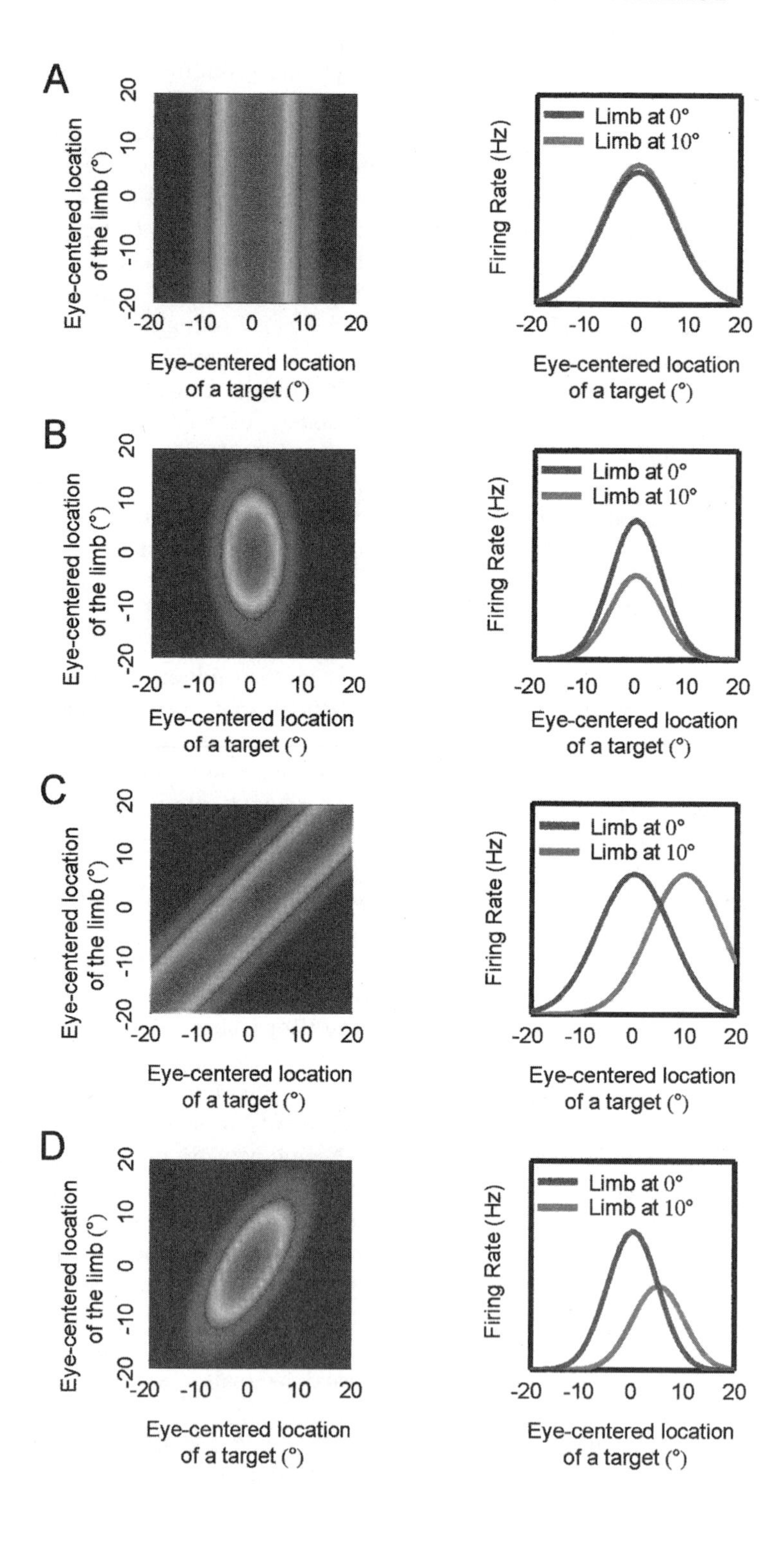
A
B
C
D
Eye-centered location of the limb (°)
Eye-centered location of a target (°)
Firing Rate (Hz)
Limb at 0°
Limb at 10°
-20 -10 0 10 20

case for transformations of auditory receptive fields from head-centered coordinates, which is typical of lower levels of the auditory system, to eye-centered coordinates, which are found in LIP and PRR. This can be surmised from the recent finding that an auditory cortical area that provides major projections into PPC (area Tpt) has auditory fields in head-centered coordinates that are gain modulated by eye position (80). On the other hand, LIP and PRR have auditory response fields in head, intermediate, and eye-centered coordinates, as well as gain field effects, suggesting that the sites of integration and transformation include these two parietal areas.

Direct Transformations

Although the preceding results suggest that there are intermediate steps in coordinate transformations, the number of steps may be limited. An example considered here is that of visually guided reaching movements. In transforming the location of a visual target for a reach from retinal to limb-centered coordinates, must the brain also use intermediate representations in head and body-centered coordinates, or can this transformation be performed more directly?

A scenario that uses explicit head- and body-centered representations, which we call the *sequential* model, is illustrated in Figure 10-3A. First visual signals in retinal coordinates are combined with eye position signals to represent targets in head-centered coordinates. Next, head position is combined with the representation of target location in head-centered coordinates to form a target representation in body-centered coordinates. In the last step, the current location of the limb, in body-centered coordinates, is subtracted from the location of the target, in body-centered coordinates, to generate the motor vector, in limb-centered coordinates. A drawback of this approach is that it requires several stages and separate computations, which would likely require a large number of neurons and cortical areas. Moreover, although there are some reports of cells in the PPC coding visual targets in extraretinal, perhaps head-centered, coordinates (51,81), the vast majority of PPC cells code visual targets in eye-centered coordinates or coordinates intermediate between eye and other reference frames (47,74,77).

One alternative scenario that uses only a single intermediate step is referred to as the *combinatorial* model (41). In this model (Fig. 10-3B), retinal target location, eye, head, limb position, and other body position signals are all combined at once, and the target location in limb-centered coordinates is then read out from this representation. A potential problem with this model is the "curse of dimensionality," which results from a combinatorial explosion when tiling a space for a large number of parameters. For example, if it takes 10 cells to tile each dimension in visual space, and 10 for each dimension of eye position, head position, and so on, the number of cells required to represent all possible combinations of such signals quickly becomes larger than the number of cells in PPC. One method the PPC ap-

FIG. 10-2. A–D: Responses of idealized posterior parietal cortex neurons. In the left column, responses are plotted for a range of target locations (x-axis) and initial limb locations (y-axis) along the horizontal, in eye-centered coordinates. In the right column, two tuning curves are shown, representing slices though each response field at initial hand locations of 0 degrees and 10 degrees. **A:** Responses of a neuron encoding target location in eye-centered coordinates. The tuning curves corresponding to the different starting locations are identical. **B:** Responses of a neuron encoding target location and limb location in eye-centered coordinates. The tuning curves are gain-modulated, but this gain effect is also represented in eye coordinates. **C:** Neuron encoding target location in hand-centered coordinates. The tuning curves are shifted with respect to one another, with the magnitude of the shift corresponding to the difference between the two starting locations. No gain modulation is observed. **D:** Neuron encoding target location in hand-centered and eye-centered coordinates. Tuning curves are partially shifted and gain-modulated.

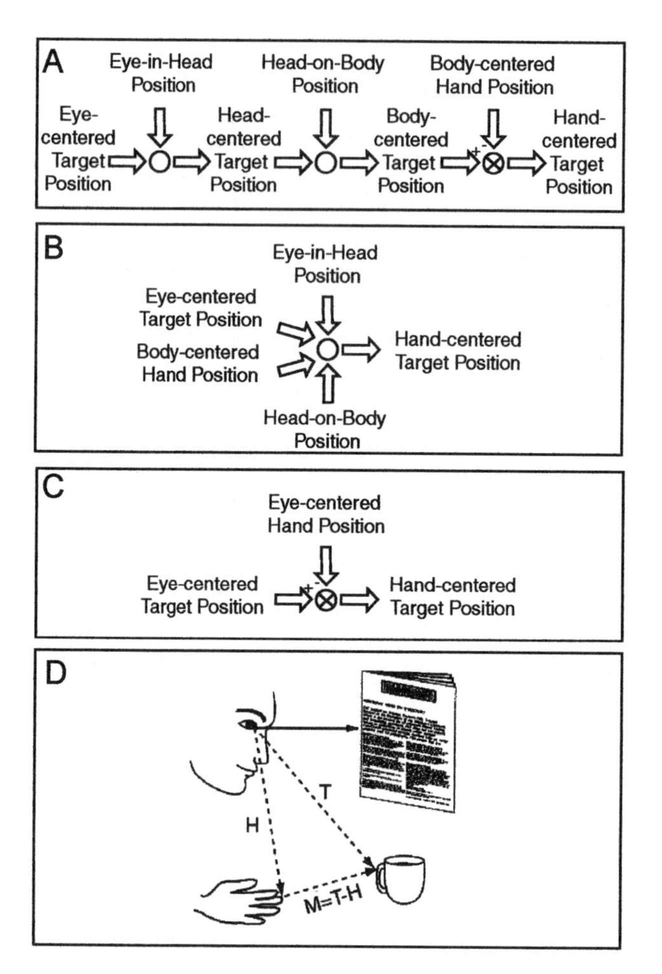

FIG. 10-3. A–C: Schemes for transforming target position from eye-centered to hand-centered coordinates. **A:** Sequential method. **B:** Combinatorial method. **C:** Direct method. **D:** Illustration of reaching for a cup while fixating a newspaper, using the direct method. The position of the cup with respect to the hand (M) is obtained by directly subtracting hand position (H) from target position (T), both in eye coordinates. (Adapted from: Andersen RA, Buneo CA. Intentional maps in posterior parietal cortex. *Annu Rev Neurosci* 2002;25: 189–220; Buneo CA, Jarvis MR, Batista AP, et al. Direct visuomotor transformations for reaching. *Nature* 2002;416:632–636, with permission.)

parently used to avoid this curse of dimensionality is to code only a limited number of variables in each of its subdivisions (35). In fact, knowledge of the set of variables encoded by an area may provide clues to its specific functions.

Another scenario that also uses one intermediate step is the *direct* model. Figure 10-3C,D shows how this model can produce a transformation in one step by vector subtraction: The current position of the hand, in eye coordinates is subtracted from the position of the target, also in eye coordinates, to directly generate the movement vector in limb coordinates. This computation is accomplished with neural circuits by representing a limited number of parameters in the same coordinate frame, and by multiplying them together (i.e., gain fields). An advantage of this approach over the sequential method is that it requires fewer computational stages. In addition, the computation is restricted to only dimensions in eye coordinates, and does not suffer the "curse of dimensionality" of the combinatorial approach.

A recent study provides evidence for the direct transformation mechanism. Single cells in area 5 of PPC, classically considered a somatomotor region, were found to code target locations simultaneously in eye- and limb-centered coordinates (74). This result is consistent with the PPC transforming target locations directly between these two frames of reference. Moreover, cells in the adjacent PRR code target location in eye-centered coordinates, and there is gain modulation by initial hand position, also coded in eye-centered coordinates (62). These two findings suggest a simple gain field mechanism underlies the transformation from eye- to limb-centered coordinates. A convergence of input from gain-modulated cells in PRR onto area 5 neurons can perform such a transformation directly without having to resort to additional stages or to a large combination of retinal, eye, head, and limb position signals (65). Note that, although explicit head- and body-centered representations are not needed in this scheme, eye and head position information would still be needed at later stages to convert the movement vector into muscle coordinates. It remains to be seen whether these signals are carried by cells in PRR and/or area 5, or by different populations of neurons.

Psychophysical evidence supporting a sequential-type model has been provided by Flanders et al. (82) and McIntyre et al. (83, 84). These results, as well as the physiologic study of Buneo et al. (74) supporting an alternative direct scheme, may reflect an underlying context dependence in the coordinate transformations that subserve visually guided reaching (85). For instance, direct transformations may be the preferred scheme when both target location and the current hand position are simultaneously visible, even for a brief instant. On the other hand, a sequential scheme may be used when visual information about the current position of the hand is not available.

Lesions

The deficits observed following lesions of the PPC are consistent with the area playing an intermediate role in sensorimotor integration. Patients with PPC lesions do not suffer from primary sensory or motor deficits. However, numerous defects become apparent when they attempt to connect perception with action, for instance, during sensory-guided movements. Such patients often suffer from optic ataxia, a difficulty in estimating the location of stimuli in three-dimensional space. This deficit results in pronounced errors in reaching movements (86,87). Patients with PPC lesions often show one or more of the apraxias, a class of deficits characterized by the inability to plan movements (88). These defects can range from a complete inability to follow verbal commands for simple movements, to difficulty in performing sequences of movements. Patients with PPC damage also have difficulty correctly shaping their hands in preparation to grasp objects. This grasp deficit again points to a disconnection between the visual sensory apparatus that registers the shape of objects, and the motor systems that shape the configuration of the hand (6,54). A most thorough review detailing the neurologic deficits following parietal lobe damage can be found in Macdonald Critchley's classic book on the subject, *The Parietal Lobes* (89).

Lesions do produce different deficits depending on the location of the damage. These results provide evidence for specificity within the PPC, as outlined in the preceding. For instance, extinction (the lack of awareness of objects in the unhealthy side of the visual field when there are competing stimuli toward the healthy side) is more common with lesions of the superior parietal lobule, whereas profound neglect is more common with lesions of the inferior parietal lobule (90). However, lesions are often large and, of course, the product of accident and not experiment; as a result, it is difficult to precisely differentiate different parts of PPC based on patient data. In fact, there is even now some debate regarding whether neglect is the result of inferior parietal lobule or superior temporal gyrus damage (91).

NETWORK PROPERTIES

Parietal and frontal cortical areas are strongly interconnected via cortical–cortical connections, and via loops through subcortical structures (92). Laboratories that have directly compared the response selectivity of areas LIP and FEF in saccadic eye movement tasks have noted a surprising similarity between these two areas. This similarity has led to the proposal that they are best understood as being parts of a network, rather than separate stages in neural processing (92,93). Moreover, the parieto-frontal circuits involved in eye movements, reaching, and grasping appear to be parallel and at least partially separated, leading to the proposal of separate networks for different movement behaviors (94).

As an example of network behaviors we examine the similarities between areas LIP and PRR. These two areas are anatomically interconnected by reciprocal corticocortical connections (29), and as such form a part of a network. LIP and PRR seem to use a number of the same operational rules, even though they are involved in very different behaviors (i.e., eye and arm movements). This similarity in operations may facilitate cooperation between these areas during behaviors requiring coordinated movements of the eyes and hands.

Coordinates

Eye and limb movements require very different coordinate frames near the output stage of a sensorimotor transformation, because of the unique mechanical and geometric properties of the different end effectors. However, neurons in both LIP and PRR encode visual targets using the same, eye-centered coordinate frame (47,48,77,95). Interestingly, the response fields of LIP and PRR neurons remain eye-centered during the sensory, delay and movement epochs of memory-saccade and memory-reach tasks, indicating that the coordinate frame remains fixed over the course of the sensorimotor transformation for these neurons.

Both areas also have response fields for auditory targets that are similar. A large proportion of cells encode auditory responses in eye-centered coordinates, whereas others encode locations in some other coordinate frame (77,95). Because head positions were not varied in these experiments, it cannot be established whether the "other" coordinate frame was a head-centered one, the most likely result for these auditory targets, or some additional reference frame (e.g., body- or world-centered). Although visual response fields, and many auditory response fields, are eye-centered in both LIP and PRR, they are also gain modulated by eye and body position signals in both areas. LIP neurons are gain modulated by eye position and neck proprioceptive signals but generally not by vestibularly derived head position signals (unlike 7a cells, which do have vestibular gain fields) (34,35). PRR neurons are gain modulated by eye and limb position signals (62,63,74). LIP has not yet been tested for limb position gain fields, nor has PRR been tested for head position gain fields.

Delay Activity

A striking feature of both LIP and PRR is strong, persistent activity when animals have planned movements but are withholding their responses. This phenomenon was first observed in LIP using a "memory" saccade paradigm (22) designed to separate sensory from movement-related activity (96). Snyder et al. (26) subsequently demonstrated that the presence of delay period activity in both LIP and PRR depends on the type of movement being planned. Delay period activity is strong in LIP when eye movements are being planned and weak or nonexistent for planned limb movements and vice versa for PRR.

Next Planned Movement

The activity of LIP and PRR neurons in the delay period reflects the next movement the animal plans, and not the memory of a sensory location. This next-movement feature has been demonstrated with sequential movement tasks, in which the animal is required to remember two target locations, and then make sequential movements to the two targets after a GO instruction. In both LIP and PRR, cells are not active when the animal is planning a movement outside the response field of the cell, even though the subsequent movement target location is within the cell's response field (97,98).

Compensation for Eye Movements

Encoding saccade and reach plans in eye-centered coordinates could be problematic in instances where a movement plan is formed and an intervening saccade is made before the movement is executed. In such cases, particularly when movements are planned to remembered locations in the dark, movements could be inaccurate if no compensation for the intervening saccade occurs, with the size of the error being directly related to the size of the eye movement. Mays and Sparks (99) were the first to probe the effects of intervening saccades on eye movements and perceptual stability while recording in the superior colliculus (SC), which contains an eye movement map in retinal coordinates (100). They found that, under these circumstances, activity shifts within the eye movement map of the SC to compensate for the intervening saccade and still codes the correct motor vector. Gnadt and Andersen (22) reported a similar result for saccade planning in area LIP. Duhamel et al. (101) extended the LIP results by showing that it is not necessary to make an eye movement for this updating to take place.

The same compensation for intervening saccades has been observed during reach planning in PRR also (47). When monkeys plan a reach to a remembered location in the dark, and the animals are required in the task to make an intervening saccade prior to the reach, all PRR cells recorded showed a shift in activity within the eye-centered map to compensate for the eye movement. A remapping of reach plans in eye coordinates has been demonstrated psychophysically in humans as well (102), consistent with this physiologic finding.

Intention

A number of studies have demonstrated that a significant component of both LIP and PRR activity is related to the intention to make a movement, an eye movement in the case of LIP, and a reach movement in the case of PRR. Evidence for an intentional role for LIP and PRR activity (as opposed to purely sensory or attentional roles) comes from studies employing "antireach" tasks, in which animals are trained to make a reach in the opposite direction to the location of a visual stimulus. Such tasks have revealed that activity within area MIP (a part of PRR) is related mostly the direction of the movement, and not the location of the stimulus (44,103). Gottlieb et al. (104) reported that the reverse is true in the lateral intraparietal area (LIP) for "antisaccade" tasks; that is, LIP neurons responded to the stimulus and not the direction of planned movement. However, a subsequent report by Zhang and Barash (105) indicated that, after a brief transient linked to the stimulus, most LIP neurons code the direction of the planned eye movement.

Experiments have been specifically designed to separate the effects of spatial attention from those of intention also (26,106, 107). In one experiment, recordings were made from areas LIP and PRR while animals attended to a flashed target and planned a movement to it during a delay period; however, in one case they were instructed to plan a saccade and in the other a reach. If neurons are selective for attention, then they should be active in both conditions; but if they are selective for intention, then they should be active in only one of the two conditions. Snyder et al. (26) found that the latter was true; the large majority of PRR neurons were active in the delay only when the monkeys were planning reaches, and conversely LIP neurons were active only when planning saccades. A subsequent experiment showed that activity in the PPC is also related to the shifting of movement plans, when spatial attention is held constant (107). Cells with a preference for a particular type of movement (reach or saccade) showed increased activity if a plan was changed from the nonpreferred to the preferred movement type (for the same target location) but not when the nonpreferred (or preferred) plan was simply reaffirmed. This result is reminiscent of proposals that the PPC plays a role in shifting attention (108), but in this case it is the intended movement that shifts, and not the spatial locus of attention. Finally, when a monkey is cued as to whether an upcoming trial will be a reach or a saccade, activity in LIP and PRR increases selectively for saccades (LIP) and reaches (PRR) (106). Thus, these two areas show differential responses depending on the type of movement that is planned, even before the location of the target has been specified.

Default Planning

Both areas LIP and PRR show default or covert planning; that is, stimuli that are often behaviorally relevant can result in planning activity, even though this plan may not be executed or may change. This issue was directly addressed in the study of Snyder et al. (26), which, found a high degree of movement specificity for LIP and PRR, as mentioned. However, this specificity was not complete—68% of recorded neurons were significantly modulated in the delay period by one movement plan (reach or saccade) but not the other. (Interestingly, even during the cue period 44% showed this specificity.) We reasoned that the remaining cells showing significant activity for both movement plans might reflect covert plans for movement, because it is very natural to look to where you reach. To control for this possibility the animals also performed a "dissociation" task, in which they simultaneously planned an eye and an arm movement in different directions, with one movement into the response field and the other outside. Sixty-two percent of the cells that were not specific for single movements were specific in the dissociation task, bringing to 84% the number of cells that showed movement planning specificity in the delay period. Interestingly, more cells also revealed specificity for the cue response as well in the dissociation task, with a total of 45% being specific for reaches and 62% for saccades. These results suggest that, when given a single target for a reach (or saccade), a proportion of activity in LIP (and PRR) reflects a default plan to also make a saccade (or reach) to the target.

Default planning may explain activity that is seen in GO/NO-GO tasks. In these tasks a stimulus appears in the response field and the animal is later cued whether to make a movement to it or not. Reach activity in area 5 (109), and saccade activity in LIP (110) continues when the animal is cued not to move. This result is not consistent with attention or intention activity, because the target is no longer important to the animal's behavior (110). However, it is consistent with a covert or default plan, which remains if no new movement plans are being formed. Evidence for this alternative explanation comes from experiments where the plan is canceled, but a new movement plan is put in place (111). When an eye movement plan is canceled by instructing that the target location has changed, LIP neurons that coded the canceled plan fall silent, unlike the NO-GO results, and cells coding the new plan become active. A similar result is found even when the location of the planned movement is held constant, but the type of movement (from reach to saccade or saccade to reach) is changed (107). The data from these various studies indicate that default plans are formed in both PRR and LIP to stimuli of behavioral significance when no alternative plan is provided, but are erased if alternative plans are formed.

Movement Decisions

Both LIP and PRR show activity related to the decisions of an animal. Experiments in LIP by Platt and Glimcher (112) and by Shadlen and colleagues (113,114) have found activity related to the decision of a monkey to make eye movements. Both the prior probability and the amount of reward influence the effectiveness of visual stimuli in LIP, consistent with a role for this area in decision making. As monkeys accumulate sensory information to make a movement plan, activity increases for neurons in LIP and the prefrontal cortex (113–115). Similar results have been found in PRR recently, where the activity of cells reflects which of two targets monkeys select for a reach (38). The fact that decision-related activity is found in LIP for saccades, and in PRR for reaches, suggests that decisions are not made by a single brain area, but rather that decision making is a more distributed phenomenon with different networks and areas responsible for different decisions.

Dynamic Evolution of Intention Activity

LIP and PRR activity evolves dynamically, reflecting sensory, cognitive, and motor variables as the demands of a task change. For instance, the temporal dynamics of PRR activity differ depending on whether monkeys plan reaches to auditory targets or visual targets in a memory-reach task (116). At cue onset, activity for visually cued trials carried more information about spatial location than activity for auditory cued trials. However, as the trials progressed and the animal was preparing a movement, the amount of information regarding spatial location increased for the auditory cued trials, so that by the time of the reach movement it was not significantly different from information carried during visually cued trials. This result suggests that the spatial location of the target is represented in the early phases in the task, and is more poorly specified for auditory compared to visual targets (which is consistent with poorer sound localization compared to visual localization in primates). Later in the task, activity reflects the plan of the animal, which is the same for both sensory types, and thus explains the similarity in spatial information at this later period.

In another study, animals were trained to make saccades to a specific location cued on an object. After the presentation of the cue, and before the onset of the saccade, the orientation of the object was changed. Early in the task area LIP cells carried information about the location of the cue and the orientation of the object, both pieces of information being important for solving the task. However, near the time of the eye movement the same neurons coded primarily the direction of the intended movement (117).

Platt and Glimcher (112) showed in a delayed eye movement task that the early activity of LIP neurons varied as a function of the expected probability that a stimulus was a target for a saccade, as well as the amount of reward previously associated with the target. However, during later periods of the trial the cells coded only the direction of the planned eye movement. A similar evolution of activity has been shown in LIP and dorsal prefrontal cortex in eye movement tasks instructed by motion signals. The strength of the motion signal is an important determinant of activity at the beginning of the trial, but at the end of the trial the activity codes the decision or movement plan of the animal (114,115). The preceding studies show that activity in LIP and PRR evolves in time to reflect sensory, cognitive, and movement components of behavior.

GENERAL FUNCTIONS: LEARNING AND ATTENTION

Learning

Few neuroscientists would argue for the existence of a single center in the brain that directs all learning. Rather, learning is a function that is distributed throughout the nervous system. What appears to distinguish different types of learning, at least at the cognitive level, is where it occurs in the nervous system. Consistent with this viewpoint, recent prism adaptation studies have shown that visuomotor learning occurs in the PPC. As demonstrated initially by Held and Hein (118), when human subjects reach to visual targets while wearing displacing prisms, they initially miss-reach in the direction of target displacement but gradually recover and reach correctly if provided with appropriate feedback about their errors. Using positron emission tomography (PET) to monitor changes in cerebral blood flow, Clower and colleagues (119) showed that this prism adaptation process results in selective activation of the PPC contralateral to the reaching arm, when confounding sensory, motor, and cognitive effects are ruled out.

Similarly, it has been found that hemispatial neglect resulting from damage to the right hemisphere can be at least partially ameliorated by first having affected patients make reaching movements in the presence of a prismatic shift, then removing the prisms. These effects are as resulting from the stimulation of neural structures responsible for sensorimotor transformations, including the PPC as well as the cerebellum. A recent electrophysiologic study employing a prism adaptation paradigm suggests that the ventral premotor cortex plays a role in this process as well (120).

Another example of the effects of learning in the PPC was revealed in a recent electrophysiologic study of LIP (36). The responses of LIP neurons to auditory stimuli in a passive fixation task were examined before and after animals were trained to make saccades to auditory targets. Before such training, the number of cells responding to auditory stimuli in LIP was statistically insignificant. After training, however, 12% showed significant responses to auditory stimuli. This indicates that at least some LIP neurons become active for auditory stimuli only after an animal has learned that these stimuli are important for oculomotor behavior. As with the learning effects discussed in the preceding, effects of this nature have been reported in other areas of cortex (e.g., area 3a for tactile discrimination, and the FEF for visual search training) (121,122), highlighting the distributed nature of learning in cortex.

Recently, rapid learning effects have been observed in PRR in experiments where the activity of a single cell is used to position a cursor on a computer screen, in the absence of

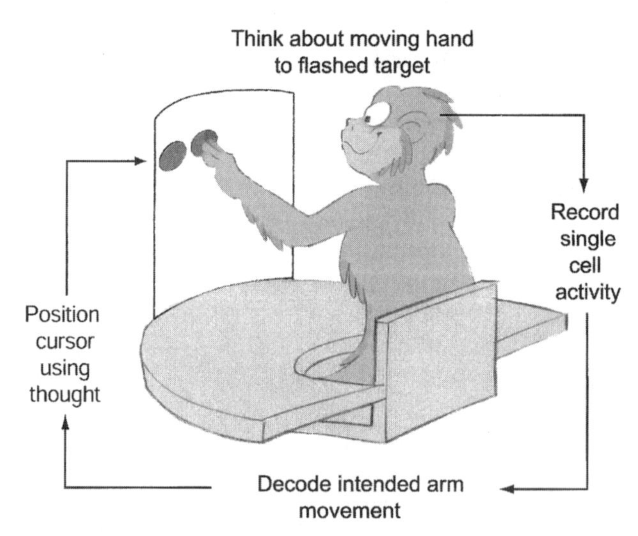

FIG. 10-4. Schematic of an experiment in "closing the loop." In the "baseline" phase, the activity of a single parietal reach region neuron is isolated and a database is constructed of the neuron's responses for reaches in the preferred and nonpreferred direction. In the "cursor-control" phase, the animal fixates and touches a central cursor on a display screen and a target is then presented either in the preferred or nonpreferred direction of the cell. Based on the response of the neuron for this one trial, and the statistics of the previously constructed database, we then predict whether the animal is intending to reach in either the preferred or nonpreferred direction. Importantly, the animal never actually makes the intended movement on these trials. A cursor moves to the location predicted from the cell activity and the animal is rewarded if the prediction corresponds to the cued location. (Adapted from: Andersen RA, Buneo CA. Intentional maps in posterior parietal cortex. *Annu Rev Neurosci* 2002;25:189–220, with permission.)

any arm movement (123). In this experiment, a database of the tuning properties of a neuron is first established while the animal performs reaching movements in the cell's preferred and nonpreferred directions ("baseline" phase). This database is then used in a subsequent "cursor control" phase to predict, on single trials, where the monkey is intending to reach, even though no reach actually occurs (Fig. 10-4). Once the cursor was put under control of the monkey's neural activity, it was found that about one half of the PRR neurons improved their selectivity within 10 to 50 trials, presumably to optimize performance in the task. This rapid learning did not occur in the baseline phase, because the animal was always being rewarded for the correct reach. It was only during the cursor control phase of the task that learning was observed, and this learning was of a similar, rapid time scale as that seen in prismatic adaptation experiments. These experiments suggest that the PPC operates to align and calibrate sensory and motor inputs as part of the sensorimotor transformation process, and that rapid learning occurs in this area in order to maintain these alignments.

Attention

Although learning is largely recognized to be a general property of cortex, the idea of attention as a general property, at least for sensory cortex, is a concept that remains unresolved. The alternative possibility is that a single center for attention exists that resides in the posterior parietal or parietal-prefrontal cortex. The idea that a central attention mechanism resides in PPC may result from the rather dramatic results of lesions to this region, which produce neglect in humans. Patients with neglect appear as if they have a primary deficit in attention, and are completely unaware of the space contralateral to the lesion (89). In addition, experiments manipulating attention produce strong activation of PPC in human fMRI studies (124) and tasks that require attention in monkeys lead to an increase in the discharge of PPC neurons (reviewed in detail in Chapter 7). All of these observations lend support to the idea of PPC being a center for attention, and in directing activity in other cortical areas. On the other hand, attention tasks tailored to the presumed functional role of areas outside the PPC have revealed attentional effects that cannot be easily attributed to PPC. For instance, attention tasks that involve fine spatial scales and contour analysis activate area V1, and tasks that examine feature and object perception show attentional modulation in areas V4 and the inferotemporal cortex (125–128).

Thus, it is possible that attention is similar to learning in that it is a general phenomenon that obtains its specificity only through the underlying functional specificity of the particular brain areas in which it is observed. In this view, attention in PPC would operate in the context of sensorimotor processes, for example, during target selection for movements. More research is required to determine if attention in PPC is restricted to the functions of this region, or if it has a more paramount role for all of sensory cortex.

SYNTHESIS

The two cortical areas focused on in this chapter, areas LIP and PRR, appear to embody the traits of task specificity and intermediate coding while also exhibiting properties consistent with being nodes of a distributed network. It is likely that these properties will generalize to other cortical regions within the PPC. Perhaps the most apparent attribute in the literature is specificity, in part because neuroscientists tend to adapt an approach of "dividing and conquering" different cortical areas in their investigations. Thus, numerous studies reviewed in this chapter point to LIP as specialized for saccades, and PRR for reaches. Within each of these areas is a map of visual space that can be considered a map of potential movements to those locations (i.e., intentional maps).

However, considerable research also points to the intermediate nature of these two cortical fields in sensorimotor transformation. Response fields are gain modulated by body position signals, and can even represent spatial locations in two distinct reference frames, suggesting that these areas occupy an intermediate stage in coordinate transformations and multisensory integration (34,35,62,63, 74). Moreover, Barash and colleagues found a small class of neurons in LIP that code both the location of a stimulus and the location of a planned movement in antisaccade tasks where the cue and the movement are opposite to one another. They proposed that these cells play an intermediate role in the transformation process in this informative task (105). Finally, even plans appear to be intermediate in PPC, being cognitive and high level at this stage and requiring further elaboration in motor areas prior to execution. One seminal indication of the cognitive nature of the PPC plans is the finding that reaches are coded in visual coordinates (47).

The similarity of coding strategies in PRR and LIP are very dramatic, considering these two areas process very different types of movements. These similarities suggest that they are parts of a single network, with the two areas speaking a common language for the purpose of coordinating movements of the hands and eyes.

Finally, attention and learning play a major role in sensorimotor processing in PPC. Perhaps because most behaviors have a spatial component and require orienting by the subject (e.g., eye movements) the PPC may be engaged in most behaviors, accounting for its apparent universal activation in fMRI experiments (124). In fact, sensorimotor transformations can be considered one of the major and essential functions of all nervous systems. With hand-eye coordination, along with language, being one of the main specializations to contribute to the success of humans as a species, it is no wonder that we have such elaborate neural structures for this exquisite ability.

REFERENCES

1. Mishkin M, Ungerleider LG. Contribution of striate inputs to the visuospatial functions of parieto-preoccipital cortex in monkeys. *Behav Brain Res* 1982;6:57–77.
2. Mountcastle VB, Lynch JC, Georgopoulos A, Sakata H, Acuma C. Posterior parietal association cortex of the monkey: command functions for operations within extrapersonal space. *J Neurophysiol* 1975;38:871–908.
3. Goldberg ME, Bushnell MC. Behavioral enhancement of visual responses in monkey cerebral cortex. II. Modulation in frontal eye fields specifically related to saccades. *J Neurophysiol* 1981;46:773–787.
4. Robinson DL, Goldberg ME, Stanton GB. Parietal association cortex in the primate: sensory mechanisms and behavioral modulations. *J Neurophysiol* 1978;41: 910–932.
5. Andersen RA, Essick GK, Siegel RM. Neurons of area 7a activated by both visual stimuli and oculomotor behavior. *Exp Brain Res* 1987;67:316–322.
6. Goodale MA, Milner AD. Separate visual pathways for perception and action. *Trends Neurosci* 1992;15:20–25.
7. Andersen RA, Buneo CA. Intentional maps in posterior parietal cortex. *Annu Rev Neurosci* 2002;25:189–220.
8. Colby CL, Goldberg ME. Space and attention in parietal cortex. *Annu Rev Neurosci* 1999;22: 319–349.
9. Brodmann K. *Vergleichende Lokalisationslehre der grosshirnrinde in ihren prinzipien dargestellt auf grund des zellenbaues.* Leipzig: JA Barth, 1909.
10. von Economo C, Koskinas GN. Die Cytoarchitektonik der Hirnrinde. Berlin: Springer, 1925.
11. von Bonin G, Bailey P. The neocortex of macaca mulatta. Urbana, Illinois: University of Illinois Press, 1947.
12. Andersen RA, Asanuma C, Essick C, et al. Corticocortical connection of anatomically and physiologically defined subdivisions within the inferior parietal lobe. *J Comp Neurol* 1990a;296:65–113.
13. Colby CL, Gattass R, Olson CR, et al. Topographical organization of cortical afferents to extrastriate visual area PO in the macaque: a dual tracer study. *J Comp Neurol* 1988;269:392–413.
14. Felleman DJ, Van Essen DC. Distributed hierarchical processing in the primate cerebral cortex. *Cerebral Cortex* 1991;1:1–47.
15. Lewis JW, Van Essen DC. Mapping of architectonic subdivisions in the macaque monkey, with emphasis on parieto-occipital cortex. *J Comp Neurol* 2000;428: 79–111.
16. Pandya DN, Seltzer B. Intrinsic connections and architectonics of posterior parietal cortex in the rhesus monkey. *J Comp Neurol* 1982;204:196–210.
17. Preuss TM, Goldman-Rakic PS. Architectonics of the parietal and temporal association cortex in the Strepsirhine primate Galago compared to the anthropoid primate Macaca. *J Comp Neurol* 1991;310:475–506.
18. Seltzer B, Pandya DN. Converging visual and somatic sensory cortical input to the intraparietal sulcus of the rhesus monkey. *Brain Res* 1980;192:339–351.
19. Seltzer B, Pandya DN. Posterior parietal projections to the intraparietal sulcus of the rhesus monkey. *Exp Brain Res* 1986;62:459–469.
20. Barash S, Andersen RA, Bracewell RM, et al. Saccade-related activity in the lateral intraparietal area. I. Temporal properties: comparison with area 7a. *J Neurophysiol* 1991a.66:1095–1108.
21. Barash S, Bracewell RM, Fogassi L, et al. Saccade related activity in the lateral intraparietal area. II. Spatial properties. *J Neurophysiol* 1991b.66:1109–1124.
22. Gnadt JW, Andersen RA. Memory related motor planning activity in posterior parietal cortex of macaque. *Exp Brain Res* 1988;70:216–220.

23. Li CS, Mazzoni P, Andersen RA. Effect of reversible inactivation of macaque lateral intraparietal area on visual and memory saccades. *J Neurophysiol* 1999;81: 1827–1838.
24. Li CSR, Andersen RA. Inactivation of macaque lateral intraparietal area delays initiation of the second saccade predominantly from contralesional eye positions in a double-saccade task. *Exp Brain Res* 2001;137: 45–57.
25. Lynch JC, McLaren JW. Deficits of visual attention and saccadic eye movements after lesions of parietooccipital cortex in monkeys. *J Neurophysiol* 1989; 61:74–90.
26. Snyder LH, Batista AP, Andersen RA. Coding of intention in the posterior parietal cortex. *Nature* 1997;386:167–170.
27. Thier P, Andersen RA. Electrical microstimulation suggests two different forms of representation of head-centered space in the intraparietal sulcus of rhesus monkeys. *Proc Natl Acad Sci USA* 1996;93: 4962–4967.
28. Asanuma C, Andersen RA, Cowan WM. The thalamic relations of the caudal inferior parietal lobule and the lateral prefrontal cortex in monkeys: divergent cortical projections from cell clusters in the medial pulvinar nucleus. *J Comp Neurol* 1985;241:357–381.
29. Blatt G, Andersen RA, Stoner G. Visual receptive field organization and cortico-cortical connections of area LIP in the macaque. *J Comp Neurol* 1990;299: 421–445.
30. Lynch JC, Graybiel AM, Lobeck LJ. The differential projection of two cytoarchitectonic subregions of the inferior parietal lobule of macaque upon the deep layers of the superior colliculus. *J Comp Neurol* 1985; 235:241–254.
31. Connolly JD, Menon RS, Goodale MA. Human frontoparietal areas active during a pointing but not a saccade delay. *Soc Neurosci* 2000;26:1329(abstr).
32. Rushworth MFS, Paus T, Sipila PK. Attention systems and the organization of the human parietal cortex. *J Neurosci* 2001;21:5262–5271.
33. Li C-SR, Mazzoni P, Andersen RA. Reversible inactivation of area LIP disrupts saccasic eye movements. *Soc Neurosci* 1995;21:281.1(abstr).
34. Brotchie PR, Andersen RA, Snyder LH, et al. Head position signals used by parietal neurons to encode locations of visual stimuli. *Nature* 1995;375:232–235.
35. Snyder LH, Grieve KL, Brotchie P, et al. Separate body- and world-referenced representations of visual space in parietal cortex. *Nature* 1988b;394:887–891.
36. Grunewald A, Linden JF, Andersen RA. Responses to auditory stimuli in macaque lateral intraparietal area. I. Effects of training. *J Neurophysiol* 1999;82:330–342.
37. Mazzoni P, Bracewell RM, Barash S, et al. Spatially tuned auditory responses in area LIP of macaques performing delayed memory saccades to acoustic targets. *J Neurophysiol* 1996b;75:1233–1241.
38. Scherberger H, Andersen RA. Neural activity in the posterior parietal cortex during decision processes for generating visually-guided eye and arm movements in the monkey. *Soc Neurosci* 2001;27:237.7(abstr).
39. Thier P, Andersen RA. Electrical microstimulation distinguishes distinct saccade-related areas in the posterior parietal cortex. *J Neurophysiol* 1998.
40. Cavada C, Goldman-Rakic PS. Topographic segregation of corticostriatal projections from posterior parietal subdivisions in the macaque monkey. *Neuroscience* 1991;42:683–696.
41. Battaglia-Mayer A, Ferraina S, Mitsuda T, et al. Early coding of reaching in the parietooccipital cortex. *J Neurophysiol* 2000;83:2374–2391.
42. Ferraina S, Battaglia-Mayer A, Genovesio A, et al. Early coding of visuomanual coordination during reaching in parietal area PEc. *J Neurophysiol* 2001;85: 462–467.
43. Ferraina S, Johnson PB, Garasto MR, et al. Combination of hand and gaze signals during reaching: activity in parietal area 7 m of the monkey. *J Neurophysiol* 1997;77:1034–1038.
44. Kalaska JF. Parietal cortex area 5 and visuomotor behavior. *Can J Physiol Pharmacol* 1996;74:483–498.
45. MacKay WA, Riehle A. Planning a reach: spatial analysis by area 7a neurons. In: Stelmach G, Requin J, eds. *Tutorials in motor behavior,* 2nd ed. New York: Elsevier, 1992.
46. Marconi B, Genovesio A, Battaglia-Mayer A, et al. Eye-hand coordination during reaching. I. Anatomic relationships between parietal and frontal cortex. *Cerebral Cortex* 2001;11:513–527.
47. Batista AP, Buneo CA, Snyder LH, et al. Reach plans in eye-centered coordinates. *Science* 1999;285: 257–260.
48. Cisek P, Kalaska JF. Modest gaze-related discharge modulation in monkey dorsal premotor cortex during a reaching task performed with free fixation. *J Neurophysiol* 2002;88:1064–1072.
49. Bremmer F, Schlack A, Shah NJ, et al. Polymodal motion processing in posterior parietal and promotor cortex: a human fMRI study strongly implies equivalencies between humans and monkeys. *Neuron* 2001;29: 287–296.
50. Duhamel JR, Colby CL, Goldberg ME. Ventral intraparietal area of the macaque: Congruent visual and somatic response properties. *J Neurophysiol* 1998;79: 126–136.
51. Duhamel JR, Bremmer F, Ben Hamed S, et al. Spatial invariance of visual receptive fields in parietal cortex neurons. *Nature* 1997;389:845–848.
52. Sakata H, Taira M, Murata A, Mine S. Neural mechanisms of visual guidance of hand action in the parietal cortex of the monkey. *Cereb Cortex* 1995;5(5):429–438.
53. Sakata H, Taira M, Kusunoki M, Murata A, Tanaka Y. The TINS Lecture. The parietal association cortex in depth perception and visual control of hand action. *Trends Neurosci* 1997;20(8):350–357.
54. Perenin MT, Vighetto A. Optic ataxia: a specific disruption in visuomotor mechanisms. I. Different aspects of the deficit in reaching for objects. *Brain* 1988; 111:643–674.
55. Newsome WT, Wurtz RH, Komatsu H. Relation of cortical areas MT and MST to pursuit eye movements. II. Differentiation of retinal from extraretinal inputs. *J Neurophysiol* 1988;60: 604–620.
56. Dursteler MR, Wurtz RH. Pursuit and optokinetic deficits following chemical lesions of cortical areas Mt and Mst. *J Neurophysiol* 1988;60:940–965.
57. Duffy CJ, Wurtz RH. Response of monkey MSTd neurons to optic flow stimuli with shifted centers of motion. *J Neurosci* 1995;15:5192–5208.
58. Bradley DC, Maxwell M, Andersen RA, et al. Mecha-

nisms of heading perception in primate visual cortex. *Science* 1996;273:1544–1547.
59. Shenoy KV, Bradley DC, Andersen RA. Influence of gaze rotation on the visual response of primate MSTd neurons. *J Neurophysiol* 1999;81:2764–2786.
60. Binkofski F, Dohle C, Posse S, et al. Human anterior intraparietal area subserves prehension: a combined lesion and functional MRI activation study. *Neurology* 1998;50:1253–1259.
61. Andersen RA, Essick GK, Siegel RM. Encoding of spatial location by posterior parietal neurons. *Science* 1985;25:456–458.
62. Buneo CA, Batista AP, Andersen RA. Frames of reference for reach-related activity in two parietal areas. *Soc Neurosci* 1998;24:262(abstr).
63. Cohen YE, Andersen RA. The parietal reach area (PRR) encodes reaches to auditory targets in an eye-centered reference frame. *Soc Neurosci* 1998;24: 262(abstr).
64. Andersen RA, Bracewell RM, Barash S, et al. Eye position effects on visual, memory, and saccade-related activity in areas LIP and 7a of macaque. *J Neurosci* 1990b;10:1176–1196.
65. Salinas E, Abbott LF. Transfer of coded information from sensory to motor networks. *J Neurosci* 1995;15: 6461–6474.
66. Zipser D, Andersen RA. A back-propagation programmed network that simulates response properties of a subset of posterior parietal neurons. *Nature* 1988; 331:679–684.
67. Pouget A, Snyder LH. Computational approaches to sensorimotor transformations. *Nat Neurosci* 2000;3: 1193–1198.
68. Xing J, Andersen RA. Models of the posterior parietal cortex which perform multimodal integration and represent space in several coordinate frames. *J Cogn Neurosci* 2000b;12:601–614.
69. Boussaoud D, Jouffrais C, Bremmer F. Eye position effects on the neuronal activity of dorsal premotor cortex in the macaque monkey. *J Neurophysiol* 1998;80: 1132–1150.
70. Connor CE, Gallant JL, Preddie DC, et al. Responses in area V4 depend on the spatial relationship between stimulus and attention. *J Neurophysiol* 1996;75: 1306–1308
71. Trotter Y, Celebrini S. Gaze direction controls response gain in primary visual-cortex neurons. *Nature* 1999;398:239–242.
72. Van Opstal AJ, Hepp K, Suzuki Y, et al. Influence of eye position on activity in monkey superior colliculus. *J Neurophysiol* 1995;74:1593–1610.
73. Salinas E, Their P. Gain modulation: a major computational principle of the central nervous system. *Neuron* 2000;27(1):15–21.
74. Buneo CA, Jarvis MR, Batista AP, et al. Direct visuomotor transformations for reaching. *Nature* 2002;416: 632–636.
75. Deneve S, Latham PE, Pouget A. Efficient computation and cue integration with noisy population codes. *Nat Neurosci* 2001;4:826–831.
76. Linden JF, Grunewald A, Andersen RA. Responses to auditory stimuli in macaque lateral intraparietal area. II. Behavioral modulation. *J Neurophysiol* 1999;82: 343–358.
77. Cohen YE, Andersen RA. Reaches to sounds encoded in an eye-centered reference frame. *Neuron* 2000;27: 647–652.
78. Toth LJ, Assad JA. Dynamic coding of behaviourally relevant stimuli in parietal cortex. *Nature* 2002;415: 165–168.
79. Xing J, Andersen RA. Memory activity of LIP neurons for sequential eye movements simulated with neural networks. *J Neurophysiol* 2000a;84:651–665.
80. Wu S, Andersen RA. The representation of auditory space in temporo-parietal cortex. *Soc Neurosci* 2001; 27:166.15(abstr).
81. Galletti C, Battaglini PP, Fattori P. Parietal neurons encoding spatial locations in craniotopic coordinates. *Exp Brain Res* 1993;96:221–229.
82. Flanders WD, DerSimonian R, Freedman DS. Interpretation of linear regression models that include transformations or interaction terms. *Ann Epidemiol* 1992;2(5):735–744.
83. McIntyre J, Stratta F, Lacquaniti F. Viewer-centered frame of reference for pointing to memorized targets in three-dimensional space. *J Neurophysiol* 1997;78 (3):1601–1618.
84. McIntyre J, Stratta F, Lacquaniti F. Short-term memory for reaching to visual targets: psychophysical evidence for body-centered reference frames. *J Neurosci* 1998; 18(20):8423–8435.
85. Carrozzo M, McIntyre J, Zago M, et al. Viewer-centered and body-centered frames of reference in direct visuomotor transformations. *Exp Brain Res* 1999;129: 201–210.
86. Balint R. Seelenlahmung des "Schauens," optische Ataxie, raumliche Storung der Aufmerksamkeit. *Monatsschr Psychiatr Neurol* 1909;25:51–81.
87. Rondot P, Recondo J, de Ribadeau Dumas J. Visuomotor ataxia. *Brain* 1977;100:355–376.
88. Geshwind N, Damasio AR. Apraxia. In: Vinken PJ, Bruyn GW, Klawans HL, eds. *Handbook of clinical neurology.* Amsterdam: Elsevier, 1985:423–432.
89. Critchley M. *The parietal lobes.* London: Arnold, 1953.
90. Milner AD. Neglect, extinction, and the cortical streams of visual processing. In: Thier P, Karnath H-O, eds. *Parietal lobe contributions to orientation in 3D space.* Heidelberg: Springer, 1997.
91. Karnath H, Ferber S, Himmelbach M. Spatial awareness is a function of the temporal not the posterior parietal lobe. *Nature* 2001;411:950–953.
92. Goldman-Rakic PS. Topography of cognition-parallel distributed networks in primate association cortex. *Annu Rev Neurosci* 1998;11:137–156.
93. Pare M, Wurtz RH. Progression in neuronal processing for saccadic eye movements from parietal cortex area LIP to superior colliculus. *J Neurophysiol* 2001;85: 2545–2562.
94. Caminiti R, Ferraina S, Johnson PB. The sources of visual information to the primate frontal lobe: a novel role for the superior parietal lobule. *Cerebral Cortex* 1996;6:319–328.
95. Stricanne B, Andersen RA, Mazzoni P. Eye-centered, head-centered, and intermediate coding of remembered sound locations in area LIP. *J Neurophysiol* 1996;76:2071–2076.
96. Hikosaka O, Wurtz RH. Visual and oculomotor functions of monkey substantia nigra pars reticulata. III. Memory-contingent visual and saccade responses. *J Neurophysiol* 1983;49:1268–1284.

97. Batista AP, Andersen RA. The parietal reach region codes the next planned movement in a sequential reach task. *J Neurophysiol* 2001;85:539–544.
98. Mazzoni P, Bracewell RM, Barash S, et al. Motor intention activity in the macaque's lateral intraparietal area. I. Dissociation of motor plan from sensory memory. *J Neurophysiol* 1996a;76:1439–1456.
99. Mays LE, Sparks DL. Dissociation of visual and saccade-related rsponses in superior colliculus neurons. *J Neurophysiol* 1980;43(1):207–232.
100. Klier EM, Wang H, Crawford JD. The superior colliculus encodes gaze commands in retinal coordinates. *Nat Neurosci* 2001;4:627–632.
101. Duhamel JR, Colby CL, Goldberg ME. The updating of the representation of visual space in parietal cortex by intended eye movements. *Science* 1992;255(5040): 90–92.
102. Henriques DY, Klier EM, Smith MA, et al. Gaze-centered remapping of remembered visual space in an open-loop pointing task. *J Neurosci* 1998;18:1583–1594.
103. Eskandar EN, Assad JA. Dissociation of visual, motor and predictive signals in parietal cortex during visual guidance. *Nat Neurosci* 1999;2:88–93.
104. Gottlieb J, Goldberg ME. Activity of neurons in the lateral intraparietal area of the monkey during an antisaccade task. *Nat Neurosci* 1999;2(10):906–912.
105. Zhang M, Barash S. Neuronal switching of sensorimotor transformations for antisaccades. *Nature* 2000;408: 971–975.
106. Calton JL, Dickinson AR, Snyder LH. Non-spatial, motor-specific activation in posterior parietal cortex. *Nat Neurosci* 2002;5:580–588.
107. Snyder LH, Batista AP, Andersen RA. Change in motor plan, without a change in the spatial locus of attention, modulates activity in posterior parietal cortex. *J Neurophysiol* 1998a;79:2814–2819.
108. Steinmetz MA, Constantinidis C. Neurophysiologic evidence for a role of posterior parietal cortex in redirecting visual attention. *Cerebral Cortex* 1995;5: 448–456.
109. Kalaska JF, Crammond DJ. Deciding not to go: neuronal correlates of response selection in a go/nogo task in primate premotor and parietal cortex. *Cerebral Cortex* 1995;5:410–428.
110. Pare M, Wurtz RH. Monkey posterior parietal cortex neurons antidromically activated from superior colliculus. *J Neurophysiol* 1997;78:3493–3497.
111. Bracewell RM, Mazzoni P, Barash S, et al. Motor intention activity in the macaque's lateral intraparietal area. II. Changes of motor plan. *J Neurophysiol* 1996; 76:1457–1464.
112. Platt ML, Glimcher PW. Neural correlates of decision variables in parietal cortex. *Nature* 1999;400(6741): 233–238.
113. Kim JN, Shadlen MN. Neural correlates of a decision in the dorsolateral prefrontal cortex of the macaque. *Nat Neurosci* 1999;2:176–185.
114. Shadlen MN, Newsome WT. Motion perception: seeing and deciding. *Proc Natl Acad Sci USA* 1996;93: 628–633.
115. Leon MI, Shadlen MN. Effect of expected reward magnitude on the response of neurons in the dorsolateral prefrontal cortex of the macaque. *Neuron* 1999; 24:415–425.
116. Cohen YE, Batista AP, Andersen RA. Comparison of neural activity preceding reaches to auditory and visual stimuli in the parietal reach region. *NeuroReport* 2002;13:891–894.
117. Sabes PN, Breznen B, Andersen RA. Parietal representation of object-based saccades. *J Neurophysiol* 2002; in press.
118. Hein A, Held R. Dissociation of the visual placing response into elicited and guided components. *Science* 1967;158(799):390–392.
119. Clower DM, Hoffman JM, Votaw JR, Faber TL, Woods RP, Alexander GE. Role of posterior parietal cortex in the recalibration of visually guided reaching. *Nature* 1996;383(6601):618–621.
120. Kurata K, Hoshi E. Reacquisition deficits in prism adaptation after muscimol microinjection into the ventral premotor cortex of monkeys. *J Neurophysiol* 1999; 81:1927–1938.
121. Recanzone GH, Merzenich MM, Jenkins WM. Frequency discrimination training engaging a restricted skin surface results in an emergence of a cutaneous response zone in cortical area 3a. *J Neurophysiol* 1992; 67:1057–1070.
122. Bichot NP, Schall JD, Thompson KG. Visual feature selectivity in frontal eye fields induced by experience in mature macaques. *Nature* 1996;381:697–699.
123. Meeker D, Cao S, Burdick JW, et al. Rapid plasticity in the parietal reach region demonstrated with a brain-computer interface. *Soc Neurosci* 2002;28(abstr).
124. Culham JC, Kanwisher NG. Neuroimaging of cognitive functions in human parietal cortex. *Curr Opin Neurobiol* 2001;11:157–163.
125. Chelazzi L, Duncan J, Miller EK, et al. Responses of neurons in inferior temporal cortex during memory-guided visual search. *J Neurophysiol* 1998;80: 2918–2940.
126. Crist RE, Li W, Gilbert CD. Learning to see: experience and attention in primary visual cortex. *Nat Neurosci* 2001;4:519–525.
127. McAdams CJ, Maunsell JHR. Attention to both space and feature modulates neuronal responses in macaque area V4. *J Neurophysiol* 2000;83:1751–1755.
128. Roelfsema PR, Lamme VAF, Spekreijse H. Object-based attention in the primary visual cortex of the macaque monkey. *Nature* 1998;395:376–381.
129. Cohen YE, Andersen RA. A common reference frame for movement plans in the posterior parietal cortex. *Nat Rev Neurosci* 2002;3:553–562.
130. Andersen RA. *The role of the inferior parietal lobule in spatial perception and visual-motor integration.* Bethesda, MD: American Physiologic Society, 1987: 483–518.

11

Somatosensory and Motor Disturbances in Patients with Parietal Lobe Lesions

Hans-Joachim Freund

Department of Neurology, University of Dusseldorf, Moorenstrasse 5, Dusseldorf, Germany

THE IMPACT OF LESION ON FUNCTION

Before discussing data mainly derived from lesion studies, a few aspects of their interpretation are considered. Many of the classical and more recent studies on parietal lobe (PL) function are based on large and sometimes even bilateral lesions not restricted to the PL. In the early reports, the lesions were assessed by subsequent pathology, neurosurgery, or (later) by computed tomography (CT) scans with poor resolution. In spite of the highly sophisticated clinical and neuropsychological analyses, therefore, often it remains unclear, how far the observed deficits represent disturbances of parietal or other adjacent cortical areas or of disconnection syndromes caused by subcortical damage to long fiber tracts. The problem is the allocation of the lesion to the functional anatomy.

Another major variable determining the emergence of functional deficits is the temporal development of the local pathology (Jackson's "lesion momentum"). In contrast to strokes, tumors often show no disturbances, even when most of the PL is invaded (see Chapter 22). This indicates the efficacy of the reorganizational processes even in eloquent areas when enough time is allotted.

The redundancy of the damaged system is another variable for its liability to functional impairment. The term eloquent brain areas addresses this from a neurosurgical perspective, thereby contrasting those regions where removals produce major and persisting loss of function with others where no clinically detectable deficits ensue.

Because many of the parietal deficits are transient, the early phase after stroke or neurosurgical excision is most suitable for detailed analysis, particularly when the lesions are small and restricted to distinct functional zones. The collection and meticulous analysis of such cases is now possible by means of high-resolution magnetic resonance images that even allow us to recognize purely cortical infarcts in cases with selective neuronal necrosis. When the structural defects are small, the corresponding functional disturbances sometimes may be clinically inapparent and only be detectable by special testing. However, these dysfunctions then can be precisely attributed to the cortical landmarks. With these new possibilities provided by refined structural imaging, the lesion approach continues to constitute a powerful tool for the investigation of the functional anatomy of the human cortex. It will further benefit from recent sophisticated methods such as volumetric voxel-based morphometry (1) and fiber imaging by means of diffusion tensor imaging. Along with advances in the experimental design and analysis of activation studies (2), the apparent inconsistencies between lesion and functional neuroimaging data may be overcome.

SOMATOSENSORY REPRESENTATIONS

The somatosensory system mediates the perception of cutaneous, kinesthetic, and noxious sensations. Its cortical processing modules are located in the parietal lobe in nonhuman primates and humans. The human anterior parietal lobe shows a similar cytoarchitectonic organization to the monkey, but its functional parcellation is less clear. The somatotopy in primary somatosensory cortex (SI) of the human postcentral gyrus shows a similar cortical representation (3–7) to previous intraoperative cortical stimulation patient studies (8–10), and reveals a high intersubject consistency of the somatotopic map. A hierarchical processing of somatosensory information with an anterior-posterior gradient in the postcentral gyrus has been described in the monkey (11). In the human, a statistical analysis of cases with lesions scattering around the postcentral gyrus showed that the impairment of discriminative functions that can not be attributed to the disturbance of elementary sensation also followed such an anterior-posterior gradient (12,13).

In spite of the clear and consistent layout of the body map on each hemisphere, the observations on patients with hemispherectomy are surprising because they indicate a bilateral representation of the axial and proximal body parts. A pronounced proximal-distal gradient characterizes the disturbance of executive motor functions with a surprisingly good preservation of axial and proximal movements. In some patients with infantile hemiparesis who underwent hemispherectomy because of intractable seizures, proximal movements remained almost normal, whereas distal movements (particularly of the fingers) were severely compromised (14). This pronounced gradient was mimicked on the somatosensory side where cutaneous sensation and kinesthesia were near normal above the elbow and knee but successively disturbed further distally. Light touch was severely reduced but not absent at the contralateral finger pads and toes and there were no dystonic features as in deafferentated patients. One patient could precisely locate with her right fingers any point touched on her left hand. She could further recognize numbers of 1.5-cm size when written with sufficient pressure on her finger pads. Position sense at the wrist and finger joints was severely reduced but not absent. Also, vibration was perceived normally even distally. Passive roughness discrimination was still possible, whereas stereognosis was completely abolished.

Unlike the effects of lesions of the secondary somatosensory area (SII) in monkeys (15,16), lesions of this area in the human did not interfere with the detection of threshold stimuli, kinesthesia or the microgeometric or macrogeometric features of tactile objects (12). A disturbance of manipulation was similar in the two species. The direct connections between SII and primary motor cortex (17) and the direct nociceptive inputs to SII have been implicated in the processing of fast reactions to harmful stimuli (18). Dorsomedial lesions, including the supplementary sensory area, acutely caused disturbances of somesthetic processing and apraxia, whereas ventrolateral lesions, including SII, induced persistent tactile agnosia (19).

SOMATOSENSORY DEFICITS FOLLOWING PARIETAL LOBE LESIONS

Lesions of the Anterior Parietal Lobe

Regarding the originality and elaboration of the prime features of somatosensory deficits in parietal lobe disease, the classical descriptions by Déjerine (20), Head and Holmes (21), Delay (22), Critchley (23), Foerster (8), and other pioneers are unsurpassed. Verger (24) and Déjerine and Mouzon (25) emphasized the preferential involvement of tactile discrimination and stereognosis as the hallmark of the disturbance of sensation after parietal lobe damage, whereas pain, temperature, and touch were better preserved. A similar pattern was seen in cases with dorsal column lesions. However, surgical excision of the postcentral gyrus clarified the fact that damage restricted to the anterior PL does not show such a differential pattern (8). The acute stage was characterized by a complete anesthesia, including pain, thermal sensation, and areflexia and hypotonia of the contralateral side. These sensory deficits were associated with a disturbance of motility, slightly diminished force, and slowing of movement associated with ataxia, dysmetria, the assumption of dystonic postures, and deficient fine motor control. The execution of precision movements or selective muscle activations required for fractionated finger movements or the maintenance of static forces was not possible. Jeannerod and coworkers (26) reported a case with complete and persisting distal anesthesia including cold, warm, and painful stimuli after an infarct of the postcentral area that extended deep into the white matter down to the level of the thalamus.

The neurosurgical cases with well-defined excisions of the postcentral gyrus demonstrate that elementary sensation is so severely impaired that the elaboration of the more complex feature extractions required for the recognition of the microgeometry and macrogeometry, of the spatial attributes and of the identity of tactile objects is not possible. Therefore, the clinical picture characterizing damage of the anterior parietal lobe resembles that of patients with peripheral nerve or dorsal root damage and does not show distinctive features that would allow one to identify the sensory deficits as parietal.

Lesions of the Posterior Parietal Lobe

The pattern of sensory deficits in patients with lesions of the superior parietal lobule (SPL) is distinct and resembles that seen after dorsal column lesions. The apparent similarity of the preferential impairment of tactile discrimination and stereognosis in the two conditions goes along with a similar pattern of disturbance of postural sensibility. In contrast to the effects of lesions of the anterior parietal lobe where the affection of the cognitive and motoric aspects is secondary to sensory loss, complex somatosensory functions such as the identification of surface texture, the recognition of shape and its spatial attributes are preferentially disturbed, whereas elementary sensation remains preserved (27). Correspondingly, complex motor dysfunctions as described below are typically observed rather than the ataxic disturbances typically seen after more anterior lesions. As in the cases with damage of the postcentral gyrus, the deficits are restricted to the contralesional side. The impairment of complex somatosensory functions can not be distinguished from that following damage to the lemniscal projections. Sensory deficits that are solely seen in parietal cases can be recognized when particular features of sensory information processing are selectively affected. The hallmark of these deficits is the disturbance of specific attributes of feature extraction and elaboration.

The use of the hand as a sense organ requires a set of functions subserving the recognition of tactile objects. Semmes and Turner (28) and more recently De Renzi and his group (29) emphasized the role of spatial cognition for the identification of object shape. They reported a case with large bilateral posterior lobe damage and impairment of tactile form recognition—morphagnosia—who showed normal discrimination of size, texture, weight, and simple shapes, but a disturbance of complex shape discrimination and spatial abilities (line orientation, pin positioning, and rod test).

However, other cases demonstrate that shape discrimination can be compromised in spite of normal spatial functions. A patient reported by Reed et al. (30) had a selective impairment of object recognition with the hand contralesional to a small infarction of the inferior parietal lobule (IPL) that deteriorated as shape complexity increased. Sensation and manual shape exploration were normal, as was judgment of metric length. Tests of spatial abilities revealed normal functions so that tactile agnosia was not secondary to the impairment of spatial cognition, but rather represented a more specific impairment of high level, modality-specific tactile shape perception, leaving the ability to associate tactually defined objects with episodic memory intact. It was concluded that tactile shape recognition is represented as a separate module distinct from amodal or modality-specific spatial perception.

Mauguière and Isnard (31) proposed that only "tactile asymbolia" should be considered as a "true" tactile agnosia. The recording of early components of the somatosensory evoked potentials (SEPs) was used as an estimator of parietal input functions in 309 subjects with focal hemispheric lesions presenting with a somatosensory deficit of any type. Astereognosis referable to impaired discrimination of textures and shapes in the absence of impaired elementary sensation was observed in only 12 patients, all with abnormal early SEPs. The only condition combining a failure of tactile recognition of objects with normal early SEPs was observed in patients with "tactile anomia" and callosal disconnections. These results were regarded as challenging the interpretation of the cases with tactile object recognition and inferior parietal damage (19,32). The problem regarding this reasoning is that changes in early SEP components do not implicate that not enough information reaches the later processing stages.

The selective impairment of object categorization and object naming that represents further stages in the processing network required for object recognition was shown for two patients reported by Endo et al. (33). Case 1 presented with a right-hand tactile agnosia. He could distinguish tactile qualities of objects flawlessly but could neither name nor categorize the objects. Case 2 suffered from a bilateral tactile aphasia characterized by a severe tactile naming deficit, but could recognize and categorize objects. Case 1 was regarded as tactile agnosia clearly distinct from tactile aphasia. CT scans of case 1 revealed lesions in the left angular gyrus, and in the right parietal, temporal, and occipital lobes. Case 2 had lesions in the left angular gyrus and posterior callosal radiation. The authors suggested that tactile agnosia appears when the somatosensory association cortex is disconnected by a subcortical lesion of the angular gyrus from the semantic memory store located in the temporal lobe, whereas tactile aphasia represents a tactual-verbal disconnection.

These selective disturbances of higher-order processing of somatosensory information fit well with the classical concepts regarding astereognosis as a disturbance of mental operations; the use of this term in deafferentated patients is a misnomer (22). In this context, Critchley addressed Wernickes' (34) distinction between the disturbance of primary identification (referring to the description of object size, weight, roughness, and the like), and the secondary identification of its meaning—a distinction that corresponds to the apperceptive and associative agnosias in vision.

THE MOTOR SIDE OF TACTILE COGNITION: ACTIVE TOUCH

Because object exploration is mainly accomplished by manipulation, the use of the hand as a sense organ also depends on its motor capacities. The sensory information obtained by the hand comprises two fundamentals: exteroceptive and proprioceptive elements. According to Gibson (35), the combination of two inputs is the important variable between touching and being touched. Active touch can elaborate the unity, stability, plasticity, and shape of phenomenal objects. The subject perceives one object only when a single object is grasped with several fingers, although several cutaneous receptor sheets are engaged. The percept is the object form and not skin form and the continuous movements of the fingers are not perceived.

Effective grasping and object manipulation are based on three fundamental properties of the motor system: the capacity to generate independent finger movements, the ability to transform sensory information concerning the object to be grasped into an appropriate hand configuration, and a sophisticated somatosensory control of finger movements (36). It has been shown by psychophysical studies that not only the perceptual goals constrain exploratory action, but also, conversely, exploration constrains what is perceived (37). Thus, higher-level knowledge about objects produces patterns of manual exploratory movements that lead to haptically driven object representations. Such knowledge-driven use of preprogrammed hand movements was proposed by early Gestalt psychologists such as Katz (38) and Hippius (39).

The interdependence of the sensory and motor processes required for object recognition and manipulation is demonstrated by cases with lesions of the SPL where the impairment of tactile gnosis is associated with a severe disturbance of the exploratory finger movements. Figure 11-1 shows the kinematic analysis in such a case. It reveals the derangement of the digital trajectories along with a breakdown of their spatial and temporal exploratory patterns (27). The essence of the dysfunction lies in the impairment of the conception of those movements necessary to shape the required sensory input.

This disturbance of active touch is characteristic for an essential feature of apraxia as defined by Liepmann (40), the loss of the purposive nature of motor behavior. Following Liepmann's definition of the three major forms: ideational, ideomotor, and limb-kinetic apraxia, the extension of this concept to disturbances of tactile exploration was first established by Klein (41). He adopted the term tactile apraxia for a purely contralesional disturbance of active touch in a patient who could not produce adequate, purposive finger movements required for the exploration of objects given into his hand. However, when he was set in front of an object, a correct reaching for grasping movement was initiated and his intransitive, expressive movements were well

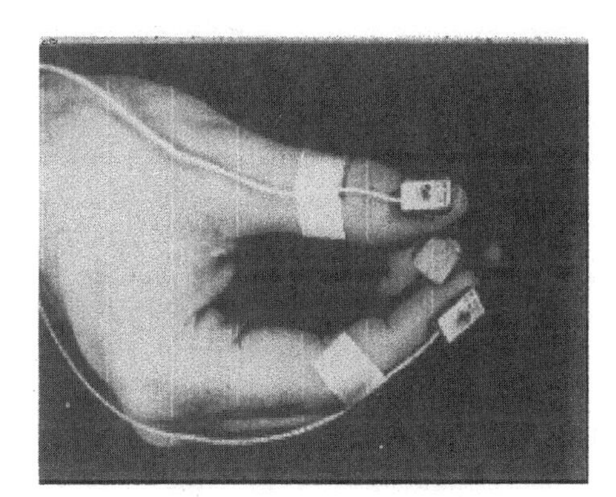

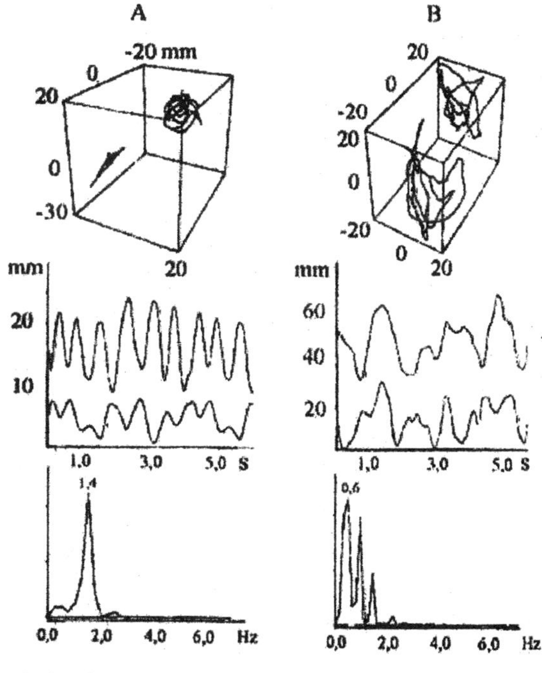

FIG. 11-1. Exploratory finger movements of the index finger and thumb during tactile exploration of the small object. **A:** Kinematic analysis of exploratory finger movements in a normal subject and **(B)** in a patient with a posterior parietal lesion. In the upper row, the movement trajectories of the forefinger and thumb are shown as viewed from the front, so that the vertical and horizontal movement components are displayed. The workspace of the scanning movements is larger and more irregular for the affected hand. The temporal profiles in the middle row show the breakdown of the finely tuned, regular motion, further illustrated by the power spectra below the respective recordings.

preserved. The unimodal nature of this dysfunction was illustrated by the fact that there was no apraxia when the patient saw the object, but there was apraxia when he was blindfolded and then started active touch. Patients with tactile apraxia usually are so severely affected that they regard their hand as useless.

Many two-dimensional (2D) object features can be recognized by passive touch, whereas the identification of complex three-dimensional (3D) properties requires active touch. Valenza and coworkers (42) reported a patient with a right hemispheric infarction who, in spite of intact sensorimotor functions, had impaired tactile object recognition with the left hand. Recognition of 2D shapes and objects was severely deficient under the condition of spontaneous exploration. Tactile exploration of shapes was disorganized and exploratory procedures, such as the contour-following strategy necessary to identify the precise shape of an object, were severely disturbed. However, recognition of 2D shapes under manually or verbally guided exploration and the recognition of shapes traced on the skin were intact, indicating a dissociation in shape recognition between active and passive touch. Functional MRI (fMRI) during sensory stimulation of the left hand showed preserved activation of the spared primary sensory cortex in the right hemisphere. The deficit was considered as representing a pure tactile apraxia without tactile agnosia, that is, a specific inability to use tactile feedback to generate the exploratory procedures necessary for tactile shape recognition.

The frequent association of tactile recognition and tactile apraxia reveal the intricate interdependence between the processes linking perception to action and the anatomic proximity of their processing modules (43). However, the cases with isolated somatosensory deficits affecting different levels of complexity further demonstrate that the respective modules are separate. Patients with tactile apraxia without astereognosia (44) or the reverse condition—a near normal exploratory manual pattern but severe disturbances of tactile recognition—reveal the selective impairment of the processes required for the elaboration of motor concept formation or cognition. Cases with tactile apraxia without astereognosis argue against the view that the motor deficit is secondary to the disturbance of complex somatosensory processing. They also show that the sensorimotor transformations required for the elaboration of motor concepts engage different neural repertoires.

The situation is strikingly different in patients where finger movements are compromised by executive motor deficits such as paresis, ataxia, bradykinesia, or tremor but tactile recognition is preserved. Kinematic recordings in such cases revealed abnormal parameters of the exploratory finger movements (43). However, in contrast to the changes of the digital palpation pattern in cases with SPL damage they affect the amplitude, speed, or regularity of the finger movements, whereas the configuration of the explorative pattern remained preserved. Obviously, these abnormalities barely interfere with active touch. Correspondingly, kinematic recordings have demonstrated that the derangement of the spatial-temporal pattern of whole limb movements in patients with limb apraxia is different from the disruption of interjoint coordination that characterized the failure to control the interactive forces arising among limb segments during multijoint movements in deafferentated patients (45).

The relationship between perception and movement in patients with infantile hemiplegia and astereognosis but otherwise normal sensation constitutes an interesting contrast to congenital abnormalities of the arm and hand. Déjerine (20) used the term "virgin hand" in order to emphasize that the sensory deficit represents an astereognosis through ignorance or inexperience and not immobility. Critchley (23) emphasized the difference of this virgin hand condition and patients with congenitally absent or rudimentary hands. He referred to a patient with Alport's agyrocephalosyndactyly with no free phalanges but stumped and deformed metacarpals. When the attenuated thumb was freed by plastic surgery, the patient acquired a previously inexperienced degree of dexterity with the stumps and a high level of texture recognition and tactile discrimination comparable to that in normal subjects.

The meticulous analysis of cases with selective disturbances of different aspects of somatosensory and sensorimotor functions following lesions of the posterior parietal lobe discloses the operations of corresponding processing modules. They illustrate that virtually every aspect of somatosensory processing can be selectively disturbed. Their precise topological allocation to particular cortical areas is only possible in those cases where the lesions are small enough. Information gained from functional imaging studies provide complementary evidence for parietal involvement in a wide range of tasks (see Chapter 18). A principal difference between lesion and activation studies lies in the unilateral versus bilateral nature of the observed effects. For the case of active touch, the deficits of object recognition and manipulation are purely contralesional, whereas the activations elicited by unilateral active touch paradigms are bilateral and often symmetrical. This is illustrated in Figure 11-2, where the activations are compared with the common overlap zone of lesions that produce active touch and grasping deficits. The center of activations associated with the respective tasks coincide around the anterior intraparietal sulcus (IPS), thus stressing the key role of this area for tactile exploration and manipulation (46).

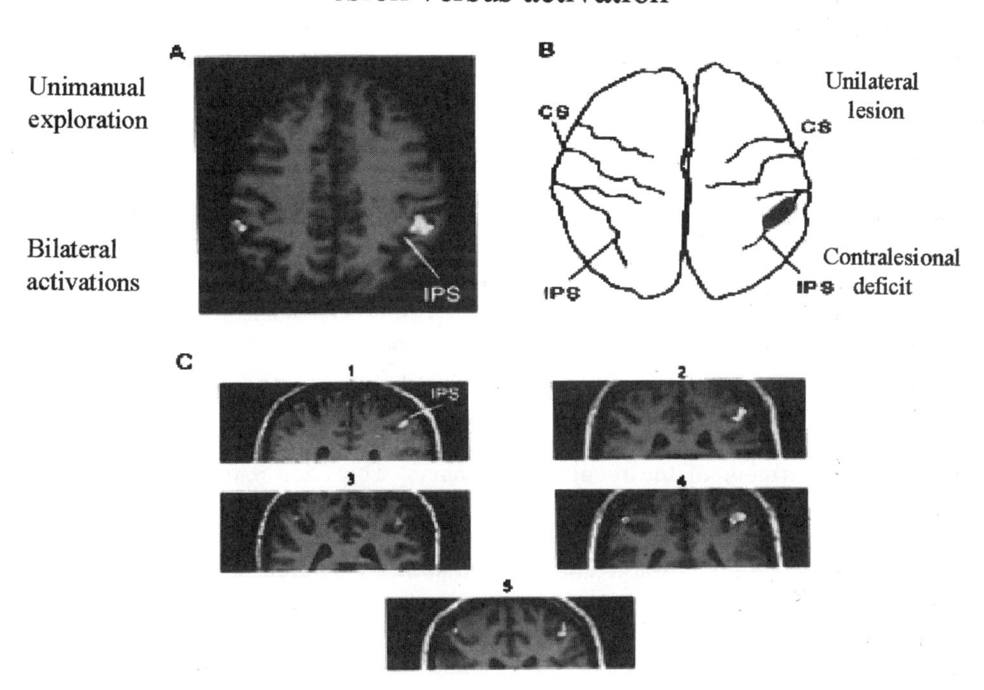

FIG. 11-2. Functional magnetic resonance imaging study during grasping compared with pointing shows a significant activation area in the anterior lateral bank of the intraparietal sulcus (IPS) in a normal subject **(A).** The activated area corresponds to the location of the common zone of overlap in the patients with lesions, including anterior IPS and prehension deficits **(B).** The stereotactic coordinates were x = –45, y = –35, z = 43. CS, central sulcus. The localization of the activation foci in the coronal sections of the five control subjects occurs despite the obvious interindividual variation in sulcal anatomy consistent across subjects **(C).** (From: Binkofski, *Neurology* 1998, with permission.)

Other constituents of the functional network involved in specific tasks components can be elaborated as shown for the creation of self-generated figures out of an amorphous mass. Whereas the finger movements during the exploration of 3D objects are predetermined by the given object shape, manual modeling requires the complex interplay between memory retrieval of 3D objects, their representation in working memory and continuous update, and the conception of the motor act guided by the internal model (47). This complex task additionally recruited an area within the superior and inferior parietal lobule and in ventral premotor cortex.

THE DEFICIENT CONTROL OF STATIC FORCE AND POSTURE

Patients with parietal lobe damage are not only affected with respect to the sophisticated sensorimotor behavior involved in active touch but also in elementary functions such as the adjustment of muscle force and synergies required for the provision of the continuous automatic control of posture. The assumption of dystonic postures and pseudoathetotic wandering in deafferentated patients illustrate the inability to sustain constant force levels required for the maintenance of posture. Foerster (8) addressed this deficiency as static ataxia interfering with the patient's ability to adjust his force for steady postural tasks, such as holding an egg, that would either be squeezed or dropped. The impairment of the regulation of force is an important aspect of parietal lobe function because it reveals that not only the more complex spatial postural control mechanisms are affected, but also basic aspects of force control, such as the maintenance of tension at particular joints. The preservation of afferent input to motor and premotor areas via the cerebellar route obviously is not sufficient to provide effective compensatory postural control because recovery remains incomplete even in chronic cases.

Figure 11-3 shows an example of a steady hold task in a patient with a lesion of the left posterior parietal lobe. The task was to match the force exerted on a strain gauge to a target force level presented at a screen. Whereas the ipsilesional hand behaved normally and showed only smaller force fluctuations after eye closure, the deficient somatosensory control of the contralesional hand could be compensated for to some extent, as long as visual feedback was provided. But as soon as the eyes were closed, force tracking became chaotic. In contrast to the regulation of static force, step tracking was surprisingly good on both sides. This corresponds to the clinical picture with better preservation of rapid predictive movements compared to slower sensory-guided movements. The patient described by Jeannerod and coworkers (26) could not force track a slow sinusoidal oscillatory signal, although this was a highly predictive task.

These observations indicate a differential liability to disturbance of sensory-guided pursuit as compared to predictive fast ramp movements of the saccadic type in parietal lobe disease. Support for such a view comes from data on the postural control of the fingers examined by a system analytic input-output approach in patients with parietal lesions (48). Whereas the responses to short torque pulse perturbations were normal, patients had difficulty returning to given preload levels when additional step torque loads had to be compensated for by their fingers. The control offset measured 0.5 s after the torque step was significantly larger in the patient group. The two types of stimuli activate different motor subsystems. Impulse-load tasks evoke spinally mediated short latency responses that are subject to modification by cerebellar and transcortical loops. The compensation of step torque loads requires a static readjustment of the force level to preload conditions, a task involving more complex functions such as memorizing preload level, step torque, and response metrics.

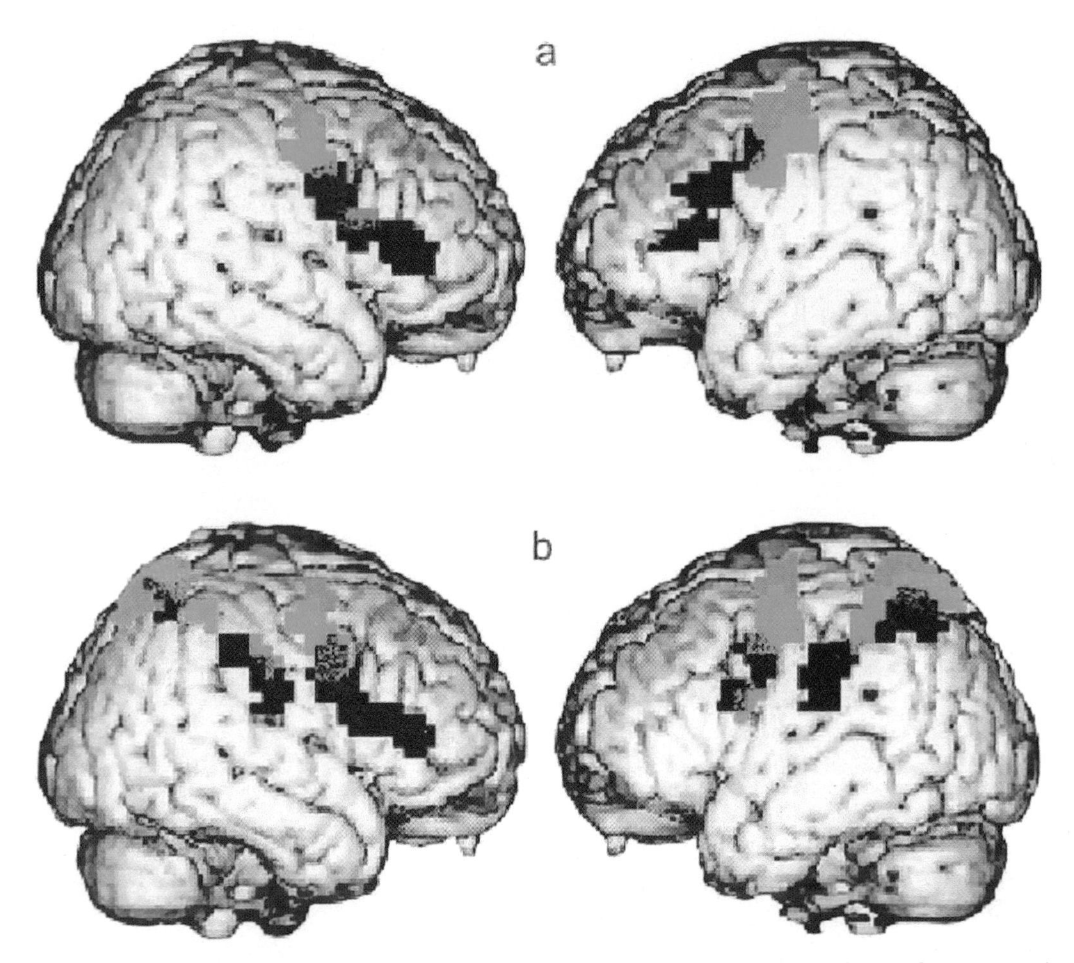

FIG. 11-4. Somatotopy of premotor and parietal cortices as revealed by action observation. **A:** Observation of non–object-related actions. **B:** Observation of object-related actions. Activation foci are projected on the lateral surface of a standard brain (MNI). Overlap of colors indicates foci present during observation of actions made by different effectors.

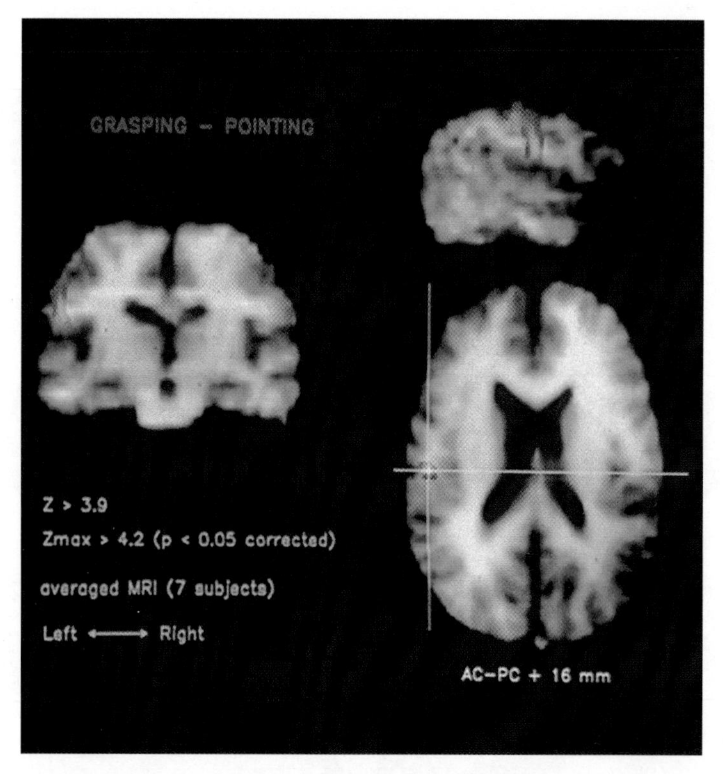

FIG. 13.3. Topography of parietal activation during grasping movements of the right hand. Average regional cerebral blood flow (rCBF) change in normal subjects as studied by positron emission tomography. Note the activation in the ventral part of the left parietal cortex, corresponding to the anterior tip of the intraparietal sulcus.

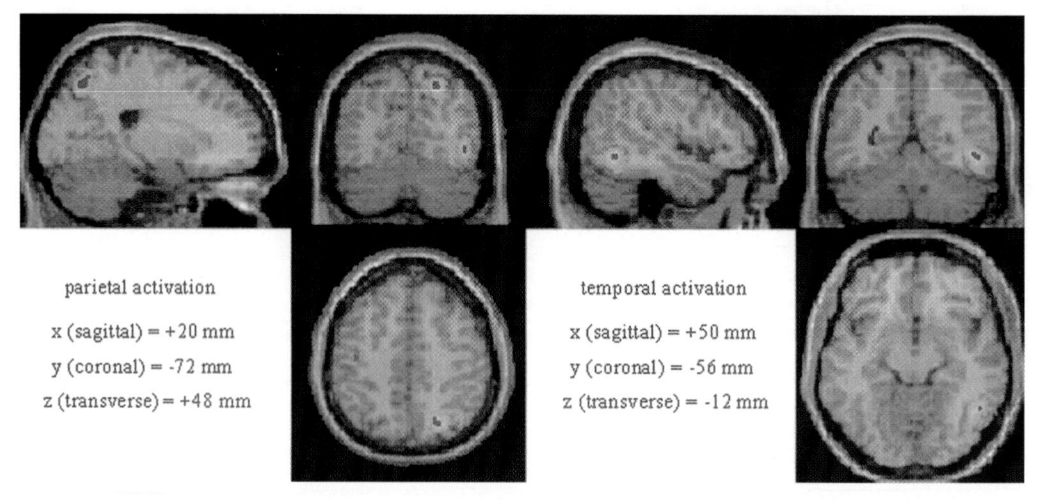

FIG. 13.4. The topography of cortical areas involved in spatial perception. A positron emission tomography study in normal subjects. The subjects were requested to judge the similarity or difference of meaningless shapes for their size or orientation. Note the activation in the right posterior parietal cortex and right inferotemporal cortex. (From: Faillenot I, Decety J, Jeannerod M. Human brain activity related to the perception of spatial features of objects. *NeuroImage* 1999;10:114–124, with permission.)

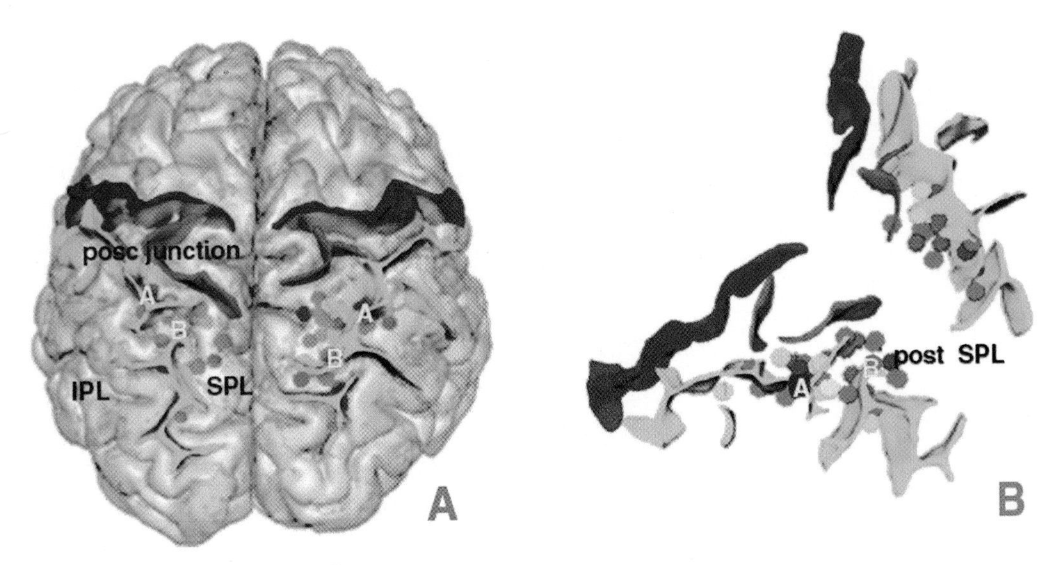

FIG. 17.1. Three-dimensional reconstruction of a brain in the Talairach and Tournoux coordinates system. The principal sulci have been extracted and underlined; the central sulcus in red, the superior branch of the post-central sulcus in pink, the intraparietal sulcus in green. Each dot represents the center of activity as reported by one of the studies listed in Table 17.1, with a color code corresponding to the type of eye movements: dark blue, saccades in darkness; red, visually guided saccades (single or sequences); turquoise, memory guided saccades (single or sequences); yellow, pursuit. **A:** Putative LIP according to Müri et al. (Müri R, Iba-Zizen MT, Cabanis EA, et al. Location of the human posterior eye field with functional magnetic resonance imaging. *J Neurol Neurosurg Psychiatry* 1996;60:445–448). **B:** Putative LIP according to Sereno et al. (Sereno MI, Pitzalis S, Martinez A. Mapping of contralateral space in retinotopic coordinates by a parietal cortical area in humans. *Science* 2001;294:1350–1354) and Simon et al. (Simon O, Mangin JF, Cohen L, et al. Topographical layout of hand, eye, calculation, and language-related areas in the human parietal lobe. *Neuron* 2002;33:475–487). **A:** Upper view. **B:** View from an upper left angle of the extracted sulci of the same brain.

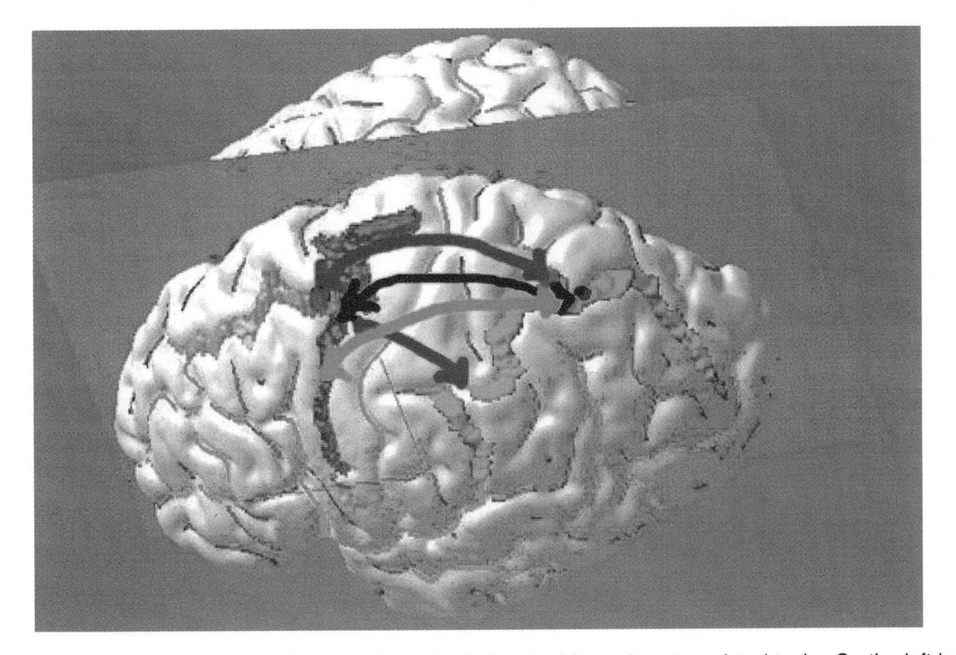

FIG. 17.3. Schematic examples of frontoparietal circuits involved in oculomotor-related tasks. On the left hemisphere of a brain transformed into Talairach and Tournoux space (MNI template, see text), the intraparietal sulcus is underlined in green, the precentral sulcus in purple, and the superior frontal sulcus in yellow. The arrows represent putative circuits as derived from fMRI results. In blue, circuits for memorized sequences of saccades; in red, visually guided saccades; in green, voluntary saccades in darkness.

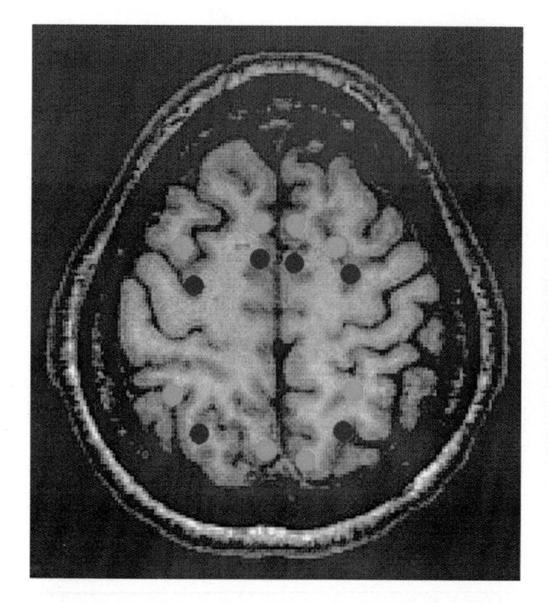

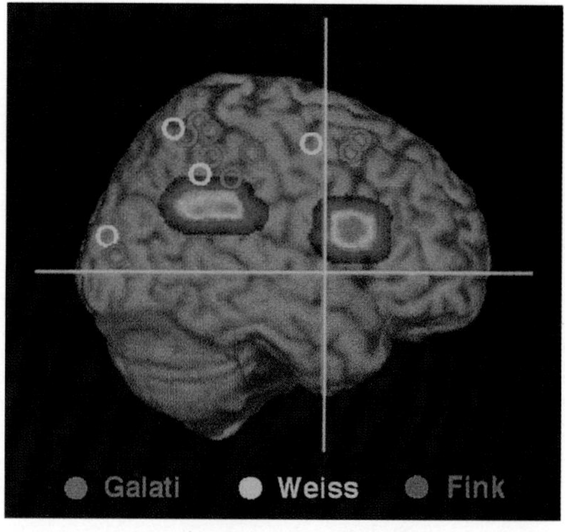

FIG. 17.4. (Top Left). Two different networks for executing sequences of saccades, in the dorsal frontoparietal cortex. In red we represented the peaks of activity observed when the volunteers executed any kind of memorized sequence of saccades; in turquoise, the peaks of activity observed when the sequence was newly learned.

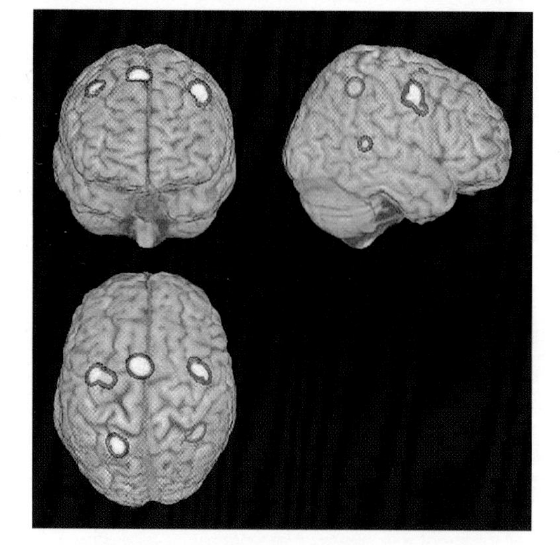

FIG. 19.7. (Top Right). Brain areas associated with spatial neglect (shaded colored areas) and metaanalysis of three functional neuroimaging experiments on line bisection. The temporoparietal area is redrawn from Vallar and Perani (see ref. 51 in Chapter 19), the frontal area from Husain and Kennard (see ref. 68 in Chapter 19), and reproduces the regions of maximal overlap of frontal lesions in that patient series. Circles represent the location of activation peaks expressed in stereotactic space in three functional neuroimaging experiments on line bisection. Only the activations on the lateral surface of the right hemisphere are reported. The horizontal line represents the plane passing through the two commissures, the vertical line the coronal plane passing through the anterior commissure. Mesial ventral occipital activations are not shown.

FIG. 19.8. (Bottom Left). Metaanalysis of two functional magnetic resonance imaging studies in which covert orienting of attention and generation of saccades were compared in the same subjects. The metaanalysis represents only brain regions found by both studies. The *areas in yellow* represent brain regions that were activated by both eye movements and covert attention. These include the dorsal premotor cortex, the supplementary motor region, the region of the left intraparietal sulcus/superior parietal lobule, and the right superior temporal sulcus. The right dorsal premotor region and the cortex of the right intraparietal sulcus/superior parietal lobule *(areas in blue)* represent cortical regions more active for covert attention. There was also another such region at the junction between the right ventral mesial extrastriate cortex and the cerebellum (not shown). There was also one area in left ventral occipital cortex and left cerebellum more active for eye movements in both studies (not shown). One study reported separate comparisons of the two experimental tasks against central fixation baseline, and direct comparisons of the two experimental tasks. The other study reported conjunctions of experimental conditions against baseline and differences across conditions (as interaction effects). To accommodate for the different statistical approaches, the metaanalysis was generated as follows: (a) each peak of activation was transformed into a 12-mm diameter sphere in stereotactic space, which approximates spatial resolution of group studies using fMRI; (b) the intersection of attention and eye movement maps was calculated for one study; (c) intersections for three set of maps from each study were then calculated: (1) areas of shared activation for covert attention and eye movements; (2) areas of greater activations for covert attention; and (3) areas of greater activation for eye movements.

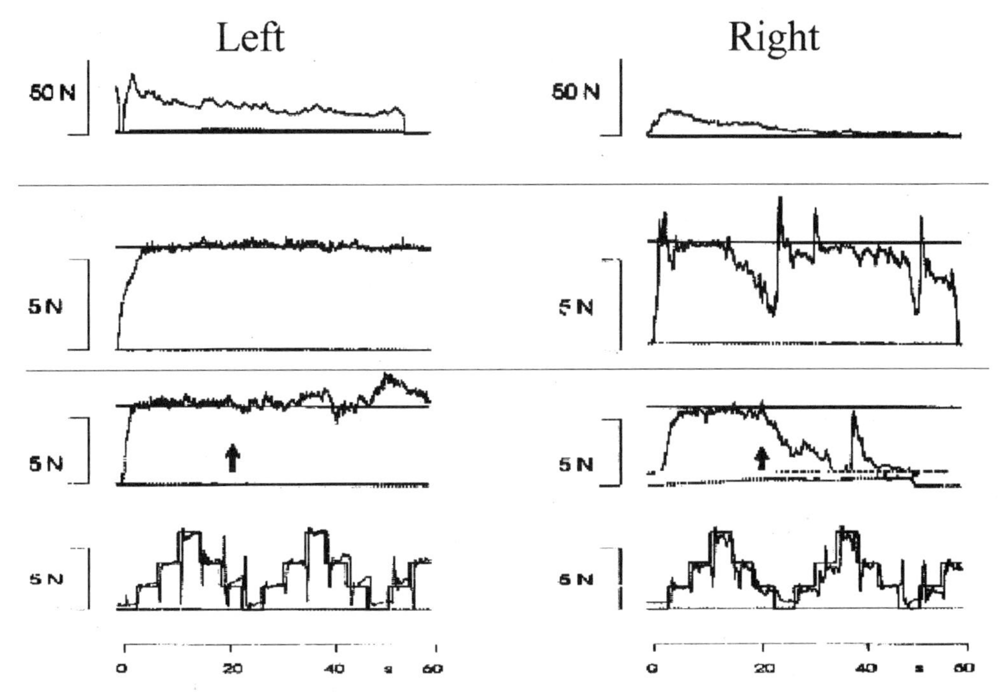

FIG. 11-3. Force control in a normal subject *(left)* and a patient with a posterior parietal lesion *(right)*. **Upper row:** maintenance of a maximal voluntary contraction over 1 minute. The typical decay of maximal force is seen in the normal subject and the patient. The second row shows that the normal subject is easily capable of maintaining a steady force level at a submaximal contraction (note the different ordinate). For the patient, the maintenance of a steady force level shows coarse force fluctuations from the beginning, but becomes impossible when vision is blocked *(arrow)*. Step tracking is near normal, in contrast to the maintenance of steady force levels.

DISTURBANCES OF BODY IMAGE

An important facet of the relation between the control of force and posture and the constitution of body image was illuminated by the description of a patient with an SPL lesion who presented with an unusual sensorimotor deficit (49). The clinical picture was characterized by her inability to maintain internal representations of her body parts. She perceived her right extremities to drift and then fade unless she could see them. Her perception of light touch was only slightly diminished and she could discriminate sharp and blunt stimuli at her right arm. However, when these stimuli were applied constantly, the sensation faded over seconds when she could not see them. Along with tactile extinction she had astereognosia and agraphesthesia of the affected hand. Without vision, postural control was poor with pronounced drifts of the affected limbs or of their perceived location and with pronounced pointing errors. Force control showed major instability so that she could not maintain steady force levels.

As compared to the permanent unawareness of their limbs in typical parietal neglect patients, the unique feature of the disturbance of this patient was that she only became unaware of her limbs for seconds and could regain awareness by looking at them. On the basis of this case the authors chose a new theoretical approach derived from the "observer framework" in engineering. This requires a system that monitors both the input and output in order to establish online state estimates to provide internal representations of the body and its environment. According to this model, the internal state is maintained by a recursive process, updated by current sensory and motor signals and stored. Because the state estimate of this patient was perceptually decaying over time, it was suggested that this storage occurs in the SPL.

Taken together, these results support and further extend the view that the parietal lobe plays a pivotal role in the constitution and maintenance of body image. However, in contrast to the long-standing alterations of body scheme in other conditions, such as neglect, anosognosia, denial of hemiparesis with or without confabulation, asomatognosia, finger agnosia, and right-left disturbances (see Chapter 20), the changes in bodily awareness were transient. The phantom third limb represents another transient symptom of disturbed body image, but in this case the lesion can be outside the parietal lobe. McGonigle et al. (50) described a patient with a right frontomedial lesion who sporadically experienced a supernumerary "ghost" left arm occupying the previous position of the real left arm after a delay of 60 to 90 s. Using a delayed response paradigm with functional MRI to examine the hemodynamic correlates of the patient's illusion, they identified an activation on the right medial wall in the supplementary motor area. The authors suggested that motor areas can influence the conscious perception of the body in "action space" while continuing to be aware of the true position of the real limb on the basis of afferent somatosensory information. Obviously, information from different sources must be integrated in order to constitute a unitary perception of the body. Its transient nature illustrates the temporal dynamics of the interplay among multiple dissociated conscious representations.

A study of a haptically deafferentated patient (51) further illustrates the differential role of afferent and efferent signals for bodily awareness. The patient could solve the conflict produced among motor intention, proprioception, and visual feedback. She could also generate accurate movements to a target with very limited feedback from her movements. However, she had no conscious awareness of that conflict, nor of her actual performance, indicating that information derived from efferent processes is not sufficient for establishing the conscious awareness of one's own actions.

In this context, it is interesting that subjects with congenitally absent limbs still have vivid phantom sensations. Functional imaging data showed that the body parts that have never been developed are represented in somatosensory cortex and therefore can not be considered as perceptual-motor memory traces of once functioning body parts (52). The emergence of a body scheme in these cases was regarded as evidence for the role of genetic factors for the formation of an innate "neuromatrix" (53).

ARE SIMPLE AND MEANINGLESS MOVEMENTS DISTURBED IN PARIETAL LOBE DISEASE?

The isolated somatomotor and visuomotor disturbances (see Chapter 13) illustrate that unimodal information processing is compromised before integration into polymodal reference frames occurs. Impairment of motor behavior at the supramodal level is represented by the apraxias as discussed in Chapter 15 by Leiguarda. Liepmann (40) defined the apraxias as an impairment of purposive motor behavior affecting both sides of the body irrespective of the modalities involved. Because they are typically seen after damage of the left hemisphere, Liepmann concluded that the left hemisphere is not only dominant for language, but also for praxis.

In the context of this chapter it is of interest whether the principle of sensorimotor transformation disorders discussed so far also applies for these "cognitive" disturbances of movement semantics. From which level of complexity or by which criteria can apractic disturbances be defined and distinguished from other nonapractic disturbances of motor behavior? Historical definitions serve this purpose, as is often the case in neurology. However, new results and quantitative recordings provide new vistas as they reveal that the spectrum of motor dysfunctions is continuous rather than qualitatively divided into different categories.

A suitable example to illustrate this point is a kinematic study of five meaningless movement sequences of increasing complexity that had to be performed by imitation or on verbal command by patients with parietal lesions (54). The performance was scored as omissions, additions, or temporal and spatial errors. Patients with left parietal damage produced more temporal and spatial errors than those with right parietal lesions, whereas additions and omissions occurred about equally in both groups. Only the left parietal patients showed an increase of error rate with task complexity. Both hands were equally affected. Instruction modality had no influence on performance and those patients clinically classified as apractic did not show a different pattern of disturbance.

The higher incidence and severity of spatial errors in the left parietal patients seems inconsistent with the well-known superiority of the right hemisphere for spatial functions. However, this may result from the fact that spatial errors were scored on the basis of kinematic recordings rather than on the production of complex spatial configurations. In another kinematic study on imitation behavior of left- and right-hemisphere damaged patients, Hermsdörfer et al. (55) also found a higher incidence of "spatial parapraxias" in patients with left-sided lesions. Haaland et al. (56) observed disturbances of hand posture and location in apractic patients with left hemisphere lesions and uncoupling of the temporal and spatial features of the movement. Complementary, Clark et al. (57) described errors in the plane of motion and trajectory shape in cases with left parietal damage.

The data by Weiss et al. (54) showed that the performance of meaningless movement sequences in patients with left parietal damage is more frequently disturbed than the various clinically apparent apraxias. The study further revealed that the cumulative effects of the disturbance of the spatial and temporal characteristics that exist already at the level of simple movements compromise the production of more complex motor behaviors. Therefore, the disorder of purposive motor behavior is, at least in part, a consequence of the derangement of its constituent elements that may be complemented by cognitive deficits.

ACTION OBSERVATION

The wide range of motor deficits ensuing from parietal damage in the human emphasizes the pivotal role of the parietal lobe for sensorimotor transformation and integration. The functional equivalence among the intention, simulation, imitation, observation, or performance of actions revealed by functional activation studies (58) adds an important dimension to our understanding of parietal lobe function and dysfunction because it demonstrates the significance of parietal–premotor interactions. As parietal sensory association cortex is relevant for motor behavior, premotor areas are involved in action observation. Experimentally, the "mirror neurone" framework (59) showed that a cluster of neurons in monkey ventral premotor cortex area F5 is active not only during movement execution but also when watching another monkey performing the same action. This concept has been extended by an imaging study in the human (60), demonstrating that action observation activates not only the possible homolog in ventral premotor, but also dorsolateral premotor cortex. The activated premotor area extended in a somatotopic manner along a strip anterior to the motor homunculus in primary motor cortex.

The other new finding in the human was that whenever the observed movement was object-directed (e.g., chewing an apple) there was strong concomitant parietal activation that was also bilateral and somatotopically organized. Thus, this system constitutes the neural substrate for a matching mechanism mapping the observed actions on the observer's motor representations (Fig. 11-4).

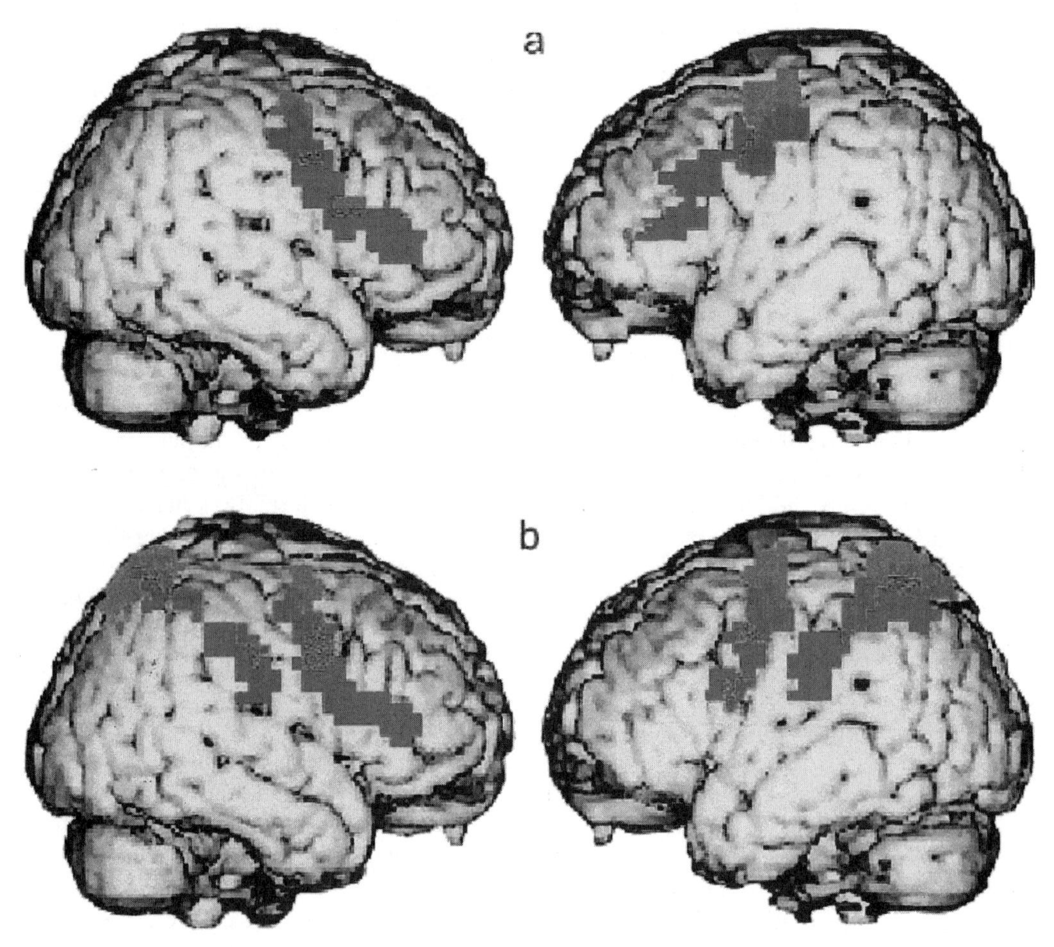

FIG. 11-4. Somatotopy of premotor and parietal cortices as revealed by action observation. **A:** Observation of non–object-related actions. **B:** Observation of object-related actions. Activation foci are projected on the lateral surface of a standard brain (MNI). Overlap of colors indicates foci present during observation of actions made by different effectors. (From: Buccino, with permission.) (See color plate of this figure after page 186.)

The object related parietal activations are in accordance with the well-known pragmatic role of parietal cortex that is dysfunctional in patients with parietal lesions. They further support the view that the parietal lobe is implicated in action-oriented object description by showing that a similar pragmatic analysis takes place when passively watching such an action.

If the perceptual correlates of action observation were mapped on the neuronal repertoires conceiving the corresponding motor acts, damage could interfere with both the motor behavior and the comprehension of meaning conveyed by motor acts. Examining the recognition and imitation of pantomimed motor acts in patients with parietal and premotor lesions showed that left parietal damage compromised the imitation but not the comprehension of gestures or tool use (61). The patients with premotor lesions did not reveal any deficits of movement imitation or recognition. This is interesting in view of the distinct activations of the premotor areas during action observation. What patients with premotor lesions did show were disturbances in the sensory cueing of movement, whereas non–motor conditional association tasks were normal (62). However, the understanding of the meaning of a given motor behavior was preserved, in contrast to patients with temporal lesions. This is in agreement with activation studies (63), which show the significance of temporal lobe structures for the knowledge of action.

THE FUNCTIONAL ORGANIZATION OF SOMATOSENSORY AND VISUOMOTOR INFORMATION PROCESSING IS DIFFERENT

The impairment of the organization of actions in the surrounding extrapersonal space in cases where the perception of the world remains preserved illustrates the well-known processing dichotomy of visual information flow into a pragmatic parietal and a cognitive temporal route (see Chapter 16). There is no such dissociation for somatosensory functions. Rather, the concomitant impairment of the cognitive and pragmatic aspects of somatosensory processing in cases with parietal lobe damage indicate that in this modality both aspects are processed in the parietal lobe. For somatsenation the parietal lobe is a unimodal sensory analyzer and a sensorimotor interface.

This important difference in the functional architecture of the two modalities is also demonstrated by activation studies of the respective channels. Visual cognitive tasks show ventral route, and tactile recognition paradigms reveal dorsal route activation. Visumotor and tactumotor tasks activate different parietal areas. During cross-modal matching tasks, distinct activations in the anterior intraparietal sulcus emphasize the role of this area for integrating polymodal sensory information and the elaboration of object features and object-related actions (64).

SUMMARY

Lesion studies show that a wide range of integrative sensorimotor functions can be selectively disturbed in patients with parietal lobe damage. Lesions restricted to the somatosensory representations on the anterior parietal lobe produce somatosensory deficits that resemble deafferentated states, including the secondary effects on motor control. Slightly more posterior lesions often are associated with impairment of more complex synthetic somatosensory functions similar to those observed after dorsal column lesions. Damage of the posterior parietal lobe can selectively interfere with virtually every aspect of somatosensory function. These perceptive and cognitive disturbances may or may not be associated with complex motor disturbances of the apractic type. The frequent association of astereognosia and tactile apraxia illustrate the mutual interdependence of the sensorimotor processes involved in active touch and the proximity of the respective processing modules.

Parietal lobe function is critical for the control of force and posture, and for the formation of the body image and its relation to external space (the guidance of movements, including the eyes, to external objects). Imaging studies underscore the prominent role of the parietal cortex as a sensorimotor interface and provide complementary information about the interrelationship between perception and action. Action observation activates premotor cortex, but parietal cortex is also recruited whenever an action involves objects, thus emphasizing the significance of parietal cortex for object-directed motor behavior.

In contrast to the ventral-dorsal route processing dichotomy in the visual system, both the perceptual-cognitive and motor aspects of somatosensory processing are compromised after parietal damage, demonstrating a different functional architecture of the two sensory systems. The preservation of the comprehension of the meaning of gestures or of object use in patients with lesions restricted to the parietal lobe reveals that the semantic aspects of motor behavior are mediated in the temporal lobe.

REFERENCES

1. Good CD, Johnsrude IS, Ashburner J, et al. A voxel-based morphometric study of ageing in 465 normal adult human brains. *NeuroImage* 2001;14:21–36.
2. Devlin JT, Moore CJ, Mummery CJ, et al. Anatomic constraints on cognitive theories of category specificity. *NeuroImage* 2002;15:675–685.
3. Hari R, Reinikainen K, Kaukoranta E, et al. Somatosensory evoked cerebral magnetic fields from SI and SII in man. *Electroencephalogr Clin Neurophysiol* 1984;5:254–263.
4. Hari R, Karhu J, Hamalainen M, et al. Functional organization of the human first and second somatosensory cortices: a neuromagnetic study. *Eur J Neurosci* 1993;5: 724–734.
5. Narici L, Modena I, Opsomer RJ, et al. Neuromagnetic somatosensory homunculus: a non-invasive approach in humans. *Neurosci Lett* 1991;21:51–54.
6. Sutherling WW, Crandall PH, Darcey TM, et al. The magnetic and electric fields agree with intracranial localizations of somatosensory cortex. *Neurology* 1988; 38:1705–1714.
7. Sakai K, Watanabe E, Onodera Y, et al. Functional mapping of the human somatosensory cortex with echoplanar MRI. *Magn Reson Med* 1995;33:736–743.
8. Foerster O. Motorische Felder und Bahnen: sensible corticale Felder. In: Bumke H, Foerster O, eds. *Handbuch der Neurologie.* Berlin: Springer-Verlag, 1936:358–448.
9. Penfield W, Rasmussen T. *The cerebral cortex in man.* New York: Macmillan, 1950.
10. Lüders H, Lesser RP, Dinner DS, et al. The second sensory area in humans: evoked potential and electrical stimulation studies. *Ann Neurol* 1985;17:177–184.
11. Iwamura Y. Hierarchical somatosensory processing. *Curr Opin Neurobiol* 1998;8:522–528.
12. Roland E. Somatosensory detection of microgeometry, macrogeometry and kinesthesia after localized lesions of the cerebral hemispheres in man. *Brain Res Rev* 1987;12:43–94.
13. Bodegard A, Geyer S, Grefkes C, et al. Hierarchical processing of tactile shape in the human brain. *Neuron* 2001;31:317–328.
14. Müller F, Kunesch E, Binkofski F, et al. Residual sensorimotor functions in a patient after right-sided hemispherectomy. *Neuropsychologia* 1991;29:125–145.
15. Murray EA, Mishkin M. Relative contributions of S II and area 5 to tactile discrimination in monkeys. *Behav Brain Res* 1984;11:67–83.
16. Garcha HS, Ettlinger G, Maccabe JJ. Unilateral removal of the second somatosensory projection cortex in the monkey: evidence for cerebral predominance. *Brain* 1982;105:787–810.
17. Friedman DP, Jones EG, Burton H. Representation pattern in the second somatic sensory area of the monkey cerebral cortex. *J Comp Neurol* 1980;192:21–41.
18. Ploner M, Schmitz F, Freund H-J, et al. Differential organization of touch and pain in human primary somatosensory cortex. *J Neurophysiol* 2000;83:1770–1776.
19. Caselli RJ. Ventrolateral and dorsomedial somatosensory association cortex damage produces distinct somesthetic syndromes in humans. *Neurology* 1993;43: 762–771.
20. Déjerine J. A propos l'agnosie tactile. *Rev Neurol (Paris)* 1907;15:781–784.
21. Head H, Holmes G. Sensory disturbances from cerebral lesions. *Brain* 1911;34:102–254.
22. Delay J. *Les astéréognosies, pathologie du toucher.* Paris: Masson & Cie, 1935.
23. Critchley M. *The parietal lobes.* New York: Hafner, 1953.
24. Verger H. Sur les troubles de la sensibilité générale consecutifs aux lésions des hémisphères cérébraux chez l'homme. *Arch gén de Méd* 1900;6:641–713.
25. Déjerine J, Mouzon J. Deux cas de syndrome sensitif cortical. *Rev Neurol (Paris)* 1914–1915;28:388–392.
26. Jeannerod M, Michel F, Prablanc C. The control of hand movements in a case of hemianesthesia following a parietal lesion. *Brain* 1984;107:899–920.
27. Pause M, Kunesch E, Binkofski F, et al. Sensorimotor disturbances in patients with lesions of the parietal cortex. *Brain* 1989;112:1599–1625.
28. Semmes J, Turner B. Effects of cortical lesions on somatosensory tasks. *J Invest Dermatol* 1977;69:181–189.
29. Saetti MC, De Renzi E, Comper M. Tactile amorphagnosia secondary to spatial deficits. *Neuropsychologia* 1999;37:1087–1100.
30. Reed CL, Caselli RJ, Farah MJ. Tactile agnosia. Underlying impairment and implications for normal tactile object recognition. *Brain* 1996;119:875–888.
31. Mauguière F, Isnard J. Tactile agnosia and dysfunction of the primary somatosensory area. Data of the study by somatosensory evoked potentials in patients with deficits of tactile object recognition. *Rev Neurol (Paris)* 1995;151:518–527.
32. Caselli RJ. Rediscovering tactile agnosia. *Mayo Clin Proc* 1991;66:129–142.
33. Endo K, Miyasaka M, Makishita H, et al. Tactile agnosia and tactile aphasia: symptomatological and anatomic differences. *Cortex* 1992;28:445–469.
34. Wernicke C. Die neueren Arbeiten über Aphasie. *Fortschritte der Medizin* 1885;3:824–830.
35. Gibson JJ. Observations on active touch. *Psychol Rev* 1962;69:477–491.
36. Jeannerod M, Arbib MA, Rizolatti G, et al. Grasping objects: the cortical mechanisms of visuomotor transformation. *Trends Neurosci* 1995;18:314–320.
37. Lederman SJ, Klatzky RL. Hand movements: a window into haptic object recognition. *Cogn Psychol* 1987;19: 342–368.
38. Katz D. Der Aufbau der Tastwelt. *Z Psychol Physiol Sinnesorg* 1925;2:1–270.
39. Hippius R. Erkennendes Tasten. *N Psychol Stud* 1934; 12:83–98.
40. Liepmann H. Das Krankheitsbild der Apraxie ("motorische Asymbolie"), auf Grund eines Falles von einseitiger Apraxie. *Monatsschr Psychiatr Neurol* 1900;8: 182–197.
41. Klein R. Zur Symptomatologie des Parietallappens. *Zeitschrift für Gesamte Neurologie und Psychiatrie* 1931;135:589–608.
42. Valenza N, Ptak R, Zimine I, et al. Dissociated active and passive tactile shape recognition: a case study of pure tactile apraxia. *Brain* 2001;124:2287–2298.
43. Binkofski F, Kunesch E, Classen J, et al. Tactile apraxia.

Unimodal apractic disorder of tactile object exploration associated with parietal lobe lesions. *Brain* 2001;124: 132–144.
44. Yamadori A. Palpatory apraxia. *Eur Neurol* 1982;21: 277–283.
45. Sainburg R L, Poizner H, Ghez C. Loss of proprioception produces deficits in interjoint coordination. *J Neurophysiol* 1993;70:2136–2147.
46. Binkofski F, Dohle C, Posse S, et al. Human anterior intraparietal area subserves prehension. A combined lesion and functional MRI activation study. *Neurology* 1998;50:1253–1259.
47. Jäncke L, Kleinschmidt A, Mirzazade S, et al. The role of the inferior parietal cortex in linking the tactile perception and manual construction of object shapes. *Cerebral Cortex* 2001;11:114–121.
48. Scholle HC, Bradl U, Hefter H, et al. Force regulation is deficient in patients with parietal lesions: a system-analytic approach. *Electroencephalogr Clin Neurophysiol* 1998;109:203–214.
49. Wolpert DM, Goodbody SJ, Husain M. Maintaining internal representations: the role of the human superior parietal lobe. *Nat Neurosci* 1998;1:529–533.
50. McGonigle DJ, Hanninen R, Salenius S, et al. Whose arm is it anyway? An fMRI case study of supernumerary phantom limb. *Brain* 2002;125:1265–1274.
51. Fourneret P, Paillard J, Lamarre Y, et al. Lack of conscious recognition of one's own actions in a haptically deafferentated patient. *NeuroReport* 2002;13:541–547.
52. Brugger P, Kollias SS, Müri RM, et al. Beyond re-membering; phantom sensations of congenitally absent limbs. *Proc Natl Acad Sci USA* 2000:97:6167–6172.
53. Melzack R, Israel R, Lacroix R, et al. Phantom limbs in people with congenital limb deficiency or amputation in early childhood. *Brain* 1997;120:1603–1620.
54. Weiss PH, Dohle C, Binkofski F, et al. Motor impairment in patients with parietal lesions: disturbances of meaningless arm movement sequences. *Neuropsychologia* 2001;39:397–405.
55. Hermsdorfer J, Goldenberg G, Wachsmuth C, et al. Cortical correlates of gesture processing: clues to the cerebral mechanisms underlying apraxia during the imitation of meaningless gestures. *NeuroImage* 2001;14: 149–161.
56. Haaland KY, Harrington DL, Knight RT. Spatial deficits in ideomotor limb apraxia. A kinematic analysis of aiming movements. *Brain* 1999;122:1169–1182.
57. Clark MA, Merains AS, Kothari A, et al. Spatial planning deficits in limb apraxia. *Brain* 1994;117: 1093–1106.
58. Grezes J, Decety J. Functional anatomy of execution, mental simulation, observation, and verb generation of actions: a meta-analysis. *Hum Brain Mapp* 2001;12: 1–19.
59. Rizzolatti G, Fadiga L, Gallese V, et al. Premotor cortex and the recognition of motor actions. *Cog Brain Res* 1996;3:131–141.
60. Buccino G, Binkofski F, Fink GR et al. Action observation activates premotor and parietal areas in a somatotopic manner: an fMRI study. *Eur J Neurosci* 2001;13: 400–404.
61. Halsband U, Schmitt J, Weyers M, et al. Recognition and imitation of pantomimed motor acts after unilateral parietal and premotor lesions: a perspective on apraxia. *Neuropsychologia* 2001;39:200–216.
62. Halsband U, Freund H-J. Premotor cortex and conditional motor learning in man. *Brain* 1990;113:207–222.
63. Martin A, Haxby JV, Lalonde FM, et al. Discrete cortical regions associated with knowledge of color and knowledge of action. *Science* 1995;270:102–105.
64. Grefkes C, Weiss PH, Zilles K, et al. Crossmodal processing of object features in human anterior intraparietal cortex: an fMRI study implies equivalencies between humans and monkeys. *Neuron* 2002;35:173–184.

12

Remapping Somatosensory Cortex After Injury

Herta Flor

Department of Neuropsychology, University of Heidelberg, Central Institute of Mental Health, Mannheim, Germany

INTRODUCTION

Until recently the leading neuroscientific opinion was that the neuronal organization of the sensory and motor maps of the cortex develops early in life and remains stable throughout adulthood. Experience-dependent plasticity of the primary sensory and motor areas seemed to be limited to certain so-called critical phases in development. The results of animal experiments as well as experiments in humans with acquired neuronal damage led to a revision of this view. Plastic changes of the somatosensory and motor maps also take place in the adult central nervous system. Today it is accepted that the ability for structural and functional reorganization is a basic principle of brain function. This ability also is often referred to as neuronal or brain plasticity. "Brain plasticity refers to the adaptive capacities of the central nervous system—its ability to modify its own structural organization and functioning. It is an adaptive response to functional demand (1)."

In this chapter we present an overview of the results of animal and human studies on the reorganization of the adult somatosensory cortex related to injury. In addition to the description of functional and structural changes related to injury, we also address the potential mechanisms of these changes, their functional significance, and potential clinical applications.

ANIMAL STUDIES

Remapping of somatosensory cortex refers to the alteration of the location or size of the homuncular representation of somatosensory inputs. Although the Penfield and Boldrey (2) description established the idea of a somatosensory homunculus, current research on the anatomy of the somatosensory cortex has questioned this assumption and rather convincingly showed multiple representations of the body in different subsections of the somatosensory cortex (3,4). Thus, the discussion of reorganization must be viewed with respect to these rather complicated representations that also seem to change based on attentional factors and situational demands (5). Therefore, the somatosensory cortex must be viewed as a very flexible dynamic and adaptive system, and specific representations may be just one expression of the variety of states that may characterize somatosensory cortex. As a consequence, topographic representations are manifestations of physiologic interactions rather than anatomic constructs (6).

Experiments on lesion-induced plasticity of the adult somatosensory cortex were performed in adult monkeys in whom the afferent neuronal input was interrupted by severing a nerve or by amputating a digit (7–10). In these studies, an invasion of the representation zone of the deafferentated or amputated body part in primary somatosensory cortex by neighboring areas was reported. For example, after the amputation of the third or both the second and third digits in owl monkeys, the representation zones of the neighboring digits and palmar zones expanded topographically into the cortical zone that formerly represented the amputated digit(s) in area 3b of the primary somatosensory cortex as revealed by microelectrode recordings. Accompanying changes in receptive field sizes showed an expansion into the deafferentated territory and the authors suggested that there might be parallel changes in sensory acuity. The changes observed in these studies were in the range of millimeters and could be explained by alterations in synaptic strength of existing connections. Pons et al. (11) reported changes of a magnitude of 1 to 2 cm in adult macaque monkeys who had been subjected to sectioning of the dorsal roots in the area C2 to T4 12 years earlier. In these monkeys, neuronal input from the face now activated the cortical zone that had previously represented the hand and arm region (Fig. 12.1). Neurons in a large section of areas 3b and 1 that had been deafferentated now responded to touch and brushing of the lower face. The question was raised to what extent this reorganizational change might be related to the growth of new connections because it could no longer be explained by alterations in synaptic strength. Changes of similar magnitude also were reported by Florence and Kaas (12) in monkeys with longstanding hand amputations.

Substantial cortical reorganization also occurs after central lesions. For example, Jain et al. (13) performed complete unilateral transsection of the dorsal columns at the C3/C4 level and showed that after a period of several months the representation of the upper limb was invaded by the face representation and responded to input from the face. Jenkins and Merzenich (14) reported that small infarcts of SI led to a takeover of inputs of this zone by adjacent territory in conjunction with an enlargement of receptive fields.

HUMAN STUDIES

The results on adult cortical plasticity in animals were soon replicated in humans. Ramachandran, Stewart, and Rogers-Ramachandran (15) observed in three amputees that slight tactile stimuli applied with a Q-tip or strong pressure on the face led to referred sensations in the amputated (phantom) hand or arm with a point to point correspondence of the sites of referred sensation and stimulation sites in the face. They saw similarities to the findings of Pons et al. (11) and postulated

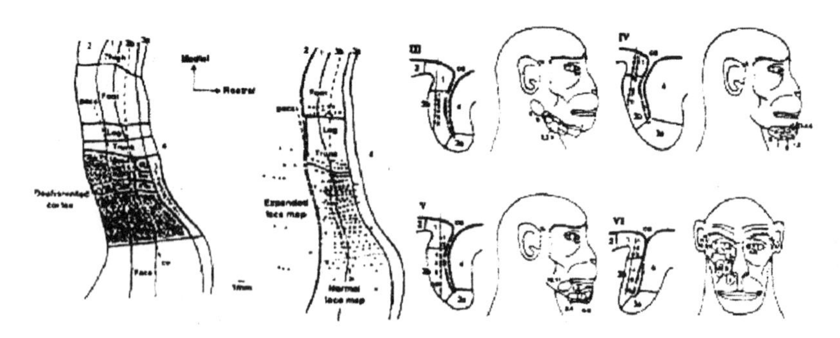

FIG. 12.1. I to II —This shows two flattened maps of SI, the left showing the deafferentated zone, and the right the recording site density in one animal. Note that the face representation has expanded over the entire zone of deafferentated cortex. **III to VI:** These show receptive field data from the cortex investigated. **III to V:** These are sections of cortex in the deafferentated zone. **VI:** This is next to the deafferentated zone. Note that there are receptive fields in the chin area in the portion of cortex that normally represents the arm and hand. (From: Pons TP, Garraghty PE, Ommaya AK, et al. Massive cortical reorganization after sensory deafferentation in adult macaques. *Science* 1991; 252:1857–1860, with permission.)

that the phenomenon of referred phantom sensation was a perceptual correlate of the reorganization that occurs subsequent to amputation and called it "facial remapping." Elbert et al. (16,17) used neuromagnetic source imaging (a combination of magnetoencephalographic recordings and magnetic resonance imaging) to determine the homuncular organization of the primary somatosensory cortex in upper extremity amputees. In all amputees a shift of the neuronal activity that was elicited by facial stimulation into the area that formerly represented the now-amputated hand and arm in primary somatosensory cortex was observed. The authors used the Euclidean distance between the hand and mouth representation of the intact side and compared it to the amputated side by mirroring the representation of the intact hand onto the other hemisphere (to replace the representation of the amputated hand, assuming symmetric representation on both hemispheres). In some cases this shift of the mouth toward the hand region spanned a distance of several centimeters.

Similar changes were also reported by Kew et al. (18) who used transcranial magnetic stimulation and positron emission tomography in a sample of arm amputees. Plastic alterations also occur in motor cortex probably owing to its close association with somatosensory cortex (19). Mogilner et al. (20) used magnetic source imaging to follow the changes that occurred subsequent to the separation of a syndactyly ("webbed hand"). Whereas the hand representation was "smeared" before surgery, after surgery the separation of the fingers was mirrored in separate representation of the individual fingers in primary somatosensory cortex.

This change in cortical organization is most likely only present when the deafferentation occurs after the completion of cortical development. Persons who lost a limb in the prenatal phase and very young child amputees did not report phantom sensation or phantom limb pain in two studies (21,22). In addition, they showed no reorganization of somatosensory cortex. However, contrary to these results Brugger et al. (23) reported phantom sensation and reorganization in a sample of subjects with congenital aplasias in accordance with reports on phantom sensations in congenital amputees by Melzack (24). The role of early experience and learning in injury-related plasticity thus needs to be further clarified.

Transient deafferentation achieved by nerve blocks or regional anesthesia was studied in humans. For example, Rossini et al. (25) showed that somatosensory evoked magnetic fields can be influenced by transient ischemic deafferentation. A shift of the representation of the stimulated finger toward the representation of the deafferentated finger was observed. Attentional factors were found to modulate the extent of transient deafferentation (26).

FUNCTIONAL SIGNIFICANCE

Surprisingly, the reorganization observed in amputees was not associated with the presence of topographic referred sensations as postulated by Ramachandran et al. (15). In several subsequent studies less than 20% of the amputees were found to show topographic referred sensation, whereas a much larger percentage (60% to 80%) of the amputees was significantly reorganized (27,28). Rather, phantom limb pain was soon identified as a perceptual correlate of cortical reorganization. A larger shift of neighboring representation zones into the zone that formerly represented the amputated limb was associated with more phantom limb pain (21,22,27–31). Whereas amputees with phantom limb pain showed a mean reorganization of 1.5 cm, the pain-free amputees revealed less than 4 mm reorganization (29) (Fig. 12.2).

As noted, nonpainful phantom sensations have consistently not been related to changes in the somatotopy of somatosensory cortex (28). Upper extremity amputees who experienced nonpainful phantoms when stimulated at the finger or the mouth showed increased activation levels in primary somatosensory cortex, posterior parietal cortex and reduced activation in ipsilateral secondary somatosen-

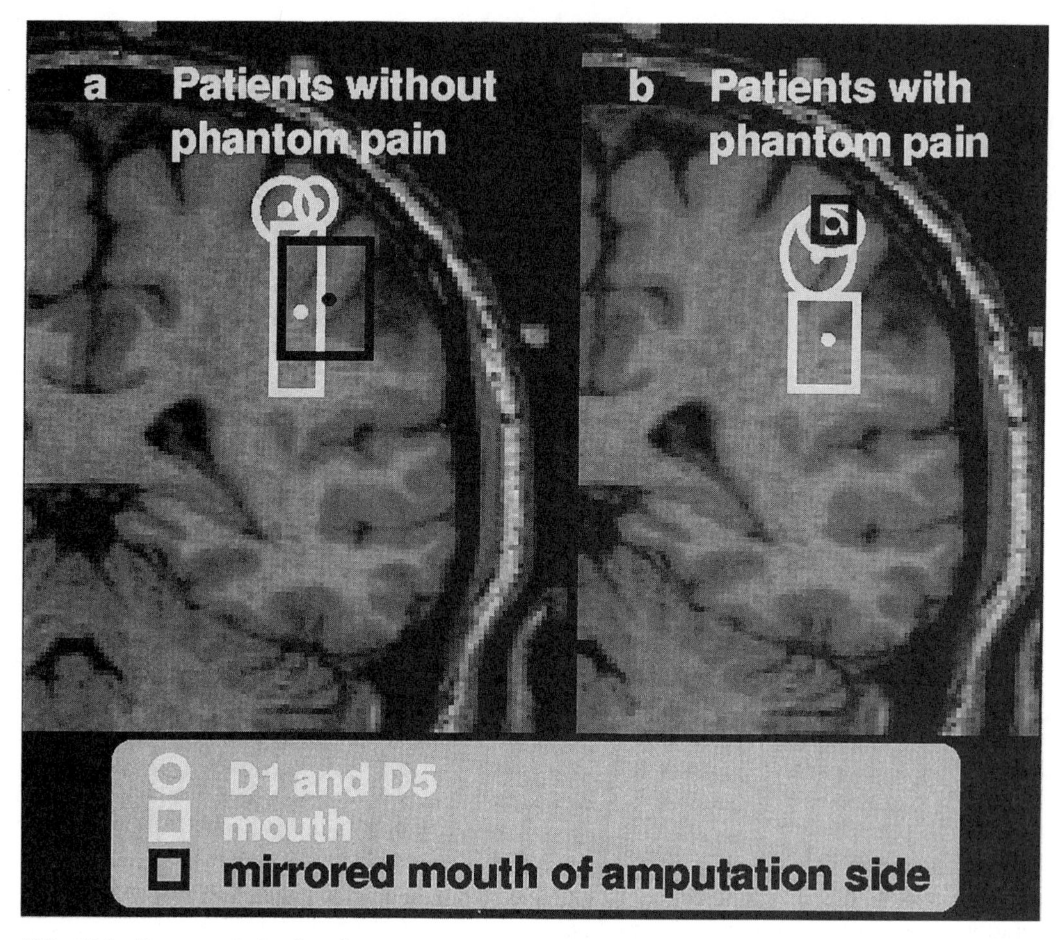

FIG. 12.2. Representation in primary somatosensory cortex of the mouth and two digits in upper extremity amputees with and without phantom limb pain. Note how the mouth representation of the amputation side has shifted into the hand region in the amputees with pain.

sory cortex (32) rather than a change in homuncular organization. A similar reduction in SII activity was also reported by Hari et al. (33) when a supernumerary limb was present in a patient who had suffered brain damage. Thus, phantom limb pain seems to be associated with changes in primary somatosensory cortex, whereas nonpainful phantom phenomena seem to be related to alterations in other brain areas such as the posterior parietal cortex.

Anesthesia of the brachial plexus in upper limb amputees suffering from phantom limb pain revealed that about half of the amputees achieved a temporary reduction of phantom limb pain whereas the other half continued to display phantom limb pain (30). The neuroelectric source analysis of activity in primary somatosensory cortex revealed that the pain-free amputees no longer showed cortical reorganization; that is, their lip representation had shifted back to the location where the lip representation of the intact side was located. The amputees without a change in phantom limb pain showed a small increase in reorganization; that is, the lip representation shifted even further into the region of the hand and arm. These data confirmed the hypothesis of a close association of phantom limb pain and changes in the map of the somatosensory cortex.

Similar alterations were found in patients with chronic low back pain (34). Electric stimulation of the back and finger resulted in enhanced activity in primary somatosensory cortex in response to the back but not the finger stimulation in the chronic back pain group as compared to healthy controls and subchronic patients. When the magnetic fields related to this stimulation were localized a shift and expansion of the back representation toward the leg area was found in the chronic group that was the larger the more chronic the pain has become. This type of "somatosensory pain memory" was assumed to contribute to the hyperalgesia and allodynia that is often found in states of chronic pain (35).

Flor (36) suggested a model of phantom limb pain that assigns a special role to cortical reorganization (Fig. 12.3). It has long been known that the primary somatosensory cortex is involved in the processing of pain and that it may be important for the sensory-discriminative aspects of the pain experience. There have been reports that phantom limb pain was abolished after the surgical removal of portions of the primary somatosensory cortex and that stimulation of somatosensory cortex evoked phantom limb pain. The model postulates that longstanding or intense acute pain in the limb prior to or during amputation may lead to the establishment of a somatosensory pain memory (37). These pain memories do not have to be explicit and thus open to conscious perception by the patient but can be implicit and may merely consist of a physiologic alteration of the somatosensory cortex. If such an implicit pain memory has been established with an important neural correlate in primary somatosensory cortex, subsequent deafferentation and an invasion of the amputation zone by neighboring input may activate preferentially cortical neurons coding for pain. Because the cortical area coding input from the periphery seems to stay assigned to the original zone of input the activation in the cortical zone representing the amputated limb is referred to this limb and the activation is interpreted as phantom sensation and phantom limb pain.

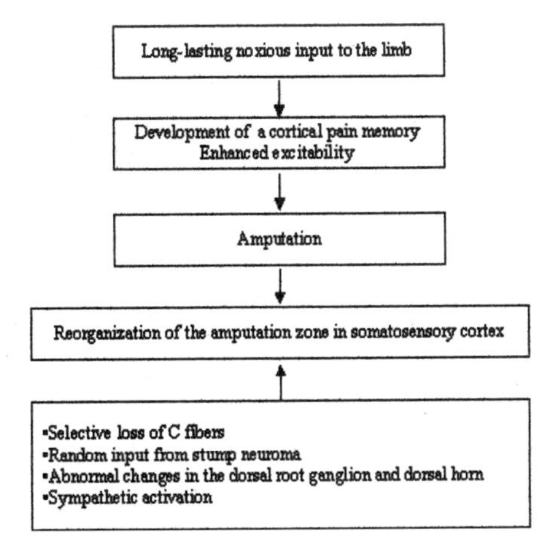

FIG. 12.3. Schematic diagram of the development of phantom limb pain. (From: Flor H. Phantom limb pain: characteristics, aetiology, and treatment. *Lancet*, in press, with permission.)

MECHANISMS

In the studies on reorganization of somatosensory cortex several phases of reorganization seem to be present. A first, very fast phase with an expansion of neighboring areas into the representation zone of the deafferentated area without necessarily completely filling this region was observed. Next was an intermediate phase that extended over weeks with a consolidation and new formation of the topographic organization and finally a third phase with a continued expansion and use-dependent changes (38). Phantom limb pain and phantom sensations can be present already in the very early stages of reorganization after amputation (39). In addition, transient alterations of somatosensory input such as the application of painful stimulation can also lead to immediate changes in the somatosensory map (40).

The original interpretations of the changes in the median-nerve sectioned monkeys were based on the assumption that week and ineffective synapses in the fringes of overlapping thalamocortical axon arbors would be potentiated. Since the discovery of massive reorganization in somatosensory cortex a number of mechanisms have been viewed as basic for these reorganizational processes. However, it is so far not clear which of these mechanisms are decisive, if they are different in the early and late stages of cortical plasticity, and if they differ between lesion- and stimulation-induced plasticity. In addition, there have been extensive discussions about the site of remapping, specifically, to what extent there is genuine cortical plasticity and to what extent cortical changes merely reflect alterations at lower levels of the neuraxis (41).

Any alteration of the pattern of input from a sensory surface immediately leads to map changes in the brain regions activated by these inputs. These map changes, however, are far from straightforward because normally an intricate pattern of inhibitory and excitatory circuits is present. The alteration of the balance of excitatory and inhibitory interactions has been described as unmasking of normally silent connections (42). Unmasking seems to depend on GABAergic mechanisms (43). Arckens et al. (44) showed with respect to the visual cortex that a GABA decrease also may be related to changes in synaptic efficiency and thus still be present long after unmasking has occurred.

The activity-dependent upregulation of glutamate at the NMDA receptor is another important mechanism of reorganization. Garraghty and Muja (45) showed that an NMDA receptor blockade after the sectioning of the median nerve led to substantially less reorganization than is normally observed. The small number of neurons that was responsive to skin stimulation was thought to be active because of unmasking that is not NMDA receptor dependent. Myers et al. (46) showed that although the later stage of reorganization is NMDA receptor dependent, the maintenance of reorganization depends on the activation of AMPA receptors.

Another mechanism that has been discussed is sprouting of new connections on several levels of the central nervous system. In monkeys with hand and forearm amputations, Florence and Kaas (12) showed that afferents from the remaining forelimb extended into the hand areas in both the dorsal horn and cuneate nucleus in the brainstem; thus suggesting that sprouting may occur on lower levels. The sprouting of new intracortical connections was shown by Florence, Taub, and Kaas (47) in monkeys subsequent to amputation or bone fractures. Injection of tracers revealed normal thalamocortical connections in area 1 but sprouting in the lateral connections in areas 3b and 1, suggesting cortical rather than thalamic changes (Fig. 12.4). Jones and Pons (48), however, examined macaque monkeys with longstanding dorsal rhizotomies and reported extensive transneuronal atrophy at the thalamic level. Merzenich (49) suggested that the level of deafferentation might explain these divergent findings because deafferentation at the spinal level is accompanied by neuronal degeneration in the dorsal root, whereas this is not the case in peripheral deafferentation. In this context the results of Ergenzinger et al. (50) are important; they showed that acute and chronic suppression of activity in SI can result in massive thalamic changes. This is related to the anatomic fact that the efferent connections from cortex are about tenfold stronger than those to cortex if one considers thalamocortical connections.

Changes in sensory maps as a consequence of deafferentation can occur at the cortical or subcortical level. Reorganization also has been observed on the thalamic level. For example, Garraghty and Kaas (51) recorded from the ventroposterior lateral nucleus of the thalamus of deafferentated monkeys and found extensive reorganization. Local anesthesia-induced reorganization in the ventral posterior medial nucleus (52) was observed as well. In monkeys with longstanding dorsal rhizotomies, Rausell et al. (53) found a selective degeneration of non-nociceptive pathways and increased activity of nociceptive cells in the thalamus. In a thalamic stimula-

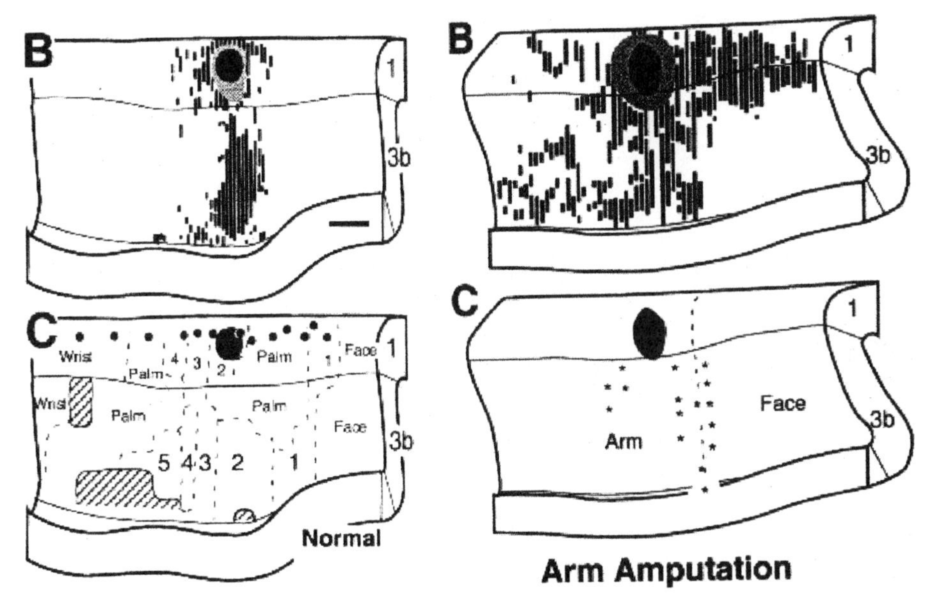

FIG. 12.4. Summary reconstruction of the hand representation in area 3b and a partial representation of the hand in area 1 based on electrophysiologic recordings using a retrograde tracer. On the left side the normal connections are shown, on the right side connections in an amputated monkey are displayed. (From: Florence SL, Taub HB, Kaas JH. Large-scale sprouting of cortical connections after peripheral injury in adult macaque monkeys. *Science* 1998;282:1117–1120, with permission.)

tion and recording study in human amputees, Davis et al. (54) reported that they could elicit phantom sensations by stimulating in the thalamic representation zone of the amputated limb and also record activity in the amputation zone when the skin adjacent to the amputation line was stimulated.

A change in synaptic connections related to Hebbian learning and long-term potentiation (LTP) is another equally important mechanism that may be involved in reorganizational changes (55). Long-term potentiation has been shown in the hippocampus mainly. In the cortex it is more complex because of the less well segregated and classified cell types and afferent pathways (42).

IMPLICATIONS FOR REHABILITATION

Based on the findings from neuroelectric and neuromagnetic source imaging, it is possible that changes in cortical reorganization might influence phantom limb pain. Animal work on stimulation-induced plasticity suggests that extensive behaviorally relevant (but not passive) stimulation of a body part leads to an expansion of its representation zone (56). Thus, the use of a myoelectric prosthesis might be one method to influence phantom limb pain. It was shown that intensive use of a myoelectric prosthesis was positively correlated with both reduced phantom limb pain and reduced cortical reorganization (19). When cortical reorganization was partialed out, the relationship between prosthesis use and reduced phantom limb pain was no longer significant, suggesting that cortical reorganization mediates this relationship. An alternative approach in patients where prosthesis use is not viable is the application of behaviorally relevant stimulation. A 2-week training that consisted of a discrimination training of electric stimuli to the stump for 2 hours per day led to significant improvements in phantom limb pain and a significant reversal of cortical reorganization. A control group of patients who received standard medical treatment and

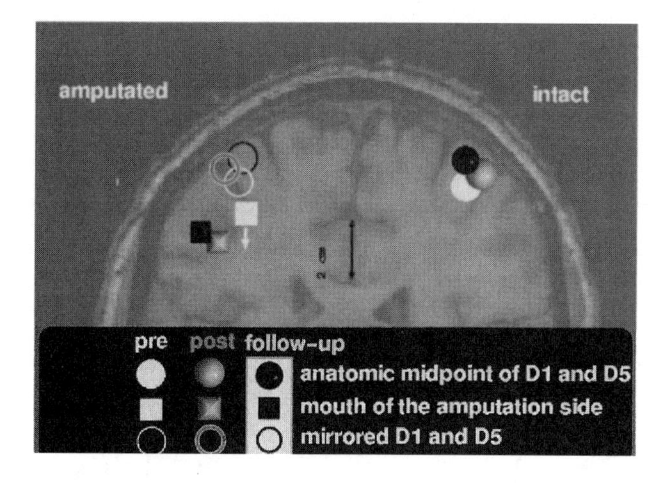

FIG. 12.5. Reversal of the mouth representation in primary somatosensory cortex to its original position subsequent to sensory discrimination training. (Based on: Flor H, Denke C, Schaefer M, et al. Sensory discrimination training alters both cortical reorganization and phantom limb pain. *Lancet* 2001; 357:1763–1764, with permission.)

general psychologic counseling in this time period did not show similar changes in cortical reorganization and phantom limb pain (57) (Fig. 12.5). The basic idea of the treatment was to provide input into the amputation zone and thus undo the reorganizational changes that occurred subsequent to the amputation. Ramachandran (58), who used a virtual reality box to train patients to move the phantom and reduce phantom limb pain, described another behaviorally oriented approach. A mirror was placed in a box and the patient inserted both his or her intact arm and the stump. The patient was then asked to look at the mirror image of the intact arm, which was perceived as an intact hand where the phantom used to be. The patient was then asked to make symmetric movements with both hands, thus suggesting real movement from the lost arm to the brain. This procedure seems to reestablish control over the phantom and reduce phantom limb pain in some patients.

Pharmacologic interventions also might be useful in the alteration of cortical plasticity and the amelioration of phantom limb pain related to central changes. Animal studies have shown that both spinal sensitization and cortical reorganization can be prevented or reversed by the use of NMDA receptor antagonists (42,59). Reorganization also was found to be related to reduced GABAergic activity and increased cholinergic activity. Thus, these substances should be beneficial in the treatment of phantom limb pain. However, studies in patients are scarce and the data are controversial. For example, whereas some studies reported a positive effect of the NMDA receptor antagonist ketamine on phantom limb pain, others found no effects of the NMDA receptor antagonist memantine on chronic phantom pain phenomena (60).

A recent study used the NMDA receptor antagonist memantine versus placebo in addi-

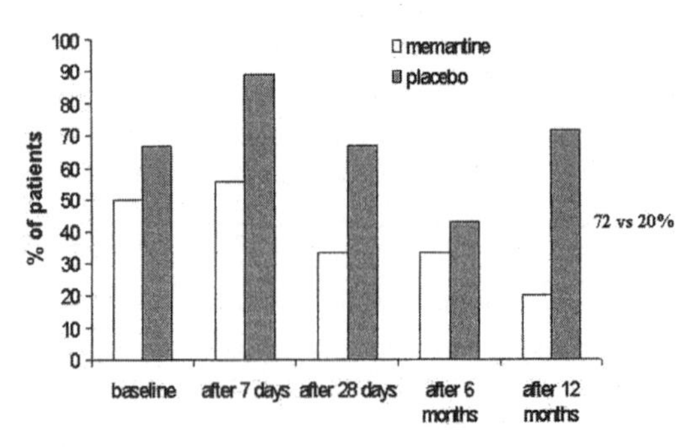

FIG. 12.6. Change in the incidence of phantom limb pain related to the application of memantine in the perioperative and postoperative phase. (From: Wiech K, Preissl H, Kiefer T, et al. Prevention of phantom limb pain and cortical reorganization in the early phase after amputation in humans. *Soc Neurosci Abstr* 2001;28:163, with permission.)

tion to brachial plexus anesthesia in patients undergoing traumatic amputations of individual fingers or a hand (61). It was found that memantine significantly reduced the incidence of phantom limb pain 1 year after the surgery, whereas placebo failed to show a similar effect (Fig. 12.6). In this study, very low levels of cortical reorganization and phantom limb pain were observed as well.

SUMMARY

The results reported here show convincingly that the adult somatosensory cortex alters its maps subsequent to injury. Studies with amputees and chronic pain patients have shown that pain may be one important perceptual correlate of the changes that were observed in primary somatosensory cortex. These results also have led to new approaches to rehabilitation. Both pharmacologic and behavioral interventions designed to alter cortical reorganization were found to not only alter the organization of primary somatosensory cortex but also maladaptive perceptual phenomena that accompany these changes.

ACKNOWLEDGMENTS

This research was supported by the Deutsche Forschungsgemeinschaft and the Max Planck Research Award for International Cooperation.

REFERENCES

1. Bach-y-Rita P. Brain plasticity as a basis for therapeutic procedures. In: Bach-y-Rita P, ed. *Recovery of function: theoretical considerations for brain injury rehabilitation.* Bern: Hans Huber, 1980:225–236.
2. Penfield W, Boldrey E. Somatic motor and sensory representations in the cerebral cortex of man as studied by electrical stimulation. *Brain* 1937;60:389–443.
3. Kaas JH, Nelson RJ, Sur M, et al. Multiple representation of the body within primary somatosensory cortex of primates. *Science* 1979;204:511–512.
4. Nelson RN, Sur M, Felleman DJ, et al. Representations of the body surface in postcentral parietal cortex of Macaca fascicularis. *J Comp Neurol* 1980;192:611–643.
5. Braun C, Schweizer R, Elbert T, et al. Differential activation in somatosensory cortex for different discrimination tasks. *J Neurosci* 2000;20:446–450.
6. Calford MB. Dynamic representational plasticity in sensory cortex. *Neuroscience* 2002;111:709–738.
7. Kaas, JH, Merzenich MM, Killackey HP. The reorganization of the somatosensory cortex following peripheral nerve damage in adult and developing mammals. *Ann Rev Neurosci* 1983;6:325–356.
8. Merzenich MM, Kaas JH, Wall J, et al. Topographic reorganization of somatosensory cortical areas 3b and 1 in adult monkeys following restricted deafferentation. *Neuroscience* 1983;8:33–56.
9. Merzenich MM, Kaas JH, Wall JT, et al. Progression of change following median nerve section in the cortical representation of the hand in areas 3b and 1 in adult owl and squirrel monkeys. *Neuroscience* 1983;10:639–666.
10. Merzenich MM, Nelson RJ, Stryker MP, et al. Somatosensory cortical map changes following digit amputation in adult monkeys. *J Comp Neurol* 1984;224: 591–605.
11. Pons TP, Garraghty PE, Ommaya AK, et al. Massive cortical reorganization after sensory deafferentation in adult macaques. *Science* 1991;252:1857–1860.
12. Florence SL, Kaas JH. Large-scale reorganization at multiple levels of the somatosensory pathway follows therapeutic amputation of the hand in monkeys. *J Neurosci* 1995;15:8083–8095.
13. Jain N, Catania MV, Kaas JH. Deactivation and reactivation of somatosensory cortex after dorsal spinal cord injury. *Nature* 1997;386:495–498.
14. Jenkins WM, Merzenich MM. Reorganization of neocortical representations after brain injury: a neurophysiologic model of bases of recovery from stroke. *Progr Brain Res* 1987;71:249–266.
15. Ramachandran VS, Stewart M, Rogers-Ramachandran DC. Perceptual correlates of massive cortical reorganization. *NeuroReport* 1992;3:583–586.
16. Elbert T, Flor H, Birbaumer N, et al. Extensive reorganization of the somatosensory cortex in adult humans after nervous system injury. *NeuroReport* 1994;5: 2593–2597.
17. Yang TT, Schwartz B, Gallen C, et al. Sensory maps in the human brain. *Nature* 1994;368:592–593.
18. Kew JJ, Ridding MC, Rothwell JC, et al. Reorganization of cortical blood flow and transcranial magnetic stimulation maps in human subjects after upper limb amputation. *J Neurophysiol* 1994;72:2517–2524.
19. Lotze M, Grodd W, Birbaumer N, et al. Does use of a myoelectric prosthesis prevent cortical reorganization and phantom limb pain? *Nat Neurosci* 1999;2:501–502.
20. Mogilner A, Grossman JA, Ribary U, Joliot et al. Somatosensory cortical plasticity in adult humans revealed by magnetoencephalography. *Proc Natl Acad Sci USA* 1993;90:3593–3597.
21. Flor H, Elbert T, Mühlnickel W, et al. Cortical reorganization and phantom phenomena in congenital and traumatic upper-extremity amputees. *Exp Brain Res* 1998; 119:205–212.
22. Montoya P, Ritter K, Huse E, et al. The cortical somatotopic map and phantom phenomena in subjects with congenital limb atrophy and traumatic amputees with phantom limb pain. *Eur J Neurosci* 1998;10: 1095–1102.
23. Brugger P, Kollias SS, Muri M, et al. Beyond re-membering: phantom sensations of congenitally absent limbs. *Proc Natl Acad Sci USA* 2000;97:6167–6172.
24. Saadah ES, Melzack R. Phantom limb experiences in congenital limb-deficient adults. *Cortex* 1994;30: 479–485.
25. Rossini PM, Martino G, Narici L, et al. Short-term

'brain plasticity' in humans: transient finger representation changes in sensory cortex somatotopy following ischemic anesthesia. *Brain Res* 1994;642: 169–177.
26. Buchner H, Richrath P, Grunholz J, et al. Differential effects of pain and spatial attention on digit representation in the human primary somatosensory cortex. *NeuroReport* 2000;27:1289–1293.
27. Knecht S, Henningsen H, Höhling C, et al. Plasticity of plasticity? Changes in the pattern of perceptual correlates of reorganization after amputation. *Brain* 1998; 121:717–724.
28. Grüsser SM, Winter C, Mühlnickel W, et al. The relationship of perceptual phantom phenomena and cortical reorganization in upper extremity amputees. *Neuroscience* 2001;102:263–272.
29. Flor H, Elbert T, Knecht S, Wienbruch C, et al. Phantom-limb pain as a perceptual correlate of cortical reorganization following arm amputation. *Nature* 1995;375: 482–484.
30. Birbaumer N, Lutzenberger W, Montoya P, et al. Effects of regional anesthesia on phantom limb pain are mirrored in changes in cortical reorganization. *J Neurosci* 1997;17:5503–5508.
31. Karl A, Birbaumer N, Lutzenberger W, et al. Reorganization of motor and somatosensory cortex in upper extremity amputees with phantom limb pain. *J Neurosci* 2001;21:3609–3618.
32. Flor H, Mühlnickel W, Karl A, et al. A neural substrate of nonpainful phantom limb phenomena. *NeuroReport* 2000;11:1407–1411.
33. Hari R, Hanninen R, Makinen T, et al. Three hands: fragmentation of human bodily awareness. *Neurosci Lett* 1998;240:131–134.
34. Flor H, Braun C, Elbert T, et al. Extensive reorganization of primary somatosensory cortex in chronic back pain patients. *Neurosci Lett* 1997;224:5–8.
35. Flor H. The functional organization of the brain in chronic pain. *Progr Brain Res* 2000;129:313–322.
36. Flor H. Phantom limb pain: characteristics, aetiology, and treatment. *Lancet Neurology* 2002;3:182–189.
37. Katz J, Melzack R. Pain ('memories' in phantom limbs: review and clinical observations. *Pain* 1990;43: 319–336.
38. Churchill JD, Muja N, Myers WA, et al. Somatotopic consolidation: a third phase of reorganization after peripheral nerve injury in adult squirrel monkeys. *Exp Brain Res* 1998;119:189–196.
39. Borsook D, Becerra L, Fishman S, et al. Acute plasticity in the human somatosensory cortex following amputation. *NeuroReport* 1998;9:1013–1017.
40. Soros P, Knecht S, Bantel C, et al. Functional reorganization of the human primary somatosensory cortex after acute pain demonstrated by magnetoencephalography. *Neurosci Lett* 2001;298:195–198.
41. Kaas JH, Florence SL, Jain N. Subcortical contributions to massive cortical reorganization. *Neuron* 1999;22: 657–660.
42. Buonomano DV, Merzenich MM. Cortical plasticity: from synapses to maps. *Ann Rev Neurosci* 1998;21: 149–186.
43. Jones EG. GABAergic neurons and their role in cortical plasticity in primates. *Cereb Cortex* 1993;3:361–372.
44. Arckens L, Schweigart G, Qu Y, et al. Cooperative changes in GABA, glutamate and activity levels: the missing link in cortical plasticity. *Eur J Neurosci* 2001;12:4222–4232.
45. Garraghty PE, Muja N. NMDA receptors and plasticity in adult primate somatosensory cortex. *J Comp Neurol* 1996;367:319–326.
46. Myers WA, Churchill JD, Muja N, et al. Role of NMDA receptors in adult primate cortical somatosensory plasticity. *J Comp Neurol* 2000;418:373–382.
47. Florence SL, Taub HB, Kaas JH. Large-scale sprouting of cortical connections after peripheral injury in adult macaque monkeys. *Science* 1998;282:1117–1120.
48. Jones EG, Pons TP. Thalamic and brainstem contributions to large-scale plasticity of primate somatosensory cortex. *Science* 1998;282:1121–1125.
49. Merzenich MM. Long-term change of mind. *Science* 1998;282:1062–1063.
50. Ergenzinger ER, Glasier MM, Hahm JO, et al. Cortically induced thalamic plasticity in the primate somatosensory system. *Nat Neurosci* 1998;1:226–229.
51. Garraghty PE, Kaas JH. Functional reorganization in adult monkey thalamus after peripheral nerve injury. *NeuroReport* 1991;2:747–750.
52. Nicolelis MAL, Lin RCS, Woodward DC, et al. Induction of immediate spatiotemporal changes in thalamic networks by peripheral block of ascending cutaneous information. *Nature* 1983;361:533–536.
53. Rausell E, Cusick CG, Taub E, et al. Chronic deafferentation in monkeys differentially affects nociceptive and non-nociceptive pathways distinguished by specific calcium-binding proteins and down-regulates gamma-aminobutyric acid type A receptors at thalamic levels. *Proc Natl Acad Sci USA* 1989;89:2571–2575.
54. Davis KD, Kiss ZHT, Luo L, et al. Phantom sensations generated by thalamic microstimulation. *Nature* 1998; 391:385–387.
55. Rauschecker JP. Mechanisms of visual plasticity: Hebb synapses, NMDA receptors, and beyond. *Physiol Rev* 1991;71:587–615.
56. Jenkins WM, Merzenich MM, Ochs MT, et al. Functional reorganization of primary somatosensory cortex in adult owl monkeys after behaviorally controlled tactile stimulation. *J Neurophysiol* 1990;63:82–104.
57. Flor H, Denke C, Schaefer M, et al. Sensory discrimination training alters both cortical reorganization and phantom limb pain. *Lancet* 2001;357:1763–1764.
58. Ramachandran VS, Rogers-Ramachandran D. Synaesthesia in phantom limbs induced with mirrors. *Proc R Soc Lond B Biol Sci* 1996;263:377–386.
59. Woolf CJ, Mannion RJ. Neuropathic pain: aetiology, symptoms, mechanisms, and management. *Lancet* 1999;353:1959–1964.
60. Nikolajsen L, Jensen TS. Phantom limb pain. *Br J Anaesthesiol* 2001;87:107–116.
61. Wiech K, Preissl H, Kiefer T, et al. Prevention of phantom limb pain and cortical reorganization in the early phase after amputation in humans. *Soc Neurosci Abstr* 2001;28:163.

13

The Visuomotor Functions of Posterior Parietal Areas

Marc Jeannerod* and Alessandro Farnè†

*†*Institut des Sciences Cognitives, Bron, France; †Department of Psychology, University of Bologna, Bologna, Italy*

INTRODUCTION

The discovery of the role of the parietal cortex in visuomotor functions owes much to neurophysiologists. In 1975, Mountcastle and colleagues published an influential paper on neuron properties in the posterior parietal cortical areas in the monkey. Quoting from a paper written by this author 20 years later (1), these neurons were active

> if and only if the animal 'had a mind' to deal with the stimulus in a behaviorally meaningful way! In area 5, there were neurons active when the animal projected his arm toward a target of interest, not during random arm movements. . . . Quite different sets of neurons appeared to drive the transport and grasping phases of reaching movements . . . (p 377)

In area 7, there were neurons active during visual fixation of objects of interest, not during casual fixations; neurons active during visually evoked but not during spontaneous saccadic eye movements . . . (p 378)

In the last two decades, Mountcastle et al.'s experiment was followed by a long series of studies aimed at elucidating, at the level of posterior parietal neurons, the mechanisms for acting in visual space. Research has focused on two main problems: first, what are the basic operations for transforming the spatial coordinates of an object relative to visual space into coordinates relative to the subject? Second, how is visual processing of the geometrical properties of an object—like its shape—transformed into motor commands?

POSTERIOR PARIETAL LESION IN THE MONKEY

The main neuronal mechanisms for coordinate transformation and generation of visuomotor commands in the monkey are localized within the intraparietal sulcus (IPS), at the border between the superior and inferior parietal lobules. Some of the IPS areas directly project onto the premotor cortex (area 6) and from there, to primary motor cortex. The dorsal part of area 6, where neurons coding for reaching movements are found, is connected with the rostralmost part of area 7a (2), with area MIP (3) and with area PO (either directly or via the superior parietal lobule) (5). In the more ventral part of area 6, neurons related to grasping movements are found: they are highly selective for different types of hand grasps (6). This part of area 6 receives direct connections from another IPS area (area anterior intraparietal area [AIP]) (7). Thus, the connection linking the parietal area AIP and ventral premotor cortex represents a specialized visuomotor system for encoding objects' geometric primitives and for generating the corresponding hand-and-finger configuration (8) (Fig. 13.1).

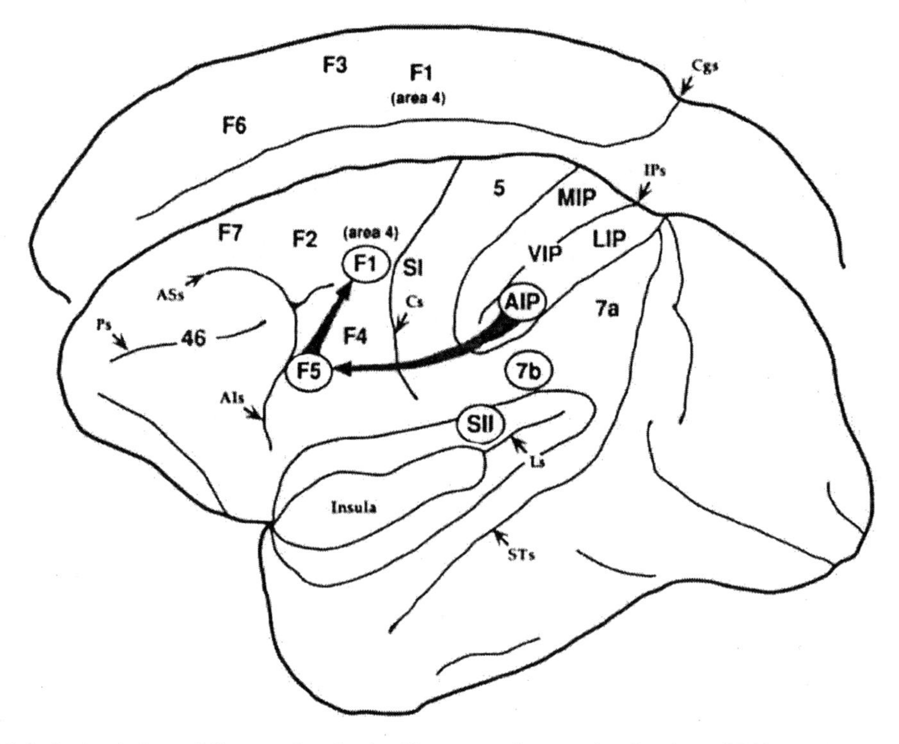

FIG. 13.1. Lateral view of the monkey brain. Note open intraparietal sulcus (IPS) and lateral sulcus (LS) to show buried areas. AIP, anterior area; F1 to F7, cytoarchitectonic fields of motor and premotor cortex; LIP, lateral area; MIP, medial intraparietal area; VIP, ventral area. (From: Jeannerod M, Arbib MA, Rizzolatti G, et al. Grasping objects. The cortical mechanisms of visuomotor transformation. *Trends Neurosci* 1995;18:314–320, with permission.)

The Mountcastle et al.'s paper (9) had prompted a series of lesion experiments with the site of the lesion focused on visuomotor mechanisms in area 7 on the convexity of the hemisphere. Lamotte and Acuna (10) and Faugier Grimaud et al. (11) both found clearcut deficits, mostly limited to visually goal directed movements performed with the hand contralateral to the lesion. During the first postoperative days, the monkey's contralateral hand remained unused, although it was in no way paralyzed, as the animal could use it, albeit awkwardly, when the other hand was attached. When the affected hand was used by the animal, severe visuomotor deficits became apparent: The animal misreached visual food targets located anywhere in the workspace. In addition, the hand was not shaped according to objects' shape and precision grip was abolished (Fig. 13.2).

More recently, selective disruption of parietal function was performed by using transient inactivation (by local injection of a GABA agonist such as muscimol) of different areas in the IPS region. Inactivation of AIP produces a subtle deficit in the monkey's performance of visually guided grasping. The hand does not preshape in anticipation to the grasp and grasping errors are observed in tasks requiring a precision grip. Such a local inactivation can be as selective as to dissociate the mechanisms for grasping and reaching, respectively: The former is altered, whereas the latter remains normal (12).

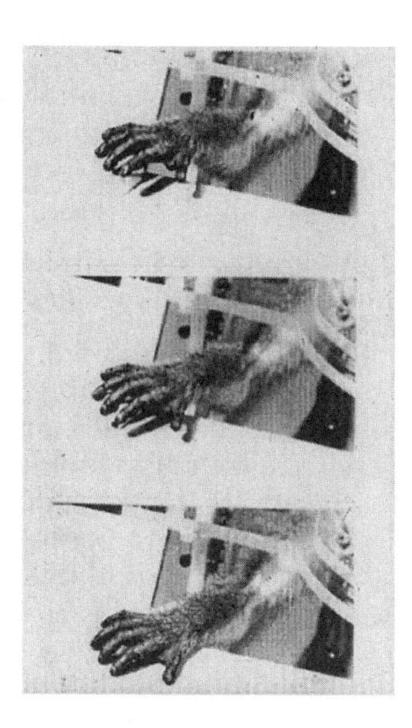

FIG. 13.2. Reach and grasp movement in a monkey following unilateral left lesion of area 7 in parietal cortex. Note the absence of hand preshaping. (From: Faugier-Grimaud S, Frenois C, Stein DG. Effects of posterior parietal lesions on visually guided behavior in monkeys. *Neuropsychologia* 1978;16:151–168, with permission.)

IMPAIRMENTS IN VISUALLY GUIDED BEHAVIOR FOLLOWING LESION OF THE PARIETAL LOBE IN HUMANS

The effects of pathologic lesions of the parietal lobes were first described by R. Balint and G. Holmes. The two authors, however, differed in the interpretation of the symptoms they observed in their patients. Whereas Holmes emphasized the "visuospatial" nature of his patients' difficulty (13), Balint, on the contrary, discussed the failures of his patient in terms of visuomotor, orienting and attentional functions (14). As expressed by Harvey and Milner (15), this difference in interpretation

> is particularly clear in relation to the phenomenon of faulty reaching for targets in visual space, which they both described in their patients. Whereas Balint concluded from his experiments that this disorder was visuomotor in Nature Holmes concluded . . . that it was due to a disorder in localizing objects in space. Balint called it 'optic ataxia' . . . whereas Holmes subsumed it as a 'disturbance of visual orientation.' (p 262)

In the following sections we describe impairments in visual function, which may fulfill both interpretations.

Optic Ataxia, a Specific Disorder of Visuomotor Transformation

Following the lead of the initial Balint's description, optic ataxia is now considered as a specific deficit attached to posterior parietal lesion. Optic ataxia can be characterized as a deep alteration of reaching movements directed toward a visual target, in the absence of any motor impairment. Three main characteristics have been described. First, the kinematics of the movements are changed (movements have a lower peak velocity, the duration of their deceleration phase is increased), a deficit that cannot be of a motor origin, as the same movements can be executed with a normal kinematic profile in non visual conditions. Second, directional coding is impaired: Movements are not directed in the proper direction, and large pointing and reaching errors are observed. A directional bias is frequently observed, with reaching or pointing deviated to the side of the lesion. This misreaching is more marked when the visual targets are presented outside the fixation point, and tends to increase when the target distance from the fixation point increases (16). Third, alteration of the movements is not limited to the reaching phase. Distal aspects of the movements are affected as well. Patients fail to orient their hand properly when they have to insert it through a slit (17). During the action of grasping an object, the finger grip aperture is increased and the usual correlation of maximum grip aperture (MGA) to object size is lost (18).

The lesion responsible for optic ataxia was localized by different authors at the convexity of the posterior parietal cortex (17). Both reaching and grasping are affected, because the lesions usually are relatively large. The question arises, however, whether more limited lesions could only affect the reaching or grasping phase of the movements, respectively. In a selected series of patients presenting a deficit limited to the grasping phase, Binkofski et al. (19) were able to circumscribe the responsible area to the anterior part of the intraparietal sulcus on the side contralateral to the impaired hand. Patients with lesions sparing the intraparietal sulcus showed intact grasping. The same authors (19,20) confirmed this localization in a functional magnetic resonance imaging analysis of prehension movements in normal subjects. The activated area was limited to the lateral bank of the anterior part of the intraparietal sulcus. This same area also had been found to be specifically activated during grasping in a positron emission tomography study by Faillenot et al. (21) (Fig. 13.3). It is noteworthy that this exquisite localization for grasping movements also is found to be activated by mere observation of grasping performed by another agent (22), or even by observation of graspable objects (23).

Case AT: A Visuomotor Impairment with Normal Object Recognition

Optic ataxia appears as a disorder limited to transforming visual properties of objects into motor commands for action directed at these objects. It is not caused by misperception of shape, orientation, or size of the objects. This point was clearly established in a patient (AT), examined by F. Michel and his colleagues. AT's lesion was an infarction of a relatively large occipitoparietal zone on both sides. Areas 18 and 19 were destroyed, as well as parietal areas 7 and 39. At the early stage of her disease, AT had presented a bilateral optic

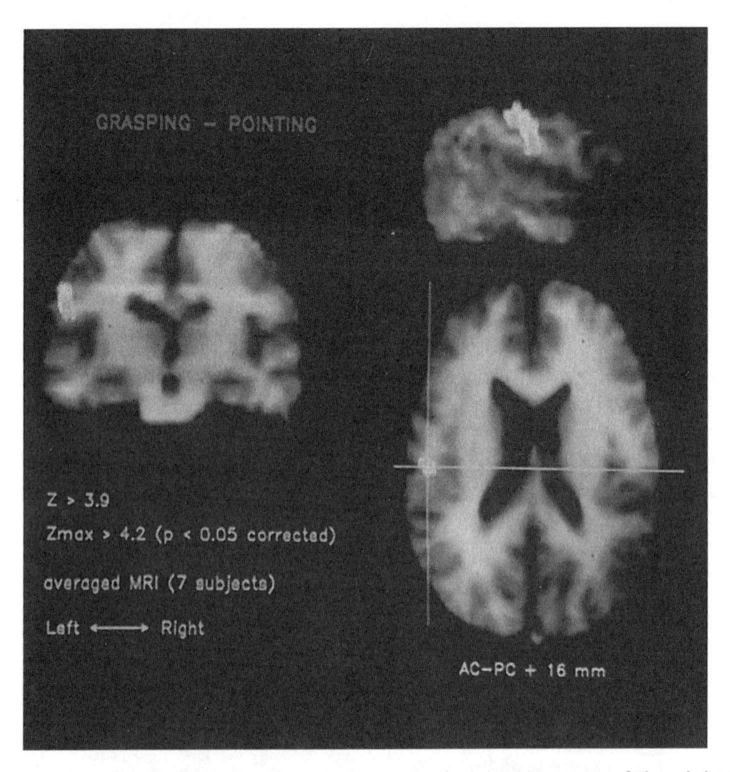

FIG. 13.3. Topography of parietal activation during grasping movements of the right hand. Average regional cerebral blood flow (rCBF) change in normal subjects as studied by positron emission tomography. Note the activation in the ventral part of the left parietal cortex, corresponding to the anterior tip of the intraparietal sulcus. (See color plate of this figure after page 186.)

ataxia as part of a typical Balint's syndrome. When she was examined several years later, reaching accuracy had became close to normal in her central field of vision, although it remained impaired in her peripheral visual field (see the following). In addition, she was normal in recognizing, describing, and naming visual forms and objects. AT still presented a severe visuospatial disorder corresponding to what is described in the following as "dorsal simultagnosia." As a consequence of this disorder, she was hampered in her everyday life for actions such as dressing, cooking, ironing, sewing, or driving.

AT was tested on two tasks, the grasping and the size matching tasks (24). In the grasping task, she was instructed to reach and grasp with a precision grip target objects (vertical plastic cylinders, 1.5 to 7.5 cm in diameter) with or without visual control of her hand. Because these objects all had the same visual aspect, except for their size, they were considered "neutral" objects. During the grasping of neutral objects, AT's movements offered a striking dissociation: Whereas in all cases the reach was correctly oriented, the grasp was incorrect, especially with the smaller objects. The patient ended the trial with the object in contact, not with the fingertips, but with the palmar surface of more proximal phalanxes or even with the palmar surface of the hand itself. This behavior (which was bilateral but more marked with the right hand) resulted in awkward and inaccurate grasps with the thumb and index finger curled around the object. On a few occasions, the grasp was not possible, because the object was pushed down by the palm of the hand. Closer examination revealed an exaggerated aperture of the finger grip, such that the end of the finger closure, which should normally bring the fingertips in contact with the object at the time when the reach stops, was delayed and the reach tended to overshoot target position. Maximum grip size correlated poorly with object size, because grip aperture was grossly exaggerated for the smaller objects. This correlation was not improved by vision of the hand during the grasp.

In contrast with the impaired pattern of grasping directed at visual objects, AT's performance appeared to be normal in the size matching task; that is, a task exploring her ability to judge the size of objects. When asked to match with the thumb and index finger of her hidden right hand the size of the same objects as those used for the grasping task, she gave accurate estimates that correlated positively with object size. (See reference 25 for a very similar case.)

This dissociation between impaired grasping and normal perceptual judgment on the same objects or shapes (demonstrated with the same motor output) stands exactly opposite to the dissociation between the impaired perceptual judgment and preserved grasping observed in patients with inferotemporal lesion (26). As such, this double dissociation lends support to the notion that different modalities of object-oriented responses are differently distributed in two cortical visual systems: Impaired grasping is a consequence of parietal lobe damage in the dorsal system, whereas the intact ventral system temporal lobes still allow normal perceptual judgment for object size.

Is this view of task-related processing of visual information compatible with a distribution of function across specific pathways? Object-oriented behavior can hardly be conceived as composed either of pure "action" or pure "perception" tasks, in which only one part of the visual system works at a time. Competent action toward objects requires more than simply encoding their size or orientation, and cannot be limited to mere mechanical problem solving, enabling the efficient use of objects in a manner consistent with their physical or geometric properties. Instead, as emphasized by Hodges et al. (27), "Competent conventional use of objects depends on additional conceptual knowledge for which inferotemporal brain structures appear to be critical" (p 9447). Indeed, the collaboration of the two systems is clearly suggested by further experiments with AT. Jeannerod et al. (24) observed that AT's grasping performance was improved when familiar objects were used as targets instead of neutral objects. In their experiment, these objects were of approximately the same size and shape as the neutral ones, but were clearly recognizable (a lipstick, a reel of thread, etc.). With these objects, the correlation between maximum grip size and object size increased up to normal values, indicating that semantic knowledge about these objects was used to calibrate the grip. Finally, and not surprisingly, when AT was tested for her ability to estimate the size of the same imagined familiar objects with her fingers, her judgment strongly correlated with object size. These striking findings demonstrate that the two visual systems are not independent of each other. In fact, AT's behavior suggests that the functions of the impaired dorsal system were substituted by information processed at the level of the intact ventral system. When semantically identifiable objects were used, the ventral system was activated and the relevant information was transferred to the motor system through a pathway bypassing the impaired parietal cortex. In contrast, when neutral objects were used, correct grasping could not be achieved because the visual primitives related to object size could no longer be transferred to the parietal areas. No other cues could be used, because all objects looked alike.

A further example of a substitution of the deficient dorsal pathway by an alternative pathway was observed in patient AT (16). She was presented with luminous targets appearing in her visual field at locations ranging from midline to 30 degrees on either side. In one set of trials, the target was on for a period of 2 seconds, after which a tone was presented, indicating that she had to point immediately at the target. In another set of trials, the tone was presented 5 seconds after the target was turned off and she had to wait until then to point at the target location. In the first set of trials, AT made inaccurate pointing responses at all peripheral targets. In addition, the degree of inaccuracy was a function of eccentricity of the target; that is, pointing errors increased for more peripheral targets. This is typical behavior in an optic ataxia patient such as AT. The surprising finding was that the pointing accuracy significantly increased in the second set of trials; that is, when pointing responses were delayed after disappearance of the target. No such effect was found in a group of control subjects; instead, their pointing accuracy significantly decreased in the delayed condition. Milner et al. (16) interpreted these findings by suggesting that the parietal lesion in AT had altered the visuomotor system normally used to reach visual targets (e.g., the system coded in egocentric coordinates), but had left intact another system coded in allocentric coordinates, which was normally used for localizing objects with respect to each other. This system becomes available only after a delay sufficient for the activation of the first, inefficient system to decay. Although in normal conditions, the alternative system is less accurate than the dedicated visuomotor system (as can be verified from the results of the control subjects), in AT its use allowed a paradoxic improvement in pointing accuracy. The problem remains to determine the anatomic location of this system. Because the lesion was restricted to parietal lobes in AT, it is likely that a more perceptual pathway through the temporal lobes was activated, which ultimately reached the motor centers without using the parietal route.

VISUOSPATIAL DISORDERS FOLLOWING LESIONS OF THE DORSAL SYSTEM

Patients with parietal lesions, with or without optic ataxia, often present visuospatial disorders. As mentioned, this fact was first recognized by Holmes, who considered that disturbances of visual orientation represented a general explanation for the impairments observed in these patients. Following the lead of Holmes, many authors reported similar observations, testifying to the existence of disorders of spatial perception or conscious manipulation of spatial information (once referred to as "blindness for places") (28) following parietal lesion, more frequently on the right side. These disorders not only include inability to localize an object in its absolute position in space, but also the loss of topographic concepts (patients cannot orient themselves on a map; they are unable to indicate on a map appropriate directions to travel from one point to another) and the loss of topographic memory (they cannot evoke an image of familiar places; they cannot find their way home) (29,30).

Dorsal Simultagnosia

Another striking phenomenon observed in these patients is the so-called "dorsal simultagnosia" (31). Simultagnosia is a condition in which a patient accurately perceives the individual elements or details of a complex picture, but cannot appreciate the overall meaning of the picture. Farah added the term "dorsal" to refer to the fact that it occurs after bilateral parieto-occipital damage. Typically, a dorsal simultagnosic patient recognizes most objects but is unable to see more than one at a time, irrespective of their size. As a consequence of this condition, such patients cannot count objects, and their description of complex scenes is slow and fragmentary; in a visual environment, they behave like blind people, groping for things and walking into obstacles; if asked to look for a particular object, they make random searching eye movements. Note that this form of simultagnosia was initially described by Balint (14) as part of a syndrome that also included disorganized search with eye movements and optic ataxia. The forthcoming patient AT presented a typical picture of dorsal simultagnosia, in addition to her optic ataxia. However, as stressed in the following, the fact that these disorders occasionally occur separately from one another raises the possibility that they pertain to different mechanisms.

Dorsal simultagnosia is interpreted as a disorder of visual attention. Along with Posner's hypothesis, Farah proposed a deficit in disengaging attention: Attention must first be disengaged from its current location for a new stimulus to be able to engage it. Parietal lobes play a critical role in this mechanism (32). Thus, people with bilateral parietal lesion should present a "sticky" attention on the current object without the possibility to shift to another one. If one assumes, in an allocentric frame of reference, that the location of an object can only be specified relative to another object, then simultagnosia, which does not allow seeing more than one object at a time, becomes a highly plausible explanation for most of the disorders in spatial orientation found in parietal patients. The same explanation is valid for the difficulties in drawing met by these patients: They are frequently unable to copy drawings that they have recognized, and their attempts at copying have a characteristic "exploded" look because the parts of the copied object are not related to each other (the so-called constructional apraxia). According to Farah (31), "this can be understood in terms of the necessity, in copying a picture, for drawing a part at a time, which requires patients to shift their attention to the individual parts of the shape and to position the parts with respect to one another" (p 45).

Unilateral Spatial Neglect

Another dramatic illustration of the visuospatial disorders produced by parietal lesions is unilateral spatial neglect. Neglect usually is bound to right-sided lesions; therefore, it affects the contralateral left hemispace. Typically, a patient with neglect of the left side of space ignores visual stimuli presented on that side and only reports items presented on his or her right. This behavior is illustrated by the well-known "cancellation" test, where the patient is instructed to cross out lines of various orientations distributed across a sheet of paper: Only the lines on the right side of the sheet are crossed. In copying simple drawings, such as a house, clock face, or flower, the patient copies only the right side of the model.

Neglect for the left hemispace not only affects the external visual world, but also the world as it is mentally represented. Consider a neglect patient imagining that he is standing in front of the Milan duomo. This patient is asked to verbally describe the surroundings that he can mentally see from his present vantage point. As Bisiach and Luzzatti discovered in 1978, he correctly lists the shops and buildings on the right side of the duomo, but ignores those on the left. If the patient now imagines the same area while he takes the opposite vantage point, he has a clear mental representation of those buildings he previously ignored, and fails to report those that he previously reported. This observation (33) carries important new information; namely, that the representation of visual space affected by neglect is coded in a frame of reference that originates from the subject (an egocentric frame). This fits with the notion that neglect is a spatial attentional disorder, in other words, a deficit of attention localized to part of space. Insofar as attention to visual stimuli (and to stimuli in other modalities as well) is a goal-directed process, it is not surprising that a lesion in areas that process visuospatial relationships results in impossibility to direct attention to the corresponding area of space. A recent experiment by Rushworth et al. (34) in normal subjects provides some insight into this mechanism. Subjects were asked to respond to the appearance of a target by a key press. In most trials, the target was precued; that is, attention was directed to the target (valid trials). In some trials, the precue was invalid and the subject had to respond to a noncued target (invalid trials). Normal subjects successfully disengaged their attention from the cue in invalid trials, and their reaction time was little affected with respect to the valid trials. Rushworth et al.'s experiment involved application of electrical shocks (via transcranial magnetic stimulation) at different sites of the right and left parietal lobes during the presentation of the visual stimuli. The reaction time was significantly increased when the shock was applied on a site corresponding to the most posterior part of the right angular gyrus during invalid trials. The shock produced no effect if applied on another part of the right parietal lobe, or the left angular gyrus. The interpretation of this result is that the right angular gyrus is the site responsible for disengaging attention from a visual stimulus: Disengagement of attention is impaired when its activity is disrupted by an intervening electrical shock. The effective site in the right angular gyrus corresponds to the main focus of the lesion observed in neglect patients (35). Hence, the hypothesis of an impossibility to disengage attention for directing attention to the left as an explanation for the unilateral spatial neglect

Lesions responsible for unilateral spatial neglect usually affect the inferior and posterior part of the right parietal lobe, in the opercular region connecting the angular gyrus with the uppermost part of the temporal lobe (35,36). This region must be clearly distinguished from the more dorsal part of the parietal lobe, which we have found involved in visuomotor functions. Indeed, a recent study by Farnè et al. (in preparation) demonstrated that reaching and grasping movements directed at visual objects presented in the left hemispace were not more affected in patients presenting neglect than in patients with right brain damage without neglect. This finding reinforces the notion of a dissociation of perceptual and motor functions within the parietal lobe. Pritchard et al. (37) found that perceptual size underestimation, commonly observed for objects presented in the left hemispace in neglect patients, was not paralleled by a corresponding change in grip size during the action of grasping directed to those objects. Perceptual and visuomotor processing of the same objects led to different responses: Only the grasping response reflected the veridical size of the objects. The problem of a dissociation between perceptual/attentional and visuomotor functions is discussed again in the last section of this chapter.

Do Visuomotor and Visuospatial Disorders Correspond to Dissociable Functions of the Parietal Lobes?

The question of whether visuomotor transformation and perception of spatial relationships are one and the same thing or two separate entities is a critical one for the validity of the dual model of visual functions. In its radical version, where vision for perception and vision for action pertain to two parallel visual systems, the model predicts that the parietal, dorsal system should subserve a single function ultimately dedicated to acting on the visual world (38). Under this conception, visuospatial disorders following parietal lesion should reflect the processing of spatial information as a precondition for action. This point was made by Bayliss and Bayliss (39). They reported on the visual behavior of one patient (RM) with a Balint syndrome following a bilateral parietal lesion. This patient showed typical optic ataxia when attempting to reach for targets in visual space with either hand. However, RM made many more errors than in the reaching task when the location of the same targets was to be named (not reached) by using location descriptors (e.g., left, front, etc.). Furthermore, the errors in describing and reaching were similar when the task consisted in first describing the location, and then reaching for it immediately afterward. This difficulty was likely related to a problem with visual space processing, because errors tended to disappear when the blindfolded subject was asked to reach for the same targets on verbal instruction (e.g., touch the left target). The interpretation given by Bayliss and Bayliss (39) to these results is that the difficulty faced by RM is caused by a severely disrupted visual representation of space that captures and degrades his otherwise good ability to reach. However, this interpretation seems in contradiction to observations reported in other patients (see the following).

In the attenuated version of the model, where the dorsal system subserves both spatial cognition (in the Mishkin sense) and object-oriented action, the two sets of disorders should reflect an impairment of two distinct functions. One of these functions is visuomotor transformation, a largely automatic function coded in egocentric coordinates; another is perception of spatial relationships, a conscious function that obeys the same rules as perception in general. Available neurophysiologic data tend to speak in favor of the latter version of the model. For example, optic ataxia can be observed in the absence of visuospatial disorders. Also, optic ataxia frequently is limited to one hand following unilateral parietal lesion: In this situation, the visuomotor deficit cannot be explained by a disorganized spatial perception. Finally, optic ataxia can be observed following lesion on either side, whereas visuospatial disorders are more likely to be observed after a right-sided lesion.

Neuroimaging data obtained from normal subjects can shed new light on this old problem. The use of experimental paradigms specifically designed for isolating visuomotor processing from perception of spatial relationships conversely allows mapping of the functional anatomy of the parietal lobes in each condition. As mentioned, pure visuomotor transformation, as studied in a task of grasping objects of varying size and shape, activates areas confined to the anterior part of the intraparietal sulcus at the junction between the two parietal lobules (19,20,40). In addition, the activation is located contralateral to the hand used in the task. The perception of spatial relationships was studied with a variety of spatial matching tasks (e.g., comparing displays where the same objects or shapes are arranged in same or different spatial locations). The results of these studies consistently showed activation of relatively posterior and ventral parietal areas, in the fundus of the intraparietal sulcus (41), and in the area of the angular gyrus in the inferior parietal lobule (42) (Fig. 13.4).

The crucial point is that the activation foci were predominantly located within the right parietal lobe in all the preceding studies. Thus, neuroimaging findings generally are consistent with the clinical data, which also emphasize the role of the right hemisphere in space perception.

The conclusion to be drawn from this section is that parietal lobes play a critical role both in visuomotor transformation and perception of spatial relationships. Parietal function might be to localize and select objects in the visual world that will be the goal for visual

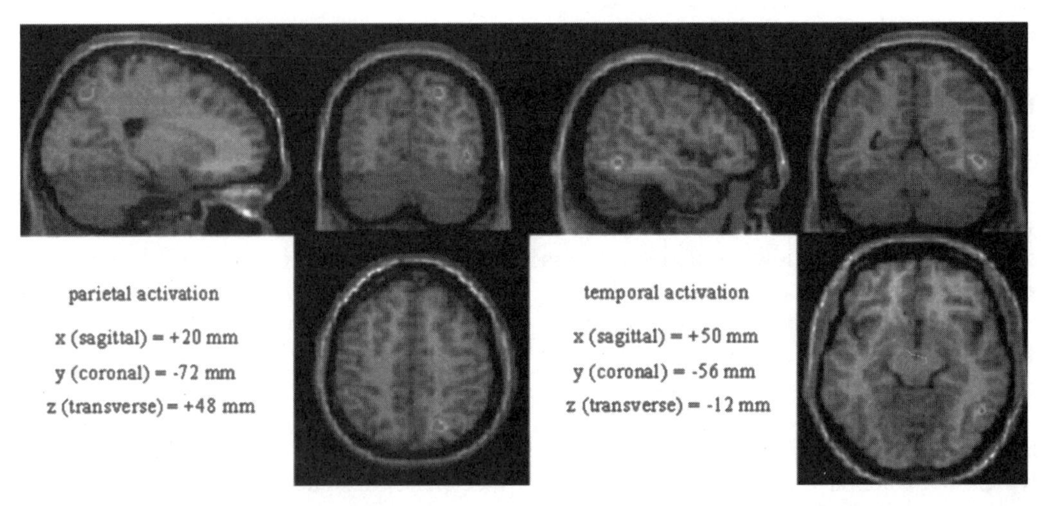

FIG. 13.4. The topography of cortical areas involved in spatial perception. A positron emission tomography study in normal subjects. The subjects were requested to judge the similarity or difference of meaningless shapes for their size or orientation. Note the activation in the right posterior parietal cortex and right inferotemporal cortex. (From: Faillenot I, Decety J, Jeannerod M. Human brain activity related to the perception of spatial features of objects. *NeuroImage* 1999;10:114–124, with permission.) (See color plate of this figure after page 186.)

perception or for visually goal–directed action. As such, parietal lobes must carry information about object identity and the context in which this visual behavior will take place. As long as external objects remain in extrapersonal space, their localization relative to other objects has to be processed in order to build a coherent representation of the environment using allocentric coordinates. When objects are to be transferred to the intrapersonal space (for manipulation, handling, use, transformation, or further identification by tactile cues), their absolute localization has to be determined in egocentric space. This transformation of coordinates is effected in parietal areas where not only visual, but also somatosensory cues and eye movement cues are available (see Chapter 12). The use of egocentric coordinates then is central to visuomotor transformation, where action is focused on one object at a time and only the cues relevant to action on that object are to be taken into consideration. Relative localization of an object by contrast, is a context-dependent operation. Relative localization may appear as a mere precondition for action in the context of object-oriented action. The full set of operations, including transfer of object position in egocentric space and visuomotor transformation, are performed rapidly and automatically. In a different context, however, such as counting objects on a table or drawing a complex scene with many objects (but with no attempt at interacting with them), the same process of relative localization operates on a conscious, perceptual mode.

A Comparison of Visuomotor Performance Between Neglect and Non-neglect Right Brain–Damaged Patients

In the mentioned experiment by Farnè et al. (43), right brain–damaged (RBD) patients with (RBD+) or without (RBD–) unilateral spatial neglect were examined for their ability to reach for visual targets presented on the two sides of a midline. Objects were vertical plastic dowels that the subjects were requested to grasp with their right (ipsilesional) hand, using a precision grip. The dowels could be presented either in a fixed position (midline, 10 degrees, or 20 degrees to the right and to the left), or they could unexpectedly move at onset of the hand movement (e.g., from midline to 10 or 20 degrees on either side, randomly). Temporal parameters and kinematics of the movement were recorded using a motion analyzer.

As a rule, RBD patients showed asymmetric visuomotor behavior, whether they reached and grasped objects located in the right or left (affected) hemispace. The problem here is whether patients' hemispatial asymmetries were differentially affected by the presence of visual neglect. In the fixed condition, RBD+ patients tended to react more slowly to left- than right-sided positions, but this effect was not significant. Movements directed to the left contralesional hemispace were slower, and all the main kinematic parameters were altered when compared to movements directed to the right ipsilesional space. However, neglect did not significantly influence any of the parameters considered. In both groups, MGA during grasping the dowels located at 20 and 10 degrees on the left did not differ from that for the dowel at the central position; grasping the dowels on the right side yielded MGA changes similar to those observed in normal subjects (Fig. 13.5).

In the perturbed condition, all perturbed movements took a longer time to be executed than control movements (directed to the central object) with homologous perturbed positions not differing across hemispaces. The kinematic parameters for the reaching phase, such as time to peak acceleration and velocity (TPA, TPV), showed a significant lengthening for objects located in the left, compared to the right hemispace. In addition, the TPV of rightward perturbed movements largely led that of control movements, thus confirming the notion of fast reactions to a sudden object's displacement (44), within the limits of the ipsilesional space.

As in the fixed condition, the maximum grip aperture was also asymmetric, increasing only from the central to right-sided objects' locations (see Fig. 13.5). However, the asymmetric pattern in perturbed reach-to-grasp movements was manifest, again, both in neglect and non-neglect RDB patients. The only differences that could be selectively ascribed

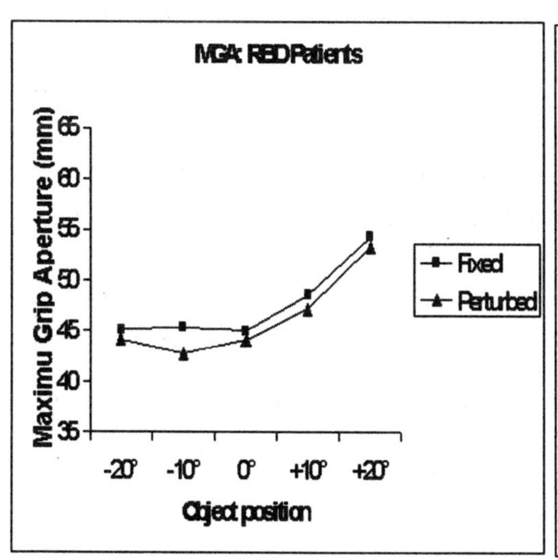

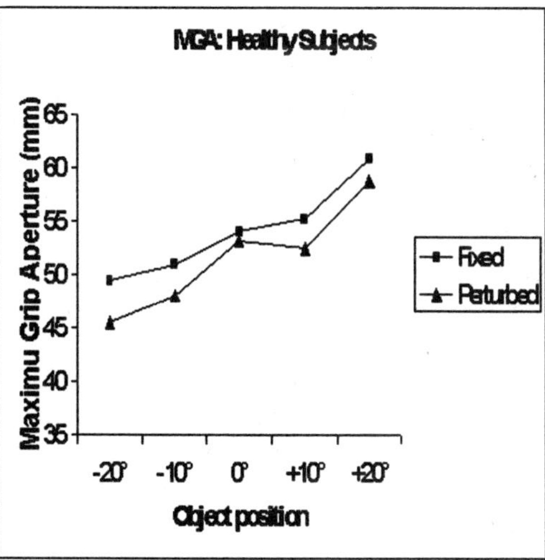

FIG. 13.5. A: Mean values of maximum grip aperture (MGA) as a function of object position and experimental condition (fixed, perturbed) in right brain–damaged patients (with and without neglect collapsed). **B:** For comparative purposes, mean values from normal subjects are reported.

to the presence of neglect were constituted by a reduced amplitude of peak acceleration, and a prolonged grip aperture phase (TGA).

In other words, the kinematics of both groups were similarly affected by object position in both conditions. Therefore, the presence of neglect (although severe in most cases) in addition to a right brain lesion does not seem responsible for patients' asymmetric motor performance. It is important to note that clear asymmetries, revealed for temporal and spatial parameters as a function of the hemispace in which the perturbation occurred, were consistently present irrespective of presence or absence of neglect.

These findings provide clear evidence that when visually guided prehension is impaired following a right hemispheric lesion, the visuomotor deficits can be immune to (and dissociable from) the additional presence of a visuospatial deficit such as neglect.

CONCLUSION

The human parietal lobe carries out several distinct functions. First, it is responsible for visuomotor transformation, a function achieved by neuronal populations in and around the intraparietal sulcus and connected with frontal premotor areas. Pathologic destruction of these areas produces the typical symptom of optic ataxia. Second, it is responsible for visuospatial perception, a function achieved by a broad system that spreads from the posterior part of the intraparietal sulcus down to the angular gyrus in the inferior parietal lobule. A lesion of this system (particularly when it affects the right parietal lobe) produces visuospatial disorders, including unilateral spatial neglect.

The two functions seem to coexist without necessarily interfering with each other. Optic ataxia may manifest itself in isolation as a unilateral deficit following a lesion on either side. Unilateral neglect does not preclude an almost correct reach and grasp of objects presented on the neglected left side: Subjects' movements are not more altered than after a right-sided lesion that does not produce neglect. The relative independence of visuomotor and visuospatial disorders occurring after a parietal lesion makes problematic the assignment of parietal functions en bloc to a dorsal system only concerned with automatic visually directed actions.

REFERENCES

1. Mountcastle VB. The parietal system and some higher brain functions. *Cereb Cortex* 1995;5:377–390.
2. Matelli M, Camarda R, Glickstein M, et al. Afferent and efferent projections of the inferior area 6 in the Macaque monkey. *J Comp Neurol* 1986;251:281–298.
3. Johnson PB, Ferraina S, Caminiti R. Cortical networks for visual reaching. *Exp Brain Res* 1993;97:361–365.
4. Tanné J, Boussaoud D, Boyer-Zeller N, et al. Direct visual pathways for reaching movements in the macaque monkey. *NeuroReport* 1995;7:267–272.
5. Caminiti R, Ferraina S, Johnson PB. The sources of visual information to the primate frontal lobe. A novel role for the superior parietal lobule. *Cereb Cortex* 1996;6:319–328.
6. Rizzolatti G, Camarda R, Fogassi L, et al. Functional organization of area 6 in the macaque monkey. II. Area F5 and the control of distal movements. *Exp Brain Res* 1988;71:491–507.
7. Matelli M, Luppino G, Rizzolatti G. Patterns of cytochrome oxydase activity in the frontal agranular cortex of the macaque monkey. *Behav Brain Res* 1985;18:125–136.
8. Jeannerod M, Arbib MA, Rizzolatti G, et al. Grasping objects. The cortical mechanisms of visuomotor transformation. *Trends Neurosci* 1995;18:314–320.
9. Mountcastle VB, Lynch JC, Georgopoulos A, et al. Posterior parietal association cortex of the monkey: command functions for operations within extra-personal space. *J Neurophysiol* 1975;38:871–908.
10. Lamotte RH, Acuna C. Defects in accuracy of reaching after removal of posterior parietal cortex in monkeys. *Brain Res* 1978;139:309–326.
11. Faugier-Grimaud S, Frenois C, Stein DG. Effects of posterior parietal lesions on visually guided behavior in monkeys. *Neuropsychologia* 1978;16:151–168.
12. Gallese V, Murata A, Kaseda M, et al. Deficit of hand preshaping after muscimol injection in monkey parietal cortex. *NeuroReport* 1994;5:1525–1529.
13. Holmes G. Disturbances of vision by cerebral lesions. *Br J Ophthalmol* 1918;2:353–384.
14. Balint R. Seelenlähmung des "Schauens", optische Ataxie, raümliche Störungen des Aufmerksamkeit. *Monatschrift für Psychiatrie und Neurologie* 1909;25: 51–81.
15. Harvey M, Milner DA. Balint's patient. *Cog Neuropsychol* 1995;12:261–281.
16. Milner AD, Paulignan Y, Dijkerman H, et al. A paradoxical improvement of misreaching in optic ataxia. New evidence for two separate neural systems for visual localization. *Proc R Soc London Biol* 1999;266: 2225–2229.
17. Perenin MT, Vighetto A. Optic ataxia: a specific disruption in visuomotor mechanisms. I. Different aspects of

the deficit in reaching for objects. *Brain* 1988;111: 643–674.
18. Jeannerod M. The formation of finger grip during prehension. a cortically mediated visuomotor pattern. *Behav Brain Res* 1986;19:99–116.
19. Binkofski F, Dohle C, Posse S, et al. Human anterior intraparietal area subserves prehension. A combined lesion and functional MRI activation study. *Neurology* 1998;50:1253–1259.
20. Binkofski F, Buccino G, Posse S, et al. A frontoparietal circuit for object manipulation in man: evidence from an fMRI study. *Eur J Neurosci* 1999;11:3276–3286.
21. Faillenot I, Toni I, Decety J, et al. Visual pathways for object-oriented action and object recognition. Functional anatomy with PET. *Cereb Cortex* 1997;7:77–85.
22. Iacoboni M, Woods RP, Brass M, et al. Cortical mechanisms of human imitation. *Science* 1999;286:2526–2528.
23. Chao LL, Martin A. Representation of manipulable man-made objects in the dorsal stream. *NeuroImage* 2000;12:478–484.
24. Jeannerod M, Decety J, Michel F. Impairment of grasping movements following a bilateral posterior parietal lesion. *Neuropsychologia* 1994;32:369–380.
25. Goodale MA, Meenan JP, Bülthoff HH, et al. Separate neural pathways for the visual analysis of object shape in perception and prehension. *Curr Biol* 1994;4:604–610.
26. Goodale MA, Milner AD, Jakobson LS, et al. Perceiving the world and grasping it. A neurological dissociation. *Nature* 1991;349:154–156.
27. Hodges JR, Spatt J, Patterson K. "What" and "how": evidence for the dissociation of object knowledge and mechanical problem-solving skills in the human brain. *Proc Natl Acad Sci USA* 1999;96:9444–9448.
28. Kleist K. *Gehirnpathologie.* Leipzig: Barth, 1934.
29. Hécaen H, Albert MA. *Human neuropsychology.* New York: Wiley, 1974.
30. Von Cramon D, Kerkhoff G. On the cerebral organization of elementary visuo-spatial perception. In: Gulyas B, Ottoson D, Roland PE, eds. *Functional organization of the human visual cortex.* Oxford, UK: Pergamon, 1993:211–231.
31. Farah MJ. *Visual agnosia. Disorders of object recognition and what they tell us about normal vision.* Cambridge, MA: MIT Press, 1995.
32. Posner MI, Walker JA, Friedrich FJ, et al. Effects of parietal lobe injury on covert orienting of visual attention. *J Neurosci* 1984;4:1863–1874.
33. Bisiach E, Luzzatti C. Unilateral neglect of representational space. *Cortex* 1978;14:129–133.
34. Rushworth MFS, Ellison A, Walsh V. Complementary localization and lateralization of orienting and motor attention. *Nat Neurosci* 2001;4:656–661.
35. Vallar G, Perani D. The anatomy of unilateral neglect after right hemisphere stroke lesion. A clinical/CT-scan correlation study in man. *Neuropsychologia* 1986;24: 609–622.
36. Vallar G. Extrapersonal visual unilateral spatial neglect and its neuroanatomy. *NeuroImage* 2001;14:52–58.
37. Pritchard LC, Milner AD, Dijkerman HC, et al. Visuospatial neglect: veridical coding of size for grasping but not for perception. *Neurocase* 1997;3:437–443.
38. Milner AD, Goodale MA. *The visual brain in action.* Oxford, UK: Oxford University Press, 1995.
39. Bayliss GC, Bayliss LL. Visually misguided reaching in Balint's syndrome. *Neuropsychologia* 2001;39:865–875.
40. Grafton ST, Mazziotta JC, Woods RP, et al. Human functional anatomy of visually guided finger movements. *Brain* 1992;115:565–587.
41. Faillenot I, Decety J, Jeannerod M. Human brain activity related to the perception of spatial features of objects. *NeuroImage* 1999;10:114–124.
42. Köhler S, Kapur K, Moscovitch M, et al. Dissociation of pathways for object and spatial vision. A PET study in humans. *NeuroReport* 1995;6:1865–1868.
43. Farnè A, Roy AC, Paulignan Y, et al. Right hemisphere visuomotor control of the ipsilateral hand. Evidence from right brain-damaged patients. *Neuropsychologica* (in press).
44. Paulignan Y, MacKenzie C, Marteniuk R, et al. Selective perturbation of visual input during prehension movements. I. The effects of changing object position. *Exp Brain Res* 1991a;83:502–512.

14

Multimodal Spatial Representations in the Human Parietal Cortex: Evidence from Functional Imaging

Emiliano Macaluso and Jon Driver

Institute of Cognitive Neuroscience, University College London, London, United Kingdom

INTRODUCTION

Spatial location can be extracted for stimuli in multiple modalities, each of which can provide targets for spatial actions. Neurophysiologic studies in animals have uncovered several brain areas containing multimodal neurons, that can respond to stimulation in more that one modality (e.g., vision and touch), and that typically have related spatial selectivity in the different modalities that can drive the cell (e.g., if the tactile receptive field falls on a particular body part, the visual receptive field may fall in a corresponding sector of space). Neurons of this general type have been found cortically in areas of parietal, superior temporal, and premotor cortex (1–4), and also at the subcortical level, in structures such as the superior colliculus (5,6) and putamen (7). Some of these neurons are also modulated by postural signals, such as those deriving from proprioception or the vestibular sense (3,8,9), and many show motor-related activity (3,9) for movements with particular effectors or for intentions to move. Such populations of spatially tuned multimodal neurons may be involved in constructing higher levels of spatial representation; enabling spatial interactions between inputs from different modalities; producing multimodal effects of spatial attention; and transforming sensory inputs into motor plans (3,10,11).

However, rather less is known about the brain structures involved in generating multimodal spatial representations for these various purposes in humans. Some clinical evidence from effects of brain damage, as in the syndrome of spatial neglect (12), suggests that multimodal spatial representations may be impaired, leading to deficits that affect the contralesional side of space in multiple modalities (e.g., producing auditory, tactile, and exploratory aspects of neglect in addition to visual neglect, in some cases) (10,13). However, the exact lesions for multimodal as compared with unimodal spatial neglect remain uncertain at present. A very different source of evidence on the neural basis of multimodal spatial representations in the human brain comes from recent functional imaging studies of normal humans.

MAPPING PARIETAL MULTIMODAL BRAIN AREAS WITH FUNCTIONAL MAGNETIC RESONANCE IMAGING

In recent years, neuroimaging has been used to map brain areas that respond not just to stimulation from one particular sensory modality (as for "unimodal" brain areas), but to stimulation from several different sensory modalities ("multimodal" or "polysensory" areas) (14,15). As might be expected given previous electrophysiologic evidence from nonhuman primates (1–4), some parietal regions show such multimodal responses in humans. Bremmer and colleagues (15) provide a recent example of mapping multimodal brain areas with neuroimaging. During functional magnetic resonance imaging (fMRI), subjects were presented with blocked stimulation in one or another of three sensory modalities: vision, touch, or audition. The visual stimulation was a pattern of random dots moving coherently; tactile stimulation was air flowing across the subject's forehead; and auditory stimulation was moving pure tones. These moving stimuli were chosen because similar stimulation has been shown to strongly activate certain parietal multimodal areas in monkeys (4). Comparison of each of these conditions individually with baseline conditions (no stimulation for touch or audition; a stationary random dot pattern for vision) revealed activation in sensory-specific unimodal areas (i.e., tactile, auditory, or visual cortices, as expected). Critically, a test for brain responses common to all three types of stimulation (i.e., multimodal responses) revealed activation of both intraparietal sulcus and inferior parietal lobule (plus ventral premotor cortex). Linking their finding to previous primate electrophysiologic studies (1,4), Bremmer and colleagues (15) suggested that the activated intraparietal region might correspond to the human homolog of ventral intraparietal area (VIP), lying deep in the fundus of the intraparietal sulcus.

Although the Bremmer et al. (15) study, and other independent but related imaging work (14) has addressed the basic issue of possible multimodal responses in parietal areas, a further issue is whether such responses depend on the location of the activating stimulus. This might be expected on the basis of electrophysiologic recording from single units in monkey parietal cortex, showing spatial tuning for several modalities in certain regions (4,16,17), although the spatial selectivity for such multimodal neurons at a population level (e.g., their distribution across the visual, auditory, and somatosensory fields for each hemisphere) have rarely been characterized in full (but see 3,4,18). It should be noted that single-unit spatial tuning is typically somewhat less acute (i.e., with larger receptive fields) in parietal areas than in early occipital visual areas (19,20). Moreover, unlike visual occipital areas that display systematic retinotopic organization (20–22), the topographic layout of neurons in parietal areas often appears to be less clearly defined in terms of spatial tuning (but see 4,16,18 for single-cell evidence; and 23 for a human imaging study on this issue).

We used fMRI in humans to investigate possible spatial specificity of multimodal responses for vision and touch (24), using simple comparisons of stimulation in the left versus right hemifield for touch or vision (with the visual stimuli delivered near to the hand on either side). We expected that if any multimodal parietal area preferentially represents the contralateral side of space (i.e., with more neurons having contralateral than ipsilateral receptive fields), then direct comparison of brain activity during stimulation of one hemifield versus the other would reveal this. This simple comparison showed segregation between multimodal responses that were spatially specific (i.e., found when comparing stimulation of the contralateral minus ipsilateral hemifield), versus areas where the signal increase during stimulation of either modality was independent of stimulus side. Spatially specific responses were found in and around the intraparietal sulcus contralateral to the stimulated side, but irrespective of stimulated modality (Fig. 14.1A,B, vertically and horizontally hatched areas and bars). In both hemispheres this multimodal spatial activation lay in the anterior part of the intraparietal sulcus (at the interception with the postcentral sulcus) and extended dorsally to the anterior part of the superior parietal gyrus. In these regions, activity increased for stimulation of the contralateral hemifield, but importantly it did so multimodally, that is irrespective of the stimulated modality (vision or touch). More

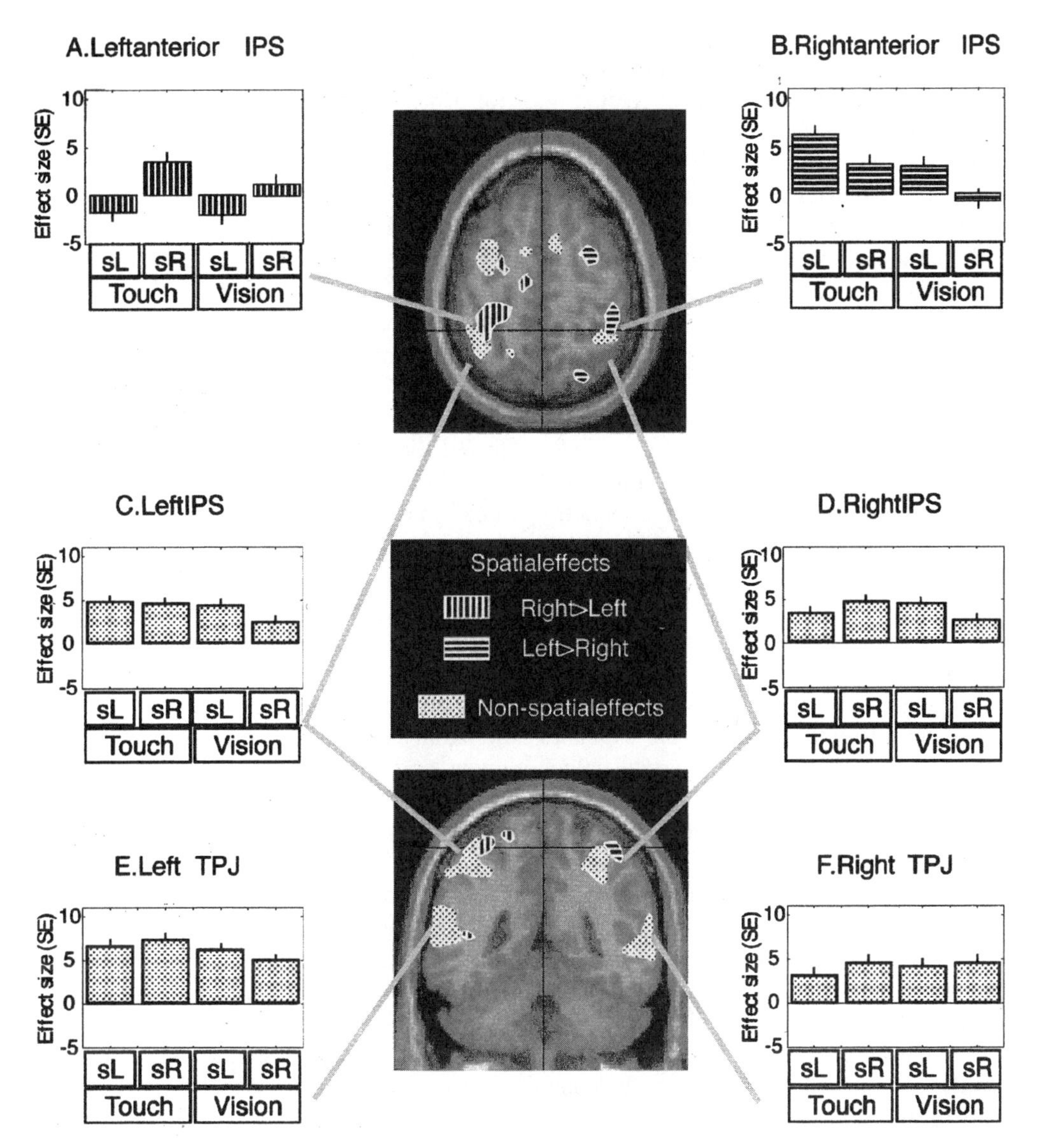

FIG. 14.1. Multimodal responses for unilateral passive stimulation in vision or touch, rendered on transverse *(top panel)* and coronal *(bottom)* sections of the Montreal Neurological Institute (MNI) brain template. (Adapted from: Macaluso E, Driver J. Spatial attention and crossmodal interactions between vision and touch. *Neuropsychologia* 2001;39:1304–1316, with permission.) **A,B:** Multimodal responses showing spatial specificity for stimulus location, in the sense of greater activity for stimulation specifically of the contralateral hemifield. **C–F:** Multimodal responses that were unaffected by stimulus hemifield. All effects are plotted in standard errors units (SE) as comparisons between stimulation condition and rest (no stimulation). (IPS, intraparietal sulcus; sL/sR: left/right stimulation; TPJ, temporal-parietal junction. SPM thresholds set to *p*-uncorrected = 0.001.)

posteriorly and in the depth of the intraparietal sulcus, multimodal responses appeared to be less specific to stimulus side (see Fig. 14.1C,D, dotted areas). Additionally, a region in the inferior parietal lobule also showed multimodal responses that did not depend on the stimulated hemifield (Fig. 14.1E–F, inferior dotted areas and bars). The latter activation extended to posterior/superior temporal cortex, where electrophysiologic studies in animals have also found multimodal neurons (25,26). Unlike regions showing spatially specific responses (that were contralateral to the stimulated side), the "nonspatial" (i.e., independent of stimulus hemifield) multimodal activations in our study were found symmetrically in the two hemispheres (see Fig. 14.1, dotted areas and bars).

Thus, the results of this simple fMRI stimulation-experiment (24), employing visual or tactile stimulation on the left or right side, were in general agreement with Bremmer et al. (15) in identifying multimodal responses in several parietal regions (intraparietal sulcus plus inferior parietal lobule). By stimulating either the left or right hemifield in vision or touch, we were able to extend previous human imaging findings, to show that a relatively anterior region of the intraparietal sulcus displayed preferential activation for stimulation in either modality but specifically within the contralateral hemifield. This dependence on stimulus location, regardless of stimulus modality, may provide one neural mechanism for integration of spatial information across sensory modalities.

MOTOR AND ATTENTIONAL INFLUENCES ON MULTIMODAL SPATIAL REPRESENTATIONS IN THE HUMAN INTRAPARIETAL SULCUS

One important aspect of spatial coding in parietal areas concerns the relation between sensory and motor properties of neurons. In particular, many single-cell studies have shown that neurons in several intraparietal regions not only respond to stimulation at one particular location, but are active when that region of space becomes the intended target for a particular spatial movement (3,27–29). For example, neurons in area LIP (in the lateral bank of the intraparietal sulcus) show responses to visual stimulation, but also display enhanced activity prior to saccadic eye movements, even in the absence of visual stimulation (27,28). A more medial region (parietal reach region, PRR) contains cells that fire for reaching movements to specific locations (29), showing increased activity prior to initiation of such a hand movement to the preferred location. These observations have led to influential proposals that posterior parietal areas may serve as sensorimotor interfaces, mapping particular positions of sensory stimuli onto the initial specification of an intention to make a particular movement to that location (3,29,30). Several functional imaging studies in humans have also shown that parietal activations can relate to motor aspects of a spatial task (31–33; plus further examples given later in this chapter).

In a recent fMRI study, unpublished as yet, we examined the dependence or independence of multimodal activations (i.e., common to visual and tactile stimulation) in parietal cortex on the type of spatial motor response (with the eye or hand) toward the stimulation. The results provide preliminary evidence that for posterior parietal regions, the type of motor response is critical (consistent with recent physiologic evidence from primates) (29); whereas in more anterior intraparietal regions, the activations that depended on the task-relevant location may apparently transcend not only stimulus modality but also response modality. We orthogonally manipulated the modality of stimulation (vision or touch, at corresponding locations in either the left or right hemifield), and the type of movement required toward that stimulation (a saccade, or an unseen manual button press on the corresponding side). In different blocks subjects performed either prosaccades to the left or right, or button presses with the left or right

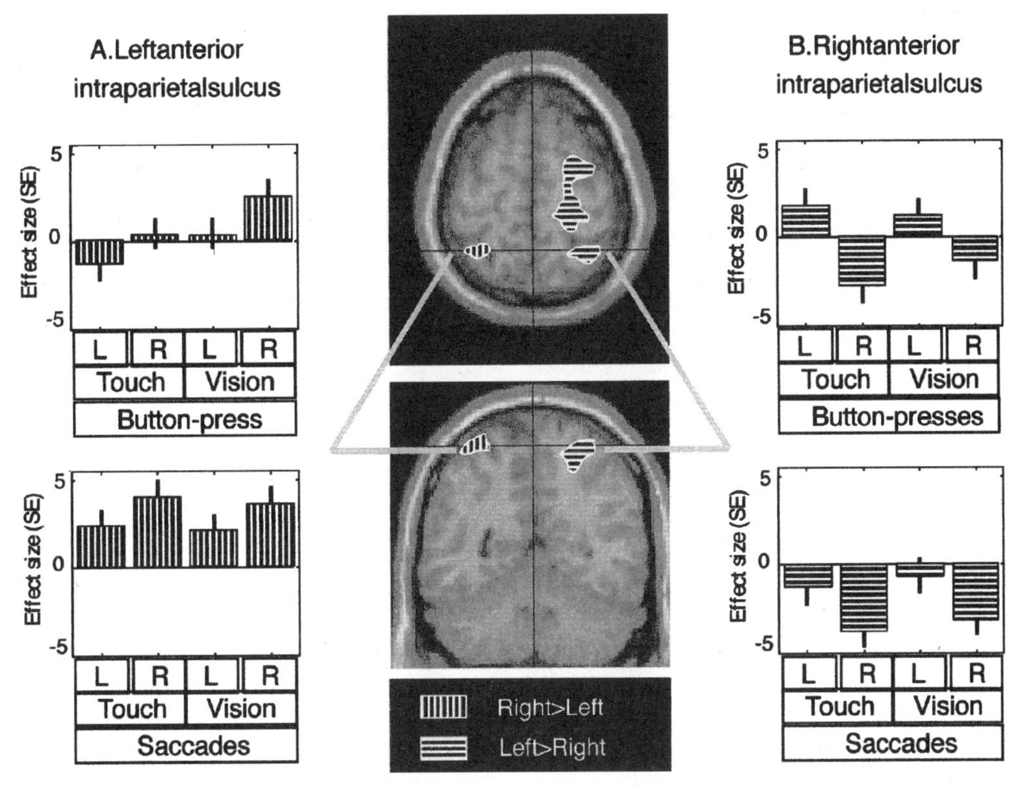

FIG. 14.2. Impact of the type of motor response to peripheral stimuli on multimodal spatial effects in the anterior intraparietal sulcus. In this experiment, subjects were asked to make either a localizing button press (with left or right hand), or a saccade to the stimulated hemifield. The results showed that the differential effect in anterior intraparietal cortex between stimulation of one versus the other hemifield transcended not only the sensory modality of stimulation but also the type of spatial motor response required (i.e., same contralateral effects detected in both button press *[upper plots]* and saccade *[lower plots]* conditions). The multimodal nature of these spatial effects replicates Figure 14.1A,B. (SPM thresholds set to p-uncorrected = 0.001.)

hand, according to the position of the target stimulus, regardless of its modality.

Figure 14.2 illustrates the results. Multimodal spatial responses contralateral to the stimulus were again found in the anterior-superior section of the intraparietal sulcus (c.f. Fig. 14.1A,B). As shown in the signal plots of Figure 14.2, activity increased here for stimulation of the contralateral hemifield regardless of stimulus modality. Thus, higher activity was observed for right stimulation (now leading to a right movement in this study, unlike the passive stimulation results of Fig. 14.1) in the left anterior intraparietal sulcus (IPS) (Fig. 14.2A), with activation in the right hemisphere for left stimulation/movement (Fig. 14.2B). Activation in this specific anterior intraparietal region was independent not only of the stimulation modality (vision or touch) but also of the type of movement performed toward the stimulated location (button presses or saccades). Thus, this anterior-dorsal region of the intraparietal sulcus may represent locations for the contralateral hemifield in a manner that transcends both sensory modality and response modality. By contrast, more posterior intraparietal regions showed specificity for the type of movement (eye versus hand), consistent not only with single-cell data from primates (3,29,30) but also with other human imaging studies that have used manual versus saccadic responses (31–33).

A further important point to consider in relation to multimodal spatial representations in parietal cortex concerns the task relevance of the stimulus. Electrophysiologic evidence shows that task relevance can affect the cellular responses to a given stimulus in several regions of parietal cortex (11,34–36). For example, Bushnell and colleagues (34) measured responses to peripheral visual stimuli while the monkey's covert attention was either directed to the location of a stimulus within the receptive field of the cell or to a location outside. Activity for area 7 parietal neurons increased when the stimulus in the receptive field was attended, providing one of the first demonstrations of spatially specific effects of covert attention on neural responses to stimulation. More recently, many functional imaging studies have implicated areas of human parietal cortex as part of a wide network involved in spatial attention, as we describe in the following. However, the majority of such imaging studies typically have considered spatial attention for only a single modality at a time (most often for vision), unlike the multimodal perspective of the present chapter.

In the first of our imaging studies to specifically address the multimodal issue for spatial attention, we presented subjects with bilateral stimulation in vision or touch during PET scanning (37). In different blocks subjects attended either the left or right hemifield to perform a discrimination task for just the attended location. We directly compared brain activity during leftward versus rightward attention, or vice versa. In order to address the multimodal issue, the bilateral peripheral stimulation could either be just visual or just tactile. Multimodal effects were highlighted by testing for any main effects of attended side irrespective of the stimulated modality. Activity in anterior intraparietal sulcus increased for attention to the contralateral hemifield, and critically these activations were independent of the stimulated modality (i.e., were found during visual stimulation and also during tactile stimulation). Thus, multimodal spatial representations in parietal cortex, in similar regions to those that can be multimodally activated "bottom-up," by passive stimulation on the contralateral side in vision or touch (see Fig. 14.1), also can be influenced "top-down" by endogenous covert spatial attention, again in a multimodal fashion.

More recently, we performed a related study using event-related fMRI (38). On a trial-by-trial basis, subjects received a central auditory cue (pure tone) whose pitch instructed them to attend either the left or right hemifield for an upcoming discrimination on that side. Bimodal, bilateral, visual-tactile stimulation usually followed the cue. According to blocked instructions given at the beginning of each scanning session, subjects discriminated just the visual or just the tactile stimulus, at the attended location specified by the symbolic auditory cue. The results replicated our previous PET findings (37,39) in showing increased activity for intraparietal areas contralateral to the attended side. Again, the spatial effects of endogenous attention here were multimodal, being independent of the modality (vision or touch) that was currently task relevant (Fig. 14.3A,B, upper plots).

This fMRI study also included some trials in which the symbolic auditory cue directing attention to the left or right side was not in fact followed by any bilateral-bimodal stimulation. This allowed us to investigate any preparatory effects of endogenous spatial attention, irrespective of subsequent target processing. Figure 14.3C,D shows that activity in the intraparietal sulcus contralateral to the attended side increased during these "cue-only" trials also, thus demonstrating preparatory activity in this region. Note that this preparatory activity was both spatially specific (i.e., higher activity for attention to the contralateral versus ipsilateral side) and multimodal (i.e., found regardless of whether the subject was preparing to judge a tactile or a visual stimulus on the attended side). This finding of increased activity in intraparietal areas when a given location becomes behaviorally rele-

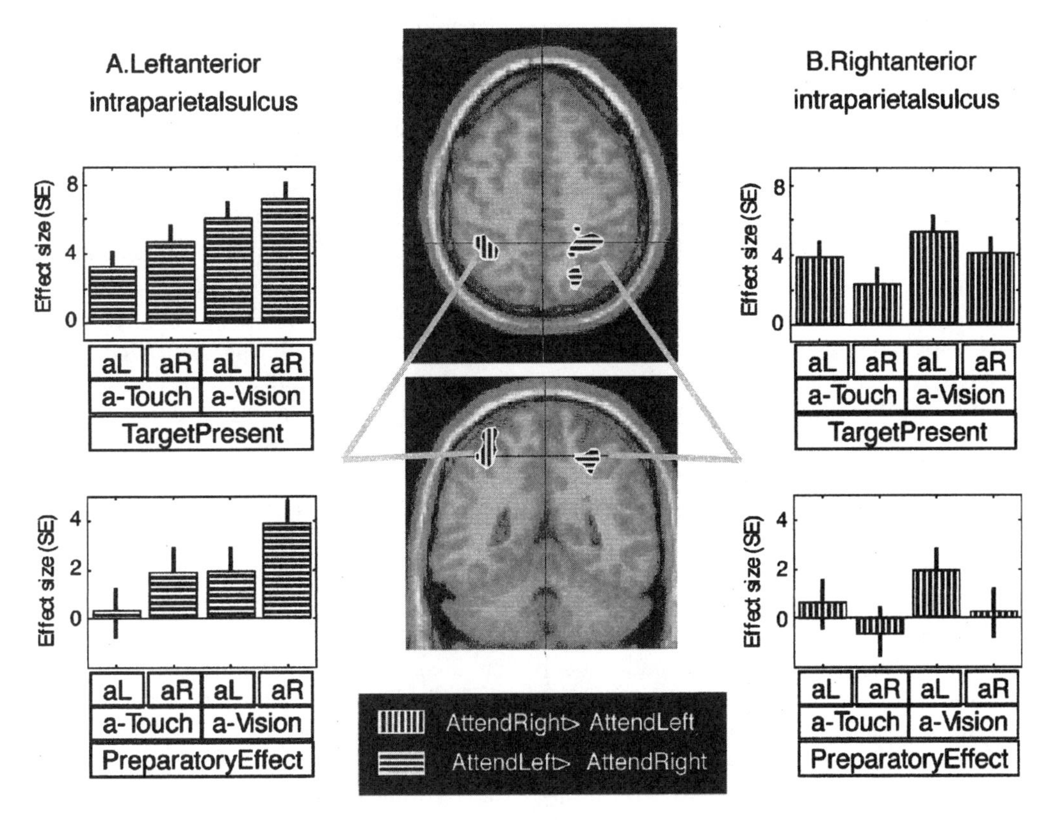

FIG. 14.3. Attentional influences and preparatory activity in the intraparietal sulcus. Here we show that activity in an anterior intraparietal region was affected by endogenous attention to one or other side of space. **Top:** Attention to the left or right hemifield followed by discrimination of visual or tactile target (according to prescanning instructions) on the attended side, during bimodal-bilateral stimulation. **Bottom:** Preparatory effects for attention to the left or right (in vision and touch), on trials where subjects directed attention to one hemifield, but no peripheral stimulation was subsequently delivered. Activity for these trials should thus reflect preparatory activity, independent of any target processing. (aL/aR, attend left/right hemifield. SPM thresholds set to *p*-uncorrected = 0.001.)

vant—even in the absence of any sensory stimulation there—is in agreement with previous electrophysiologic recording in monkeys (28,40). Our results extend previous findings to show that these preparatory spatial effects in intraparietal cortex can be observed multimodally, independent of the sensory modality of the anticipated target stimulus. It should be noted that electrophysiologic data have consistently indicated that the majority (around two thirds) of neurons showing preparatory effects in posterior regions of intraparietal cortex (the lateral intraparietal area [LIP] and the parietal reach region [PRR]) show increased responses specific to the type of intended movement associated with the task-relevant location (40,41). Future imaging studies might use different types of motor responses (e.g., planned reaches versus saccades) to search for spatial preparatory activity in human parietal cortex that depends on motor intention (31–33).

In summary, the recent imaging evidence considered herein indicates some analogies between multimodal spatial representations in the human brain and previous electrophysiologic findings from the monkey brain, in and around the intraparietal sulcus. We have shown that a region in anterior-dorsal human intraparietal sulcus shows preferential responses for stimulation of the contralateral hemifield. These spatially specific effects were found for both visual and tactile stimulation, indicating the presence of multimodal spatial representations here. More posterior intraparietal regions can also be activated for multiple modalities in humans, but their activation depends more on the type of motor response to the stimulus (eye versus hand). Our preliminary data indicate that the more anterior intraparietal region shown in Figure 14.2 can be activated in a manner that transcends not only sensory modality but also response modality (saccades versus button press). Moreover, modulation of this region by spatial attention is found, with activity increasing when the contralateral hemifield is task relevant. As might be expected for a multimodal area, these spatial attention effects are independent of the modality stimulated (37) or attended (39). Finally, our results also show that multimodal spatial attention effects can be observed here even in the absence of any peripheral sensory stimulation, thus highlighting preparatory spatial activity under purely "top down" influences (38).

These results from our own studies should be placed in the broader context of an increasing body of research that seeks to relate well-defined and well-studied areas within monkey parietal cortex to the less understood organization of human parietal cortex (15,31, 33,42–45). As in monkey parietal cortex, human imaging studies have already demonstrated some functional segregation between different areas of the intraparietal sulcus (IPS), particularly in relation to motor specificity. Activity related to grasping movements has been localized in the most anterior part of IPS (42), whereas saccade-related responses have been found more posteriorly (31,33,46, 47). However, such different types of movement have been found to activate overlapping regions of IPS across some studies (32,33,45); for example, with saccadic eye movements sometimes activating regions in anterior IPS as well (48). Generally, it appears that direct within-experiment comparisons of different movement types under closely matched conditions are needed to reveal convincing movement selectivity within human IPS, whereas comparisons against lower-level baselines produce more diffuse activation of an extensive network.

Our own work has focused on spatially specific activations (i.e., those depending on the location that is stimulated, or attended, or responded to), which are found multimodally for both vision and touch. As Figures 14.1 through 14.3 show, this has consistently highlighted a region in the anterior part of the intraparietal sulcus, at the junction with the postcentral gyrus. Although it would be premature to assert any definite homology between the anterior intraparietal area activated here, and specific intraparietal areas as defined in monkey studies, we may briefly consider some possibilities. Multimodal spatial responses have been studied in several parietal areas in monkeys at the single-cell level, in particular in the context of sensorimotor spatial transformations (3,4,11,49,50). As mentioned, neurons in LIP and MIP (or PRR) show strong preferences for one or another type of movement (e.g., saccade versus reach), whereas our anterior intraparietal activation apparently can arise regardless of the type of motor response (see Fig. 14.2) as well as of the stimulus modality, and is found even for passive stimulation (see Fig. 14.1). Anatomically, its location in anterior-dorsal IPS could perhaps suggest some possible correspondence with AIP, and/or part of area 5. Single-cell work in monkey AIP has highlighted the role of this region in grasping and processing of visual information used to guide grasping movements (51). Several imaging studies have now exploited these known characteristics of monkey AIP to identify a potentially corresponding region in humans (42,44), and indeed have activated the anterior part of IPS with grasping. Our own studies never required grasping, but activation of AIP might nevertheless be possible, given that monkey AIP seems involved in representing peripersonal space near the hands (52), while our own experimental setup always used visual or tactile stimuli in close spatial proximity to the hands, and a grasp-related region might also be concerned with visual-tactile integration of information at or near the hands.

Another potential candidate for homology with nonhuman primates may be area VIP, where multimodal responses have been detected both at the single-cell level (4) and using neuroimaging (15), as described. Future imaging studies could combine some of the paradigms presented here (e.g., perhaps including different types of intended movements in tests for preparatory spatial-attention effects), under various different combinations of stimuli and responses (e.g., using moving targets in different modalities, and pointing or grasping movements), in order to further investigate the possible relation between the anterior-dorsal intraparietal region activated here in the human brain (Figs. 14.1 to 14.3), and the various parietal areas previously identified in monkeys.

In addition to the anterior intraparietal region that we have discussed, Figure 14.1 also shows multimodal responses in several other human parietal regions (indicated in Fig. 14.1, by dotted hatching) (15). Unlike the anterior-dorsal intraparietal region, these additional multimodal areas did not show any spatial specificity for left versus right stimulation. The next section considers several manipulations that have been found to affect activity in these further parietal areas.

POSSIBLE PARIETAL INVOLVEMENT IN CONTROL OF SPATIAL ATTENTION

As briefly mentioned, several lines of evidence point to a role of parietal cortex in the direction of spatial attention. Electrophysiologic evidence has shown how task relevance can strongly affect the activity of single cells in parietal cortex (34–36), whereas many recent imaging studies employing spatial attention tasks (usually in just the visual modality) have consistently activated parietal regions (53–55), along with frontal and superior temporal areas in some cases. Some of the earliest imaging studies used "low-level" control conditions that might make interpretation of results somewhat problematic (53,54), but recent studies have employed increasingly subtle comparisons (55,56) and still found activation in several parietal areas. For example, Gitelman et al. (55) compared shifting covert visual attention between peripheral locations minus sustaining attention at a central location, in a study where sensory inputs and motor components were well matched across these conditions. The results showed activation of intraparietal regions (extending to superior parietal cortex) and also of more inferior temporal-parietal regions (46,57).

Corbetta et al. (56) were able to dissociate activity in superior and inferior parietal cortex using an adaptation of a classic spatial cueing paradigm in a purely visual study (58). On a trial-by-trial basis, a central symbolic cue indicated the most likely location for a subsequent visual target. The majority of trials had "valid" cues, with the target subsequently presented at the cued location. A minority of trials had "invalid" cues, with the visual target being presented on the side opposite to that indicated by the preceding central cue. Such invalid trials have been argued to trigger an additional shift of spatial attention from the cued location, to the actual location of the target when it arrives on the uncued side (58). Corbetta et al.'s (56) results showed that although superior parietal cortex was unaffected by the validity of the cue, inferior parietal areas (and superior temporal cortex) were particularly active for the invalid trials, when attention should be shifted by the onset of the visual target at the uncued location. This led to proposals that superior intraparietal regions were involved in preparatory top-down control (59) of endogenous attention (possibly involving covert motor plans toward the cued location); whereas the more inferior regions are engaged by exogenous (bottom-up or stimulus-driven) factors triggering an interruption of the endogenous attentional set, as when a target appears unexpectedly at the uncued location (60).

Given our previous results showing multimodal activations in both intraparietal sulcus and in inferior parietal cortex (see Fig. 14.1, dotted hatching), it seemed possible that the subsystems proposed by Corbetta and colleagues (56,60) might operate supramodally rather than only for the visual modality that they had studied. To address this, we recently used a modified version of the spatial cueing paradigm to investigate any possible common neural substrate for control of spatial attention in both vision and touch (61). Unlike previous unimodal studies, the targets could now be either visual or tactile (with visual targets located near either hand). On a trial-by-trial basis, an informative auditory cue (a voice saying "left" or "right") instructed the subjects to direct attention endogenously to one particular hemifield. Subsequently, a target appeared either at the cued location (80% of the trials) or in the opposite hemifield (20%) for discrimination. Consistent with the recent results within the visual modality (56), we found segregation between responses in intraparietal cortex versus more inferior temporal-parietal regions. Figure 14.4A shows a coronal section illustrating both superior and inferior activations, plus plots of activity for two clusters in the right hemisphere. (Significant effects for invalidly cued minus validly cued trials were detected only in the right hemisphere, but strong trends were present also in the left hemisphere.) As shown in Figure 14.4A.1, superior parietal cortex (including several maxima along the intraparietal sulcus) responded during both valid (dark shaded bars) and invalid trials (light shaded bars) alike. Moreover, these responses were independent of the sensory modality (and side) of the visual or tactile target, highlighting once again the multimodal nature of this parietal region. The finding of activation common to both valid and invalid trials may suggest a possible role for this region in initial voluntary orienting toward the cued side. This would fit with proposals from previous, purely visual, imaging studies that implicated superior parietal cortex in voluntary control of spatial attention (56,59), but would extend this to the multimodal context (61,62).

The direct comparison of invalid minus valid trials in our study (61) was matched for sensory stimulation and motor responses, plus the requirement for peripheral discrimination, thus allowing us to isolate processes specific to stimulus-triggered shifts of spatial attention when the target appeared on the uncued side. This revealed activation of the inferior parietal lobule, extending to superior temporal cortex. Critically, signal increase here was also independent of the modality of the target (see Fig. 14.4A.2), indicating that a common neural system operated whenever unexpected attention-shifts were required during this task,

FIG. 14.4. Parietal activations for tasks thought to engage attentional control processes. Plots shown for right hemisphere only. **A:** Areas engaged during a multimodal variant (61) of a classical spatial cueing paradigm. **A1:** Common activation for valid and invalid trials (SPM display threshold: *p*-corrected = 0.001). All trial-types in this spatial attention study engaged several sites in the parietal lobe, especially in and around the intraparietal sulcus. These areas were unaffected by modality and side of the target, strongly activating above baseline for all these subconditions, as shown in the plot for the right intraparietal maxima. **A2:** Comparing invalid to valid trials showed increased activation in more inferior regions of the parietal lobule (SPM display threshold: *p*-corrected = 0.05). This activation was specific for invalid trials (i.e., when a stimulus-driven reorienting of attention was triggered by the target, from the cued location to the uncued location where the target actually appeared). It was found independent of the target side (sL/sR), and critically independent of the target modality also. **B:** Activation of the superior-posterior temporal lobe for sustained endogenous attention to peripheral locations during bimodal-bilateral stimulation (39). (SPM display threshold: *p*-uncorrected = 0.001). Again activation was detected irrespective of attended modality and side. (Compare bars 1 through 4 against bar 5.) (aL/aR, attend left/right hemifield; US, control condition, with unattended bimodal-bilateral stimulation, and attention now directed to the central fixation cross.

for both visual and tactile targets. These results are again consistent with previous, purely visual studies (56), but extend them to show that the role of inferior parietal cortex in such spatial attention shifts extends across different sensory modalities (63).

Further evidence for a supramodal involvement of inferior-parietal/superior-temporal regions in control of spatial attention is available from imaging studies of endogenous sustained spatial attention. For example, we recently compared conditions of sustained

A.SHIFTING SPATIAL ATTENTION

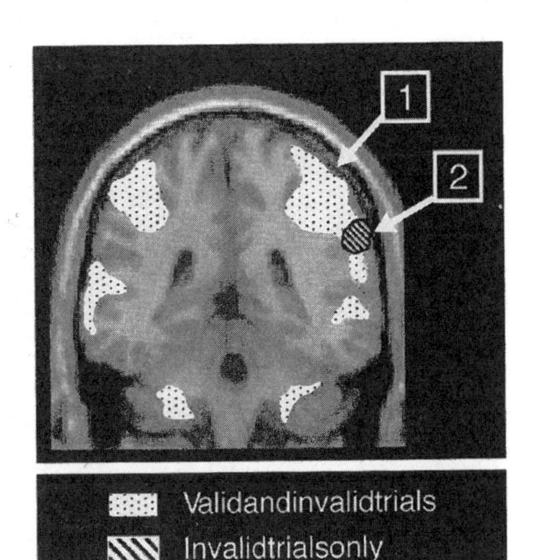

1.Rightintraparietalsulcus:
Commonforvalidandinvalidtrials

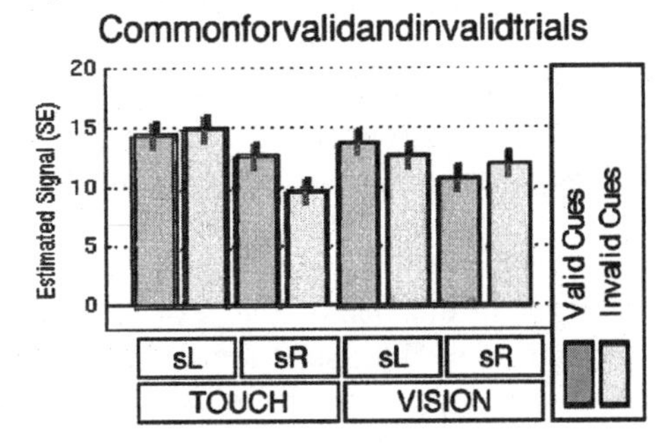

2.Rightinferior parietallobule:
Specifictoinvalidtrials

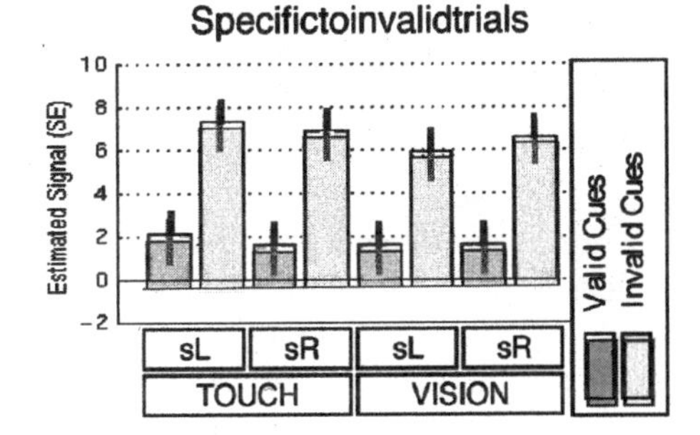

B.PERIPHERAL SUSTAINED ATTENTION

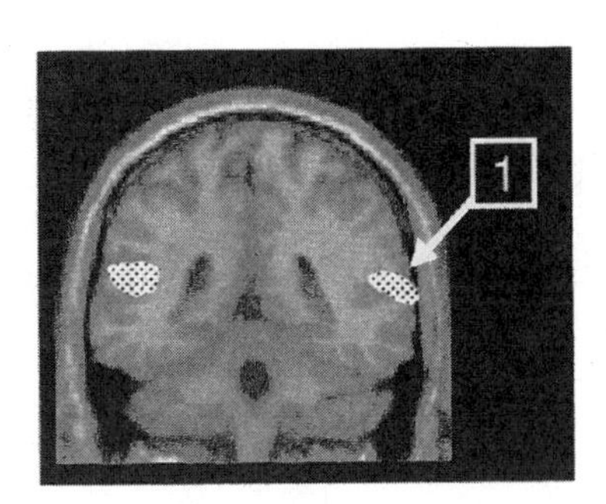

1.Rightsuperior temporalsulcus

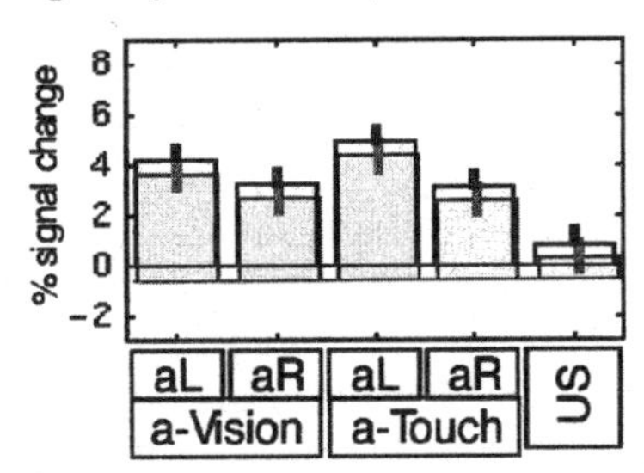

attention to a peripheral location on either side with attention to a central location (39). In conditions of attention to the periphery, subjects discriminated targets in one modality on just one side, during bimodal and bilateral stimulation. Direct comparison of attention toward one or the other side again showed activation in the anterior intraparietal sulcus contralateral to the attended side, irrespective of attended modality. (See also Figure 14.3, for related findings.) During the control condition, the same bilateral-bimodal stimulation was delivered, but subjects now performed a discrimination task on the central fixation cross (detecting its occasional dimming), ignoring any peripheral visual-tactile stimulation. Subtracting this control condition allowed us to test for brain areas commonly activated by both leftward and rightward endogenous covert spatial attention (for both vision and touch). This revealed multimodal activation of the superior temporal sulcus (and superior premotor cortex) when attending spatially to one or other side. (Compare bars 1 through 4 against 5 in Figure 14.4B.) Critically, this effect of sustained attention to peripheral locations was independent of relevant modality, once again highlighting the multimodal role of the temporal-parietal junction in control of spatial attention (see Figs. 14.4A2 and 14.4B).

Our activations of anterior-dorsal intraparietal sulcus for multimodal effects that are spatially specific (i.e., depending on the stimulated or attended side) were found to be anatomically consistent across several different types of comparison (Figs. 14.1 to 14.3). By contrast, the precise locus of activation for multimodal effects in more inferior temporal-parietal regions appeared to vary somewhat across studies and comparisons (Fig. 14.4A2,B). For instance, effects associated with stimulus-triggered reorienting of attention on invalidly cued trials (61) were centered in the inferior parietal lobule, in correspondence with the supramarginal gyrus (Fig. 14.4A2); whereas the effects of endogenous spatial attention to the periphery on either side, as found in Macaluso et al. (39), were primarily restricted to the posterior part of the superior temporal sulcus (Fig. 14.4B). Both superior temporal and inferior parietal (area 7b) cortex are known to contain multimodal neurons from cellular recordings (2,25). On the basis of the imaging data presented here, one might speculate that some distinction could exist between endogenous (superior-posterior temporal) and stimulus-driven exogenous (inferior parietal) control of spatial attention. This suggestion receives partial support from a purely visual study that directly compared endogenous and exogenous shifts of visual-spatial attention (64). Although this study found very similar networks activated for the two tasks, the results also indicated higher activity in the occipital-temporal junction (just posterior to the activation depicted in Fig. 14.4B) for the endogenous task. However, further evidence may be less consistent with such a proposal. For example, in another purely visual study Rosen and colleagues (65) failed to reveal any difference between endogenous and exogenous shifts of visual-spatial orienting; whereas Hopfinger et al. (57) reported robust activation of inferior parietal areas (centered around the angular gyrus) in an endogenous spatial cueing paradigm. Thus, one objective for future studies will be to dissociate activations of human inferior parietal cortex versus superior temporal sulcus, as a function of the attentional conditions, within the same experiment. Despite such uncertainties about the possible specialization of different areas situated around the temporal-parietal junction, the data considered here demonstrate the multimodal involvement of the temporal-parietal junction region in control of covert spatial attention, for both vision and touch

CONCLUDING REMARKS

This chapter has considered several recent imaging studies investigating the role of human parietal cortex in multimodal spatial processing for vision and touch. First, we presented a distinction between multimodal parietal areas that show differential responses depending on the location of the stimuli (in one or other hemifield) for both modalities; versus those that respond to both vision and touch irrespective of the stimulated hemifield. Hemifield-specific responses were localized in the anterior-dorsal part of the intraparietal sulcus, contralateral to stimulation. More posterior areas in the intraparietal sulcus, plus regions in the inferior parietal lobule, did not show such specificity for stimulus side, but still responded multimodally for both vision and touch.

Given that many parietal areas are increasingly considered as sensorimotor interfaces between perception and action, that also may be involved in spatial attention, further experiments investigated the possible influences of different motor responses, or of selective spatial attention, on activity in multimodal parietal areas. Whereas posterior intraparietal multimodal areas can show strong dependence on the type of movement executed or intended (with eye versus hand), our preliminary data suggest that activations in anterior intraparietal sulcus may transcend both sensory and response modalities, despite their specificity for the hemifield of the task-relevant location. The anterior intraparietal region showing multimodal activation for contralateral stimuli also was found to be influenced by endogenous covert spatial attention (even in the absence of any sensory stimulation), with stronger activation when attending visually or tactually to the contralateral hemifield.

More posterior areas in intraparietal sulcus and inferior parietal cortex also were influenced by spatial attention, again irrespective of sensory modality (vision or touch), but now also regardless of the attended side. Activation of these areas can be found in the context of both endogenous spatial attention and also stimulus-triggered (exogenous) reorienting of spatial attention. Superior parietal areas appear more involved in voluntary, anticipatory allocation of attention to peripheral locations (which may involve covert motor planning); whereas more inferior regions were activated for stimulus-driven reorienting of attention (although we also found activation of a more inferior—temporal—region for sustained endogenous attention to peripheral locations). In all these cases, we found that the attentional activations were multimodal (that is, found in common for vision and touch). The data presented here thus illustrate that parietal regions are involved in the representation of space across sensory modalities, have differing dependencies on motor factors, and roles in the control of spatial attention across different senses.

ACKNOWLEDGMENTS

The authors were supported by a Programme Grant from the Medical Research Council (UK). JD holds a Royal Society–Wolfson Research Merit Award. The Functional Imaging Laboratory at the Wellcome Department of Imaging Neuroscience, where the imaging experiments were conducted, is supported by the Wellcome Trust. Our thanks to Chris Frith for his past and ongoing collaboration on the work reported here.

REFERENCES

1. Colby CL, Duhamel JR, Goldberg ME. Ventral intraparietal area of the macaque: anatomic location and visual response properties. *J Neurophysiol* 1993;69: 902–914.
2. Graziano MS, Gross CG. The representation of extrapersonal space: a possible role for bimodal, visual-tactile neurons. In: Gazzaniga MS, ed. *The cognitive neurosciences.* Cambridge, MA: MIT Press, 1995: 1021–1034.
3. Andersen RA, Snyder LH, Bradley DC, et al. Multimodal representation of space in the posterior parietal cortex and its use in planning movements. *Ann Rev Neurosci* 1997;20:303–330.
4. Duhamel JR, Colby CL, Goldberg ME. Ventral intraparietal area of the macaque: congruent visual and somatic response properties. *J Neurophysiol* 1998;79: 126–136.
5. Stein BE, Meredith MA. *The merging of the senses.* Cambridge, MA: MIT Press, 1993.
6. Wallace MT, Meredith MA, Stein BE. Multisensory integration in the superior colliculus of the alert cat. *J Neurophysiol* 1998;80:1006–1010.
7. Graziano MS, Gross CG. A bimodal map of space: somatosensory receptive fields in the macaque putamen with corresponding visual receptive fields. *Exp Brain Res* 1993;97:96–109.
8. Graziano MS. Where is my arm? The relative role of vision and proprioception in the neuronal representation of limb position. *Proc Natl Acad Sci USA* 1999;96: 10418–10421.

9. Buneo CA, Jarvis MR, Batista AP, et al. Direct visuomotor transformations for reaching. *Nature* 2002;416:632–636.
10. Driver J, Spence C. Attention and the crossmodal construction of space. *Trend Cog Sci* 1998;2:254–262.
11. Colby CL, Goldberg ME. Space and attention in parietal cortex. *Ann Rev Neurosci* 1999;22:319–349.
12. Driver J, Vuilleumier P. Perceptual awareness and its loss in unilateral neglect and extinction. *Cognition* 2001;79:39–88.
13. Pavani F, Ladavas E, Driver J. Selective deficit of auditory localisation in patients with visuospatial neglect. *Neuropsychologia* 2002;40:291–301.
14. Lewis JW, Beauchamp MS, DeYoe EA. A comparison of visual and auditory motion processing in human cerebral cortex. *Cereb Cortex* 2000;10:873–888.
15. Bremmer F, Schlack A, Shah NJ, et al. Polymodal motion processing in posterior parietal and premotor cortex: a human fMRI study strongly implies equivalencies between humans and monkeys. *Neuron* 2001;29:287–296.
16. Blatt GJ, Andersen RA, Stoner GR. Visual receptive field organization and cortico-cortical connections of the lateral intraparietal area (area LIP) in the macaque. *J Comp Neurol* 1990;299:421–445.
17. Bremmer F, Schlack A, Duhamel JR, et al. Space coding in primate posterior parietal cortex. *NeuroImage* 2001;14:46–51.
18. Ben Hamed S, Duhamel JR, Bremmer F, et al. Representation of the visual field in the lateral intraparietal area of macaque monkeys: a quantitative receptive field analysis. *Exp Brain Res* 2001;140:127–144.
19. Motter BC, Mountcastle VB. The functional properties of the light-sensitive neurons of the posterior parietal cortex studied in waking monkeys: foveal sparing and opponent vector organization. *J Neurosci* 1981;1:3–26.
20. Tootell RB, Mendola JD, Hadjikhani NK, et al. Functional analysis of V3A and related areas in human visual cortex. *J Neurosci* 1997;17:7060–7078.
21. Tootell RB, Silverman MS, Switkes E, et al. Deoxyglucose analysis of retinotopic organization in primate striate cortex. *Science* 1982;218:902–904.
22. DeYoe EA, Carman GJ, Bandettini P, et al. Mapping striate and extrastriate visual areas in human cerebral cortex. *Proc Natl Acad Sci USA* 1996;93:2382–2386.
23. Sereno MI, Pitzalis S, Martinez A. Mapping of contralateral space in retinotopic coordinates by a parietal cortical area in humans. *Science* 2001;294:1350–1354.
24. Macaluso E, Driver J. Spatial attention and crossmodal interactions between vision and touch. *Neuropsychologia* 2001;39:1304–1316.
25. Bruce C, Desimone R, Gross CG. Visual properties of neurons in a polysensory area in superior temporal sulcus of the macaque. *J Neurophysiol* 1981;46:369–384.
26. Hikosaka K, Iwai E, Saito H, et al. Polysensory properties of neurons in the anterior bank of the caudal superior temporal sulcus of the macaque monkey. *J Neurophysiol* 1988;60:1615–1637.
27. Andersen RA, Essick GK, Siegel RM. Encoding of spatial location by posterior parietal neurons. *Science* 1985;230:456–458.
28. Colby CL, Duhamel JR, Goldberg ME. Visual, presaccadic, and cognitive activation of single neurons in monkey lateral intraparietal area. *J Neurophysiol* 1996;76:2841–2852.
29. Snyder LH, Batista AP, Andersen RA. Coding of intention in the posterior parietal cortex. *Nature* 1997;386:167–170.
30. Snyder LH. Coordinate transformations for eye and arm movements in the brain. *Curr Opin Neurobiol* 2000;10:747–754.
31. Kawashima R, Naitoh E, Matsumura M, et al. Topographic representation in human intraparietal sulcus of reaching and saccade. *NeuroReport* 1996;7:1253–1256.
32. Desouza JF, Dukelow SP, Gati JS, et al. Eye position signal modulates a human parietal pointing region during memory-guided movements. *J Neurosci* 2000;20:5835–5840.
33. Simon O, Mangin JF, Cohen L, et al. Topographical layout of hand, eye, calculation, and language-related areas in the human parietal lobe. *Neuron* 2002;33:475–487.
34. Bushnell MC, Goldberg ME, Robinson DL. Behavioral enhancement of visual responses in monkey cerebral cortex. I. Modulation in posterior parietal cortex related to selective visual attention. *J Neurophysiol* 1981;46:755–772.
35. Robinson DL, Bowman EM, Kertzman C. Covert orienting of attention in macaques. II. Contributions of parietal cortex. *J Neurophysiol* 1995;74:698–712.
36. Gottlieb JP, Kusunoki M, Goldberg ME. The representation of visual salience in monkey parietal cortex. *Nature* 1998;391:481–484.
37. Macaluso E, Frith C, Driver J. Selective spatial attention in vision and touch: unimodal and multimodal mechanisms revealed by PET. *J Neurophysiol* 2000;83:3062–3075.
38. Macaluso E, Eimer M, Frith C, et al. Preparatory states in cross-modal spatial attention: spatial specificity and possible control mechanisms. *Exp Brain Res* 2003;149:62–74.
39. Macaluso E, Frith CD, Driver J. Directing attention to locations and to sensory modalities: multiple levels of selective processing revealed with PET. *Cereb Cortex* 2002;12:357–368.
40. Snyder LH, Batista AP, Andersen RA. Change in motor plan, without a change in the spatial locus of attention, modulates activity in posterior parietal cortex. *J Neurophysiol* 1998;79:2814–2819.
41. Snyder LH, Batista AP, Andersen RA. Intention-related activity in the posterior parietal cortex: a review. *Vis Res* 2000;40:1433–1441.
42. Binkofski F, Buccino G, Stephan KM, et al. A parietopremotor network for object manipulation: evidence from neuroimaging. *Exp Brain Res* 1999;128:210–213.
43. Connolly JD, Goodale MA, Desouza JF, et al. A comparison of frontoparietal fMRI activation during antisaccades and anti-pointing. *J Neurophysiol* 2000;84:1645–1655.
44. Grefkes C, Weiss PH, Zilles K, et al. Crossmodal processing of object features in human anterior intraparietal cortex: an fMRI study implies equivalencies between humans and monkeys. *Neuron* 2002;35:173–184.
45. Culham JC, Kanwisher NG. Neuroimaging of cognitive functions in human parietal cortex. *Curr Opin Neurobiol* 2001;11:157–163.
46. Perry RJ, Zeki S. The neurology of saccades and covert shifts in spatial attention: An event-related fMRI study. *Brain* 2000;123:2273–2288.
47. Nobre AC, Gitelman DR, Dias EC, et al. Covert visual

spatial orienting and saccades: overlapping neural systems. *NeuroImage* 2000;11:210–216.
48. Corbetta M, Akbudak E, Conturo TE, et al. A common network of functional areas for attention and eye movements. *Neuron* 1998;21:761–773.
49. Hyvarinen J. Regional distribution of functions in parietal association area 7 of the monkey. *Brain Res* 1981; 206:287–303.
50. Stricanne B, Andersen RA, Mazzoni P. Eye-centered, head-centered, and intermediate coding of remembered sound locations in area LIP. *J Neurophysiol* 1996;76: 2071–2076.
51. Murata A, Gallese V, Luppino G, et al. Selectivity for the shape, size, and orientation of objects for grasping in neurons of monkey parietal area AIP. *J Neurophysiol* 2000;83:2580–2601.
52. Kalaska JF, Scott SH, Cisek P, et al. Cortical control of reaching movements. *Curr Opin Neurobiol* 1997;7: 849–859.
53. Corbetta M, Miezin FM, Shulman GL, et al. A PET study of visuospatial attention. *J Neurosci* 1993;13: 1202–1226.
54. Nobre AC, Sebestyen GN, Gitelman DR, et al. Functional localization of the system for visuospatial attention using positron emission tomography. *Brain* 1997;120:515–533.
55. Gitelman DR, Nobre AC, Parrish TB, et al. A large-scale distributed network for covert spatial attention: further anatomical delineation based on stringent behavioural and cognitive controls. *Brain* 1999;122: 1093–1106.
56. Corbetta M, Kincade JM, Ollinger JM, et al. Voluntary orienting is dissociated from target detection in human posterior parietal cortex. *Nat Neurosci* 2000;3: 292–297.
57. Hopfinger JB, Buonocore MH, Mangun GR. The neural mechanisms of top-down attentional control. *Nat Neurosci* 2000;3:284–291.
58. Posner MI. Orienting of attention. *Q J Exp Psychol* 1980;32:3–25.
59. Kastner S, Pinsk MA, De Weerd P, et al. Increased activity in human visual cortex during directed attention in the absence of visual stimulation. *Neuron* 1999;22: 751–761.
60. Corbetta M, Shulman GL. Control of goal-directed and stimulus-driven attention in the brain. *Nat Rev Neurosci* 2002;3:215–229.
61. Macaluso E, Frith CD, Driver J. Supramodal effects of covert spatial orienting triggered by visual or tactile events. *J Cog Neurosci* 2002;14:389–401.
62. Bushara KO, Weeks RA, Ishii K, et al. Modality-specific frontal and parietal areas for auditory and visual spatial localization in humans. *Nat Neurosci* 1999;2: 759–766.
63. Downar J, Crawley AP, Mikulis DJ, et al. A multimodal cortical network for the detection of changes in the sensory environment. *Nat Neurosci* 2000;3:277–283.
64. Kim YH, Gitelman DR, Nobre AC, et al. The large-scale neural network for spatial attention displays multifunctional overlap but differential asymmetry. *NeuroImage* 1999;9:269–277.
65. Rosen AC, Rao SM, Caffarra P, et al. Neural basis of endogenous and exogenous spatial orienting: A functional MRI study. *J Cog Neurosci* 1999;11:135–152.

15

Apraxias and the Lateralization of Motor Functions in the Human Parietal Lobe

Ramón C. Leiguarda

Department of Neurology, Rául Carrea Institute of Neurological Research, Buenos Aires, Argentina

INTRODUCTION

The majority of right-handed patients with left-hemisphere damage exhibit a disruption in the control of complex motor skills executed by both the contralateral and ipsilateral hands. Apraxias, as clinically defined by Liepmann, are the most dramatic disorders of learned, skilled movements (1–3). Thus, their study provides a special window into complex motor functions as well as into the role played by the left hemisphere in planning, programming, and executing such functions.

Limb apraxia comprises a wide spectrum of cognitive-motor disorders owing to acquired brain disease affecting the performance of a skilled, learned, purposeful movement with or without preservation of the ability to perform the same movement outside the clinical setting in the appropriate situation, or environment, which cannot be accounted for by elementary motor or sensory deficits (4,5). There are several types of limb apraxia, which at present are basically classified according to the nature of errors committed by patients during movement performance, as well as the pathway through which the movement is evoked (i.e., verbal, visual, tactile) (4,6,7).

In 1905, Liepmann studied 42 patients with right hemiplegia and 41 with left hemiplegia for the presence of apraxia. He found apraxia in 20 of those with left-hemisphere damage but in none of those with right-hemisphere damage. Therefore, he inferred that in definitely right-handed individuals the left hemisphere controls praxis of both halves of the body (8). The meticulous clinical and pathologic investigation of the Ochs case, which he carried out together with Mass in 1907, provided him with definitive evidence supporting the leading role of the left hemisphere in praxis control (9).

Liepmann considered apraxia as a higher-order motor disorder and advanced fundamental concepts of its underlying cognitive and motor mechanisms. Basically, he believed that the idea of the action, motor engrams, or movement formulae containing the space–time form picture of the movement (information such as the purpose of the action, participating body parts, timing, sequencing, and path or spatial instructions) were stored or represented in the left parietal lobe. In order to carry out a skilled movement, the space–time plan has to be retrieved and associated via cortical connections with the innervatory pattern stored in the left sensorimotorium (precentral and postcentral gyri and pes of the superior, middle, and inferior frontal convolutions), which conveys the information on formulae to the appropriate left primary motor areas. When the left limb performs the movement, the information has to be transmitted from the left to the right sensor motorium through the corpus callosum, thereafter to activate the right motor cortex (1–3). Liepmann conceived ideational apraxia as a disruption of the space–time plan or its proper activation to such an extent that it was impossible to construct the idea of the movement; the disorder becomes particularly evident when a sequence of motions with different objects is to be made. In contrast, in ideomotor apraxia, the space–time plans are intact but they can no longer guide the innervatory engrams that implement the movements because they are disconnected from one another; limb performance is no longer subject to guidance by the ideational component or movement plans. The patient knows what to do but not how to do it. Lastly, limb-kinetic apraxia appears when the disruption of the innervatory engrams interferes with the selection of the required muscle synergies to perform the skilled movement (2,3). Although the division of the apraxias into limb-kinetic, ideomotor, and ideational types is by no means undisputed (10), this terminology is still used owing to the lack of a better one based on the pathophysiologic mechanisms underlying the various clinical patterns of these higher-order motor disturbances.

According to Liepmann, limb apraxia ensues when the pathologic process affects the dominant posterior parietal (mainly the supramarginal gyrus and superior parietal lobe) and frontal (mainly the premotor) cortices, the corpus callosum, and the intrahemispheric white matter bundles that connect frontal and parietal association areas (3). Liepmann's original postulate regarding the crucial role played by the posterior parietal lobe in the genesis of apraxia has been largely confirmed by subsequent studies (11–15).

Later on, Heilman et al. and Rothi et al. proposed that the "movement formulae" or the "visuokinesthetic motor engrams" were stored in the left inferior parietal lobule, whose damage would cause not only a production deficit but also a comprehension-discrimination disorder (16,17). Thereafter, Rothi et al. suggested that the visuokinesthetic motor engrams represented in the left inferior parietal lobule were stored in a three-dimensional (3D) supramodal code and transcoded to the innervatory pattern mainly through the supplementary motor area (SMA) (18) rather than in the convexity of the premotor cortex (PMC), as previously postulated (19).

Roy and Square have proposed a model for the organization of actions based on the operation of a two-part system involving a conceptual component that encompasses an abstract knowledge base for actions and a production component that provides the mechanism for movement. According to these authors, the conceptual system (or action semantics) incorporates three types of knowledge relevant to limb praxis: knowledge of the functions that tool and object may serve, knowledge of actions independent of tools and objects, and knowledge relevant to the organization of single actions into sequences. On the other hand, the production system involves the sensorimotor component of action knowledge that includes the structural (space–time) information contained in action (motor) programs and the translation of these programs into actions. The performance of a given action involves a balance between action programs subserved at higher level systems that demand attention to keep the action sequence directed toward the intended goal and the more autonomous operations subserved at lower levels with minimal, if any, attentional demands. Thus, acting in the world depends on the interaction of abstract knowledge related to tools, objects, and actions and the structural information contained in motor programs (4). The dysfunction of the praxis production system causes ideomotor and limb-kinetic apraxias, whereas damage to the conceptual system gives rise to ideational or conceptual apraxias.

Freund stressed the role of polymodal and unimodal sensory association areas in motor control, classifying posterior apraxias as supramodal (ideomotor and ideational) and unimodal (e.g., visuomotor and tactile apraxia) (10). More recently, Leiguarda and Marsden have attempted to interpret the praxis deficits observed in patients with ideomotor and limb-kinetic apraxias as resulting from disruption of parallel parietofrontal circuits, together with their subcortical connections, supporting the processes necessary for translating an action goal into a movement by integrating sensory input with central movement representations, based on prior experience (20).

This chapter focuses on the hemispheric asymmetries of complex skilled movements, with special emphasis on the role of the dominant parietal lobe. In particular, we deal with the clinical and kinematic deficits observed in apraxia patients with ideomotor and ideational apraxia secondary to left parietal lobe damage and the participation of the left parietal lobe in the imitation and the recognition of actions and the perception of self-generated movements, as well as the role of the parietal lobe in motor imagery and its relationship with praxis.

IDEOMOTOR TYPE OF PRAXIS ERRORS

As described, Liepmann originally postulated that the "movement formulae" controlling purposeful skilled movement of the limbs on both sides of the body were stored in the left parietal lobe (3). Later, Heilman et al. localized the visuokinesthetic motor engrams in the inferior parietal lobule (16). However, apraxia as tested by the imitation of gestures and object use pantomime has been found in about 50% of patients with left hemisphere damage, as well as in less than 10% of those with right hemisphere damage (5), which means that in many subjects praxis functions have bilateral representations, and that in some right-handed subjects the right hemisphere also may possess some praxis skills, especially for the left half of the body.

Freeman suggested that the main function of the left hemisphere is to exert a priority control in situations appropriate to its actions (21). As a matter of fact, most of the spatial and temporal errors exhibited by patients with ideomotor apraxia are equally seen in left- or right-hemisphere–damaged patients when they pantomime nonrepresentative and representative-intransitive gestures, but are observed predominantly in left-hemisphere–damaged patients when they pantomime tool/object use movements (transitive), because it is this action type that is performed outside the natural context (22). The improvement in performance observed when the patient actually uses the tool/object may conceivably result from the advantage provided by visual and tactile-kinesthetic cues emanating from the tool/object and/or by the fact that in this condition, the patient is performing the movement in a more natural context and therefore is less dependent on the left (dominant) hemisphere.

Rapcsak et al. exhaustively studied a right-handed patient with a massive left-hemisphere stroke in order to explore the putative right-hemisphere praxis skills. The patient was almost unable to pantomime transitive gestures or reproduce novel nonsymbolic movements with the left hand, but intransitive gestures and actual object use were relatively unimpaired. Gesture recognition and discrimination were preserved as well. The author proposed that the left hemisphere was dominant for "abstract" or context-independent performance of transitive movements and in learning novel movement sequences, whereas either hemisphere controlled the execution of concrete or context-dependent familiar and action routines (23). Schnider et al. further emphasized that the left hemisphere motor dominance reflected by ideomotor apraxia refers to spatially and temporally complex movements performed in an artificial context (24). Furthermore, Rushworth et al. have suggested that the left hemisphere is also dominant for the selection of learned actions (25). Thus, clinical studies strongly suggest that interhemispheric differences in the control of praxis skills largely depend on the context in which the movement is performed and the cognitive requirements of the task; that is, when goal-oriented movements are performed outside the usual context and depend on higher-level cognitive abilities for planning and self-monitoring the action, the left hemisphere emerges as the dominant one (24–29).

In a series of landmark papers, Poizner and coworkers described the application of new technological advances in movement analysis to the study of patients with limb apraxia (30–32). Using 3D motion analysis, they demonstrated that patients with ideomotor apraxia mainly caused by left parietal lesions, as well as patients with asymmetric cortical degenerative syndromes, show deficits in the spatial and temporal attributes of wrist trajectory, in space–time relationships as well as the coordination of joint motion (30–33). We further demonstrated similar abnormalities in patients with ideomotor apraxia in the context of corticobasal degeneration and primary progressive aphasia (34).

Movement in apraxia patients usually shows a slow and hesitant build-up of hand velocity not only initially but also within the movement, at points of transition between movement components. The velocity profile is often irregular and nonsinusoidal, reflecting the awkwardness and coarseness of the movement. Several aspects regarding the spatial organization of the gestural movement are also disrupted in apraxia patients. The amplitude of the movement is quite variable and the shape of the wrist trajectory, the reversal phase at the end of the forward wrist stroke, and the spatial orientation of the plane of motion when performing a gesture such as slicing a loaf of bread are all abnormal. The shape of the wrist trajectory usually is curved and irregular with a blunted phase reversal, instead of being linear, tightly organized and with a sharp phase reversal. The movement is faultily oriented; in contrast to normal subjects who orient the movement in a sagittal plane, perpendicular to the object, the patient's movement is predominantly performed in a frontal plane, parallel to the object, and with variable changes of planes in successive movements instead of producing closely overlapping planes as normally made. The normal relationship between hand velocity and curvature or degree of bending is impaired or decoupled in apraxia patients, who also show deficits of interjoint coordination, such as improper joint synchronization, distorted angle–angle relationships and abnormally apportioned arm angles (30–34). However, when patients with left-hemisphere damage and limb apraxia are compared to those with right-hemisphere damage, it was found that only the former showed impaired coupling of spatial and temporal aspects of wrist trajectories and deficits in interjoint coordination, whereas both patient groups exhibited abnormalities in 3D wrist motion planes (35).

Kinematic deficits in apraxia patients are not restricted to gestures alone. Carving a turkey and slicing a loaf of bread are repetitive movements, which mainly explore the transport or reaching phase. However, the large majority of transitive gestures included in most apraxia batteries are prehension movements reflecting proximal (transport) and distal limb control (grasping), as well as coupling of transport and grasping movements.

Although scanty, studies designed to explore these components of the action system in patients with apraxia have provided consistent results. Charlton et al. evaluated an apraxia patient and demonstrated that the coordination of the transport and grasp component, together with the grasp component itself, were markedly abnormal (36). Caselli et al. as well as Leiguarda et al. studied patients with progressive apraxia owing to corticobasal degeneration and Alzheimer's disease. Compared to control subjects, apraxia patients show disruption of both the transport and grasp phases of the movement together with transport-grasping uncoupling (37,38). Lastly, Binkofsky et al. studied three patients with left (two of whom had ideomotor apraxia) and two with right hemisphere lesions involving the anterior lateral bank of the intraparietal sulcus (IPS). Patients displayed selective temporal and spatial kinematic deficits in the coordination of finger movements required for grasping a switch. A delayed time to achieve maximal hand aperture and a prominent disturbance of hand shaping was observed in all five patients, regardless of the side of the lesion and the presence or absence of ideomotor apraxia (39). A recent functional magnetic resonance imaging (fMRI) study further confirms the crucial role of the parietal lobe in grasping and manipulation, because it demonstrated bilateral activation during actual right-hand grasping and manipulation of complex objects (40).

Contrasting with the limited number of reports on prehension movements in apraxia patients with parietal involvement, there is a large amount of studies assessing the performance of the ipsilesional hand of brain-damaged patients when aiming or pointing to a target. On the whole, findings suggest that the left hemisphere would be specialized for ballistic, open-loop components of movements that are preprogrammed and largely independent of feedback mechanisms. On the other hand, the right hemisphere would be dominant during closed-loop processing, which is mainly dependent on sensory feedback (41). However, this claim has been questioned (29,42), because in a reciprocal aiming task, left-hemisphere–damaged patients showed deficits in selecting and implementing an optional movement velocity as movement amplitude increased, which would indicate a problem in adjusting the motor program for changes in the movement context. Furthermore, only apraxia patients showed impaired accuracy for movements to small but not larger targets, which may point to a degraded program for movements that require a high degree of spatial accuracy (43). In a later study, Haaland et al. employed kinematic analysis of simple aiming movements in patients with left-hemisphere damage with and without limb apraxia and in normal controls, in order to investigate preprogramming and response implementation deficits in apraxia. Kinematic abnormalities in apraxia patients were restricted to the spatial aspects of movements; indeed, spatial errors were mainly found during the secondary adjustment phase and in particular when visual cues were unavailable. No differences in arm movement deficits were observed between apraxia patients with frontal or parietal lesions using this relatively simple aiming task (44).

PLANNING AND SEQUENCING MOVEMENTS

Performance of a motor sequence requires the execution of preprogrammed temporal and spatial movement patterns. Functional brain imaging studies have shown that distinct neural systems are actively engaged in the preparation and generation of a sequential action, depending on whether a sequence has been prelearned or is a new one, and contingent to the complexity of the attentional demands of the task (45–47). The SMA, primary sensorimotor cortex, basal ganglia (midposterior putamen) and cerebellum may be mainly involved in the execution of automatic, overlearned, sequential movements, whereas the prefrontal, premotor, and posterior parietal cortices and the anterior part of the caudate/putamen, would be particularly recruited—in addition to such areas engaged in the execution of simple movement sequences—when a complex or newly learned sequence, which requires attention, integration of multimodal information and working memory processes for its appropriate selection and monitoring has to be performed (46–48).

Dysfunction of sequential movements has been reported more commonly after left- than right-hemisphere damage. Abnormalities in sequencing movements have been described in patients with parietal, frontal, and basal ganglia involvement (13,14,28,49–52). Luria emphasized the role of the frontal lobes in controlling actions requiring sequencing of different movements over time (49). Kolb and Milner studied meaningless movement sequences in a group of epileptic patients with cortical ablations, and found that those with left parietal lesions were more impaired than those with right parietal or left frontal lesions

(13). Similar results were documented by Kimura and De Renzi et al., using a multiple hand movement task and the imitation of three unrelated movements carried out in a sequence (14,28). Patients with frontal and parietal lesions had deficits in sequencing movements, but the impairment in those with frontal damage became evident only with more complex sequences (28).

Several clinical studies have shown that impairment in sequencing is particularly apparent for the left-hemisphere–damaged patients when the tasks place demands on memory (53,54). However, when the temporal aspects of sequencing, reflecting response preparation and programming, are considered, left-hemisphere–damaged patients exhibit deficits in movement sequencing even though they are not required to select the movement from memory; moreover, patients with ideomotor apraxia were found to be specifically impaired in planning and implementing sequences of various hand movements, which depend on developing and retrieving diverse motor programs for each movement in the sequence (51). Rushworth et al. have further proposed that the left hemisphere is not only dominant for learning to select movements in a sequence but also for learning to select a limb movement that is appropriate for the use of an object. These learning processes depend on distinct but adjacent response selection systems, both lateralized to the left hemisphere. The system made up by lateral premotor and parietal cortex, basal ganglia, thalamus, and white matter fascicles participate in the selection of limb movement responses, whereas an adjacent system integrated by lateral area 8 and possibly interconnected parietal regions, thalamus, striatum, and white matter fascicles are concerned with the selection of object-oriented responses (25). Finally, the process of motor attention has also been lateralized to the left hemisphere, in particular to the parietal lobe, so that left-hemisphere–damaged patients exhibit abnormalities in the sequencing of movements owing to inability to shift the focus of motor attention from one movement in the sequence to the next (26).

Weiss et al. studied the execution and imitation of meaningless arm movement sequences in patients with left and right parietal lesions, in comparison with control subjects. Patients with left parietal lesions showed more temporal and spatial errors than those with right parietal lesions, whereas additions or omissions of movement components occurred almost equally in both groups. In addition, only the left parietal-damaged patients showed a significant increase in error rates with greater movement complexity (55). These findings are consistent with those of Haaland et al., using a simple aiming task, and Clark et al., using a gestural movement, thus corroborating the role of the left parietal lobe in the spatiotemporal organization of simple as well as complex movements (32,44).

IDEATIONAL APRAXIA DISORDERS

Pick coined the term ideational apraxia to denote the inability to carry out a series of acts involving the use of several objects (e.g., prepare a letter for mailing) (56). Liepmann (2) and others (12,57,58) advanced a similar conception and attributed it to damage to the parieto-occipital or parieto-temporal regions of the left hemisphere. However, other authors use the term to denote a failure to use single tools appropriately. Denny-Brown considered ideational apraxia as an agnosia for object use (59), and De Renzi interpreted it as an inability to remember the general configuration of the action when attempting to use a tool (5). To overcome this confusion, Ochipa et al. suggested restricting the term ideational apraxia to the failure to sequence correctly a series of acts leading to an action goal, and introduced the term conceptual apraxia to denote the loss of different types of tool–action knowledge (60).

Patients with conceptual apraxia primarily exhibit content errors in the performance of transitive movements (e.g., the patient pantomimes shaving for a target of tooth brushing or uses the toothbrush as if it were a shaver), because they are unable to associate tool and objects with their corresponding actions. They may also lose the ability to associate tools with the object that receives their action; thus, when a partially driven nail is shown, the patient may select a pair of scissors rather than a hammer from an array of tools to perform the action. In addition, patients with conceptual apraxia lose the mechanical advantage afforded by tools (mechanical knowledge). For example, when asked to complete an action and the appropriate tool is not available (e.g., a hammer to drive a nail), they may not select the most adequate tool for that action (e.g., a wrench) but rather one that is inadequate (e.g., a screwdriver) (60,61). Patients with ideational or conceptual apraxia are disabled in everyday life, because they use tools/objects improperly, they misselect tools/objects for an intended activity, perform a complex sequential activity in a mistaken order, or do not complete the task at all (62).

Although Poeck and Lehmkuhl (63) maintained that patients with ideational apraxia "quite often are able to correctly manipulate single objects", when these patients are systematically investigated, they have also been found to be impaired in demonstrating the use of single objects (64). De Renzi and Lucchelli tested 20 left-brain–damaged patients with a single object and a multiple object test and found that performance of both tests strongly correlated; thus, a strict difference between ideational and conceptual apraxia may not be possible in every patient. Parietal damage predominates in these patients, either in isolation or in combination with temporal and frontal involvement, although some patients had only frontal, frontotemporal, or even basal ganglion lesions (64).

Heilman et al. evaluated patients with focal hemisphere lesions for deficit in the conceptual praxis system. They found that patients with conceptual apraxia had damage exclusively to the left hemisphere, but only about half of them exhibited apraxia. Most patients had involvement of the parietal and frontal association areas, together or separately, and with or without subcortical involvement (61).

Goldenberg and Hagmann assessed pantomime of tool use, actual tool use, and mechanical problem solving, with tests of selection and application of novel tools, in patients with left and right brain damage and in control subjects. They found that only left-brain–damaged patients differed from controls in all tests, although right-brain–damaged patients may have difficulties with the use but not the selection of novel tools. Analysis of left-brain–damaged patients' lesions suggested that, whereas frontal and parietal lesions may cause abnormal pantomime of tool use, only parietal lesions cause defective tool selection. Thus, the ability to infer function from structure and the subsequent capacity to solve mechanical problems is mainly subserved by the dominant parietal lobe (65).

IMITATION OF MOVEMENTS AND POSTURES

Although performance in ideomotor apraxia usually improves on imitation, there are patients who are more impaired when imitating than when pantomiming to command, so-called conduction apraxia (66), or cannot imitate but perform flawlessly under other modalities (visual-imitative apraxia) (67–69). The deficits may be restricted solely to the imitation of meaningless gestures, with preserved imitation to meaningful ones. Genuine imitation can be achieved only if the action to be imitated is novel, lacking any established correspondence with representation; otherwise, a meaningful gesture can be comprehended and followed by its verbal retrieval and reproduction from stored knowledge in long-term memory (70).

In humans, PET experiments have demonstrated that the imitation of meaningful actions seems to be indicated by implicit knowledge about the form as well as the meaning of the gesture, which is processed by regions involved in the planning and generation of actions, plus the left temporal cortex; whereas imitation of meaningless actions mainly depends on the decoding of their spatiotemporal layout in the "dorsal" visual pathway (occipitoparietal) and dorsal PM cortex of the right hemisphere with the contribution of regions within the ventral pathway (71). In a further study, Grezes et al. also used PET to specifically explore the normal network involved in the perception of meaningless actions without any purpose or with the intent to imitate them later. As expected, the observation of hand movements irrespective of the subject's intention was associated with bilateral activation of the areas involving the motion complex, such as the intraparietal sulcus and occipitotemporal junction. The intention to imitate was reflected by increased regional cerebral blood flow (rCBF) in the dorsal pathway extending to the lateral PM and dorsolateral prefrontal cortices, which indicates the information processing for prospective actions. When the meaningless action was learned, areas located in the occipitotemporal junction and the right middle occipital gyrus were less activated but the frontopolar area 10 predominantly in the left hemisphere, and part of the angular gyrus became more active; thus, this frontoparietal network may be involved in recognition processing (72). Hermsdörfer et al. performed a PET study of brain activation during matching of hand as well as finger postures. Both conditions gave rise to activation in the lateral occipitotemporal junction and the supramarginal gyrus of the left hemisphere, whereas matching of finger postures produced additional activations of the right intraparietal sulcus and supramarginal gyrus, as well as bilaterally in the extrastriatal visual cortex (73). The activation of the left parietal region reflects its participation in body part coding, whereas the occipitotemporal region is involved in processing the motion implied by the gesture presented as a static image (70,73). Finally, an fMRI study performed by Iacoboni et al. showed that imitation of simple finger movements leads to activation mainly in the left frontal operculum, the right anterior parietal region, and the right parietal operculum. According to the authors, the activation of the first two regions supports the "direct matching hypothesis" of action imitation, whereas activation of the parietal operculum reflects body identity preservation during imitation (74).

The anatomic correlates of imitation deficits have not been specifically studied, although abnormal performance on imitation was found in patients with lesions in several brain regions of the left hemisphere, in particular the parietal lobe. Using a single- and multiple-movement imitation test, De Renzi et al. found that patients with left frontal and parietal lesions were significantly more impaired than controls and that the frequency and severity of the imitation deficits was greater following parietal than frontal damage (14). In another study, the imitation of meaningless hand position was found to be abnormal, not only in patients with left parietal and frontal lesions, but also in some with left temporal and/or subcortical lesions (75). Haaland et al. studied patients with left hemisphere stroke and ideomotor apraxia, as disclosed by spatial errors while imitating gestures, and identified a middle frontal-intraparietal network as the crucial one involved in the representation of goal-directed movements. Interestingly, these patients exhibited more errors when imitating transitive than intransitive and meaningless movements, and target errors were more common with parietal than frontal lobe damage (76). Furthermore, patients with left parietal lobe damage seem to have more difficulties when imitating meaningful transitive gestures on their own bodies than when imitating movements with reference to an external object use, displaying more pronounced deficits in the spatial domain, a finding that may suggest that the basic deficit in these subjects concerns the ability to code and comprehend movements in relation to their body scheme (77).

Imitation of meaningless hand and finger postures discloses differential susceptibility to right and left brain damage. Left-brain–damaged patients have more difficulties imitating hand than finger postures, whereas right-brain–damaged patients commit more errors with finger than with hand postures (78). It has been suggested that the right hemisphere is required for visuoperceptual exploration and gesture analysis, whereas the left hemisphere is mainly responsible for coding gestures with reference to knowledge concerning body structure (70). Single case studies of patients with visuo-imitative apraxia (67,69), who have a severe deficit restricted to the imitation of meaningless gestures, together with the result of functional imaging (71,73), further support the crucial role of the left parietal lobe in the imitation of novel actions.

RECOGNITION OF ACTIONS AND PERCEPTION OF SELF-GENERATED MOVEMENTS

Ferro et al. studied action recognition deficits in 65 left hemisphere stroke patients in different stages of the disease. In the acute stage, the more affected structures were the frontal areas 44, 45, and 46; basal ganglia; and parietotemporal areas 39, 40, and 21, 22, 37. At the recent stage (third to fourth month), involvement of parietal area 7 and temporal area 37 predominated with maximal overlapping in the angular gyrus, whereas chronic patients more commonly showed involvement and areas of maximal overlap in the parietal areas 39 and 40 (79). Varney and Damasio studied only aphasic patients with left-hemisphere damage and demonstrated that defects in pantomime recognition appeared to result mainly from lesions affecting areas 40, 39, 37, and 22 (80). Rothi et al. found that apraxia-aphasic patients with left parietal and parietofrontal damage made more errors in a nonverbal paradigm of pantomime comprehension than aphasic-nonapraxia patients and normal subjects (81). Thereafter, Rothi et al. also described two nonapraxic, nonaphasic patients with inability to recognize the meaning of an action but with preservation of gesture imitation and without generalized visual agnosia who had occipitotemporal lesions on computed tomography scans (82). Wang and Goodglass found abnormal pantomime recognition in 30 aphasic subjects with frontal, parietal, temporoparietal, and frontotemporoparietal lesions (83). Recently, Halsband et al. compared gesture comprehension and imitation in patients with left and right parietal lobe and left and right PM/SMA lesions. Gesture comprehension was slightly disturbed only in patients with left parietal lesions. The severity of comprehension errors failed to differ between aphasic and nonaphasic apraxia patients, and there was no correlation between the number of errors in gesture imitation and gesture comprehension. The lack of consistent gesture comprehension deficits in patients with left parietal damage—as the authors pointed out—could result from the relatively dorsally located lesions barely involving temporal lobe structures that seem to be crucial for the knowledge of actions (77).

Thus, the recognition of actions seems to be subserved through a distributed neural system preferentially but not exclusively dependent on the left dominant hemisphere. This system appears to be made up by several interconnected nodes mainly in the parietal, frontal, and temporal cortices, each node having preferential functions within the system. Single-unit studies in monkeys and functional neuroimaging studies tend to support these assumptions.

Di Pellegrino et al. discovered a particular subset of neurons in F5 that discharge during the time a monkey observes meaningful hand movements made by the experimenter, in particular when interacting with objects; they called them "mirror neurons" and speculated that they belonged to an observation/execution matching system involved in understanding the meaning of motor events (84). Neurons with properties similar to those of mirror neurons in F5 are also found in the superior temporal sulcus (STS) (85), as well as in the parietal lobe. Preliminary data seem to demonstrate that neurons responding to hand–object interaction are also present in area 7b (86). Area 7b in the rostral part of the convexity of the inferior parietal lobule sends its cortical output to the convexity of F5 (87) and area 7b receives projection from the STS region (88), thus closing a cortical network involved in the perception of hand–object interaction; the crucial cognitive role of the STS-7b-F5 network is the internal representation of actions that, when evoked by an action made by others, are involved in two related functions, namely, action recognition and action imitation (89).

Positron emission tomography studies in humans support neurophysiologic findings in monkeys. Grasping observation markedly increased rCBF in the cortex of the STS, the rostral part of Broca's area and in the rostral part of the left intraparietal sulcus on the left hemisphere of right-handed subjects (90–92). The recent study of Buccino et al. demonstrated that action observation activates not only ventral PM cortex but also dorsolateral PM cortex in a somatotopic manner; and, whenever the observed movement was object-oriented there was strong concomitant parietal activation that was also bilateral and somatotopically organized (93).

Thus, converging evidence from patients with focal brain lesions and functional imaging in healthy subjects suggests that action observation is processed in the dorsal stream areas in order to automatically generate an internal replica of that action but needs semantic decoding along the ventral stream to be adequately recognized.

Apraxia patients with left parietal damage may not only be unable to recognize the action made by others, but also may have difficulties when they are required to discriminate from their own an external hand that performs the same movement (94). The impairment in correctly attributing the ownership of the movement is particularly evident when complex gestures are required. Thus, the left parietal lobe seems to play an important role in generating and maintaining a kinesthetic model of ongoing movements. Damage to the left parietal cortex results in an inability to evaluate and compare internal and external feedback about movements (94).

MOTOR IMAGERY AND PRAXIS

Motor imagery may be defined as the mental rehearsal of a motor act without any overt movement (95). Several observations suggest that common neurocognitive networks subserve the ideation of motor imagery and the intentional generation of overt movements.

Functional imaging studies have demonstrated a pattern of cortical and subcortical activation with imagined movements resembling that of an intentionally executed action (71,91,96–99). Stephan et al. asked normal subjects to imagine and execute a joystick movement and found a consistent pattern of bilateral cerebral activation during mental rehearsal of movements that mainly includes rostral SMA, anterior cingulate areas, lateral PM and ventral opercular PM areas, and part of the superior and inferior parietal areas. They suggested that motor imagery involves areas associated with selection of actions and multisensory integration and differ from those involved in the execution of movements mainly by the lack of activation of the primary sensorimotor cortex (97). However, other studies have shown that the primary motor cortex is also activated during imagery of a movement, although to a lesser degree than during motor performance (91,98).

Using PET, Decety et al. and Grafton et al. investigated the neural substrate of imaging to grasp neutral and familiar objects, respectively (91,96). Decety et al. found activation of the prefrontal cortex, area 6 in the inferior part of the frontal gyrus, parietal area 40, anterior cingulate areas, caudate nucleus, and cerebellum (96), whereas Grafton et al. additionally showed increased activity of the left rostral SMA, left inferior frontal area 44, middle frontal and left dorsal PM cortices, and left caudal inferior parietal cortex. These studies demonstrated that conscious representation of a grasping action involves a pattern of brain activation that closely resembles the circuit for hand grasping recently defined with fMRI (40).

Cortical lesions may or may not dissociate execution from mentally simulated movements (100,101). A patient with a right rolandic lesion and left hemiparesis exhibited the same increase in movement time when a motor task was performed as when mentally simulated, including compliance with Fitt's law, which relates movement duration with task difficulty. On the other hand, a patient with a left superior parietal lesion was unable to match actual movement times during mental imagery with the contralateral affected arm; movement times were not modulated by changing the difficulty of the task in the mentally imaged condition, as would be expected from Fitt's law, which was the case when movements were executed. It thus appears that a cortical lesion may or may not impair mentally performed actions to the same extent as real movements depending on its localization (101).

According to Jeannerod, the usually unconscious process of motor representation can be transformed through motor imagery into a conscious one; furthermore, because imaging movements result from the use of stored representations of action, it can be used to understand the content of such representations (94). Some praxis deficits may appear when the patient shifts from a strategy where object-oriented actions are processed automatically to when the content of those actions has to be explicitly represented as a consequence of a disturbance in using stored motor representations to build up mental images of actions or in the ability to evoke actions mentally (102).

Ochipa et al. recently described a patient with progressive ideomotor apraxia and striking parietal lobe atrophy, who exhibited a parallel impairment in her ability to answer imagery questions about joint movement and spatial position of the hands during actions, but with preserved visual object imagery (103). The patient of Sirigu et al. with severe apraxia for hand postures owing to hypometabolism in posterior parietal regions may have a similar deficit because he was impaired in producing as well as in verbally describing the appropriate hand posture for tool use (104).

Thus, patients with left parietal lesions often present with difficulty in tasks that require representing an action; they may be unable to pantomime actions involving an object or a tool; they may show deficits when mentally performing a motor task; they may be unable to imitate and may not recognize pantomimed actions, and may fail to recognize their own actions from those performed by other people. These impairments, which are part of the apraxia syndrome, clearly correspond to a lack of representation and recognition of action (105).

REFERENCES

1. Liepmann H. The syndrome of apraxia (motor asymboly) based on a case of unilateral apraxia (translated from: *Monatsschrift für Psychiatrie und Neurologie* 1900;8:15–44). In: Rottenberg DA, Hockberg FH, eds. *Neurological classics in modern translation.* New York: Macmillan, 1977.
2. Liepmann H. *Drei Aufsätze aus dem Apraxiegebiet.* Berlin: Karger, 1908.
3. Liepmann H. Apraxie. *Ergeb Gesamten Medizin* 1920; 1:516–543.
4. Roy EA, Square PA. Common considerations in the study of limb, verbal, and oral apraxia. In: Roy EA, ed. *Neuropsychological studies of apraxia and related disorders.* Amsterdam: North-Holland, 1985;111–161.
5. De Renzi E. Apraxia. In: Boller F, Grafman J, eds. *Handbook of neuropsychology.* Amsterdam: Elsevier, 1989:245–263.
6. De Renzi E, Faglioni P, Sorgato P. Modality-specific and supramodal mechanisms of apraxia. *Brain* 1982; 105:301–312.
7. Rothi LJG, Heilman KM. *Apraxia: the neuropsychology of action.* Hove, UK: Psychology Press, 1997.
8. Liepmann H. The left hemisphere and action (translated from: *Münchener Medizinische Wochenschrift* 1905; 49:22–26). The University of Western Ontario, 1980.
9. Liepmann H, Maas O. Ein Fall von linksseitiger Agraphie und Apraxie bei rechtsseitiger Lähmung. *Monatsschrift für Psychiatrie und Neurologie* 1907;10: 214–227.
10. Freund HJ. The apraxias. In: Asbury AK, McKhann GM, McDonald WJ, eds. *Diseases of the nervous system. Clinical neurobiology,* 2nd ed. Philadelphia: WB Saunders, 1992:751–767.
11. Morlass J. *Contribution á l'étude de l'apraxie.* Paris: Amédee, Legrand, 1928.
12. De Ajuriaguerra J, Hecaen H, Angelergues R. Les apraxies. Variétés cliniques et latéralisation lésionelle. *Rev Neurol* 1960;102:66–94.
13. Kolb B, Milner B. Performance of complex arm and facial movements after local brain lesions. *Neuropsychologia* 1981;19:491–503.
14. De Renzi E, Faglioni P, Lodesani M, et al. Performance of left brain-damaged patients on imitation of single movements and motor sequences. Frontal and parietal-injured patients compared. *Cortex* 1983;19: 333–343.
15. Faglioli P, Basso A. Historical perspectives on neuroanatomic correlates of limb apraxia. In: Roy EA, ed. *Neuropsychological studies of apraxia and related disorders.* Amsterdam: North-Holland, 1985:3–44.
16. Heilman KM, Rothi LJG, Valenstein E. Two forms of ideomotor apraxia. *Neurology* 1982;32:342–346.

17. Heilman KM, Rothi LJG. Apraxia. In: Heilman KM, Valenstein E, eds. *Clinical neuropsychology.* New York: Oxford University Press, 1985:131–150.
18. Rothi LJG, Ochipa C, Heilman KM. A cognitive neuropsychological model of limb praxis. *Cog Neuropsychol* 1991;8:443–458.
19. Geschwind N. Disconnection syndromes in animals and man. *Brain* 1965;88:237–294, 585–644.
20. Leiguarda R, Marsden CD. Limb apraxias: higher-order disorders of sensorimotor integration (review). *Brain* 2000;123:860–879.
21. Freeman RB Jr. The apraxias, purposeful motor behavior, and left hemisphere function. In: Prinz W, Saunders AF, eds. *Cognition and motor processes.* New York: Springer-Verlag, 1984:29–50.
22. Haaland KY, Flaherty D. The different types of limb apraxia errors made by patients with left vs. right hemisphere damage. *Brain Cognit* 1984;3:370–384.
23. Rapcsak SZ, Ochipa C, Beeson P, et al. Praxis and the right hemisphere. *Brain Cognit* 1993;23:181–202.
24. Schnider A, Hanlon RE, Alexander DN, et al. Ideomotor apraxia: behavioural dimensions and neuroanatomic basis. *Brain Language* 1997;58:125–136.
25. Rushworth MFS, Nixon PD, Wade DT, et al. The left hemisphere and the selection of learned actions. *Neuropsychologia* 1998;36:11–24.
26. Rushworth MFS, Nixon PD, Renowden S, et al. The parietal cortex and motor attention. *Neuropsychologia* 1997b;35:1261–1273.
27. Kimura D, Archibald Y. Motor functions of the left hemisphere. *Brain* 1974;97:337–335.
28. Kimura D. Left hemisphere control of oral and brachial movements and their relation to communication. *Phil Trans R Soc Lond Biol Sci* 1982;298: 135–149.
29. Haaland KY, Harrington DL. Hemispheric asymmetry of movement. *Curr Op Neurobiol* 1996;6:796–800.
30. Poizner H, Mack L, Verfaellie M, et al. Three-dimensional computergraphic analysis of apraxia. *Brain* 1990;113:85–101.
31. Poizner H, Clark MA, Merians AS, et al. Joint coordination deficits in limb apraxia. *Brain* 1995;118: 227–242.
32. Clark MA, Merians AS, Kothari A, et al. Spatial planning deficits in limb apraxia. *Brain* 1994;117: 1093–1106.
33. Rapcsak SZ, Ochipa C, Anderson KA, et al. Progressive ideomotor apraxia: evidence for a selective impairment of the action production system. *Brain Cogn* 1995;27:213–236.
34. Leiguarda R, Starkstein SE. Apraxia in the syndromes of Pick complex. In: Kertesz A, Muñoz DG, eds. *Pick's disease and Pick complex.* New York: Wiley-Liss, 1998:129–143.
35. Poizner H, Merians AS, Clark MA, et al. Left hemispheric specialization for learned, skilled, and purposeful action. *Neuropsychology* 1998;12:163–182.
36. Charlton J, Roy EA, Marteniuk RG, et al. Disruption to reaching in apraxia. *Soc Neurosci Abstr* 1988;14: 1243.
37. Caselli RJ, Stelmach GE, Caviness JV, et al. A kinematic study of progressive apraxia with and without dementia. *Mov Disord* 1999;14:276–287.
38. Leiguarda R, Merello M, Balej J. Apraxia in corticobasal degeneration. In Litvan I, Goetz C, Lang A, eds. *Corticobasal degeneration and related disorders.* Philadelphia: Lippincott Williams & Wilkins, 2000: 103–121.
39. Binkofski F, Dohle CM, Posse S, et al. Human anterior intraparietal area subserves prehension: a combined lesion and functional MRI activation study. *Neurology* 1998;50:1253–1259.
40. Binkofski F, Buccino G, Posse S, et al. A fronto-parietal circuit for object manipulation in man. *Eur J Neurosci* 1999;11:3276–3786.
41. Winstein CJ, Pohl PS. Effects of unilateral brain damage on the control of goal-directed hand movements. *Exp Brain Res* 1995;105:163–174.
42. Fisk JD, Goodale MA. The effects of unilateral brain damage on visually guided reaching: hemispheric differences in the nature of the deficit. *Exp Brain Res* 1988;72:425–435.
43. Haaland KY, Harrington DL. Hemispheric control of the initial and corrective components of aiming movements. *Neuropsychologia* 1989;27:961–969.
44. Haaland KY, Harrington DL, Knight RT. Spatial deficits in ideomotor limb apraxia. A kinematic analysis of aiming movements. *Brain* 1999;122:1169–1182.
45. Jenkins IH, Brooks DJ, Nixon PD, et al. Motor sequence learning: a study with positron emission tomography. *J Neurosci* 1994;14:3775–3790.
46. Grafton ST, Hazeltine E, Ivry R. Functional mapping of sequence learning in normal humans. *J Cog Neurosci* 1995;7:497–510.
47. Catalan MJ, Honda M, Weeks R, et al. The functional neuroanatomy of simple and complex sequential finger movements: a PET study. *Brain* 1998;121: 253–264.
48. Miyachi K, Hikosak O, Miyashita K, et al. Differential roles of monkey striatum in learning of sequential hand movement. *Exp Brain Res* 1997;115:1–5.
49. Luria AR. *Higher cortical function in man,* 2nd ed. New York: Basic Books, 1980.
50. Benecke R, Rothwell JC, Dick JPR, et al. Disturbance of sequential movements in patients with Parkinson disease. *Brain* 1987;110:361–379.
51. Harrington DL, Haaland KY. Motor sequencing with left hemisphere damage. Are some cognitive deficits specific to limb apraxia? *Brain* 1992;115:857–874.
52. Halsband U, Ito N, Tanji J, et al. The role of premotor cortex and the supplementary motor area in the temporal control of movement in man. *Brain* 1993;116: 243–266.
53. Jason G. Hemispheric asymmetries in motor function. Left hemisphere specialization for memory but not performance. *Neuropsychologia* 1983;21:35–46.
54. Roy EA, Square PA. Neuropsychology of movement sequencing disorders and apraxia. In: Zaidel DW, ed. *Neuropsychology.* San Diego: Academic Press, 1994: 183–218.
55. Weiss PH, Dohle C, Binkofski F, et al. Motor impairment in patients with parietal lesions: disturbances of meaningless arm movement sequences. *Neuropsychologia* 2001;39:397–405.
56. Pick A. *Studien über motorische Apraxie und ihre nahestehende Erscheinungen:ihre Bedeutung in der Symptomatologie Psychopathologischer Symptomenkomplexe.* Leipzig: Deuticke, 1905.
57. Hécaen H. *Introduction à la neuropsychologie.* Paris Larousse, 1972.

58. Poeck K. Ideational apraxia. *J Neurol* 1983;230:1–5.
59. Denny-Brown D. The nature of apraxia. *J Nerv Mental Dis* 1958;126:9–32.
60. Ochipa C, Rothi LJG, Heilman KM. Conceptual apraxia in Alzheimer's disease. *Brain* 1992;115: 1061–1071.
61. Heilman KM, Maher LH, Greenwald L, et al. Conceptual apraxia from lateralized lesions. *Neurology* 1997;49:457–464.
62. Foundas A, Macauley BL, Raymer AM, et al. Ecological implications of limb apraxia: evidence from mealtime behavior. *J Int Neuropsychol Soc* 1995;1:62–66.
63. Poeck K, Lehmkuhl G. Das Syndrom der ideatorischen Apraxie und seine Lokalisation. *Nervenarzt* 1980;51:217–225.
64. De Renzi E, Lucchelli F. Ideational apraxia. *Brain* 1988;111:1173–1185.
65. Goldenberg G, Hagmann S. Tool use and mechanical problem solving in apraxia. *Neuropsychologia* 1998;36:581–589.
66. Ochipa C, Rothi LJ, Heilman KM. Conduction apraxia. *J Neurol Neurosurg Psychiatry* 1994;57: 1241–1244.
67. Mehler MF. Visuo-imitative apraxia (abstract). *Neurology* 1987;34:129.
68. Goldenberg G, Hagmann S. The meaning of meaningless gestures: a study of visuo-imitative apraxia. *Neuropsychologia* 1997;35:333–341.
69. Merians, A.S., Clark, M., Poizner, H., et al. Visual-imitative dissociation apraxia. *Neuropsychologia* 1997; 35:1483–1490.
70. Goldenberg G. Imitation and matching of hand and finger postures. *NeuroImage* 2001;14:S132–136.
71. Decety J, Grezes J, Costes N, et al. Brain activity during observation of action. Influence of action content and subject's strategy. *Brain* 1997;120:1763–1777.
72. Grèzes J, Costes N, Decety J. The effects of learning and intention on the neural network involved in the perception of meaningless actions. *Brain* 1999;122: 1875–1887.
73. Hermsdörfer J, Goldengerg G, Wachsmuth C, et al. Cortical correlates of gesture processing. Clues to the cerebral mechanisms underlying apraxia during the imitation of meaningless gestures. *NeuroImage* 2001; 14:149–161.
74. Iacoboni M, Woods RP, Brass M, et al. Cortical mechanisms of human imitation. *Science* 1999;286: 2526–2528.
75. Hermsdörfer J, Mai N, Spatt J, et al. Kinematic analysis of movement imitation in apraxia. *Brain* 1996;119: 1575–1586.
76. Haaland KY, Harrington DL, Knight RT. Neural representation of skilled movement. *Brain* 2000;123: 2306–2313.
77. Halsband U, Schmitt J, Weyers M, et al. Recognition and imitation of pantomimed motor acts after unilateral parietal and premotor lesions: a perspective on apraxia. *Neuropsychologia* 2001;39:200–216.
78. Goldenberg G. Matching and imitation of hand and finger postures in patients with damage in the left or right hemispheres. *Neuropsychologia* 1999;37: 559–566.
79. Ferro JM, Martins I, Mariano G, et al. CT scan correlates of gesture recognition. *J Neurol Neurosurg Psychiatry* 1983;46:943–952.
80. Varney N, Damasio H. Locus of lesion in impaired pantomime recognition. *Cortex* 1987;23:699–703.
81. Rothi LJG, Heilman KM, Watson RT. Pantomime comprehension and ideomotor apraxia. *J Neurol Neurosurg Psychiatry* 1985;48:207–210.
82. Rothi LJG, Mack L, Heilman KM. Pantomime agnosia. *J Neurol Neurosurg Psychiatry* 1986;49:451–454.
83. Wang L, Goodglass H. Pantomime, praxis and aphasia. *Brain Language* 1992;42:402–418.
84. di Pellegrino G, Fadiga L, Fogassi L, et al. Understanding motor events: a neurophysiologic study. *Exp Brain Res* 1992;91:176–180.
85. Carey DP, Perrett D, Oram M. Recognizing, understanding and reproducing action. In: Boller F, Grafman J, eds. *Handbook of neuropsychology,* vol. 11. Amsterdam: Elsevier, 1997:111–129.
86. Fogassi L, Gallese V, Fadiga L, et al. Neurons responding to the sight of goal-directed hand/arm actions in the parietal area PF (7b) of the macaque monkey. *Soc Neurosci Abstr* 1998;24:654.
87. Matelli M, Luppino G, Rizzolatti G. Patterns of cytochrome oxidase activity in the frontal agranular cortex of the macaque monkey. *Behav Brain Res* 1985; 18:125–136.
88. Seltzer B, Pandya D. Frontal lobe connections of the superior temporal sulcus in the rhesus monkey. *J Comp Neurol* 1989;281:97–113.
89. Rizzolatti G, Luppino G, Matelli M. The organization of the cortical motor system: new concepts. *Electroencephalogr Clin Neurophysiol* 1998;106:283–296.
90. Rizzolatti G, Fadiga L, Matelli M, et al. Localization of grasp representations in humans by positron emission tomography. 1. Observation versus execution. *Exp Brain Res* 1996;111:246–252.
91. Grafton ST, Arbib MA, Fadiga L, et al. Localization of grasp representations in humans by positron emission tomography. 2. Observation compared with imagination. *Exp Brain Res* 1996;112:103–111.
92. Bonda E, Petrides M, Ostry D, et al. Specific involvement of human parietal systems and the amygdala in the perception of biological motion. *J Neurosci* 1996; 16:3737–3744.
93. Buccino G, Binkofski F, Fink GR, et al. Action observation activates premotor and parietal areas in a somatotopic manner: an fMRI study. *Eur J Neurosci* 2001; 13:400–404.
94. Sirigu A, Daprati E, Pradat-Diehl P, et al. Perception of self-generated movement following left parietal lesion. *Brain* 1999;122:1867–1874.
95. Jeannerod M. The representing brain. Neural correlates of motor intention and imagery. *Behav Brain Sci* 1994;17:187–245.
96. Decety J, Perani D, Jeannerod M, et al. Mapping motor representations with PET. *Nature* 1994;371:600–602.
97. Stephan KM, Fink GR, Passingham RE, et al. Functional anatomy of the mental representation of upper extremity movements in healthy subjects. *J Neurophysiol* 1995;73:373–386.
98. Roth M, Decety J, Raybaudi M, et al. Possible involvement of primary motor cortex in mentally simulated movement. A functional magnetic resonance imaging study. *NeuroReport* 1996;7:1280–1284.
99. Gerardin E, Sirigu A, Lehériay S, et al. Partially overlapping neural networks for real and imagined hand movements. *Cereb Cortex* 2000;10:1093–1104.

100. Decety J, Boisson D. Effects of brain and spinal cord injuries on motor imagery. *Eur Arch Psychiatry Clin Neurosci* 1990;240:39–43.
101. Sirigu A, Cohen L, Duhamel JR, et al. Congruent unilateral impairments for real and imagined hand movements. *NeuroReport* 1995;6:997–1001.
102. Jeannerod M, Decety J. Mental motor imagery: a window into the representational stages of action. *Curr Opin Neurobiol* 1995;5:727–732.
103. Ochipa C, Rapcsak SZ, Maher L, et al. Selective deficit of praxis imagery in ideomotor apraxia. *Neurology* 1997;49:474–480.
104. Sirigu A, Cohen L, Duhamel JR, et al. A selective impairment of hand posture for objects utilization in apraxia. *Cortex* 1995;31:41–55.
105. Jeannerod M. Neural simulation of action: a unifying mechanism for motor cognition. *NeuroImage* 2001;14:S103–S109.

16

Interactions Between the Dorsal and Ventral Streams of Visual Processing

Melvyn A. Goodale and Angela M. Haffenden*

*CIHR Group on Action and Perception, Department of Psychology, University of Western Ontario, London, Ontario, Canada; *Psychology Department, Foothills Medical Centre, Calgary, Alberta, Canada*

INTRODUCTION

Vision not only allows us to make sense of the world at a distance, it also provides exquisite control over the movements we make in that world. Over the last 10 years, a large body of work, ranging from studies of neurologic patients to single-cell recordings in monkeys, has implicated different regions of the primate cerebral cortex in the mediation of these two rather distinct functions of vision (1–3). The so-called ventral stream of visual projections, which runs from primary visual cortex to inferotemporal cortex, has been shown to play a critical role in transforming visual information into perceptual representations that embody the enduring characteristics of objects and their relations. Such representations, which are coded in scene-based coordinates, enable us to identify objects, attach meaning and significance to them, and establish their causal relations; operations that are essential for accumulating knowledge about the world. In contrast, a separate dorsal stream of visual projections, which runs from primary visual cortex to the posterior parietal cortex, has been shown to use moment-to-moment information about the location and disposition of objects with respect to the observer for the control of goal-directed actions. However, even though the two systems transform visual information in quite different ways, they work together in the production of adaptive behavior. In general terms one could say that the selection of appropriate goal objects and the action to be performed depends on the perceptual machinery of the ventral stream, whereas the execution of a goal-directed action is carried out by dedicated online control systems in the dorsal stream. However, the functioning of the two systems is even more integrated than this.

Consider what must happen when you reach out and pick up a cup of coffee at the dinner table. You must first identify your cup of coffee—not only find a cup, but determine that it is indeed yours and not your dining companion's. To do this, the visual scene is parsed into distinct objects, and the particular cup is identified—presumably on the basis of its distinctive features and/or its relative location with respect to other objects on the dinner table (and your memory of where you last left it). But to pick up the cup, you must localize the cup (or more specifically, its handle) with respect to the hand you plan to use to grasp it. You must also compute the size of the handle and its orientation with respect to your fingers. At the same time, the initial lift and grip forces you apply to the cup's handle are planned on the basis of previous information or expectations about how full the cup might be as well as about the friction coefficients and compliance of the material from which the cup is made. As we hope to show in this chapter, although each stream (and its related networks) takes primary responsibility for different aspects of the planning and execution of this deceptively simple act, the interaction between the two systems is complex and intimate.

THE EMERGENCE OF THE TWO-VISUAL-SYSTEMS MODEL

Nearly 20 years ago, Ungerleider and Mishkin (4) identified two broad "streams" of projections from primary visual cortex to extrastriate visual areas in the macaque cerebral cortex: a ventral stream projecting eventually to the inferotemporal cortex and a dorsal stream projecting to the posterior parietal cortex. Recent evidence from functional magnetic imaging (fMRI) studies suggests that the visual projections from primary visual cortex to the temporal and parietal lobes in the human brain are organized in a similar fashion (for review, see refs. 5 and 6).

Ungerleider and Mishkin (4) argued that the two streams of visual processing play different but complementary roles in the processing of incoming visual information. According to their original account, the ventral stream plays a critical role in the identification and recognition of objects, whereas the dorsal stream mediates the localization of those same objects—a distinction that has sometimes been characterized as one between "what" versus "where." Support for this idea came from work with monkeys. Lesions of inferotemporal cortex in monkeys produced deficits in their ability to discriminate between objects on the basis of their visual features but did not affect their performance on a spatially demanding "landmark" task, in which the animal had to determine which of two goals was closer to a distinctive landmark (7,8). Conversely, lesions of the posterior parietal cortex produced deficits in performance on the landmark task but did not affect object discrimination learning. Although the evidence for the Ungerleider and Mishkin proposal initially seemed quite compelling, recent findings from a broad range of studies in both humans and monkeys has forced a reinterpretation of the division of labor between the two streams. (For a discussion of these issues and a critique of the landmark task as a probe of dorsal-stream function, see refs. 3 and 9.)

Some of the most telling evidence against a simple "what" versus "where" distinction has come from studies with neurologic patients. Since the pioneering work of Bálint (10), it has been known that patients with damage to the posterior parietal cortex have difficulty reaching in the correct direction to objects placed in different positions in the visual field, even though they have no difficulty reaching out and grasping different parts of their body indicated by the experimenter. This deficit in visually guided behavior, which Bálint called "optic ataxia," has often been interpreted as a failure of spatial vision (although interestingly not by Bálint himself; see ref. 11). Two other sets of observations in these patients, however, suggest a rather different interpretation. First, patients with damage to this region of cortex often show an inability to rotate their hand or open their fingers properly to grasp an object placed in front of them, even when it is always placed in the same location (12,13). Second, these same patients are able to describe the orientation, size, shape, and even the relative spatial location of the very objects they are unable to grasp correctly (12,13). Clearly, this pattern of deficits and spared abilities cannot be explained by appealing to a general deficit in spatial vision.

Other patients, in whom the brain damage appears to involve ventral rather than dorsal stream structures, show the complementary pattern of deficits and spared visual abilities. Such patients have great difficulty recognizing common objects on the basis of their visual appearance, but have no problem grasping objects placed in front of them. Consider, for example, the patient DF, a young woman who suffered brain damage as a result of anoxia from carbon monoxide poisoning (14). The most significant damage was in the posterior regions of the brain, particularly in ventrolateral occipital cortex (in areas corresponding to the human homolog of the ventral stream). DF developed a profound visual form agnosia and is unable to identify objects visually on the basis of their shape (although her ability to perceive color and visual texture is essentially normal). Even today, more than 10 years after the accident, she cannot discriminate between simple geometric shapes such as a triangle or circle or distinguish horizontal from vertical lines. However, despite her inability to indicate the size, shape, and orientation of an object placed before her, she shows normal preshaping and rotation of her hand when she reaches out to grasp the object (15). Appealing to a general deficit in "object vision" does not help us to understand her problem. In her case, she is able to use visual information about the location, size, shape, and orientation of objects to control her grasping movements (and other visually guided movements) despite the fact that she is unable able to perceive those same object features (for a review, see refs. 3 and 16).

It was the double dissociation between deficits and spared abilities in these two kinds of patients that led David Milner and me to propose a new division of labor between the dorsal and ventral streams of visual processing (1). It seemed to us that the most parsimonious way to explain the dissociation was to posit that processing in the ventral stream delivers our perceptual representations of objects, whereas processing in the dorsal stream mediates the visual control of actions directed at those objects. This account was also consistent with what was known at the time about the properties of single neurons in the ventral and dorsal streams of the monkey (3). Since then, evidence for the perception-action theory has continued to accumulate, including evidence from a number of neuroimaging studies in humans (for review, see refs. 5 and 16). Nevertheless, the fact that there are two distinct visual systems, one for the perception of objects and one for object-directed action, begs the question as to why there are two streams in the first place. Why is there not one general-purpose visual system to serve both perceptual representation and the visual control of action? The answer to this question can be found in the differing demands of perception on the one hand and action on the other.

DIFFERENT METRICS FOR PERCEPTION AND ACTION

Perception, in the way I am using the term here, refers to the internal representation of the world that serves as a foundation for a broad range of cognitive operations that contribute to our knowledge of the world. Indeed, the cognitive operations are themselves intimately involved in the construction of the representations on which they operate. However, even though our perception of the world certainly appears remarkably rich and detailed, it is becoming increasingly apparent that much of this perceptual representation is "virtual" and is derived from memory rather than visual input (17–19). Although perception allows us to think about the world and plan our actions, as we see later in the chapter, it offers a poor metric for the actual control of the actions that we might wish to carry out (for review, see refs. 20 and 21). The absence of action-relevant metrics in the perceptual representation is a direct consequence of a tradeoff between the costs of computing those metrics and the costs of computing all the object-based features of the scene before us. If the perceptual system were to attempt to deliver the real metrics of all the objects the visual array in observer-based frames of reference, the computational load would be astronomical. The solution that perception appears to have adopted is to use world-based coordinates in which the real metrics of that world need not be computed. Only the relative position, orientation, size, and motion of objects with respect on one another are of concern to perception. Such world-based frames of reference are sometimes called allocentric (in contrast to observer-based or egocentric frames of reference). The use of allocentric frames of reference (in which the metric properties of objects are computed with reference to each other in a scene-based frame of reference) means that we can, for example, watch the same scene unfold on television or on a movie screen without being confused by the enormous absolute change in the coordinate frame. It also explains why objects in the real world may appear larger or smaller at times, depending on the size and position of juxtaposed objects in the scene. Indeed, visual size-contrast illusions, such as those commonly illustrated in introductory psychology textbooks, arise in part because of the perceptual system's reliance on relative metrics (for review, see ref. 20).

The demands of the visuomotor system are, however, quite different from those of the perceptual system. Relative metrics and a scene-based frame of reference do not provide the information required to carry out effective visually guided movements, particularly movements directed at objects we have not seen before. To be accurate, actions such as grasping must be finely tuned to the absolute, not the relative, size of the object and to the location and disposition of the object with respect to the observer; that is, they must use an egocentric frame of reference. Moreover, because different actions engage different effectors, the egocentric frame of reference must be specific to the particular motor output required. Directing a saccadic eye movement toward an object in the visual array demands different transformations of visual input to motor output from those required to direct a manual grasping movement to that object. Grasping movements, for example, require that the goal object be coded in "arm-centered" coordinates (22,23) and presumably in hand- and finger-centered coordinates as well (for a review, see ref. 24).

As we have emphasized, the visual system could not possibly provide the absolute metric properties of all the objects in the visual array and continually update the location and disposition of these objects in the many different egocentric frames of reference that would be required for the multitude of actions we are capable of performing. This would be a computational nightmare. Instead, the required transformations from vision to action are carried out just before the action is performed. Moreover, because these computations are based on the absolute metric properties of the target—and not on its relative size, location, orientation, and shape with respect to other objects in the scene—the scaling of the actions directed at the target are not sensitive to many of the visual illusions that affect perceptual judgments of object properties (for review, see ref. 25). The insensitivity of vision-for-action to such illusions has been shown in a number of experiments in which, for example, targets have been placed in displays that generate size-contrast illusions (26–30). Although observers show robust perceptual illusions in these experiments—even in matching tasks in which they are required to open their index finger and thumb to match the perceived size of the target—their grip aperture is correlated with the real size of the target when they reach out to pick it up.

The time scale over which vision-for-perception operates is several orders of magnitude longer than that of vision-for-action. We can remember things we have seen for minutes, days, even years—and these memories can affect later perceptual processing. However, the "just-in-time" computations carried out by the visuomotor system are not stored for longer than a few seconds (if that). Indeed, grasping movements that are initiated a few seconds after the goal object disappears from view are quite different in their spatiotemporal organization than grasping movements that are directed toward a visible target—and appear to use quite different kinds of visual information. For example, grasping movements made after a delay are scaled to the relative not the absolute size of the target object and, unlike grasping movements made in real time, are sensitive to many common size-contrast illusions—findings that suggest that grasping movements made after a delay use visual information about the object that was gleaned from perceptual not visuomotor networks (for review, see refs. 28 and 29). Recent evidence from our laboratory suggests that the critical factor determining whether or not the dorsal-stream visuomotor modules are engaged when a movement is programmed is not so much the length of the delay but rather the presence (or absence) of the target on the retina (31). We showed that the scaling of a grasping movement can resist a size-contrast illusion only if the target is visible at the moment the signal to initiate the movement is given. This suggests that: (a) the computation of the absolute metrics of the target object and its location and disposition within the relevant egocentric frames of reference requires access to retinal and extraretinal signals about a visible, not a remembered, target; and (b) such computations occur only when the signal (or internal command) is given to initiate the action.

Although it is theoretically possible that a highly sophisticated "general-purpose" visual system could accommodate the different computational demands of perception and action described in the preceding, nature appears to have opted for a different and more modular solution: a dorsal "action" stream and a ventral "perceptual" stream have evolved as relatively separate sets of pathways in the primate cerebral cortex. Nevertheless, the two streams of visual processing evolved together and play complementary roles in the control of behavior. In some ways, the limitations of one system are the strengths of the other. As we have just seen, the ventral perception system delivers a rich and detailed representation of the world, but the metrics of the world with respect to the organism are not well specified. In contrast, the dorsal action system delivers accurate metric information in the required egocentric coordinates but these computations are spare and evanescent. How do the two streams interact in the production of everyday behavior?

INTERACTIONS BETWEEN THE VENTRAL AND DORSAL STREAMS: A TELE-ASSISTANCE MODEL

A useful analogy for understanding the different contributions of the dorsal and ventral stream to visually guided behavior can be found in robotic engineering. That analogy is tele-assistance, one of the general schemes that have been devised whereby human operators can control robots working in hostile environments, such as in the crater of a volcano or on the surface of another planet (32,33). In tele-assistance, a human operator identifies or "flags" the goal object, such as an interesting rock on the surface of Mars, and then uses a symbolic "language" to communicate with a semiautonomous robot that actually picks up the rock and retains it for analysis.

Tele-assistance is much more flexible than completely autonomous robotic control, which is limited to the working environment for which the robot has been programmed and cannot cope easily with novel events. Tele-assistance is also more efficient than teleoperation, in which the movements or instructions of the human operator (the master) are simply reproduced by the robot (the slave). Tele-operation (i.e., the human operator) cannot cope with sudden changes in scale or the delay between action and feedback from that action (33). In tele-assistance, the human operator does not need to know the real metrics of the workspace or the timing of the movements made by the robot; instead, the human operator has the task of identifying a goal and specifying an action on that goal in general terms. Once this information is communicated to the semiautonomous robot, the robot can use its on-board range finders and other sensing devices to work out the required movements for achieving the specified goal. In short, tele-assistance combines the flexibility of teleoperation with the precision of autonomous robotic control.

The interaction between the ventral and dorsal streams is an excellent example of the principle of tele-assistance, but in this case instantiated in biology. The perceptual-cognitive systems in the ventral stream, like the human operator in tele-assistance, identify different objects in the scene—using a system that delivers a representation that is replete with detail but is metrically imprecise. When a particular goal object has been flagged, dedicated visuomotor networks in the dorsal stream (in conjunction with related circuits in premotor cortex, basal ganglia, and brainstem) are activated to perform the desired motor act. Thus, the networks in the dorsal stream, with their precise egocentric coding of the location, size, orientation, and shape of the goal object, are like the robotic component of tele-assistance. Both systems are required for purposive behavior—one system to select the goal object from the visual array, the other to carry out the required metric computations for the goal-directed action.

In making the analogy between the two visual systems and tele-assistance, we do not wish to underestimate the progress that is being made in the design of autonomous robots. It is clear that in the not-too-distant future, the role of the "human operator" will be incorporated into the design of the machine. However, when this happens, it is unlikely that this intelligent robot will be given a general-purpose visual system, one that both recognizes objects in the world and guides the robot's movements. As we have argued, the computational demands of object recognition and scene analysis are simply incompatible with the computational demands of visuomotor control. A much more effective design for such a robot would be to emulate the division of labor between the ventral and dorsal visual streams that has evolved in the primate brain.

THE CONTRIBUTION OF THE VENTRAL STREAM TO MOTOR PROGRAMMING: THE CASE OF VISUALLY GUIDED GRASPING

The Computation of Initial Grip and Lift Forces

Perception contributes quite directly to certain aspects of motor programming, notably those that depend on information that cannot be derived directly from the retina. One example is the initial grip and lift forces that are applied to an object when we pick it up. Unlike the size, location, orientation, or even shape of an object, which can be computed from the projected image on the retina, the object's mass, compliance, and surface friction can be gleaned only through experience. Yet, when people pick up familiar objects, the force they apply to the object is scaled appropriately for the object's weight (and other characteristics) from the moment their hand makes contact, well before any somatosensory feedback is available (34). This means that they must have retrieved information about the object from memory, information that could be activated only when they recognized the object. Such recognition of course would be mediated by visual mechanisms in the ventral, not the dorsal, stream. Just how this information is communicated to the motor system is not understood.

When people reach out and pick up objects they have not encountered before, they have to make an estimate of the object's weight—an estimate that is presumably based on previous experience with similar objects. Thus, when people reach out to pick up objects that "look" heavy, such as objects that appear to be made of metal, they apply more force than they do to objects that look as though they are made of lighter materials, such as polystyrene. Moreover, when subjects are presented with objects of different sizes that are made of the same material, they typically apply more force to the larger object than they do to the smaller objects—even when the objects are actually the same weight (35). When subjects heft such objects, they typically experience what has been called the "size–weight illusion" and conclude that the smaller object is heavier than the larger one. Even though it has been often assumed that the size–weight illusion is a consequence of the mismatch between applied force and the actual weight of the object, it turns out that this is not the case (36). When subjects make repeated lifts of a set of objects that have the same weight but different sizes, they eventually apply the same initial lift force to all the objects. Nevertheless, the size–weight illusion is not at all diminished by this recalibration of lift force. In other words, subjects continue to believe that the smaller object is heavier than the larger one, even though they apply to same initial lift force to both.

Despite this apparent dissociation, there are good reasons to believe that both the calibration of the lift forces and the maintenance of the size–weight illusion depend on object recognition systems in the ventral stream. First, if the objects to which the observers have recalibrated their lift forces are replaced with new objects, which like original objects have different sizes but the same weight, the observers now revert to applying less force to the smaller object in the new set (JR Flanagan, personal communication). Second, the patient DF, who has visual form agnosia, does not show any evidence of a size–weight illusion when she picks up cylindrical objects of different sizes (but the same weight) using a handle attached to the top of the cylinders. In other words, when DF's only information about the size of the object is derived from vision, she correctly reports that the cylinders are the same weight (HC Dijkerman, personal communication). Moreover, unlike normal subjects, DF's grip and lift forces on the initial trials are not correlated with the size of the cylinder; that is, she does not apply greater force to the larger object (37). In contrast, when she picks up the cylinders across their diameter with either her eyes open or closed, she shows the size–weight illusion. In other words, when she has kinesthetic information about the size of the cylinders, she incorrectly perceives the smaller cylinders as weighing more than the larger ones. Taken together, these findings suggest that ventral-stream processing plays an essential role both in computing the required forces for a successful grasp and in generating and maintaining the size–weight illusion.

Another piece of evidence that the estimates of required lift and grip forces depend on perceptual mechanisms in the ventral stream comes from experiments with pictorial illusions. When observers reach out and pick up an object that is placed in different positions on a Ponzo Illusion background, they scale their lift and grip forces to the apparent not the real size of the object (27,38). As can be seen in Figure 16.1, the smaller the object appears to be, the less force they apply. At the same time, the size of the opening between the index finger and thumb as the observer reaches out to grasp the object is scaled to the real size of the object—and is unaffected by changes in the object's apparent size when the object is positioned in different parts of the display. In short, context affects applied force but not grip aperture. However, this is exactly what one would expect because the mass of an object (and thus the required force to pick up that object) can be discovered only through experience, whereas the scaling of the grip aperture can be determined directly from the retinal image (in combination with extraretinal information such as the vergence state of the eyes). In other words, the initial forces are determined by stored information that is activated by perceptual mechanisms in the ventral stream, whereas grip aperture can be computed in an entirely bottom-up fashion by visuomotor mechanisms in the dorsal stream.

Functional Elements of Grasping

Many visually guided actions, particularly manual actions, are directed at objects that have an obvious function. This means that the movements of the hand reflect not only the size, shape, orientation, and location of the target object in these situations, but also what we intend to do with it. For example, when we reach out to pick up a coffee mug with the intention of drinking from it, our hand and fingers adopt a very different posture in flight than they do when we reach out to pick up the mug with the intention of putting it away in the cupboard. Objects with an obvious function often solicit "use-appropriate" hand postures when we pick them up—even if we have no intention of using them at the time. For example, if we are asked to pick up a screwdriver placed in front of us with the handle facing away, we typically turn our hand right around and grasp the handle as if we were about to use the screwdriver, even when we have no intention at that moment of using it.

How are these functional elements of an action selected? Selecting the appropriate action clearly depends on knowing what an object is.

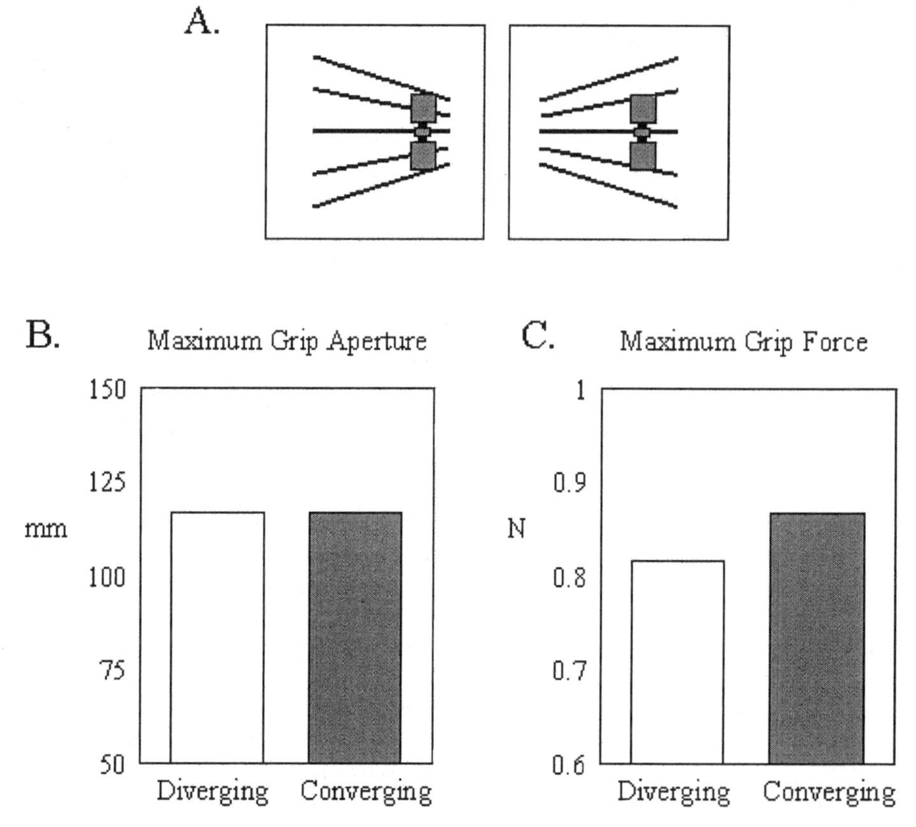

FIG. 16.1. The Ponzo illusion experiment. **A:** Participants reached to cylinders presented against the converging or diverging ends of the Ponzo or "railway track" illusion. The opening of the grasping hand was tracked using an optoelectronic recording system. The cylinders contained force transducers that recorded the grip force exerted on the object when the subject first picked up the object. **B:** Maximum grip aperture was not affected by the illusion. Grip force, however, increased with the apparent size of the object. In other words, more force was used to pick up the object when it was presented against the converging end of the illusion as compared to when it was presented against the diverging end of the illusion. (Adapted from: Jackson SR, Shaw A. The Ponzo illusion affects grip-force but not grip-aperture scaling during prehension movements. *J Exp Psychol Hum Percept Perform* 2000;26:418–423, with permission.)

The perceptual mechanisms in the ventral stream, which are known to play an essential role in object identification, presumably invoke the required posture by virtue of their connections with stored information about the function of the object. The use of semantic processes in selecting the functional elements of actions such as grasping has been shown in an elegant experiment conducted by Creem and Proffitt (39). In this experiment, subjects were presented with a series of tools and implements (e.g., a toothbrush, a frying pan, a screwdriver) with the handles pointed away from them. When the subjects reached out to pick up such objects, they adopted the same awkward, but highly functional hand posture described earlier in which they grasped the object's handle. If, however, they were required to perform a semantically demanding task at the same time (a paired-associates recall task), they no longer exhibited any functional elements in their grasping movement, and simply picked up the object in a metrically efficient but functionally inappropriate fashion. In contrast, when they performed a concurrent visuospatial task that did not tax the semantic system, they continued to pick up the objects in a functionally appropriate manner. To explain these results, Creem and Proffitt (39) quite reasonably argued that the paired-associates task put so many extra demands on the semantic system that its resources could not be properly deployed in selecting the function-appropriate hand postures in the grasping task. The semantic system is intimately connected with ventral-stream perceptual mechanisms that enable the subject to identify the object in the first place. The metric properties of the grasping movement, however, were not affected by this semantically challenging task because the visuomotor systems in the dorsal stream that mediate this aspect of grasping do not depend on the retrieval of semantic information about the object or its properties.

The behavior of the patient DF with functional objects also provides evidence for the contribution of ventral-stream perceptual mechanisms to the selection of functionally appropriate grasp postures. When DF is asked to pick up a man-made object, such as a knife, a screwdriver, or a coffee cup (objects that she cannot identify on the basis of their form), her grasp is perfectly matched to the object's size, shape, and orientation, but shows no indication that she understands its function (unpublished observations). For example, rather than picking up the coffee cup by its handle, she might grasp it deftly, but quite inappropriately, across the top, as though it were a lump of clay. In other words, because her damaged ventral stream is unable to process the cup's shape, DF has no idea what it is and is therefore unable to generate grasping movement that is appropriate for a coffee cup. Nevertheless, her intact visuomotor systems in her dorsal stream can still compute the required metrics to ensure that her grasping movement, however inappropriate, is well formed and efficient. (Of course, if she knows that the object is a cup—from having just handled it—she will adopt a hand posture appropriate for picking up a cup.)

Just how (and where in the brain) the functional aspects of grasping are integrated with the metric scaling of the movements is not yet understood. Studies of patients with damage to the dorsal stream show that functional elements can be generated in the absence of the visuomotor mechanisms that drive the metric scaling, suggesting that the integration of the movements takes place elsewhere in the brain. Consider, for example, the patient AT who has optic ataxia from bilateral lesions of the posterior parietal cortex (40). Even though AT could not configure her grasp to the dimensions of simple solid geometric objects that varied in size from trial to trial but were "neutral" with respect to function, she was able to configure her grasp to familiar objects of the same size and shape, such as a tube of lipstick or a drinking glass. It appears that AT's intact perception of the familiar object's identity allowed her to use stored information about the object's size and shape to form her grasp in flight, information that was not evoked when she was faced with neutral objects.

Evidence from another neurologic patient suggests that there are regions of the posterior parietal cortex that are not directly involved in the metric scaling of movements but play a critical role in mediating the production of a functional grasp. Sirigu et al. (41) describe a patient who demonstrated a striking dissociation between object recognition and functional control of grasping. This patient, LL, had a normal computed tomography scan but showed bilateral hypometabolism in the posterior parietal regions on a positron emission tomography (PET) scan. LL showed appropriate metric scaling for object size, shape, and orientation but could not incorporate functional elements into her grasp, even though she could readily identify the objects being grasped. Thus, she could pick up a tool quite efficiently but the posture of her hand would be quite inappropriate for the use of that tool—even though she knew exactly what the tool was and what it could be used for. Sirigu et al. (41) suggested that the lesion had either created a disconnection between object identity and those motor mechanisms that create the appropriate postures or had interfered with the function of the specialized motor mechanisms that code the range of complex postures associated with different kinds of tools and other familiar objects. Either (or both) of these hypotheses could be correct—as could a number of other possible scenarios. At present, there is little more that can be said about how function and metrics are combined in the generation of purposive action.

There is some evidence that the neural systems mediating the explicit identification of an object and its function are dissociable at some level from the neural systems mediating the functional hand postures such objects elicit. Thus, there are several reports in the literature of patients with profound object agnosia or semantic dementia who continue to exhibit relative preservation of object-appropriate hand postures (42–44). Sirigu et al. (42), for example, describe a patient who had developed multisensory associative agnosia from bilateral lesions of the temporal lobes involving medial, polar, and anterior inferotemporal structures. Even though this patient could not describe the function of an object or the context in which it would normally be used, he nevertheless could place his hand on the object in a posture that was appropriate to its use. This result appears to suggest that even though different postures come to be associated with different categories of objects, these learned associations may not require an explicit identification of the object's identity and purpose. Other investigators, however, have offered a rather different interpretation of this dissociation. On the basis of experiments with patients who have developed semantic dementia from temporal lobe atrophy, Hodges et al. (45) concluded that the ability of these patients to use everyday objects appropriately despite being unable to identify them is a consequence of their relatively spared ability to solve mechanical problems. Whether the functional aspects of object-directed actions are mediated by entirely different semantic representations from those mediating verbal descriptions and other cognitive operations or whether action and verbal outputs depend on a single semantic representation of the object is an issue well beyond the scope of this chapter. (Interested readers are directed to a recent review by Patterson and Hodges; ref. 46.) However semantics might be linked to functional hand postures, the production of such actions appears to depend on ventral stream processing.

Taken together, the evidence from studies with both normal observers and neurologic patients suggests that, for objects with known functions, the configuration of a visually guided grasping movement is a joint product of stored knowledge about the object and bottom-up metric computations of the target object's orientation, shape, size, and location. The stored knowledge about the object is activated via the ventral stream of visual processing, whereas the metric information is obtained from the dorsal stream of visual processing. In the case of novel objects, for which we have no stored information about functional postures, our grasp is largely guided by the metrics of the object and thus primarily by processing in the dorsal stream.

Contributions of Learned Associations to the Metric Scaling of Grasping Movements

Our perception of an object's features, particular its color and other surface characteristics, contribute nothing to the metric scaling of a grasping movement in flight, unless we have encountered the object (or similar objects) before. As it happens, however, novel objects do not stay novel for long. We typically pick up the same toothbrush and the same coffee cup every morning. As a consequence, "nonmetric" features of an object, such as its color, can become associated with metric properties such as its size and shape—and these reliable associations could then be used to program grip aperture. In other words, just as we apply the appropriate forces to pick up familiar objects, we can also use learned associations to scale our grasp to familiar objects.

Using perceptual information to recover a motor routine theoretically could reduce the need for bottom-up computations of the metrics of the target object, thereby increasing the overall efficiency of the programming. In other words, the incorporation of stored parameters such as grip aperture could reduce considerably the computational load on the visuomotor system. Thus, even though the grasp would have to be fine-tuned and adjusted to the particular situation, information retrieved from memory could provide the initial parameters for selecting the posture and scaling the grasp. In short, perceived object features such as color, which have no inherent link to the control of goal-directed actions, could provide a cue for the programming of such control through association.

We recently tested this idea in an experiment in which we established a relationship between the color (red or yellow) of an object and the size category (large or small) to which that object belonged (47). It was expected that in a later test, the perceived size of colored objects that were halfway in size between the sets of large and small objects would be based on relative size comparisons made with respect to the matching color category. The expected comparisons and the resulting size contrasts are illustrated in Figure 16.2. These "perceptual" effects are reminiscent of a long history of experimental and theoretic work on "adaptation level" by Helson (48). In a typical experiment in the adaptation-level tradition, previous exposure to a series of stimuli, heavy weights for example, would bias later judgments of the weight of other objects. Lighter objects that were presented after exposure to a number of heavier objects, would now be perceived as even lighter. If the same test objects had been presented after exposure to a series of extremely light objects, however, they would now feel a good deal heavier.

The particular twist in our experiment was that the subjects were adapted with objects anchored at the opposite ends of the size continuum at the same time and the size categories to which the objects belonged were color-coded. Thus, two "relative size-contrast effects" could occur simultaneously. Large and small groups of "key" blocks established the color cue to size. We presented half the subjects with a random series of large yellow and small red square wooden key blocks, and half with a series of key blocks in which the pairing of color and size was reversed. The presence of a relative-size contrast effect was tested by later inserting trials in which "test" blocks were presented that were halfway in size between the large and small key blocks. One test block was the same color as the set of large blocks and one was the same color as the set of small blocks. The direction of the relative size-contrast effect with each test block was expected, therefore, to depend on

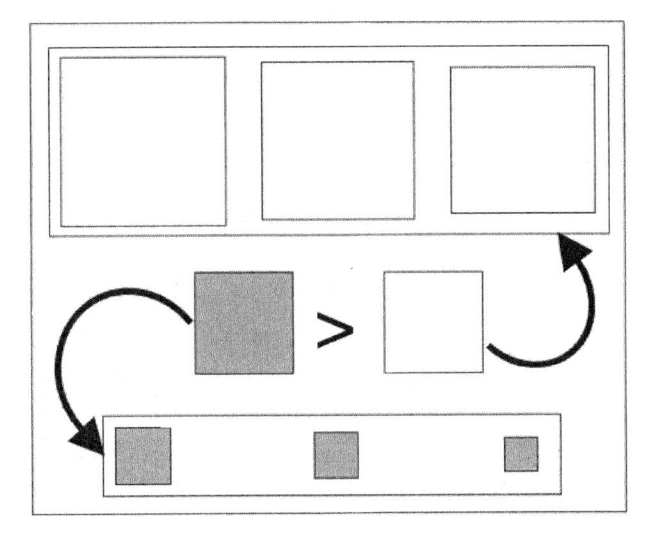

FIG. 16.2. A schematic illustration of the expected change in perceived size for probe blocks halfway in size between a set of large blocks and set of small blocks that are coded for size by their color. Relative size comparisons made within each color group would result in the red (dark-colored) probe block being perceived as being larger in size than the yellow (light-colored) probe block even though they are physically identical in size. (Adapted from: Haffenden AM, Goodale MA. The effect of learned perceptual associations on visuomotor control varies with kinematic demands. *J Cogn Neurosci* 2000;12:950–964, with permission.)

whether it was the same color as the large or the small key blocks. Thus, we could use the size-contrast effect to probe whether or not a learned perceptual association between an arbitrary cue and object size would affect anticipatory grip scaling when subjects picked up the test blocks.

We tested the effect of the learned association on subjects' perception of object size with a manual estimation task in which subjects were asked to indicate the width of each block in the series by opening their index finger and thumb a matching amount. The size of the opening between the index finger and thumb was measured using optoelectronic recording of small infrared light-emitting diodes (IREDs) attached to the tips of both digits. As Figure 16.3 illustrates, we found that a size-contrast effect was indeed operating in the subjects' estimates of the widths of the two test blocks: Subjects gave larger size estimates for the test block that was the same color as set of small key blocks than they did for the test block that was the same color as the set of large key blocks—even though the two test blocks were exactly the same size. Illusions of this kind, in which the sizes of similar objects are implicitly compared (in an automatic and obligatory fashion), have been studied by a number of investigators. Using the Ebbinghaus illusion, for example, Coren and Enns (49) found that the magnitude of the size-contrast effect was strongly influenced by the similarity between the target and surrounding stimuli: The greater the similarity, the larger the size-contrast effect. Choplin and Medin (50) have since argued that such effects are not due to conceptual similarity, but rather to the similarity between the perimeters of the inducing objects and the test objects. They postulate that the visual system "uses features that are diagnostic of, rather than isomorphic with, conceptual categories" (p. 11. ref. 50). Of course, in our experiment, color was diagnostic of the size category to which the object belonged and provided a salient cue by which to group the blocks. Thus, the conspicuous color-coding of the large and small categories of objects apparently resulted in the formation of two distinct groups, and the test blocks were preferentially compared to the size group that they were most similar to. This kind of category assignment and subsequent comparison is a quintessential example of the kinds of operations that are carried out in the perceptual domain—the domain of the ventral stream and its connections with long-term memory mechanisms in the medial temporal regions and frontal cortex.

To measure the effect of the learned association between color and size on visuomotor control, a different group of subjects was tested on a grasping task in which they were asked to reach out and pick up the blocks using their index finger and thumb. The maximum grip aperture in flight (typically achieved 70% of the way through the trajectory) was measured using the same opto-electronic recording system that was used to measure the manual estimates in the perceptual judgment task. Because the subjects had an opportunity to take advantage of the reliable relationship between object color and object size to scale their grip aperture in flight, we expected to see the same relative-size contrast effect in grip scaling that we observed in perceptual size judgments when the subjects picked up the test blocks. This is indeed what we found. As Figure 16.3 shows, subjects opened their hand wider in flight when they reached out to pick up the test block that was the same color as the set of small blocks than they did when they reached out to pick up the test block that was the same color as the set of large blocks. This result strongly suggests that repeated exposure to a correlation between object color and object size allowed subjects to set up a motor routine that could be retrieved when subjects were later confronted with objects of a particular color. The use of such learned motor routines would only be efficient for classes of objects in which shape, compliance, and surface friction were relatively constant across repeated encounters with those objects. The target objects in our experiment were designed with this in mind; they were made from the same material, they were all square in shape, and they were always

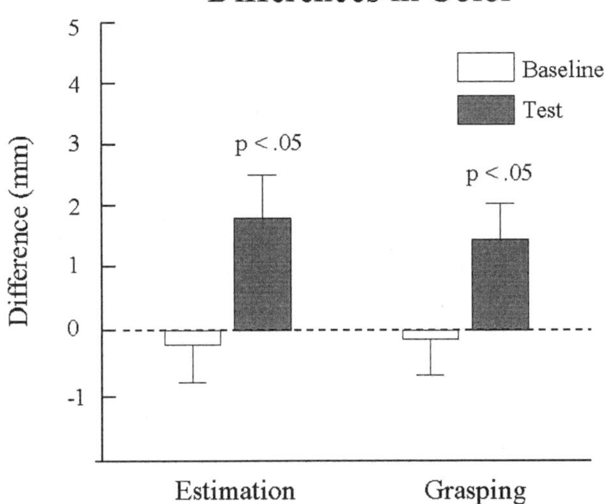

FIG. 16.3. Mean differences between responses to the probe block matched in color to the set of large key blocks and the probe block matched in color to the set of small key blocks. The difference in manual estimations for the baseline and test conditions is illustrated on the left and the difference in maximum grip aperture for the baseline and test conditions is illustrated on the right. (Adapted from: Haffenden AM, Goodale MA. The effect of learned perceptual associations on visuomotor control varies with kinematic demands. *J Cogn Neurosci* 2000;12:950–964, with permission.)

placed in the same position relative to the observer. Thus, the required kinematics of the elicited movements would be nearly equivalent across trials; therefore, motor routines that made use of learned perceptual information about object properties could be employed.

In a second experiment, illustrated in Figure 16.4, we gave subjects an opportunity to learn an association between object shape and object size by presenting them with a set of large circular key objects and a set of small hexagonal key objects (or vice versa). However, in this experiment, unlike the one with just described, the anticipatory posture of the hand during grasping would have to be adjusted to accommodate not only the varying sizes of the different objects, but also the difference in their shape. In other words, a hexagonal object demanded a somewhat different grasp from a circular one. Thus, even though subjects might learn an association between the shape of the object and its size, this learned association would not be useful

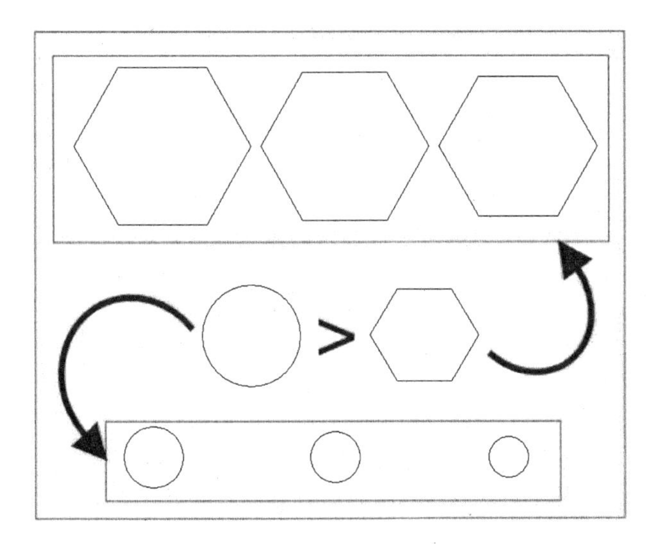

FIG. 16.4. A schematic illustration of the expected change in perceived size for probe blocks halfway in size between a set of large blocks and set of small blocks that are coded for size by their shape. Relative size comparisons made within each shape group would result in the hexagonal probe block being perceived as being larger in size than the circular probe block, even though they are physically identical in size. (Adapted from: Haffenden AM, Goodale MA. The effect of learned perceptual associations on visuomotor control varies with kinematic demands. *J Cogn Neurosci* 2000; 12:950–964, with permission.)

for setting up motor routines because of the variation in the hand and finger postures that were demanded by the different shapes. When we tested the subjects by again inserting test objects that were intermediate in size between the two sets of objects, we found the usual size-contrast effect in their manual estimates of the sizes of the test objects; in other words, subjects gave larger size estimates for the test object that was the same shape as the small key objects than they did for the test block that was the same shape as the set of large key objects (see Fig. 16.5). However, when we asked subjects to reach out and pick up the objects, we found no evidence that the perceived size-contrast effect had intruded into the scaling of their anticipatory grip aperture. Subjects did not open their hand any wider for the test object that was the same shape as small key objects than they did for the test object that was the same shape as the large key objects (see Fig. 16.5). In short, subjects scaled their grasp to the actual not the perceived size of the test objects. Thus, unlike what happened in the experiment with the colored blocks, subjects in this experiment did not appear to make use of the learned perceptual association between an arbitrary object feature and object size to set up motor routines that can be used when interacting with familiar objects.

Taken together the results of these two experiments suggest that visuomotor programming can make use of learned size information in situations where movement kinematics are quite consistent from one occasion to the next. If the kinematics requirements vary from trial to trial, however, then learned motor routines are less useful. This idea was supported by the results of a later experiment in which we varied the location of the target objects for one group of subjects (the variable-location group) and kept the location constant for another group (the constant-location group) (51). There is evidence from a number of different experiments that, although the grasp component of manual prehension is less influenced by changes in target location than is the transport component, a number of grasp parameters, including grip aperture, do change systematically with changes in the location of the target object (52–55). Both the constant-location group and variable-location group were trained with the same red and yellow key blocks that we had used in the original experiment (large red key blocks and small yellow key blocks, or vice versa). When we looked at their manual estimates of the

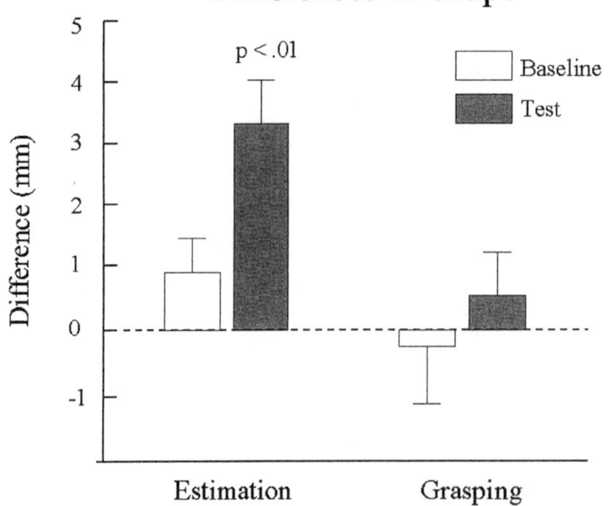

FIG. 16.5. Mean differences between responses to the probe block matched in shape to the set of large key blocks and the probe block matched in shape to the set of small key blocks. The difference in manual estimations for the baseline and test conditions is illustrated on the left and the difference in maximum grip aperture for the baseline and test conditions is illustrated on the right. (Adapted from: Haffenden AM, Goodale MA. The effect of learned perceptual associations on visuomotor control varies with kinematic demands. *J Cogn Neurosci* 2000; 12:950–964, with permission.)

perceived size of the midsized test blocks, we found that both groups showed the usual relative size contrast effect; that is, the test block that was the same color as the small key blocks was perceived to be larger than the test block that was the same color as the large key blocks. This meant that both groups had learned to associate the color of the blocks with their size. When subjects reached out and picked up the test blocks, however, only those subjects in the constant-location group made use of this perceptual learning, opening their hand wider in flight for the test block that was the same color as the small key blocks than they did for the test block that was the same color as the smaller key blocks. In contrast, the subjects in the variable-location group showed no evidence of perceptual learning in the scaling of their grasp; in other words, their preparatory grip aperture was the same for the two types of test blocks. Again, it appears, perceptual learning can be incorporated in the programming of a grasping movement only when the required kinematics of the action to the intended goal are consistent from trial to trial.

The experiments just described show that, following practice, learned associations between object metrics and readily identifiable object features can be incorporated into the programming of grasping movements provided the target object does not vary in shape and appears in the same location with respect to the observer. In other words, consistency in the kinematic requirements from one occasion to the next is a prerequisite for having learned perceptual information contribute to the programming of anticipatory hand posture. Of course, this idea of "kinematic consistency" is not novel, but has its roots in the literature dealing with movement "automatism" (for review, see ref. 56). Inconsistency in required movements has been found to slow the rate of development of automatic movements, such as those performed while driving. Indeed, in some cases, inconsistency in movement requirements has been found to completely inhibit automatization. It is possible, of course, that in the case of grasping, more practice with the same object placed in different locations would eventually result in the incorporation of learned information into the grip scaling and other grasp parameters. After all, in the real world, a familiar object can be encountered in a number of different locations with respect to the observer. However, there is a considerable range in the number of times different objects are grasped in a typical day and the number of different locations in which those objects can be encountered. An object such as a coffee cup, for example, might be repeatedly picked up and placed back on the desktop several times a day. This kind of action might therefore be more likely to call up stored motor routines than an action directed at an object such as a cereal box that is usually picked up only once a day or less and whose size and location could vary a good deal from one occasion to the next. However, whatever the constraints might be on the use of motor routines that are based on learned associations between object features and object metrics, our experiments suggest that such learning does occur and that it can be exploited by the motor system.

THE NEURAL SUBSTRATES OF LEARNED ASSOCIATIONS AMONG OBJECTS, FEATURES, AND ACTION

How such learned associations drive the motor system is not clear. As was discussed earlier in the section on functional aspects of grasping, there is neuropsychological evidence that the ventral stream of processing plays a critical role in the recognition of the objects and object features that drive the learned motor routines associated with particular tools and familiar objects. It seems quite likely, therefore, that the perceptual mechanisms in the ventral stream are also involved in the use of learned associations between arbitrary object features, such as color, and object metrics—associations that can come to control the grip scaling when we reach out to pick up such objects. What is not clear, however, is how these learned associations make contact with motor systems controlling hand

and finger movements. One plausible scenario has been put forward by Passingham and Toni (57). On the basis of a series of neuroimaging experiments in humans and lesion studies in monkeys, these authors have argued that the selection of a particular movement on the basis of its association with an arbitrary visual stimulus recruits networks involving both the dorsal and ventral streams, with the prefrontal cortex acting as an integrator for the information from the two streams. According to their proposal, prefrontal cortex is in the unique position to integrate information about the external context, the targets of action, and the success of the response. Mechanisms in the dorsal stream certainly participate in specifying the coordinates of potential actions, but the processing in prefrontal cortex acts in a top-down fashion to modulate activity in the parietal cortex as well as premotor cortex in those situations where the subjects have learned to associate particular actions with arbitrary object features. The identification of the relevant object features (and the context in which they occur) depends on an interaction between mechanisms in the ventral stream and prefrontal cortical areas.

Using PET, Toni et al. (58) directly contrasted human brain activity during the performance of a task in which a grasping movement was spatially congruent with the target with activity during the performance of a task in which a hand movement was arbitrarily associated with the stimuli used to elicit the grasping movements. Activation was observed in parietal cortex and the opercular region of the precentral gyrus when grasping movements were directly mapped onto the location and geometric properties of a target. However, when decisions about what movement to make were determined by a context independent of the spatial properties of the target, activation was seen in the ventral prefrontal cortex, striatal regions, and dorsal areas of the precentral gyrus. The authors propose that fronto-striatal loops are involved in specifying what action should be performed in a given context. This idea is similar to Murray et al.'s (59) proposal that a distributed network involving regions of the premotor and prefrontal cortex, the basal ganglia, and (in the initial stages) the hippocampus compute specific object-to-action mappings in contexts where the network must learn the action associated with a given input. Indeed, the notion that networks in the prefrontal cortex (and the striatum) permit humans and other primates to learn learned context-dependent associations between particular stimuli and particular actions is an idea that has been put forward (in various forms) by many authors (for review, see ref. 60).

The involvement of frontal-striatal loops in the selection of an action may be dependent on how arbitrary the relationship is between a given stimulus and the response it elicits. In the Toni et al. (58) experiment described in the preceding, the learned association between the stimuli and the required hand movements were completely arbitrary. In the experiments conducted by Haffenden and Goodale (51) in which subjects learned to rely on the association between color and size to program their grip aperture, the association between size and color was arbitrary, but the final movement was determined by the position and orientation of the target object. This is a variation on the arbitrary motor associations described in the preceding, and may fall in between this type of programming, and the case where the movement is specified entirely by the spatial properties of the target object. How such movements are organized is not known. What seems clear, however, is that, in both humans and nonhuman primates, networks involving the ventral stream and its projections to ventral prefrontal cortex play a crucial role in determining which action to select in a particular context. Indeed, this context-dependent selection of actions may be the norm rather than the exception. As Asaad et al. (61) point out, real-world behavior is more complicated than a simple one-to-one mapping of an action to a stimulus. The same stimulus can invoke a different response depending on the situation. On the basis of experiments showing that context has an overarching influence on the activity of neurons in monkey prefrontal cortex, Assad et al. have suggested that a major function of this region is the acquisition and implementation of the contextual information used to guide behavior. Such a system allows for the flexible mapping between objects and actions—and allows us to resist the impulsive demands of our sensory world.

In summary, it appears that prefrontal cortex allows for the integration of visuomotor transformations carried out in the dorsal stream with information about arbitrary object features processed by the ventral stream. However, how learned associations between stimuli and action might modulate the activity in the dorsal stream itself is not well understood. There is some evidence in monkey that activity of neurons in this region is modified with experience. Toth and Assad (62), for example, have shown that neurons in the lateral intraparietal sulcus (LIP), an area in the dorsal stream that is known to be involved in the visual control of saccadic eye movements, can become sensitive to a nonspatial stimulus attribute, such as color, when that attribute is directly linked to the direction of the required eye movements. Similar changes in the coding of cells in LIP has also been observed when eye movements have been arbitrarily associated with certain visual patterns (63). Thus, following training, cells in this region now show coding that is specific to the visual patterns that were used in training. It seems likely, but unproved, that prefrontal cortex plays the central role in this process, allowing arbitrary information about color and pattern (that is presumably processed in the ventral stream) to influence the activity of neurons in dorsal-stream structures that play a role in transforming visual information into action.

CONCLUSION

Let us conclude by returning now to the example of an everyday action with which we began this chapter: picking up our coffee cup from the dinner table. It is the perceptual machinery in the ventral stream that allows us to identify the objects on the table and to distinguish our cup from others that might be there. It is our ventral stream, too, that allows us to parse the handle from the rest of the cup and to select (with the help of networks in the prefrontal cortex and striatum) the appropriate hand posture for picking up the cup (by its handle) to take a drink of coffee. However, having identified the handle of our cup and the action we wish to perform, it is then up to the visuomotor machinery in the dorsal stream to impose the appropriate metric scaling on our movements to get our hand and fingers positioned efficiently on the cup's handle. Depending on the context and how familiar we are with this situation, the dorsal stream and its associated premotor networks might carry out all these computations in an entirely bottom-up fashion from the visual array—or alternatively, these networks could employ shortcuts for some of the computations (with help from links with ventral stream and prefrontal networks) and make use of stored information about the size and/or shape of the cup's handle. In any case, the scaling of the initial forces that are applied to lift the cup to our mouth is based on the stored information about the weight of the cup, which is presumably updated as we drink more of the coffee. If this information is not updated correctly, we can have the rather unpleasant experience of applying too much lift force and spilling the coffee.

Thus, everyday actions like picking up a coffee cup might appear quite simple and straightforward. The production of such actions, however, is complex and involves both online visual processing and stored information. Though the neural networks that integrate the required information remain to be decisively determined, it is clear that skilled actions, particularly actions directed at tools and other familiar objects, depend on the joint processing of visual information in the dorsal and ventral streams.

REFERENCES

1. Goodale MA, Milner AD. Separate visual systems for perception and action. *Trends Neurosci* 1992;15:20–25.

2. Goodale MA, Humphrey GK. Separate visual systems for action and perception. In: Goldstein B, ed. *Handbook of perception.* London: Blackwell Scientific, 2001:311–343.
3. Milner AD, Goodale MA. *The visual brain in action.* Oxford, UK: Oxford University Press, 1995.
4. Ungerleider LG, Mishkin M. Two cortical visual systems. In: Ingle DJ, Goodale MA, Mansfield RJW, eds. *Analysis of visual behavior.* Cambridge MA: MIT Press, 1982:549–586.
5. Culham JC, Kanwisher NG. Neuroimaging of cognitive functions in human parietal cortex. *Curr Opin Neurobiol* 2001;11:157–163.
6. Tootell RB, Dale AM, Sereno MI, et al. New images from human visual cortex. *Trends Neurosci* 1996;19: 481–489.
7. Pohl W. Dissociation of spatial discrimination deficits following frontal and parietal lesions in monkeys. *J Comp Physiol Psychol* 1973;82:227–239.
8. Ungerleider LG, Brody BA. Extrapersonal spatial orientation: the role of posterior parietal, anterior frontal, and inferotemporal cortex. *Exp Neurol* 1977;56: 265–280.
9. Goodale MA. The cortical organization of visual perception and visuomotor control. In: Kosslyn S, ed. *Visual cognition and action,* 2nd ed. Cambridge, MA: MIT Press, 1995:167–213.
10. Bálint R. Seelenlämung des Schauens, optische Ataxie, räumliche Störung der Aufmerksamkeit. *Monatschr Psychiatr Neurolog* 1909;25:51–81.
11. Harvey M, MilnerAD. Bálint's patient. *Cogn Neuropsychol* 1995;12:261–264.
12. Perenin M.-T, Vighetto A. Optic ataxia: a specific disruption in visuomotor mechanisms. I. Different aspects of the deficit in reaching for objects. *Brain* 1988;111: 643–674.
13. Jakobson LS, Archibald YM, Carey DP, et al. A kinematic analysis of reaching and grasping movements in a patient recovering from optic ataxia. *Neuropsychologia* 1991;29:803–809.
14. Milner AD, Perrett DI, Johnston RS, et al. Perception and action in "visual form agnosia." *Brain* 1991;114: 405–428.
15. Goodale MA, Milner AD, Jakobson LS, et al. A neurologic dissociation between perceiving objects and grasping them. *Nature* 1991;349:154–156.
16. Goodale MA, Humphrey GK. The objects of action and perception. *Cognition* 1998;67:179–205.
17. O'Regan JK. Solving the "real" mysteries of visual perception: the world as an outside memory. *Can J Psychol* 1992;46:461–488.
18. McConkie GW, Currie CB. Visual stability across saccades while viewing complex pictures. *J Exper Psychol Hum Percep Perf* 1996;22:563–581.
19. Rensink RA, O'Regan JK, Clark JJ. To see or not to see: The need for attention to perceive changes in scenes. *Psychol Sci* 1997;8:368–373.
20. Goodale MA, Haffenden A. Frames of reference for perception and action in the human visual system. *Neurosci Biobehav Rev* 1998;22:161–172.
21. Intraub H. The representation of visual scenes. *Trends Cogn Sci* 1997;1:217–222.
22. Graziano MS, Gross CG. Spatial maps for the control of movement. *Curr Opin Neurobiol* 1998;8:195–201.
23. Soechting JF, Flanders M. Moving in three-dimensional space: frames of reference, vectors, and coordinate systems. *Annu Rev Neurosci* 1992;15:167–191.
24. Colby CL. Action-oriented spatial reference frames in cortex. *Neuron* 1998;20:15–24.
25. Goodale MA. Different spaces and different times for perception and action. In: Casanova C, Ptito M, eds. *Vision: from neurons to cognition.* Amsterdam: Elsevier, 2001:313–331.
26. Aglioti S, DeSouza JF, Goodale MA. Size-contrast illusions deceive the eye but not the hand. *Curr Biol* 1995;5:679–685.
27. Brenner E, Smeets JBJ. Size illusion influences how we lift but not how we grasp an object. *Exp Brain Res* 1996;111:473–476.
28. Haffenden AM, Goodale MA. The effect of pictorial illusion on prehension and perception. *J Cogn Neurosci* 1998;10:122–136.
29. Hu Y, Goodale MA. Grasping after a delay shifts size-scaling from absolute to relative metrics. *J Cogn Neurosci* 2000;12:856–868.
30. Servos P, Carnahan H, Fedwick J. The visuomotor system resists the horizontal-vertical illusion. *J Motor Behav* 2000;32:400–404.
31. Westwood D, Goodale MA. Perceptual illusion and the real-time control of action. *Spatial Vision.* In press.
32. Rogers, E, Murphy RR. *Tele-assistance for semi-autonomous robots.* Proceedings of AIAA Conference on Intelligent Robots in Field, Factory, Service and Space. Hampton, Virginia: NASA Conference Publication. 1994;3251:500–508.
33. Pook PK, Ballard DH. Deictic human/robot interaction. *Robot Auton Syst* 1996;18:259–269.
34. Gordon AM, Westling G, Cole KJ, et al. Memory representations underlying motor commands used during manipulation of common and novel objects. *J Neurophysiol* 1993;69:1789–1796.
35. Gordon AM, Forssberg H, Johansson RS, et al. Visual size cues in the programming of manipulative forces during precision grip. *Exp Brain Res* 1991;83:477–482.
36. Flanagan JR, Beltzner MA. Independence of perceptual and sensorimotor predictions in the size-weight illusion. *Nat Neurosci* 2000;3:737–741.
37. McIntosh R. Seeing size and weight. *Trends Cogn Sci* 2000;4:442–444.
38. Jackson SR, Shaw A. The Ponzo illusion affects grip-force but not grip-aperture scaling during prehension movements. *J Exp Psychol Hum Percept Perform* 2000; 26:418–423.
39. Creem SH, Proffitt DR. Grasping objects by their handles: a necessary interaction between cognition and action. *J Exper Psychol Hum Percept Perf* 2001;1: 218–228.
40. Jeannerod M, Decety J, Michel F. Impairment of grasping movements following a bilateral posterior parietal lesion. *Neuropsychologia* 1994;32:369–380.
41. Sirigu A, Cohen L, Duhamel J-R, et al. A selective impairment of hand posture for object utilization in apraxia. *Cortex* 1995;31:41–55.
42. Sirigu A, Duhamel JR, Poncet M. The role of sensorimotor experience in object recognition. A case of multimodal agnosia. *Brain* 1991;114:2555–2573.
43. Buxbaum LJ, Schwartz MF, Carew TG. The role of semantic memory in object use. *Cogn Neuropsychol* 1997;14:219–254.
44. Lauro-Grotto R, Piccini C, Shallice T. Modality-spe-

cific operations in semantic dementia. *Cortex* 1997;33: 593–622.
45. Hodges JR, Spatt J, Patterson K. What and how: evidence for the dissociation of object knowledge and mechanical problem solving skills in the human brain. *Proc Natl Acad Sci USA* 1999;96:9444–9448.
46. Patterson K, Hodges JR. Semantic dementia: One window on the structure and organisation of semantic memory. In: Boller F, Grafman J, eds. *Handbook of neuropsychology,* 2nd ed. Amsterdam: Elsevier, 2000: 313–333.
47. Haffenden AM, Goodale MA. The effect of learned perceptual associations on visuomotor control varies with kinematic demands. *J Cogn Neurosci* 2000;12:950–964.
48. Helson H. *Adaptation-level theory.* New York: Harper & Row, 1964.
49. Coren S, Enns J. Size contrast as a function of conceptual similarity between test and inducers. *Percept Psychophys* 1993;54:579–588.
50. Choplin JM, Medin DL. Similarity of the perimeters in the Ebbinghaus illusion. *Percept Psychophys* 1999; 61:3–12.
51. Haffenden AM, Goodale MA. Learned perceptual associations influence visuomotor programming under limited conditions: kinematic consistency. *Exp Brain Res* 2002;147:485–493.
52. Jakobson LS, Goodale MA. Factors affecting higher-order movement planning: a kinematic analysis of human prehension. *Exp Brain Res* 1991;86:199–208.
53. Chieffi S, Gentilucci M. Coordination between the transport and the grasp components during prehansion movements. *Exp Brain Res* 1993;94:471–474.
54. Castiello U, Bennett K, Chambers H. Reach to grasp: the response to a simultaneous perturbation of object position and size. *Exp Brain Res* 1998;120:31–40.
55. Meulenbroek RG, Rosenbaum DA, Jansen C, et al. Multijoint grasping movements. Simulated and observed effects of object location, object size, and initial aperture. *Exp Brain Res* 2001;138:219–234.
56. Shiffrin RM, Dumais ST, Schneider W. Characteristics of automatism. In Long J, Baddeley A, eds. *Attention and performance IX.* Hillsdale, NJ: Erlbaum, 1981: 223–238.
57. Passingham RE, Toni I. Contrasting the dorsal and ventral visual systems: guidance of movement versus decision making. *NeuroImage* 2001;14:S125–S131.
58. Toni I, Rushworth MFS, Passingham RE. Neural correlates of visuomotor associations. Spatial rules compared with arbitrary rules. *Exp Brain Res* 2001;141: 359–369.
59. Murray EA, Bussey TJ, Wise SP. Role of prefrontal cortex in a network for arbitrary visuomotor mapping. *Exp Brain Res* 2000;133:114–129.
60. Miller EK, Cohen JD. An integrative theory of prefrontal cortex function. *Annu Rev Neurosci* 2001;24: 167–202.
61. Asaad WF, Rainer G, Miller EK. Task-specific neural activity in the primate prefrontal cortex. *J Neurophysiol* 2000;84:451–459.
62. Toth LJ Assad JA. Dynamic coding of behaviourally relevant stimuli in parietal cortex. *Nature* 2002;415: 165–168.
63. Sereno AB, Maunsell JH. Shape selectivity in primate lateral intraparietal cortex. *Nature* 1998;395:500–503.

17

Parieto-Frontal Networks and Gaze Shifts in Humans: Review of Functional Magnetic Resonance Imaging Data

Marie-Hélène Grosbras* and Alain Berthoz†

*Cognitive Neuroscience Unit, Montreal Neurological Institute, Montreal, Quebec, Canada;
†Laboratoy of Physiology, College de France–LPPA, Paris, France

INTRODUCTION

Focal lesions in the human and nonhuman primates' parietal lobes can impair the control of eye movements, most often in conjunction with other visuospatial deficits. Electrical microstimulation studies in monkeys identified saccade-related neurons in the lateral bank of the intraparietal sulcus (lateral intraparietal area, area LIP) as well as on the medial aspect of the parietal lobe (1,2). In addition, anatomic and functional connectivity studies underlined the importance of fronto-parietal circuits in the control of gaze shifts (3). However, the location and functional characterization of parietal "eye fields" in humans are still unclear.

New brain imaging methodologies, especially functional magnetic resonance imaging (fMRI), offer the potential to investigate brain function noninvasively with relatively high spatial and temporal resolution in normal human volunteers. We review a selection of recent brain imaging studies that have highlighted the oculomotor function of the parietal cortex. We particularly address the relationships between eye movements and covert shifts of visuospatial attention. We describe the organization of oculomotor behavior in the context of fronto-parietal circuits.

THE STUDY OF EYE MOVEMENTS USING BRAIN IMAGING TECHNIQUES

Primates' repertory of eye movements includes saccades that rapidly bring the center of gaze onto a peripheral target, and pursuit that maintains the center of gaze on slowly moving targets. Saccades can be reflexive (toward the location of a sudden stimulus) or voluntary (generated by an internal command and not directly by an external stimulus). Most often, reflexive saccades are triggered by the sudden appearance of a visual target (visually guided saccades). In a variant of this task, the so-called GAP paradigm, the fixation point is extinguished just before the onset of the target for the saccade, producing saccades with very short latencies. The main experimental tasks to test voluntary saccades are: (a) exploratory saccades toward fixed visual targets; (b) endogenous saccades in darkness; (c) memory guided saccades (toward the location of a target that has been extinguished for a delay); (d) antisaccades (toward a location symmetrical to the flashed target); and (e) memorization of saccade sequences. All these oculomotor paradigms involve a common final pathway, controlling the execution of eye movement, as well as higher-level selection and decision processes specific to each task to control where and when to move the eyes. In nonhuman primates, electrophysiologic recording during different types of eye movements has allowed us to identify three main cortical oculomotor areas: the frontal eye field (FEF) in the anterior bank of the arcuate sulcus, the supplementary eye field (SEF), in the dorsomedial frontal cortex anterior to the supplementary motor area, and the parietal eye field, in the LIP. These three regions are densely interconnected, and each of them presents strong connections with subcortical oculomotor regions, such as the superior colliculus, the pontine nuclei, and the pulvinar. Stimulation in all of those sites elicits eye movements, and neuronal activity is related to the oculomotor behavior (4).

Clinical electrophysiologic and lesion studies have provided evidences that a similar frontoparietal network is involved in eye movement control in humans (5). Pierrot-Deseilligny and colleagues have studied the oculomotor behavior of several patients after unilateral lesions located in the lateral part of the precentral gyrus (involving both the posterior extremity of the middle frontal gyrus and the adjacent precentral sulcus and gyrus), or in the posterior parietal cortex (including the main part of the intraparietal sulcus). They observed that in both cases the latency of reflexive visually guided or memory guided saccades are increased, and that the number of predictive saccades is reduced in a visually guided saccade task with predictive targets. In contrast in the gap paradigm, the latencies of saccades were normal after frontal lesions and less disturbed than in the case of the "fixation-remaining" paradigm after parietal lesions. This suggests that both frontal and parietal cortex are important for the disengagement of fixation. Comparison of the effects of frontal and parietal lobe lesions, respectively, suggested that the frontal lobe is more important for intentional (voluntary) saccades. (Patients with frontal lobe lesions are more impaired for voluntary exploratory saccades than for reflexive saccades.) The parietal lobe is more important for reflexive (externally triggered) saccades. (Patients with parietal lesions are more impaired for visually guided saccades [5,6].) In addition, patients with parietal lesions were particularly impaired in sequences of saccades for which the coordinates of the second saccades had to be recalculated from the land point of the first saccade (double-step paradigm). Altogether, the lesions studies show that saccadic eye movements can be generated even after lesions of the FEF region or of the posterior parietal cortex, but also that the intact parieto-frontal circuit is necessary to generate eye movements accurate in space and time. Such studies, however, gave access only to pathologic conditions and lack spatial resolution in the case of lesions. Brain imaging techniques widened the opportunities to study brain function in healthy volunteers and to eventually draw comparisons with the knowledge gathered from nonhuman primate studies.

Positron emission tomography (PET) and functional magnetic resonance imaging (fMRI) give an indirect measure of the brain activity. They allow us to compare local brain activity in different conditions, and thus give access to neural activity relative to a control condition. The intrinsic spatial resolution is 1 cm for the former and up to 1 mm for the latter. Most often, group results (five to 15 subjects) are described and reported relative to a standard stereotaxic coordinate system (7,8), as well as (sometimes) relative to gross anatomic landmarks.

IDENTIFICATION OF EYE FIELDS USING FUNCTIONAL MAGNETIC RESONANCE IMAGING: SEVERAL SITES IN THE PARIETAL CORTEX

Over the past 20 years, most of the studies that have used PET or fMRI to investigate the cerebral networks involved in the control of eye movements have reported focal activity in the parietal and frontal cortex (9). In Table 17.1, we summarize the results relative to the

TABLE 17.1. *Coordinates of Parietal Sites of Activity*

			Intraparietal sulcus + lobules						Precuneus							
	Reference		Left			Right			Left			Right			FEF	SEF
Visually guided saccades/fix																
10	**Paus, 1993**	PET		—		27	–54	45		—			—		B	R
11	**Anderson, 1994**	PET	–18	–68	36	20	–74	36		—			—		B	—
12	**Kawashima, 1996**	PET		—		25	–61	44		?			?		?	?
13	**Law, 1997**	PET	–40	–40	48	20	–48	48							B	B
			–12	–64	44	18	–66	48		—			—			
14	**Luna, 1998**	fMRI	–19	–60	56	21	–61	54	–5	–61	48		—		B (s + i)	B
15	**Thulborn, 2000**	MRI	–26	–73	33	24	–80	12					—		B	B
16	**Berman, 1999**	fMRI	–31	–52	48	22	–59	51	–7	–66	54	8.3	–68	51	B (s + i)	B
17	**Connolly, 2000**	fMRI	–41	–43	41	39	–47	42	–11	–73	39	12	–73	39	B	B
18	**Kimmig, 2001**	er fMRI	–31	–71	53	26	–72	52	–15	–79	50	15	–83	49	B	B
19	**Petit, 1999**	fMRI	–30	–58	45	37	–55	45	0	–70	46		—		B	B
20	**Heide, 2001**		–28	–56	60	20	–60	68				8	–68	44	B (s + i)	L
			–24	–68	60											
21	**Merriam, 2001**	fMRI	–21	–71	53	17	–75	54		—			—		B	B
39	**Simon, 2002**	fMRI	–16	–80	48	16	–72	56	–4	–60	56	8	–80	44	B (s + i)	B
						40	–48	48		—			—			
Voluntary saccades toward fixed visual targets																
22	**Perry, 2000**	fMRI	–36	–48	56	30	–52	54		—			—		B	R
	(single sacc.)		–38	–48	56	40	–44	48								
23	**Grosbras, 2001**	fMRI	–28	–76	32				–12	–64	64	12	–64	64	B (s + i)	B
	(sequences)		–28	–60	56	24	–56	56								
			–48	–32	44	32	–28	40								
			–44	–56	32											
Voluntary saccades in darkness/rest																
24	**Petit, 1996**	PET		—			—			—			—		B	B
25	**Law, 1998**	PET		—			—			—			—		B	B
26	**Dejardin, 1998**	PET	–24	–68	48	18	–76	44		—		8	–84	28	B	L
						50	–28	24								
Memory guided saccades/fixation																
27	**O'Sullivan, 1995**	PET		—		40	–44	36	–2	–66	44	10	–68	44	B	B
11	**Anderson, 1994**	PET	–14	–44	52											
			–18	–56	48		—		–14	–44	52		—		B	B
			–30	–34	40											
24	**Petit, 1996**	PET	–14	–72	44	30	–72	32		—		2	–76	40	B	B
	(sequence/rest)															
20	**Heide, 2001**	fMRI	–32	–56	60	24	–72	52				12	–80	48	B (i)	—
21	**Heide, 2001**		–32	–56	56	20	–60	68	–12	–64	64	12	–80	48	B (s + i)	B
	(sequence)		–20	–72	56	20	–72	52								
Memory guided/visually guided saccades																
29	**Sweeney, 1996**	PET	–18	–64	32	28	–62	32		—			—		B	B
Antisaccades/fixation or visually guided saccades																
30	**Doricchi, 1997**	PET	–24	–54	44	16	–76	44				4	–66	48	B (s + i)	R
			–36	–62	32	50	–50	32								
31	**O'Driscoll, 1995**	PET		—		15	–62	52		—			—		B	L
29	**Sweeney, 1996**	PET	–14	–68	32	20	–76	32		—			—		R	R
Pursuit eye movements/rest																
16	**Berman, 1999**	fMRI	–30	–57	52	–31	–52	48	–6	–63	51	10	–67	49	B (s + i)	B
32	**Schmidt 2001**	fMRI	–32	–54	56	30	–38	44		—			—		B (i)	B
19	**Petit, 1999**	fMRI	–30	–62	45	28	–69	39				3	–80	39	B	B
Voluntary covert shifts of attention																
33	**Corbetta, 1993**	PET	–27	–59	56	21	–61	50	–9	–53	56	15	–71	50		
			–23	–39	44	23	–47	52								
22	**Perry, 2000**	fMRI	–44	–48	36	36	–48	36								
34	**Kim, 1999**	fMRI	–30	–48	45	33	–54	51		—			—		B (s)	B
35	**Corbetta, 2000**	eR fMRI	–25	–67	48	21	–65	52								
			–25	–57	46	27	–59	52								
			–23	–67	32	29	–71	22								

TABLE 17.1. *(Continued)*

36	**Kastner, 1999**	fMRI	**−18**	**−63**	**54**	**16**	**−61**	**57**		—			—		**B**	**B**
			−37	**−32**	**59**	**38**	**−34**	**57**								
			−27	−78	**38**											
37	**Beauchamp, 2001**	fMRI	**−26**	**−57**	**50**	**27**	**−59**	**50**	−1	−65	**48**				**B (s + i)**	**L**
38	**Gitelmann, 1999**	fMRI	**−21**	**−60**	**51**	**27**	**−60**	**57**								
39	**Simon, 2002**	fMRI	**−36**	**−60**	**60**	**24**	**−64**	**60**	−4	−60	**48**	**8**	−64	**52**	**B**	**B**
						44	**−40**	**52**								

Coordinates of parietal sites of activity published in a selection of positron emission tomography (PET) or functional magnetic resonance imaging (fMRI) studies (references 9–35). The last two columns indicate if concomitant activity was observed in the frontal and supplementary eye fields (FEF and SEF), in the left (L), right (R), or both (B) hemispheres. When two sites where explicitly distinguished for the FEF we referred to them as superior (s) FEF and inferior (i) FEF. Those studies report the peak of activity after averaging in a group of volunteers.

parietal cortex involvement in eye movements from a selection of brain imaging studies (10–39). The peaks of activity, as reported in those articles, are plotted on the high-resolution anatomic image of a brain matched to the Montreal Neurological Institute template in the Talairach and Tournoux space (Fig. 17.1). The first main observation from this summary is that several parietal sites are activated even for simple oculomotor tasks. Remarkably, the peaks of activity are consistently located either along the intraparietal sulcus, or on the medial wall (precuneus). The oculomotor-related activity along the intraparietal sulcus is relatively dispersed, and several sites can be distinguished (Figs. 17.1 and 17.2). The first site lies in the main, horizontal part of the sulcus, at the junction of the supramarginal gyrus, angular

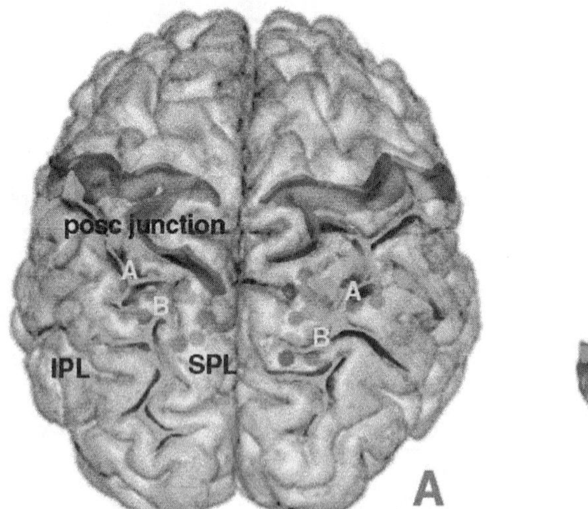

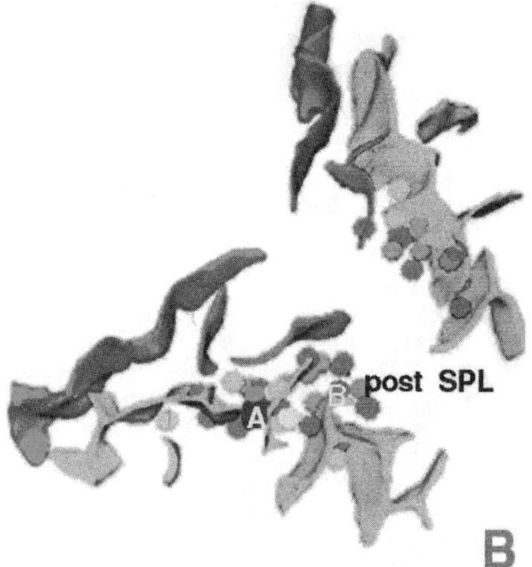

FIG. 17.1. Three-dimensional reconstruction of a brain in the Talairach and Tournoux coordinates system. The principal sulci have been extracted and underlined; the central sulcus in red, the superior branch of the postcentral sulcus in pink, the intraparietal sulcus in green. Each dot represents the center of activity as reported by one of the studies listed in Table 17.1, with a color code corresponding to the type of eye movements: dark blue, saccades in darkness; red, visually guided saccades (single or sequences); turquoise, memory guided saccades (single or sequences); yellow, pursuit. **A:** Putative LIP according to Müri et al. (41). **B:** Putative LIP according to Sereno et al. (28) and Simon et al. (39). **A:** Upper view. **B:** View from an upper left angle of the extracted sulci of the same brain. (See color plate of this figure after page 186.)

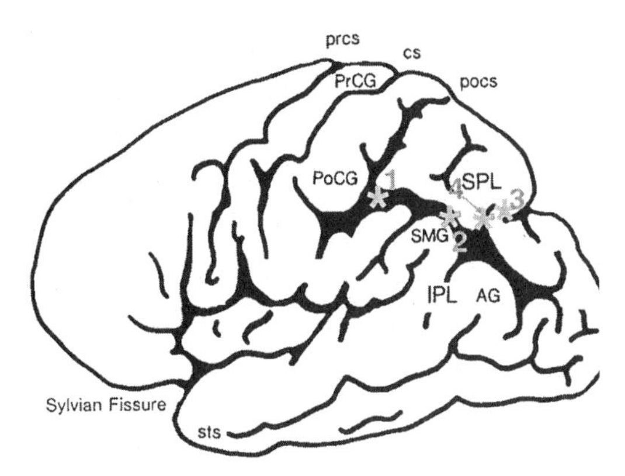

FIG. 17.2. Drawing of the human parietal lobe from an upper left view. The asterisks show the main sites where oculomotor-related activity has been described. AG, angular gyrus; CS, central sulcus; IPL, inferior parietal lobule; POCS, postcentral sulcus; POcG, postcentral gyrus; PRCS, precentral gyrus; PrCG, precentral gyrus; SMG, supramarginal gyrus; SPL, superior parietal lobule.

gyrus, and superior parietal lobule (#2 on Figure 2) (40). In most studies, it is not possible to establish whether the focus of activity is in the medial or lateral bank of the sulcus. Nevertheless, most often authors report that the activity in this region extends in the ventral-anterior direction; that is, in the direction of the lateral bank. Various oculomotor paradigms, including reflexive saccades (10–21), voluntary saccades toward fixed targets (22,23) or in darkness (24–26), memory guided saccades (9,18), and pursuit (16,19,32), as well as covert shifts of attention evoked activity in this part of the brain (33–38). Along the intraparietal sulcus, some studies also have evidenced oculomotor-related activity in the rostral part of the sulcus joining the postcentral sulcus (#1 on Figure 2) (13,23,27,32,39). Activation of this region is less frequently reported yet and seems related to paradigms involving a relatively high degree of voluntary planning, such as sequencing. In addition, a third group of foci is located medial to the posterior part of the intraparietal sulcus, in the superior parietal lobule (during visually guided saccades [12,16,19]; sequences of saccades [21,22,18]; antisaccades [29,30]; or covert shifts of attention [#4 on Figure 2] [32,33,35,39]).

The second parietal region where oculomotor activity is reported is located on the medial wall, more precisely in the precuneus. (See Table 17.1 for references.) The activation in the precuneus is typically clustered in a region approximately 1 cm behind the dorsalmost aspect of the marginal ramus of the cingulate sulcus, but can extend caudally until the dorsal portion of the parieto-occipital sulcus (see Table 17.1 and Fig. 17.1).

HOMOLOGY WITH NONHUMAN PRIMATES

In nonhuman primates, oculomotor neurons also have been recorded in different brain areas. As mentioned, the main cortical oculomotor area is the area LIP, located in the caudal third of the lateral bank of the intraparietal sulcus and strongly connected to other oculomotor regions. Saccade-related activity has been observed as well on the gyral surface of the caudal inferior parietal lobule, in area 7a, which does not present such strong oculomotor connectivity, however. It remains unclear whether a strict homology can be drawn between human and nonhuman primates and in this case which area is the homolog of LIP. Some authors have claimed that the region identified in the main (horizontal) part of the intraparietal sulcus (see Fig. 17.2) is the homolog of the LIP area identified in monkeys, and should be named "parietal eye field" (coordinates not published) (41). Kawashima and colleagues, using PET, observed activity in this region when the volunteers executed a saccade, but not when they reached toward a target with the hand without moving the eyes, which indicated a certain degree of specificity for eye movement execution (12). The arguments for such a strict homology between human and nonhuman primate LIP, however, are still incomplete and anatomic landmarks are lacking. Based on an fMRI experiment investigating saccades toward remembered targets at various visual angles, Sereno and colleagues observed that only a region in the dorsal parietal cortex showed a robust retinotopic mapping, indicating that this region might correspond to the macaque area LIP, according to the authors (42). This region was consistently (among the 12 subjects) located in a small sulcal branch just beyond the medial end of the intraparietal sulcus (B on Figure 1 and #3 on Figure 2). This is posterior to the focus described by Müri et al. or Kawashima et al. (A on Figure 1 and #2 on Figure 2). The coordinates correspond well with the focus reported by Simon et al. for both visually guided saccades and covert shifts of visuospatial attention (39). Simon et al. observed in addition another focus of activation specific to saccades in a more posterior part of the intraparietal sulcus. Beyond methodologic issues, such variability and discrepancies in locating the homolog of LIP in the human brain might result from the fact that the oculomotor representation in the parietal cortex is more complex than previously thought. Indeed, sites of oculomotor activity within the parietal lobes can be multiple and dispersed along the intraparietal sulcus as well as on the gyral surface. In monkeys, myelinization characteristics allow us to distinguish dorsal and ventral regions in the LIP area (43). Functional studies indicated some differences between the two regions (44). One possibility to explore is that in humans this dissociation evolved into clearly separated areas along the intraparietal sulcus and in the posterior superior parietal lobule (see Fig. 17.2).

Putative homology with data acquired in nonhuman primates also can help to interpret the medial wall oculomotor activity. Saccade-related activity has been recorded in the macaque medial parietal cortex (area 7m of Cavada and Goldman-Rakic or PGm of Pandya and Setzer), and several authors proposed that an additional eye field exists in this region (2). In support of this hypothesis, Leichnetz and colleagues demonstrated strong bilateral connectivity with cortical and subcortical oculomotor areas (45). Given the frequency of brain imaging reports of activity in the precuneus in humans, an eye movement representation likely exists in the human medial parietal cortex.

Moreover, another organizational difference between nonhuman and human primates is the hemispheric asymmetry for visuospatial skills control. Visuospatial impairments, including disturbance of the orienting systems, are observed commonly after right but not left hemisphere lesions, especially in the parietal cortex. Brain imaging studies have reported higher activity in the right than in the left parietal cortex when the participants made visually guided saccades, which can be considered a simple orienting task (14). Thulborn and colleagues observed the same asymmetry in healthy volunteers, but reported evidence of left hemisphere dominance in patients with Alzheimer's disease. This was accompanied by a higher activity in the prefrontal cortex and could be used as a functional marker for Alzheimer's disease (15).

In summary, brain imaging studies have allowed us to identify several sites within the parietal cortex, which present oculomotor-related activity, as is also the case in nonhuman primates. We need more data to conclude physiologic homologies between human and nonhuman primates, however. Future functional imaging studies have now to identify precise anatomic landmark, as some studies attempted to for the frontal lobe (46,47) and to further clarify the role of all the identified sites. Prior to discussing the functional significance of the parietal oculomotor activity observed in brain imaging experiments, we believe that it is important to consider these results, including the other parts of the network, particularly in the context of fronto-parietal circuits.

FRONTO-PARIETAL NETWORKS AND INTEGRATED CONTROL OF GAZE SHIFTS

In his book, Critchley (40) says that "the parietal lobes are empirical conceptions rather than autonomous entities in the anatomic and physiologic sense," and "there is something essentially artificial about a full-dress discussion limited to a sector of the cerebral hemisphere." The more our understanding of the brain progresses, the more we are able to approach the problem of cerebral function from a wider angle. In particular, the study of action control benefits greatly from the concept of fronto-parietal circuits dedicated to specific motor acts (3). In the macaque monkey, for instance, distinct areas in the parietal lobe are densely connected anatomically and functionally to distinct areas in the premotor cortex (48–51). The circuits formed by these connections control very specific movements in relation to a specific frame of reference. For example, the circuit formed by area VIP (in the fundus of the intraparietal sulcus) and area F4 (in the anterior ventral precentral gyrus) plays a crucial role in encoding the peripersonal space and transforming object locations into appropriate movements toward them. The circuit consisting of LIP, FEF, and SEF controls specifically the movements of the eyes. The existence of several eye movements related foci in the human parietal cortex mirrors the existence of several foci in the frontal lobe, suggesting that several fronto-parietal and prefronto-parietal circuits coexist to control different aspects of eye movements in human (Fig. 17.3). In the region classically defined as the FEF, two eye movement–related sites can be distinguished for saccade performance, one on the surface of the gyral inferior to the superior frontal sulcus, the other in the deep portion of the precentral sulcus, near its junction with the superior frontal sulcus (14,47,52). In addition, a distinct site is specific for the pursuit of eye movements, and the proximal cortex in the superior frontal sulcus is specific for spatial working memory (53,54). Besides, we identified an area on the dorsomedial frontal cortex, distinct and rostral to the supplementary eye field, that is active during the memory performance of unfamiliar sequences of saccades (23). A region in the rostral part of the precuneus showed a similar pattern of activation, suggesting the existence of a specific prefrontal-parietal functional circuit for the organization of novel sequences of eye movements (Fig. 17.4). Petit et al. (24) observed with PET a similar circuit

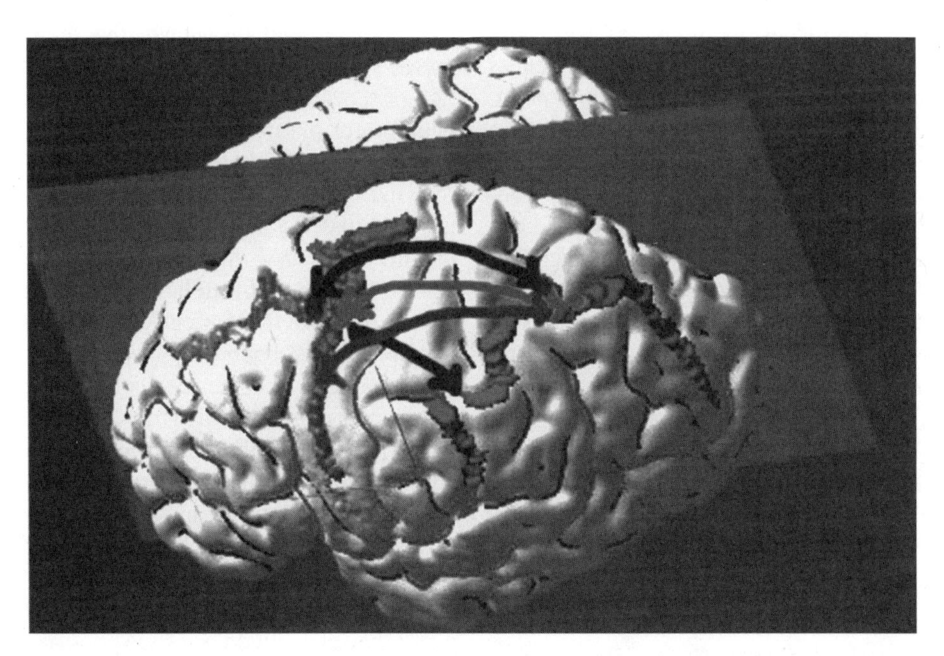

FIG. 17.3. Schematic examples of frontoparietal circuits involved in oculomotor-related tasks. On the left hemisphere of a brain transformed into Talairach and Tournoux space (MNI template, see text), the intraparietal sulcus is underlined in green, the precentral sulcus in purple, and the superior frontal sulcus in yellow. The arrows represent putative circuits as derived from fMRI results. In blue, circuits for memorized sequences of saccades (23,20); in red, visually guided saccades (14,16,19); in green, voluntary saccades in darkness (26,46). (See color plate of this figure after page 186.)

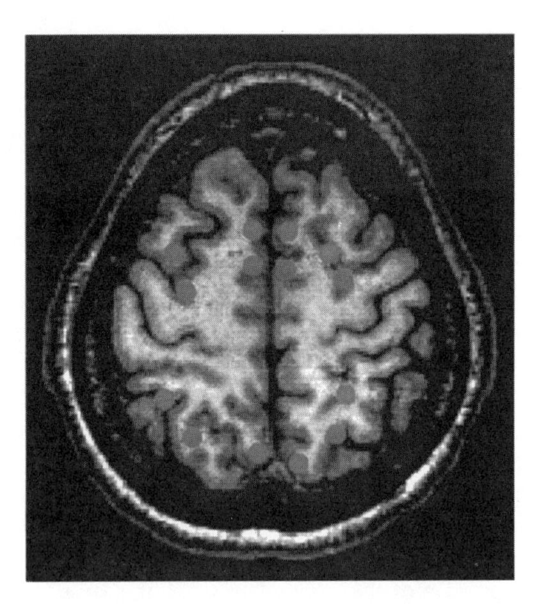

FIG. 17.4. Two different networks for executing sequences of saccades, in the dorsal frontoparietal cortex. In red we represented the peaks of activity observed when the volunteers executed any kind of memorized sequence of saccades; in turquoise, the peaks of activity observed when the sequence was newly learned (see text and ref. 23). (See color plate of this figure after page 186.)

activated during the learning of saccadic sequences; Schmid et al. did so as well with fMRI when studying the learning sequences of pursuit eye movements (32). At that point, one can only speculate about the precise fronto-parietal circuits involved in eye movements in humans, because no carefully controlled functional connectivity study has been published yet. A study combining transcranial magnetic stimulation and PET has shown that stimulating the dorsomedial frontal eye field increases blood flow in a region of the posterior dorsal parietal cortex, which indicates specific connectivity between the two regions (55). Further studies are needed to decipher the precise physiology of this connectivity (see Fig. 17.3).

NATURE OF PARIETAL ACTIVATION DURING EYE MOVEMENTS: SENSORY MOTOR INTEGRATION AND SPATIAL PERCEPTION

How the various subtasks involved in oculomotor control are divided into the different frontal and parietal oculomotor representations is still partially known. Electrophysiologic data in monkeys suggest that the parietal cortex subserves the encoding of spatial location of saccades targets, and thereby guide the oculomotor planning. In the area LIP, the visual receptive field of neurons shifts short before the eyes actually move, therefore ensuring anticipated adaptation to the new position of the eyes, and thereby space constancy across saccades (56). In humans also the integrative role of the parietal cortex in spatial perception has been emphasized, especially because the consequences of parietal cortex lesions principally involve disturbance of space perception and action guidance. Some authors proposed that the parietal oculomotor cortex is important for integrating eye movement control with visual scene perception. Among the studies listed in Table 17.1 that do not report activity in the parietal lobe, the majority have involved saccades in darkness. This might have suggested that visual stimulation is necessary to activate the parietal cortex. Other studies of saccades or covert shifts of attention in darkness, however, have observed parietal activation (26,36). In addition, given the intrinsic constraint of functional brain imaging, no strong conclusion can be drawn from a lack of activation. We prefer the working hypothesis that considers that some fronto-parietal circuits are specialized for specific aspects of the control of eye movements (together with subcortical circuits). For instance, in the macaque monkey, specific neuronal populations in the intraparietal sulcus and arcuate sulcus exhibit activity related to pursuit eye movements, whereas distinct neuronal populations are specialized for the control of visually guided eye movements (57). Tian and Lynch have demonstrated that labeling injection in the pursuit or saccade-related frontal eye fields, respectively, labeled the corresponding pursuit and saccade-related putative parietal eye fields (58). Recently, Petit and colleagues provided fMRI data that indicates such distinction between one fronto-parietal circuit for pursuit and another for saccadic eye movements, with a topologic relationship similar to that in the monkey (19). We believe that future research using brain imaging should aim at identifying such circuits in the human cortex anatomically and functionally, within the limit of the spatial and temporal resolution available.

For instance, in a recent study comparing the cerebral activity between the performance from memory of novel and familiar sequences of saccades, respectively, we provided some indication of such dissociation between different cortical circuits for different cognitive aspects of oculomotor control (see Fig. 17.4). In this experiment the subjects learned sequences of five horizontal saccades toward fixed visual targets, at a rhythm given by a metronome. When subjects executed newly learned or familiar sequences we observed activity in two foci in the precentral gyrus corresponding to the superior and inferior frontal eye field (52), in the supplementary eye field, as well as in several sites within the parietal cortex. When comparing newly learned saccades with familiar saccades, we observed higher activity in the whole common circuit, except in the ventrolateral site in the frontal cortex and in the site centered in the deep of the intraparietal sulcus at the level of boundary between the supramarginal, angular, and superior parietal gyrus, suggesting that the circuit formed by this region is more involved in the executive aspect of the sequential saccades. In addition, we observed activity for newly learned sequences only in a network comprising the precuneus, a region rostral to the frontal eye field in the superior frontal sulcus, in the medial prefrontal cortex rostral to the supplementary eye field, in the anterior part of the intraparietal sulcus near the postcentral sulcus, and in the superior parietal lobule. We suggest that this network is specific for the organization and planning of unfamiliar eye movement sequences. Recently, Merriam and colleagues (21) showed activity in a similar network when volunteers performed a task that required a relatively high cognitive decision where to make an eye movement. These authors also showed that this network was distinct from the one involved in reflexive visually guided eye movements. Interestingly, the same fronto-parietal network was also described during mental exploration of complex scenes without visual stimulation (59). This offers an example of the functional dissociation between distinct fronto-parietal circuits (see Fig. 17.4).

OVERT VERSUS COVERT SHIFTS OF ATTENTION

Lang et al. (60) asked volunteers to imagine that they were moving their eyes, but without actually moving their eyes, while he scanned them with PET. They observed that a common network was activated by overt and covert saccades. Berthoz (61) proposed a hierarchical theory to explain this property and suggested that this finding was a confirmation of the "motor theory of attention" (62): Covert shifts of attention are eye movements that are planned but not executed. Several experiments, involving various visuospatial constraints, have identified parieto-frontal couples that activity is related to covert shifts of attention (see Table 17.1). This activation includes mainly the superior frontal eye field, supplementary eye field, and posterior intraparietal sulcus. The degree of overlap and relative amount of activity induced by eye movements and covert shifts of attention, respectively, depend on the experimental paradigm. Recently, Beauchamp and colleagues (37) investigated the difference between the cortical network for overt and covert shifts of attention using a parametric design in which subjects made either overt or covert eye movements at three different rates. At every shift rate, overt eye movements elicited higher activity in the fronto-parietal network than covert shifts of attention. In addition, increasing the frequency increased the activation for overt than covert eye movement. The authors concluded that a minimum activity in the network is sufficient to orient attention, whereas activity above a certain threshold is necessary for the actual execution of an eye movement. This and other similar studies support the premotor theory of spatial attention, which stipulates that spatial shifts of attention are eye movements whose command has been blocked just before execution (63). Yet, other studies using other paradigms have found higher activity for covert than for overt shifts of attention (22,64), underscoring that the activity in the oculomotor network is highly dependent on the task and cognitive context, as has been shown in monkeys (65). Interestingly, some brain imaging studies have suggested that nonspatial or foveal attention also yield to activity in similar fronto-parietal regions (66,67). In addition, the view of the posterior parietal cortex as a sensory cortex has been challenged by observation of activity during the delay of memory guided saccade tasks as well as during the expectation of visual targets in the absence of eye movements, both in monkeys and humans (36). Therefore, the sensory, motor, and cognitive aspects driving the activity in the oculomotor cortical network need still further investigation. Hopefully, brain imaging methodologies will provide additional data in the near future.

CONCLUSION

Functional brain imaging has evidenced that several regions in the parietal cortex exhibit activity related to eye movements at the same time as do frontal regions. The use of fine-tuned paradigms has allowed investigators to start segregating specific fronto-parietal circuits from a functional point of view. For instance, pursuit- and saccade-related circuits have been dissociated; prefronto-parietal circuits are specific for oculomotor behaviors that require higher cognitive planning. The extension of this line of research in the next years will undoubtedly provide us with a better panorama of the physiology of those circuits. In particular, important outstanding questions are: Which coordinate systems are used and built in the parietal cortex? How are they related to the coordinate systems used in the premotor areas? What specific coordinate transformations are performed by the parieto-frontal circuits? How are such fronto-parieto circuits linked to limbic and memory circuits? What is specific to eye movement, and do supramodal circuits for orienting exist? How are the various aspects of spatiomotor control integrated? Are we able to identify landmarks to reliably localize the oculomotor areas in the human brain? What are the nature and significance of hemispheric specialization? To what extent does the pattern of activity in oculomotor-related circuits vary among individuals? How does the pattern of activity vary as a function of training and experience within an individual?

ACKNOWLEDGMENT

M-H Grosbras is supported by a grant from the Fyssen Foundation, Paris. The authors thank J.F. Mangin for 3D sulcus extraction in Figure 1.

REFERENCES

1. Andersen RA, Brotchie PR, Mazzoni P. Evidence for the lateral intraparietal area as the parietal eye field. *Curr Opin Neurobiol* 1992;2:840–846.
2. Thier P, Andersen RA. Electrical microstimulation distinguishes distinct saccade-related areas in the posterior parietal cortex. *J Neurophysiol* 1998;80:1713–1735.

3. Rizzolatti G, Luppino G, Matelli M. The organization of the cortical motor system: new concepts. *Electroencephalog Clin Neurophysiol* 1998;106:283–296.
4. Schall JD. Visuomotor areas of the frontal lobe. In: Rockland K, Peters A, Kaas J, eds. Extrastriate cortex of primates, volume 12 of cerebral cortex. New York: Plenum Publications, 1997;527–587.
5. Pierrot-Deseilligny C, Rivaud S, Gaymard B, et al. Cortical control of saccades. *Ann Neurol* 1995;37:557–567.
6. Heide W, Kompf D. Combined deficits of saccades and visuo-spatial orientation after cortical lesions. *Exp Brain Res* 1998;123:164–171.
7. Talairach J, Tournoux P. Co-planar stereotacic atlas of the human brain: 3-dimensional proportional system: an approach to cerebral imaging. Stuttgard, Germany: Thieme, 1988.
8. Evans AC, Kamber M, Collins DL, et al. An MRI-based probabilistic atlas of neuroanatomy. In: Shorvon D, Fish D, Andermann F, Bydder GM, Stefan A, eds. Magnetic resonance scanning and epilepsy. New York: Plenum, 1994;263–274.
9. Petit L, Orssaud C, Tzourio N, et al. Superior parietal lobule involvement in the representation of visual space: a PET review. In: Thiers P, Karnath HO, eds. Parietal lobe contributions to orientation in 3D space. Heidelberg: Springer-Verlag, 1997;77–90.
10. Paus T, Petrides M, Evans AC, et al. Role of the human anterior cingulate cortex in the control of oculomotor, manual, and speech responses: a positron emission tomography study. *J Neurophysiol* 1993;70:453–469.
11. Anderson TJ, Jenkins IH, Brooks DJ, et al. Cortical control of saccades and fixation in man: a positron emission tomography study. *Brain* 1994;117:1073–1084.
12. Kawashima R, Naito E, Matsumara M, et al. Topographic representation in human intraparietal sulcus of reaching and saccade. *NeuroReport* 1996;7:1253–1256.
13. Law I, Svarer C, Holm S, et al. The activation pattern in normal humans during suppression, imagination and performance of saccadic eye movements. *Acta Physiol Scand* 1997;161:419–434.
14. Luna B, Thulborn KR, Strojwas MH, et al. Dorsal cortical regions subserving visually guided saccades in human: an fMRI study. *Cerebral Cortex* 1998;8:40–47.
15. Thulborn KR, Martin C, Voyvodic JT. Functional MR imaging using a visually guided saccade paradigm for comparing activation patterns in patients with probable Alzheimer's disease and in cognitively able elderly volunteers. *AJNR Am J Neuroradiol* 2000;21:524–531.
16. Berman RA, Colby CL, Genovese CR, et al. Cortical networks subserving pursuit and saccadic eye movements in humans: an FMRI study. *Hum Brain Mapp* 1999;209–225.
17. Connolly JD, Goodale MA, Desouza JF, et al. A comparison of frontoparietal fMRI activation during antisaccades and anti-pointing. *J Neurophysiol* 2000;84: 1645–1655.
18. Kimmig H, Greenlee MW, Huethe F, et al. MR-eyetracker: a new method for eye movement recording in functional magnetic resonance imaging. *Exp Brain Res* 1999;126:443–449.
19. Petit L, Haxby JV. Functional anatomy of pursuit eye movements in humans as revealed by fMRI. *J Neurophysiol* 1999;82:463–471.
20. Heide W, Binkofski F, Seitz RJ, et al. Activation of frontoparietal cortices during memorized triple-step sequences of saccadic eye movements: an fMRI study. *Eur J Neurosci* 2001;13:1177–1189.
21. Merriam EP, Colby CL, Thulborn KR, et al. Stimulus-response incompatibility activates cortex proximate to three eye fields. *NeuroImage* 2001;13:794–800.
22. Perry RJ, Zeki S. The neurology of saccades and covert shifts in spatial attention: an event-related fMRI study [in process citation]. *Brain* 2000;123: 2273–2288.
23. Grosbras M-H, Leonards U, Lobel E, et al. Human cortical networks for new and familiar sequences of saccades. *Cerebral Cortex* 2001;11:936–945.
24. Petit L, Orssaud C, Tzourio N, et al. Functional anatomy of a prelearned sequence of horizontal saccades in humans. *J Neurosci* 1996;16(11):3714–3736.
25. Law I, Svarer C, Rostrup E, et al. Parieto-occipital cortex activation during self-generated eye movements in the dark. *Brain* 1998;121:2189–2200.
26. Dejardin S, Dubois S, Bodart JM, et al. PET study of human voluntary saccadic eye movements in darkness: effect of task repetition on the activation pattern. *Eur J Neurosci* 1998;10:2328–2336.
27. O'Sullivan T, Jenkins IH, Henderson L, et al. The functional anatomy of remembered saccades: a positron emission tomography study. *NeuroReport* 1995;6: 2141–2144.
28. Sereno MI, Pitzalis S, Martinez A. Mapping of contralateral space in retinotopic coordinates by a parietal cortical area in humans. *Science* 2001;294:1350–1354.
29. Sweeney JA, Mintun MA, Kwee S, et al. Positron emission tomography study of voluntary saccadic eye movements and spatial working memory. *J Neurophysiol* 1996;75:454–468.
30. Doricchi F, Perani D, Incoccia C, et al. Neural control of fast-regular saccades and antisaccades: an investigation using PET. *Exp Brain Res* 1997;116:50–52.
31. O'Driscoll G, Alpert NM, Mathysse SW, et al. Functional neuroanatomy of antisaccade eye movements investigated with positron tomography. *Proc Natl Acad Sci USA* 1995;92:925–929.
32. Schmid A, Rees G, Frith C, et al. An fMRI study of anticipation and learning of smooth pursuit eye movements in humans. *NeuroReport* 2001;12:1409–1414.
33. Corbetta M, Miezin FM, Shulman GL, et al. A PET study of visuospatial attention. *J Neurosci* 1993;13: 1202–1226.
34. Kim YH, Gitelman DR, Nobre AC, et al. The large-scale neural network for spatial attention displays multifunctional overlap but differential asymmetry. *NeuroImage* 1999;9:269–277.
35. Corbetta M, Kincade JM, Ollinger JM, et al. Voluntary orienting is dissociated from target detection in human posterior parietal cortex. *Nat Neurosci* 2000;3:292–297.
36. Kastner S, Pinsk MA, De Weerd P, et al. Increased activity in human visual cortex during directed attention in the absence of visual stimulation. *Neuron* 1999;22: 751–761.
37. Beauchamp MS, Petit L, Ellmore TM, et al. A parametric fMRI study of overt and covert shifts of visuospatial attention. *NeuroImage* 2001;14:310–321.
38. Gitelman DR, Nobre AC, Parrish TB, et al. A large-scale distributed network for covert spatial attention. Further anatomic delineation based on stringent behavioural and cognitive controls. *Brain* 1999;122: 1093–1106.

39. Simon O, Mangin JF, Cohen L, et al. Topographical layout of hand, eye, calculation, and language-related areas in the human parietal lobe. *Neuron* 2002;33:475–487.
40. Critchley M. The parietal lobes. London: Edward Arnold and Co., 1957.
41. Müri R, Iba-Zizen MT, Cabanis EA, et al. Location of the human posterior eye field with functional magnetic resonance imaging. *J Neurol Neurosurg Psychiatry* 1996;60:445–448.
42. Sereno MI, Pitzalis S, Martinez A. Possible homolog of area LIP in humans. *Soc Neurosci Abs* 2000;404.11:933.
43. Blatt G, Andersen RA, Stoner G. Visual receptive field organization and cortico-cortical connections of area LIP in the Macaque. *J Comp Neurol* 1990;299:421–445.
44. BenHamed S. Caractérisation fonctionelle de la représentation visuelle de l'aire intrapariétale latérale (LIP): spéciliatés et modulations comportementales. Doctoral thesis dissertation, University Paris VI, France, 1999.
45. Leichnetz GR. Connections of the medial posterior parietal cortex (area 7m) in the monkey. *Anat Rec* 2001;263:215–236.
46. Grosbras M-H, Lobel E, Van de Moortel P-F, et al. An anatomic landmark for the supplementary eye field revealed with fMRI. *Cerebral Cortex* 1999;9:705–711.
47. Lobel E, Kahane P, Leonards U, et al. Localization of the human frontal eye fields: anatomic and functional findings from fMRI and intracerebral electrical stimulation. *J Neurosurg* 2001;95:804–815.
48. Wise SP, Boussaoud D, Johnson PB, et al. Premotor and parietal cortex: corticocortical connectivity and combinatorial computations. *Annu Rev Neurosci* 1997;20: 25–42.
49. Matelli M, Govoni P, Galletti C, et al. Superior area 6 afferents from the superior parietal lobule in the macaque monkey. *J Comp Neurol* 1998;402:327–352.
50. Matelli M, Luppino G. Parietofrontal circuits for action and space perception in the macaque monkey. *NeuroImage* 2001;14:S27–S32.
51. Andersen RA, Asanuma C, Essick G, et al. Corticocortical connections of anatomically and physiologically defined subdivisions within the inferior parietal lobule. *J Comp Neurol* 1990;296:65–113.
52. Lobel E, Berthoz A, Leroy-Willig A, et al. fMRI study of voluntary saccadic eye movements in humans. *NeuroImage* 1996;S396.
53. Courtney SM, Petit L, Maisog JA, et al. An area specialized for spatial working memory in human frontal cortex. *Science* 1998;279:1347–1350.
54. Leonards U, Sunaert S, Van Hecke P, et al. Attention mechanisms in visual search: an fMRI study. *J Cogn Neurosci* 2000;12:61–75.
55. Paus T, Jech R, Thompson CJ, et al. Transcranial magnetic stimulation during positron emission tomography: a new method for studying connectivity of the human cerebral cortex. *J Neurosci* 1997;17:3178–3184.
56. Duhamel J-R, Colby CL, Goldberg ME. The updating of the representation of visual space in parietal cortex by intended eye movements. *Science* 1992;255:90–92.
57. Gottlieb JP, McAvoy MG, Bruce CJ. Neural responses to smooth-pursuit eye movements and their correspondence with electrically elicited smooth eye movements in the primate frontal eye field. *J Neurophysiol* 1994; 72:1634–1653.
58. Tian JR, Lynch JC. Corticocortical input to the smooth and saccadic eye movement subregions of the frontal eye field in Cibus monkeys. *J Neurophysiol* 1996;76: 2754–2771.
59. Mellet E, Briscogne S, Tzourio-Mazoyer N, et al. Neural correlates of topographic mental exploration: the impact of route versus survey perspective learning. *NeuroImage* 2000;12:588–600.
60. Lang W, Petit L, Höllinger P, et al. A positron emission tomography study of oculomotor imagery. *NeuroReport* 1994;5:921–924.
61. Berthoz A. The role of inhibition in the hierarchical gating of executed and imagined movements. *Cogn Brain Res* 1996;3:101–113.
62. Rizzolatti G, Riggio L, Dascola I, et al. Reorienting attention across the horizontal and vertical meridians: evidence in favor of a premotor theory of attention. *Neuropsychologia* 1987;25:31–40.
63. Sheliga BM, Riggio L, Rizzolatti G. Orienting of attention and eye movements. *Exp Brain Res* 1994;98: 507–522.
64. Corbetta M, Akbudak E, Conturo TE, et al. A common network of functional areas for attention and eye movements. *Neuron* 1998;21:761–773.
65. Colby CL, Goldberg ME. Space and attention in parietal cortex. *Annu Rev Neurosci* 1999;22:319–349.
66. Coull JT, Nobre AC. Where and when to pay attention: the neural systems for directing attention to spatial locations and to time intervals as revealed by both PET and fMRI. *J Neurosci* 1998;18:7426–7435.
67. Wojciulik E, Kanwisher N. The generality of parietal involvement in visual attention. *Neuron* 1999;23: 747–764.

18

Modular Organization of Parietal Lobe Functions as Revealed by Functional Activation Studies

Rüdiger J. Seitz and Ferdinand Binkofski*

*Department of Neurology, University Hospital Düsseldorf, Düsseldorf, Germany; *Department of Neurology, University Hospital Schleswig-Holstein, Lübeck, Germany*

INTRODUCTION

The parietal lobe is situated in the posterior portion of the brain between the central sulcus, lateral sulcus, and occipito-parietal sulcus. Cytoarchitectonic studies have shown that the macroanatomic structures of the parietal cortex can be parcellated into a number of subareas with a particularly rich organization in the cortex lining the intraparietal sulcus (1). These parietal cortical areas build up a group of differentiated circuits with specific links to cytoarchitectonically defined subareas in the premotor (2) and prefrontal cortex (3). In addition, there is a prominent projection from parietal cortical areas to the pontine nuclei in which information is relayed to the cerebellar cortex (4–6). The postcentral gyrus at the rostral aspect of the parietal lobe contains the representations of the somatosensory system with a well-developed somatotopic organization (7), thus providing internal information from the body. At its caudal end the parietal cortex borders the unimodal visual areas (8). From both sides there is a strong projection to the different cortical subareas in the parietal lobe. Both superior and inferior parietal lobules receive somatosensory and visual information. The anterior part of the superior parietal lobule receives mainly somatosensory and the posterior part mainly visual information. The anterior part of the inferior parietal lobule seems to integrate visual and somatosensory information (2,9). Further, for the visual input it is well established that it provides external spatial information called the "dorsal stream," which is different from the ventral stream to the temporal lobe conveying object characteristics (10). Thus, in light of the neuroanatomic and electrophysiologic data, the posterior parietal cortex emerges as a conglomerate of small, specialized areas, each of which receives specific sensory information that becomes transformed into action-relevant information (11,12). Because of the direct connectivity and indirect projections via the cerebellum to the frontal lobe, the parietal cortex is an interface between perception and action, the latter of which is mediated by the frontal lobe (13).

Neuroimaging methods open avenues to disentangle the cerebral structures involved in the different aspects underlying human behavior. Here, we report about neuroimaging studies addressing the question of how perception is relevant for action control. A major observation is that activation studies reveal functional subsystems that usually consist of interspersed patterns of closely adjacent activation foci. We argue that the function of the parietal cortex is to provide information about egocentric and allocentric spatial relations for intransitive and transitive body movements; eye movements; and higher-order processes such as mental navigation, visual imagery, and thought. We use the example of action in mirror-space to show that the combined use of lesion and corresponding activation studies allow the concept of a modular distribution of subfunctions in the human parietal lobe.

METHODS

Metaanalysis

Fifty neuroimaging studies covering different aspects of cognitive functions were analyzed. The studies were required to report activation foci in Talairach coordinates (14). These activation foci had to be the locations of peak mean change of activation found for the cohort of healthy volunteers studied. The stereotactic coordinates as reported in these publications were grouped together such that for each hot spot the coordinates of the different studies varied by maximal ±5 mm in x-, y-, or z-dimension. Right- and left-hemispheric foci were collapsed. Thereby, the coordinates were grouped according to stereotactic localization, not by the neuropsychological task nor by pseudo-anatomic labeling provided by the authors. Mean activation foci were calculated for each location from these data. They were plotted into Talairach space (14).

Action in Mirror-Space

One group of nine right-handed male volunteers (age range: 25 to 40 years old) with no history of neurologic illness participated in an activation study with positron emission tomography (PET). The study was approved by the Ethics Committee of the Heinrich-Heine-University Dusseldorf. Right handedness was assessed by the Edinburgh Inventory (15). The subjects lay in a supine position and were required to point to a visual target that was positioned by a robot such that they could see it in a mirror that was located within reaching distance in front of them (16). In the activation condition, the subjects had to point to the virtual target in the mirror, whereas in the control condition the subjects pointed to the real target location. Twelve reaches were performed during the 60 seconds of the PET activation study. The imaging data were analyzed with SPM99 software (Leopold Müller Functional Imaging Laboratory, London). After correction for head movements, signal intensity variation, images were normalized spatially into stereotactic space and smoothed with a Gaussian filter to an image resolution of 10 mm. Thereafter, group analysis of the data was performed. Significance was calculated pixel-by-pixel after analysis of covariance for global CBF changes. Only activations at $p < 0.001$ with a spatial extent of 150 pixels were accepted as significant. For visualization, the loci of the peak activations were superimposed onto the brain template of the atlas of Talairach and Tournoux (14).

The activation data were compared with the lesion location of a group of five patients suffering from mirror agnosia (17). These patients were unable to reach to the real target while seeing it in a mirror. In contrast, they could reach accurately to real targets under direct view. They had an intact egocentric coordinate system, because they could match limb positions in bilateral arm position tracking and could reach to different parts of their body with their eyes closed. Their egocentric coordinate system was craniocentric, because the patients indicated their subjective body midline in relation to their head position. The lesions of the patients were plotted into Talairach space to allow for comparison with activation studies (17).

RESULTS

The anatomic locations of the activation studies are plotted in Figure 18.1. Table 18.1 shows the broad range of activation paradigms investigated in these studies ranging from sensorimotor control to cognitive tasks. Nine different activation foci were identified along the intraparietal sulcus.

The most rostrolateral activation focus occurred in the anterior portion of the supramarginal gyrus (area 1 in Fig. 18.1). It was ac-

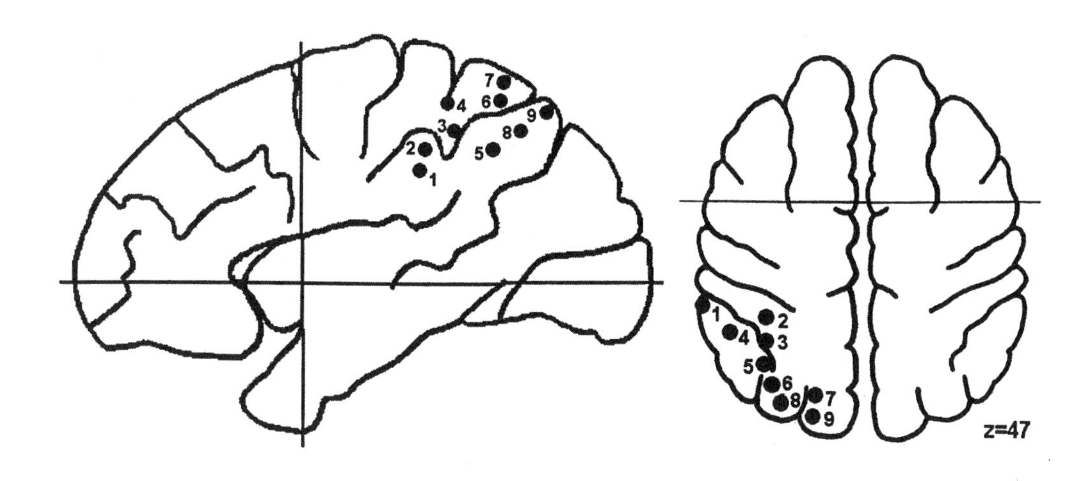

		Talairach coordinates (x,y,z in mm)		
1:	**supramarginal gyrus**	**64**	**-35**	**34**
2:	**AIP, anterior parietal area**	**32**	**-38**	**40**
3:	**VIP, ventral intraparietal area**	**35**	**-46**	**47**
4:	**LIP, lateral intraparietal area**	**44**	**-43**	**55**
5:	**IPL, inferior parietal lobule**	**31**	**-52**	**42**
6:	**pIP, posterior intraparietal area**	**32**	**-59**	**54**
7:	**pSPL, posterior portion of superior parietal lobule**	**18**	**-64**	**59**
8:	**cIPS, caudal portion of intraparietal sulcus**	**27**	**-67**	**44**
9:	**V6, visual area 6**	**18**	**-70**	**49**

FIG. 18.1. Localization of the mean activation foci bordering the intraparietal sulcus in a lateral view *(left)* and in an axial plane 47 mm dorsal to the intercommissural plane *(right).* Indicated are the origin of the coordinate system at the anterior commissure in the lateral view and the coronal plane through the anterior commissure in the axial plane.

tivated in tasks involving somatosensory information sampling (18), letter copying (19), precisely tuned grip (20), reaching (21,22), discrimination of hand orientation (23), simultaneous performance of two signal-reaction tasks (24), mental rotation (25), spatial orientation (26), and the switching of response (27,28). Most of these tasks were sensorimotor control tasks of forelimb and eye movements. Altogether, these activation paradigms have in common that the execution of the actions had to be precisely tuned with respect to the task demands. Based on the experimental isolation of outward saccades from inward saccades, Perry and Zeki (27) suggested that the function of the (right) supramarginal gyrus may be a more supraordinate one alerting another more specialized brain system for finely tuned task performance. Interestingly, a similar activation area in the left supramarginal gyrus was activated during phonological processing (29).

TABLE 18.1. *Experimental Conditions in the Evaluated Neuroimaging Studies Reporting Activation Areas Along the Intraparietal Sulcus*

Deiber et al., 1991 (43)	Auditorily cued joystick movements versus joystick movements in fixed direction
Pardo et al., 1991 (39)	Detection pause of sensory stimulation versus visual fixation
Corbetta et al., 1993 (40)	Shift of visuospatial attention versus fixation
Paulesu et al., 1993 (29)	Rhyme judgment of seen letters versus visual shape similarity judgment
Petrides et al., 1993 (48)	Self-ordering of digits from 1 to 10 versus counting from 1 to 10
Fletcher et al., 1995 (65)	Viewing words with high versus low imaginability
Bonda et al., 1995 (23)	Orientation of photographed hands versus monitoring of photographed hands
Kawashima et al., 1995 (21)	Skilled reaching versus rest
Parsons et al., 1995 (33)	Orientation of photographed hands versus visual fixation
Owen et al., 1996 (32)	Planning of sequential screen touches versus touching of targets
Vandenberghe et al., 1996 (60)	Semantic judgments on viewed words versus viewing words
Decety et al., 1997 (44)	Observation of meaningless gestures versus observation of meaningful gestures
Dolan et al., 1997 (66)	Viewing objects and faces after learning versus before learning
Faillenot et al., 1997 (47)	Match sequentially seen objects versus pointing
Kertzman et al., 1997 (45)	Reaching to visual target versus visual fixation
Seitz et al., 1997 (19)	Copying of letterlike symbols versus holding a stylus
Binkofski et al., 1998 (30)	Object grasping versus pointing to object
Coull and Nobre, 1998 (26)	Visual target detection versus rest
Binkofski et al., 1999 (18)	Object exploration versus sphere manipulation
Dehaene et al., 1999 (56)	Numerical approximation versus exact calculation
Binkofski et al., 2000 (34)	Imagery of movement trajectories versus rest
Ehrrson et al., 2000 (68)	Simultaneous hand and foot movements versus rest
Gerardin et al., 2000 (35)	Imagery of finger flexion and extension versus rest
Perry and Zeki, 2000 (27)	Saccades to peripheral stimulus versus no saccades and no peripheral stimulus
Rowe et al., 2000 (36)	Maintaining target location in memory versus interval without memory content
Shikata et al., 2000 (38)	Discrimination of orientation of visual stimuli versus color discrimination
Azari et al., 2001 (67)	Religious recite versus happy recite in religious subjects
Bodegard et al., 2001 (37)	Tactile exploration versus passive touch
Bremmer et al., 2001 (46)	Polimodal motion processing versus modality-specific baseline
Buccino et al., 2001 (50)	Observation of object-related action versus observation of static scene
Desmurget et al., 2001 (22)	Reaching to stationary targets versus looking at stationary targets
Heide et al., 2001 (31)	Triple-step saccades versus fixation
Herath et al., 2001 (24)	Detection of two stimulus characteristics versus single stimulus detection
Hermsdörfer et al., 2001 (55)	Gesture discrimination versus face discrimination
Kuhtz-Buschbeck et al., 2001 (20)	Gentle precision grip versus firm grip
Rushworth et al., 2001 (28)	Response switching versus visual stimulation switch
Sereno et al., 2001 (57)	Phase-encoded delayed saccades
Zago et al., 2001 (41)	Mental calculation versus reading numbers
Suchan et al., 2002 (25)	Imagery of visuospatial patterns versus viewing numbers in ascending order
Suchan et al., 2002 (53)	Sequential matching of matrices versus simultaneous matching of matrices

In a more medial and dorsal location we observed a further focus. According to its location at the lower bank of the intraparietal sulcus and its functional properties, this area was tentatively labeled anterior intraparietal area (AIP) (see Fig. 18.1). This focus was shown to be concerned with execution of precisely tuned finger and eye movements (19,30,31), planning of sequential screen touches (32), imagery of self-generated and visually guided movements (33–35), movement selection (36), tactile discrimination of shape (37), discrimination of visual pattern orientation (38), attention (39,40), and mental computation (41). This assembly of different functions suggest that AIP may not simply be related to tuning of finger movements. Rather, it appears that this area mediates attentive processing of fine-tuned actions with a strong relation to perception in peripersonal space.

The perceptive information can be derived from the somatosensory and visual or even cognitive domain. The strong relation to action is supported by the strong connection from AIP to the ventral premotor cortex (2).

Slightly more caudally and dorsally as compared to AIP is another focus that may correspond to the ventral intraparietal area (VIP) (see Fig. 18.1) according to the nomenclature proposed by Colby et al. (42). This area was concerned with simple and sequential saccadic eye movements (27,31), selection of forelimb movements (43), observation of meaningless gestures (33,44), letter copying (19), reaching to visual targets (45), polymodal motion processing (46), object matching (47), switching of response (28), temporal orientation of subsequent stimuli (26), and self-ordering of digits (48). This area seems to process eye movement–related information by relating the different components of an entire action to each other in head-centered coordinates (49). This function seems to work across different modalities but does not appear to be as strongly correlated to action as AIP.

The next focus (area 4 in Fig. 18.1) is located in a position lateral to the intraparietal sulcus. It is engaged during saccade performance to peripheral stimuli (27), response switching (28), letter copying (19), observation of object-related action with the arm (50), orientation toward upcoming stimuli (26), and mental calculation (41). The information is less comprehensive than for the preceding loci. Nevertheless, this is an area along the intraparietal sulcus that maps to the most lateral localization. Also, because it is activated in relation to saccades, this area may correspond to the lateral intraparietal area (LIP) as observed in the nonhuman primate (51). From animal experiments it has been stated that LIP codes space in retinotopic coordinates (51,52). The neuroimaging data in humans also point out that this area may be concerned with feedback control. The human data suggest, however, that this area may have a broader role: It may update the brain with immediately previously performed actions during the course of an ongoing composite action even for cognition.

The next focus mapped to a location inferior and lateral to the intraparietal sulcus that was labeled inferior parietal lobule (IPL) (see Fig. 18.1). It is engaged in skilled reaching (21), discrimination of hand orientation (23), planning of sequential screen touches (32), delayed matching of rotation (53), self-ordering of digits (48), and simultaneous performance of two signal-reaction tasks (24). These tasks appear quite different on the first glance. However, it is suggested that common to these activation paradigms there is integration of information over time. In fact, three of the tasks involved working memory and have recently been reported to rely on specific information processing in the parietal association cortex (54). Similarly, reaching to different targets according to a prelearned sequence and dual-task performance also share the integration of ongoing information over time. Thus, IPL seems to relate external information with reference to preexisting internal information. It is located slightly more dorsal than activation foci observed in semantic language processing (see the following).

Dorsally adjacent is a further focus that is located at the caudal superior bank of the intraparietal sulcus. We tentatively called it posterior intraparietal area (pIP) (see Fig. 18.1). This area was engaged with saccade generation (27,31), tactile exploration (18), reaching (22), selection of movement direction (36,43), gesture recognition (33,55), visuomotor imagery (34,35), planning of sequential screen touches (32), and numerical approximation (56). Thus, this area seems to be involved in generating goal-directed actions, such as forelimb and eye movements, in a spatially scaled, egocentric reference frame. Thus, this area may provide the spatial coordinates for an upcoming action but simultaneously conveys information also about the previous action.

Dorsal to pIP is a further focus that was termed posterior superior parietal lobule (pSPL) (see Fig. 18.1). This area has been shown to be involved in the generation of spatially complex saccades (31,57), discrimination of object and hand orientation (38,55), observation of object-related action (50), selection of movements in space (36,43,45), and numerical approximation (56). This area seem to subserve movement generation in spatial coordinates that is similar to coding of movement intention as suggested by Snyder et al. (58). We speculate that this area may correspond to the reach-planning area of the posterior parietal cortex (59). Neural activity in this area was found to be more consistent with eye-centered than arm-centered coding, which renders this area particularly suitable for hand-eye-coordination (59).

At the caudal end of the intraparietal sulcus the common focus of activation was called cIPS (area 8 in Fig. 18.1). This area is involved in copying of letters of different size (19), spatial movement selection (43), skilled reaching according to a prelearned sequence (21), observation of meaningless gestures (23,33,44), motor imagery (34), matching of consecutive objects (26,47), switching of response (28), generation of peripheral saccades (27), semantic judgement of visually presented words (60), and self-ordering of digits (48). This area is apparently of major importance for the control of actions. Specifically, in this area actions seem to be governed according to the relation of external object characteristics. It is tempting to speculate that the three-dimensional surface orientation of visually presented objects studied in primates (61) may represent a comparable object cue as the different stimuli in the neuroimaging studies cited in the preceding. However, the data are also compatible with the notion that this area may play a role for attention to action (49).

Finally, there is a focus most caudal and medial along the intraparietal sulcus. We presume that this area may correspond to area V6 (area 9 in Fig. 18.1). The area is involved in saccade generation (31), response switching according to visual stimuli (28), response selection from memory in relation to previously seen visual stimuli (32,36), observation of meaningless gestures (44), visuomotor imagery (34), and arithmetic calculation (41). Thus, this area apparently processes higher-order visual information during vision and in the absence of visual cues as in visuomotor imagery and arithmetic calculation. Based on comparable findings obtained with single unit recordings in primates (62–64), this area seems to guide finger and eye movements based on visual cues even without visual feedback with reference to craniotopic space.

In addition to the foci that mapped along the intraparietal sulcus, there are a number of studies in which the reported activation foci occurred at slightly different locations outside the vicinity of the intraparietal sulcus. Two of such activations were observed in relation to tasks involving memory-related imagery of visual information and occurred in the precuneus at the medial surface of the brain (Table 18.2). Fletcher et al. (65) required subjects to recall visual associations cued by object words. An almost identical activation area was observed during matching a visual matrix to a previously seen visual matrix (53). In the study by Dolan et al. (66), subjects were required to identify impoverished visual objects after they had seen the same objects with high resolution. In the study by Azari et al. (67) subjects were required to enter a previously experienced cognitive state cued by reciting the appropriate verse. Previously, Deiber and colleagues (43) reported activation in this portion of the precuneus during movement selection in relation to a previously given instruction. The activation in these latter studies occurred in a more posterior location than

TABLE 18.2. *Activation Areas in Memory Related Imagery in the Precuneus at the Mesial Surface of the Parietal Lobe*

Anterior portion	3	–52	35	Fletcher et al., 1995 (65); Suchan et al., 2002 (25)
Posterior portion	9	–78	41	Dolan et al., 1997 (66); Azari et al., 2001 (67)

Talairach coordinates in x, y, z in mm.

TABLE 18.3. *Activation Areas Related to Kinematic Limb Control in the Anterior Part of the Parietal Lobe*

aSPL, ant. superior	32	–44	67	Ehrrson et al., 2000 (68); Gerardin et al., 2000 (35); Hermsdörfer et al., 2001 (55)
SII, secondary somatosensory area	53	–26	24	Burton et al., 1993 (69); Binkofski et al., 1999 (18); Ehrrson et al., 2000 (68); Bottini et al., (70); Desmurget et al., 2001 (22); Buccino et al., 2001 (50); Suchan et al., 2002 (53)

Talairach coordinates in x, y, z in mm.

that in the former two studies. It is tempting to speculate that the latter two tasks involved more mnemonic visual information processing than the first two tasks, whereas conversely the former tasks required more intuitive action.

A further area was found in the anterior portion of the parietal lobule in a most dorsal location (i.e., 20 mm dorsal to VIP) (Table 18.3). This activation occurred in relation to flexion-extension movements of the hand (35), in gesture recognition (55), and as an area of common activation in hand and foot movements (68). Thus, this area seems to code kinematics of limb movements.

Another focus that was not included in Figure 18.1 was the secondary somatosensory area at a ventral location of the anterior portion of the parietal cortex (Table 18.3). It is activated by tactile object identification (18), observation of object-related action with the mouth (50), reaching (22), visuospatial imagery (25), and as a common area of activation in hand and foot movements (68). This activation area coincides with the location of the secondary somatosensory area as reported by Burton et al. (69) and that during caloric vestibular stimulation and neck vibration (70), but is about 14 mm caudal to the activation site found in esophageal sensory stimulation (71).

In language studies it was found that activation of the supramarginal and/or angular gyrus was located in a closely similar location as the secondary somatosensory area but about some 30 mm more caudally (72,73). Thus, it is located approximately 16 mm ventral to IPL.

Moreover, there are activation foci reported in individual papers that do not show a cross-study correspondence. One of those was the activation focus in the anterior portion of the inferior parietal cortex described by Hermsdörfer et al. (55) during gesture discrimination. It mapped closely to AIP in rostroventral coordinates but was about 20 mm more lateral. Similarly, Clower et al. (74) reported a parietal cortex activation in a prism-adaptation task at a location that was close to our area IPL, but some 20 mm more lateral to it.

Finally, the lesion and activation data in relation to action in mirror-space occurred at still different locations (Fig. 18.2). Most remarkable, the area of lesion overlap extended over a considerable portion of the inferior parietal cortex reaching from the parieto-occipito-temporal junction to the caudal aspect of the intraparietal sulcus. The anatomic structures involved were the lobulus parietal inferior, the angular gyrus, and the adjacent posterior portion of the middle temporal gyrus (see Fig. 18.2A). The activation foci during action in mirror-space in healthy subjects mapped ventrally adjacent to the lesion into the inferior temporal gyrus and caudally adjacent into the posterior portion of the angular gyrus (see Fig. 18.2B). Thus, there was no correspondence of the characteristic lesion location in the patients and of the activation area in healthy subjects as was shown previously for AIP in relation to the precision grip (30). Rather, it appeared that the lesion in mirror agnosia disconnected cortical areas that were mandatory for action in mirror-space (16). Specifically, the activity in the human visual motion area (75–77) was disconnected from a more dorsal activation area that was close to IPL and cIPS (see Figs. 18.1 and 18.2).

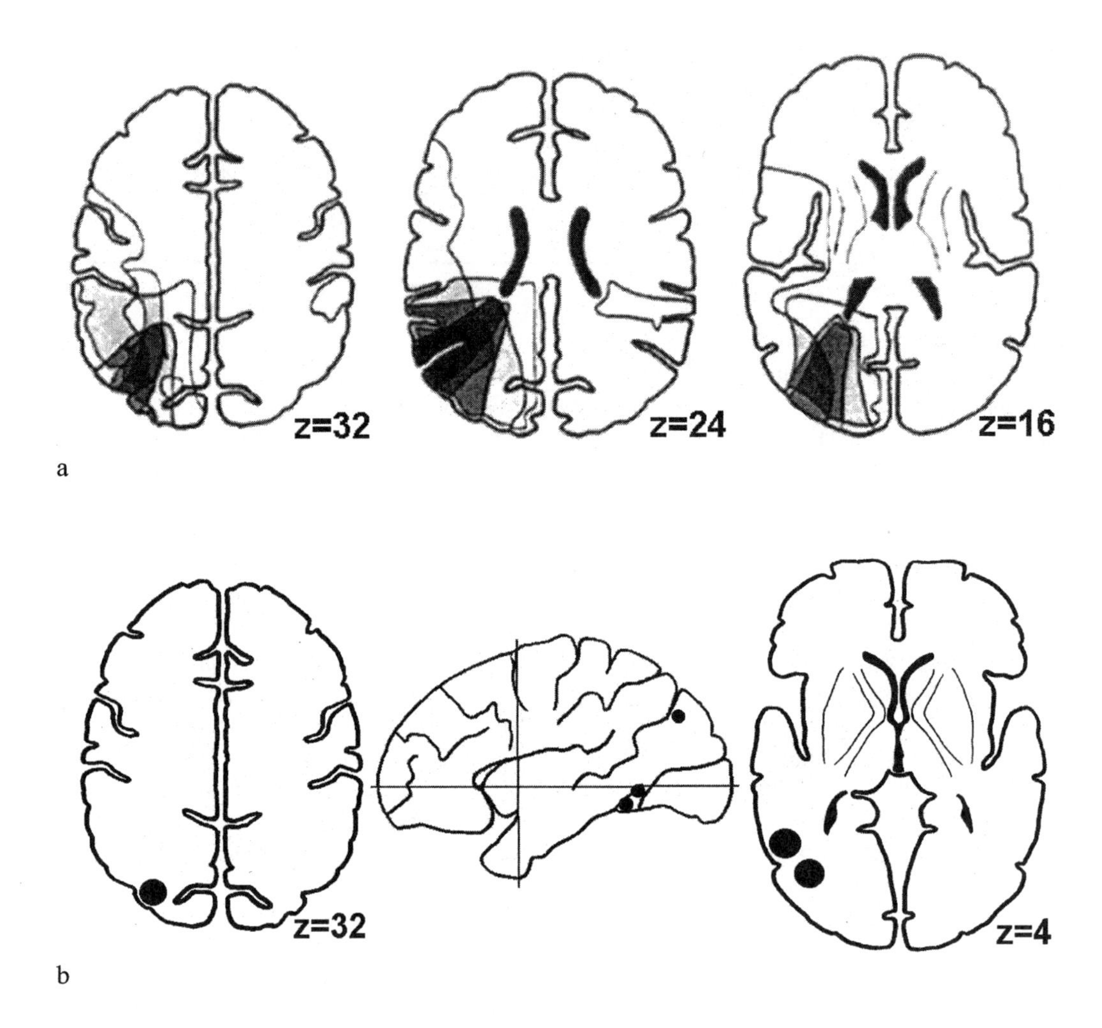

FIG. 18.2. Correspondence of lesion and activation studies for action in mirror space. **A:** Superimposition of the brain lesions in five patients with mirror agnosia into one hemisphere of Talairach space (14). The area of mean lesion overlap *(shaded)* is located at the junction of the angular gyrus, the lobulus parietalis inferior, and the middle temporal gyrus extending from z = 16 mm to z = 32 mm above the intercommissural plane. **B:** Mean activation areas related to pointing to the virtual target in the mirror as compared to pointing to the real target viewed through the mirror after superimposition into one hemisphere. Lateral view and axial planes 4 mm and 32 mm above the intercommissural plane.

DISCUSSION

In this study we performed a metaanalysis of recently published neuroimaging articles that reported mean activation areas in the human parietal lobe after transformation into stereotactic space. The studies were selected on the basis of the results of the imaging findings, not on the basis of the activation paradigms. We incorporated studies from our laboratory beyond studies of other laboratories published in the open literature. The selection is representative for different task conditions, such as sensorimotor control, attention, working memory, and cognitive computation. Obviously, this chapter cannot be comprehensive by any means. The main idea was to localize the activation areas in parietal cortex in order to find out whether different portions of the parietal cortex can be separated depending on modality of processed information or on the behavioral task. Thus, this chapter focuses on the posterior parietal cortex which is located caudal to the postcentral sulcus. Recently we described the topographically organized representations of the somatosensory system in the postcentral gyrus which is located anterior to the postcentral suclus (78).

By listing the activations according to stereotactic location it immediately became evident that the activation foci clustered at different locations rather than showing a complete mix across different studies. Consequently, we grouped the activation foci such that the coordinates reported in the publications did not exceed a distance of 10 mm in either dimension of space across the studies. This procedure minimized the spread of the peak foci around arithmetic centers. These centers represented the mean coordinate values for similarly located activation foci reported in different studies being beyond the spatial resolution of the functional images from which the activation foci were taken.

We feel that this approach is valid even if one acknowledges the fact that spatial normalization of the imaging data results in residual spatial variability (79,80). Thus, spatial normalization introduces noise into the comparison of coordinate values coming from different imaging systems and laboratories. Nevertheless, we observed an enormous consistency of the activation foci in similar paradigms across different studies. This is particularly remarkable, because different subjects were studied using different types of equipment in the different laboratories. Consequently, neuroimaging allows to provide reliable maps of human brain function. However, the nomenclature the different authors used for ascribing their activation foci to functional topographic locations was quite inconsistent unless contradictory. Only the supramarginal gyrus and AIP were labeled consistently across the studies. Thus, there is a great need to create functional-anatomic descriptions of the activated parietal subareas that can be communicated between different laboratories. We tentatively labeled the common spots according to the functional nomenclature proposed by primate studies. Consequently, our labeling may coincide with that of other authors but may also be quite different, although the stereotactic coordinates show the correspondence across studies. For example, it seems possible from this retrospective analysis that the activation in parietal cortex reported in relation to discrimination of shape occurred in AIP rather than the superior parietal lobule (81). Further, comparisons between the human and primate brain may be difficult in pure anatomic terms (1), but the similarity of function among areas found in primates and humans may help further to establish homologs (30,46).

From our metaanalysis, it also became evident that a given task may involve a number of areas in the parietal cortex. This indicates that different functions cannot be pinpointed to one center in the parietal lobe. Therefore, we suggest that there are critical local networks allowing for parallel information processing of subfunctions that are yet to be determined in detail, which subserve different aspects of behavior by their composition. It is tempting to speculate that event-related functional imaging approaches may shed light on the successive character of these operations.

This view transfers the concept of the modular organization of the parietal lobe to a higher level than that apparent from human lesion studies. Circumscribed lesions suggest that specific portions of the parietal lobe represent certain subfunctions (17,30,82). In such studies it was shown that patients with different lesion configurations may exhibit a common deficit and may show a small area of lesion overlap. This is illustrated in Figure 18.2A for mirror agnosia. The association of a specific, well-defined clinical deficit and a circumscribed common brain lesion has been the basis for suggesting a modular organization of the parietal lobe. Support for this hypothesis comes from activation studies in healthy subjects showing activation areas in the same location in a corresponding activation task (30). However, taken the evidence from mirror agnosia and action in mirror-space (see Fig. 18.2), it can be inferred that a lesion may rather damage a local functional network that mediates the corresponding normal behavior. Thus, experimental neuroimaging approaches based on psychophysical subtraction have not shown single processing nodes in the human parietal cortex as one may expect on the basis of deep electrode recordings in primates. Rather, neuroimaging appears suited to disentangle the different, but closely adjacent sites related to distributed information processing in the human parietal lobe.

CONCLUSIONS

The primate parietal cortex consists of cytoarchitectonically defined subareas that are supposed to be functionally highly specialized. In this meta-analysis of 50 neuroimaging studies in healthy subjects we found a high consistency of the mean activation foci for similar paradigms across different laboratories. The different subareas were involved in the control of limb and eye movements in egocentric and allocentric coordinates as well as in attention, short-term memory, and cognitive problem solving highlighting the role of the parietal lobe as an interface between perception and action. The combination of lesion and activation studies substantiated the concept of a modular organization of the human parietal lobe in which locally interspersed circuits mediate specialized cognitive subfunctions. Thus, neuroimaging provides a new framework for understanding the distributed information processing in the human parietal cortex.

ACKNOWLEDGMENTS

The work presented in the article was supported by the Sonderforschungsbereich 194 of the Deutsche Forschungsgemeinschaft.

REFERENCES

1. Zilles K, Palomero-Gallagher N. Cyto-, myelo-, and receptor architectonics of the human parietal cortex. *NeuroImage* 2001;14:S8–20.
2. Rizzolatti G, Luppino G, Matelli M. The organization of the cortical motor system: new concepts. *Electroencephalogr Clin Neurophysiol* 1998;106:283–296.
3. Petrides M, Pandya DN. Projections to the frontal cortex from the posterior parietal region in the rhesus monkey. *J Comp Neurol* 1984;228:105–116.
4. Glickstein M, May JG, III, Mercier BE. Corticopontine projection in the macaque: the distribution of labeled cortical cells after large injections of horseradish peroxidase in the pontine nuclei. *J Comp Neurol* 1985;235: 343–359.
5. Middleton FA, Strick PL. Basal ganglia and cerebellar loops: motor and cognitive circuits. *Brain Res Brain Res Rev* 2000;31:236–250.
6. Schwarz C, Thier P. Binding of signals relevant for action: toward a hypothesis of the functional role of the pontine nuclei. *Trends Neurosci* 1999;22:443–451.
7. Foerster O. Sensorischer Kortex. In: Bumke O, Foerster O, eds. *Allgemeine Neurologie.* Berlin: Springer-Verlag, 1936:1–357.
8. Van Essen DC, Anderson CH, Felleman DJ. Information processing in the primate visual system: an integrated systems perspective. *Science* 1992;255:419–423.
9. Caminiti R, Ferraina S, Johnson PB. The sources of visual information to the primate frontal lobe: a novel role for the superior parietal lobule. *Cerebral Cortex* 1996;6:319–328.
10. Ungerleider LG, Haxby JV. 'What' and 'where' in the human brain. *Curr Opin Neurobiol* 1994;4:157–165.
11. Jeannerod M, Arbib MA, Rizzolatti G, et al. Grasping objects: the cortical mechanisms of visuomotor transformation. *Trends Neurosci* 1995;18:314–320.
12. Milner AD, Goodale MA. Visual pathways to perception and action. *Prog Brain Res* 1993; 95:317–337.
13. Seitz RJ, Stephan KM, Binkofski F. Control of action as mediated by the human frontal lobe. *Exp Brain Res* 2000;133:71–80.
14. Talairach J, Tournoux P. *Co-planar stereotaxic atlas of the human brain.* Stuttgart: Thieme, 1988.
15. Oldfield RC. The assessment and analysis of handedness: the Edinburgh inventory. *Neuropsychologia* 1971;9:97–113.
16. Butler A, Binkofski F, Fink GR, et al. Neural correlates associated with visuo-spatial transformations using a mirror. *NeuroImage* 2001;13:S1141.
17. Binkofski F, Buccino G, Dohle C, et al. Mirror agnosia and mirror ataxia constitute different parietal lobe disorders [see comments]. *Ann Neurol* 1999;46:51–61.
18. Binkofski F, Buccino G, Posse S, et al. A fronto-parietal circuit for object manipulation in man: evidence from an fMRI-study. *Eur J Neurosci* 1999;11:3276–3286.
19. Seitz RJ, Canavan AG, Yaguez L, et al. Representations of graphomotor trajectories in the human parietal cortex: evidence for controlled processing and automatic performance. *Eur J Neurosci* 1997;9:378–389.
20. Kuhtz-Buschbeck JP, Ehrsson HH, Forssberg H. Human brain activity in the control of fine static precision grip forces: an fMRI study. *Eur J Neurosci* 2001;14:382–390.
21. Kawashima R, Roland PE, O'Sullivan BT. Functional anatomy of reaching and visuomotor learning: a positron emission tomography study. *Cereb Cortex* 1995;5:111–122.
22. Desmurget M, Grea H, Grethe JS, et al. Functional anatomy of nonvisual feedback loops during reaching: a positron emission tomography study. *J. Neurosci* 2001;21:2919–2928.

23. Bonda E, Petrides M, Frey S, et al. Neural correlates of mental transformations of the body-in-space. *Proc Natl Acad Sci USA* 1995;92:11180–11184.
24. Herath P, Kinomura S, Roland PE. Visual recognition: evidence for two distinctive mechanisms from a PET study. *Hum Brain Mapp* 2001;12:110–119.
25. Suchan B, Yaguez L, Wunderlich G, et al. Neural correlates of visuospatial imagery. *Cog Brain Res* 2002;131: 163–168.
26. Coull JT, Nobre AC. Where and when to pay attention: the neural systems for directing attention to spatial locations and to time intervals as revealed by both PET and fMRI. *J Neurosci* 1998;18:7426–7435.
27. Perry RJ, Zeki S. The neurology of saccades and covert shifts in spatial attention: an event-related fMRI study. *Brain* 2000;123:2273–2288.
28. Rushworth MF, Paus T, Sipila PK. Attention systems and the organization of the human parietal cortex. *J Neurosci* 2001;21:5262–5271.
29. Paulesu E, Frith CD, Frackowiak RS. The neural correlates of the verbal component of working memory. *Nature* 1993;362:342–345.
30. Binkofski F, Dohle C, Posse S, et al. Human anterior intraparietal area subserves prehension: a combined lesion and functional MRI activation study. *Neurology* 1998;50:1253–1259.
31. Heide W, Binkofski F, Seitz RJ, et al. Activation of frontoparietal cortices during memorized triple-step sequences of saccadic eye movements: an fMRI study. *Eur J Neurosci* 2001;13:1177–1189.
32. Owen AM, Doyon J, Petrides M, et al. Planning and spatial working memory: a positron emission tomography study in humans. *Eur J Neurosci* 1996;8:353–364.
33. Parsons LM, Fox PT, Downs JH, et al. Use of implicit motor imagery for visual shape discrimination as revealed by PET. *Nature* 1995;375:54–58.
34. Binkofski F, Amunts K, Stephan KM, et al. Broca's area subserves imagery of motion: a combined cytoarchitectonic and MRI study. *Hum Brain Mapp* 2000;11: 273–285.
35. Gerardin E, Sirigu A, Lehericy S, et al. Partially overlapping neural networks for real and imagined hand movements. *Cereb Cortex* 2000;10:1093–1104.
36. Rowe JB, Toni I, Josephs O, et al. The prefrontal cortex: response selection or maintenance within working memory? *Science* 2000;288:1656–1660.
37. Bodegard A, Geyer S, Grefkes C, et al. Hierarchical processing of tactile shape in the human brain. *Neuron* 2001;31:317–328.
38. Shikata E, Hamzei F, Glauche V, et al. Surface orientation discrimination activates caudal and anterior intraparietal sulcus in humans: an event-related fMRI study. *J Neurophysiol* 2001;85:1309–1314.
39. Pardo JV, Pardo PJ, Janer KW, et al. The anterior cingulate cortex mediates processing selection in the Stroop attentional conflict paradigm. *Proc Natl Acad Sci USA* 1990;87:256–259.
40. Corbetta M, Miezin FM, Shulman GL, et al. A PET study of visuospatial attention. *J Neurosci* 1993;13: 1202–1226.
41. Zago L, Pesenti M, Mellet E, et al. Neural correlates of simple and complex mental calculation. *NeuroImage* 2001;13:314–327.
42. Colby CL, Duhamel JR, Goldberg ME. Ventral intraparietal area of the macaque: anatomic location and visual response properties. *J Neurophysiol* 1993;69: 902–914.
43. Deiber MP, Passingham RE, Colebatch JG, et al. Cortical areas and the selection of movement: a study with positron emission tomography. *Exp Brain Res* 1991;84: 393–402.
44. Decety J, Grezes J, Costes N, et al. Brain activity during observation of actions. Influence of action content and subject's strategy. *Brain* 1997;120:1763–1777.
45. Kertzman C, Schwarz U, Zeffiro TA, et al. The role of posterior parietal cortex in visually guided reaching movements in humans. *Exp Brain Res* 1997;114: 170–183.
46. Bremmer F, Schlack A, Shah NJ, et al. Polymodal motion processing in posterior parietal and premotor cortex: a human fMRI study strongly implies equivalencies between humans and monkeys. *Neuron* 2001;29:287–296.
47. Faillenot I, Toni I, Decety J, et al. Visual pathways for object-oriented action and object recognition: functional anatomy with PET. *Cereb Cortex* 1997;7:77–85.
48. Petrides M, Alivisatos B, Meyer E, et al. Functional activation of the human frontal cortex during the performance of verbal working memory tasks. *Proc Natl Acad Sci USA* 1993;90:878–882.
49. Culham JC, Kanwisher NG. Neuroimaging of cognitive functions in human parietal cortex. *Curr Opin Neurobiol* 2001;11:157–163.
50. Buccino G, Binkofski F, Fink GR, et al. Action observation activates premotor and parietal areas in a somatotòpic manner: an fMRI study. *Eur J Neurosci* 2001; 13:400–404.
51. Colby CL, Duhamel JR, Goldberg ME. Visual, presaccadic, and cognitive activation of single *Neuron* s in monkey lateral intraparietal area. *J Neurophysiol* 1996; 76:2841–2852.
52. Duhamel JR, Bremmer F, BenHamed S, et al. Spatial invariance of visual receptive fields in parietal cortex neurons. *Nature* 1997;389:845–848.
53. Suchan B, Yaguez L, Wunderlich G, et al. Hemispheric dissociation of visuo-spatial processing and visual rotation. *Cog Brain Res* 2002;136:533–544.
54. Cohen J. Prefrontal cortex involved in higher cognitive functions. Introduction. *NeuroImage* 2000;11:378–379.
55. Hermsdorfer J, Goldenberg G, Wachsmuth C, et al. Cortical correlates of gesture processing: clues to the cerebral mechanisms underlying apraxia during the imitation of meaningless gestures. *NeuroImage* 2001;14:149–161.
56. Dehaene S, Spelke E, Pinel P, et al. Sources of mathematical thinking: behavioral and brain-imaging evidence. *Science* 1999;284:970–974.
57. Sereno MI, Pitzalis S, Martinez A. Mapping of contralateral space in retinotopic coordinates by a parietal cortical area in humans. *Science* 2001;294:1350–1354.
58. Snyder LH, Batista AP, Andersen RA. Coding of intention in the posterior parietal cortex. *Nature* 1997;386: 167–170.
59. Batista AP, Buneo CA, Snyder LH, et al. Reach plans in eye-centered coordinates. *Science* 1999;285:257–260.
60. Vandenberghe R, Price C, Wise R, et al. Functional anatomy of a common semantic system for words and pictures. *Nature* 1996;383:254–256.
61. Taira M, Tsutsui KI, Jiang M, et al. Parietal neurons represent surface orientation from the gradient of binocular disparity. *J Neurophysiol* 2000;83:3140–3146.
62. Galletti C, Fattori P, Battaglini PP, et al. Functional de-

marcation of a border between areas V6 and V6A in the superior parietal gyrus of the macaque monkey. *Eur J Neurosci* 1996;8:30–52.
63. Cavada C. The visual parietal areas in the macaque monkey: current structural knowledge and ignorance. *NeuroImage* 2001;14:S21–S26.
64. Fattori P, Gamberini M, Kutz DF, et al. 'Arm-reaching' neurons in the parietal area V6A of the macaque monkey. *Eur J Neurosci* 2001;13:2309–2313.
65. Fletcher PC, Frith CD, Baker SC, et al. The mind's eye—precuneus activation in memory-related imagery. *NeuroImage* 1995;2:195–200.
66. Dolan RJ, Fink GR, Rolls E, et al. How the brain learns to see objects and faces in an impoverished context. *Nature* 1997;389:596–599.
67. Azari NP, Nickel J, Wunderlich G, et al. The neural correlate of religious experience. *Eur J Neurosci* 2001; 13:1649–1653.
68. Ehrsson HH, Naito E, Geyer S, et al. Simultaneous movements of upper and lower limbs are coordinated by motor representations that are shared by both limbs: a PET study. *Eur J Neurosci* 2000;12:3385–3398.
69. Burton H, Videen TO, Raichle ME. Tactile-vibration-activated foci in insular and parietal-opercular cortex studied with positron emission tomography: mapping the second somatosensory area in humans. *Somatosens Mot Res* 1993;10:297–308.
70. Bottini G, Karnath HO, Vallar G, et al. Cerebral representations for egocentric space: Functional-anatomic evidence from caloric vestibular stimulation and neck vibration. *Brain* 2001;124:1182–1196.
71. Binkofski F, Schnitzler A, Enck P, et al. Somatic and limbic cortex activation in esophageal distention: a functional magnetic resonance imaging study. *Ann Neurol* 1998;44:811–815.
72. Demonet JF, Chollet F, Ramsay S, et al. The anatomy of phonological and semantic processing in normal subjects. *Brain* 1992;115:1753–1768.
73. Jessen F, Erb M, Klose U, et al. Activation of human language processing brain regions after the presentation of random letter strings demonstrated with event-related functional magnetic resonance imaging. *Neurosci Lett* 1999;270:13–16.
74. Clower DM, Hoffman JM, Votaw JR, et al. Role of posterior parietal cortex in the recalibration of visually guided reaching. *Nature* 1996;383:618–621.
75. Watson JD, Myers R, Frackowiak RS, et al. Area V5 of the human brain: evidence from a combined study using positron emission tomography and magnetic resonance imaging. *Cereb Cortex* 1993;3:79–94.
76. Buchel C, Josephs O, Rees G, et al. The functional anatomy of attention to visual motion. A functional MRI study. *Brain* 1998;121:1281–1294.
77. Goebel R, Khorram-Sefat D, Muckli L, et al. The constructive nature of vision: direct evidence from functional magnetic resonance imaging studies of apparent motion and motion imagery. *Eur J Neurosci* 1998; 10:1563–1573.
78. Schnitzler A, Seitz RJ, Freund H-J. The somatosensory system. In: Toga AW, Mazziotta JC, eds. Brain Mapping: The applications. San Diego: Academic Press, 2000, 291–329.
79. Seitz RJ, Bohm C, Greitz T, et al. Accuracy and precision of the computerized brain atlas programme for localization and quantification in positron emission tomography. *J Cereb Blood Flow Metab* 1990;10: 443–457.
80. Roland PE, Zilles K. Brain atlases—a new research tool. *Trends Neurosci* 1994;17:458–467.
81. Seitz RJ, Roland PE, Bohm C, et al. Somatosensory discrimination of shape: tactile exploration and cerebral localization. *Eur J Neurosci* 1991;3:481–492.
82. Binkofski F, Kunesch E, Classen J, et al. Tactile apraxia: unimodal apractic disorder of tactile object exploration associated with parietal lobe lesions. *Brain* 2001;124: 132–144.

19

Neglect Syndromes: The Role of the Parietal Cortex

Giuseppe Vallar, Gabriella Bottini*, and Eraldo Paulesu

*Department of Psychology, University of Milano-Bicocca, Milano, Italy; *Department of Psychology, University of Pavia, Pavia, Italy*

NEGLECT SYNDROMES

The Spectrum of Neglect

Spatial unilateral neglect refers to the defective ability of patients with unilateral brain damage to explore the side of space contralateral to the lesion (contralesional), and to report stimuli presented in that portion of space. The adjective "unilateral" denotes a main feature of the disorder, which distinguishes spatial neglect from more global deficits of spatial exploration and perception. The patients' performance is comparatively preserved in the side ipsilateral to the lesion (ipsilesional), or a gradient may be present.

Spatial unilateral neglect has long been considered a largely unitary disorder. Since the late 1970s, however, it has been increasingly clear that the clinical term "spatial unilateral neglect" refers to a number of discrete disorders, which frequently occur in association, but may also present in isolation (1–3). Table 19.1 (4,5) shows a simplistic taxonomy of the main neurologic manifestations that go under the rubric of spatial unilateral neglect and related disorders. The many neurologic observations in individual patients of specific components of the neglect syndrome (e.g., extrapersonal neglect without "personal" neglect, or vice versa) (6,7) may suggest that a unitary treatment of the disorder is meaningless, and may misdirect the experimental investigation (2). Should unilateral neglect be considered a "weak" syndrome; namely, an association of symptoms caused by the anatomic contiguity of the responsible lesions, which disrupt different and unrelated functions, or a "strong" syndrome, an association of symptoms produced by the derangement of a single unitary function (8)? Three main reasons argue for a partly unitary treatment of the disorder.

1. The laterality of neglect: The manifold manifestations of neglect summarized in Table 19.1 are more frequent and severe in the left side of extrapersonal and personal space.
2. The hemispheric side of responsible cerebral lesions: Neglect is more frequent and severe after damage in the right cerebral hemisphere.
3. The effects of a variety of sensory stimulations and modulations: Vestibular, optokinetic, transcutaneous mechanical vibration and electrical stimulation, adaptation to visually displacing prisms, and other manipulation of sensory input may improve or worsen the manifestations of neglect summarized in Table 19.1 in a basically similar fashion.

TABLE 19.1. *A Taxonomy of the Clinical Syndrome of Spatial Unilateral Neglect (USN).*

Defective Manifestations		
Extrapersonal space		Personal/bodily space
Dimension	Variety	
(Input/output)	Perceptual USN[a]	Hemiasomatognosia[b]
	Premotor/intentional USN, directional hypokinesia[d]	Anosognosia[c]
		Motor neglect[e]
Sectors of space (with reference to the body)	Lateral external USN[f]	
	Lateral internal (imaginal) USN	
	Altitudinal[g]	
Reference frames	Egocentric USN[h]	
	Allocentric/object-based USN	
Sensory modality[i]	Visual USN (pseudo-hemianopia)	Somatosensory USN
	Auditory USN	
	Olfactory USN	
Processing domain (material-specific forms of neglect)	Facial USN	
	Neglect dyslexia	
Productive Manifestations		
Extrapersonal space		Personal/bodily space
Avoidance[j]		Somatoparaphrenia[k]
Hyperattention, magnetic attraction toward ipsilesional targets		
Perseveration		

Defective manifestations (the better known and more extensively investigated aspect of USN) refer to *negative phenomena*, characterized by the *absence* of specific behavioral responses, such as the impaired exploration of the contralesional side of space, or the failure to report stimuli presented in that sector of space. *Productive* manifestations refer to *positive phenomena*, characterized by the *presence* of specific behaviors.

[a]Defective awareness of targets in the neglected sector of space.
[b]Defective awareness of the contralesional side of the body.
[c]Defective awareness or denial of contralesional motor, somatosensory and visual half-field deficits.
[d]Defective programming of movements of the ipsilesional limbs towards targets in the neglected, contralesional, sector of space.
[e] Failure to move the contralesional limbs, in the absence of primary motor impairment (hemiparesis or hemiplegia).
[f]Along a left-right axis: near, far.
[g]Along a vertical axis: upper, lower.
[h]With reference to the head, the trunk, and limbs.
[i]Defective awareness of sensory input in a particular sensory modality.
[j]Active withdrawal from contralesional targets.
[k]Delusional beliefs concerning the contralesional side of the body.

Arguments 1 and 2 are well-known and time-honored neurologic knowledge (9). Argument 3 is based on a wide number of more recent studies, initiated using vestibular stimulation in patients with extrapersonal visual neglect (10–12), and followed by the investigation of the effects of other sensory inputs and manipulations (13,14). Specifically, the similarity of the effects of these sensory modulations suggests the existence of a basic spatial medium, shared by the different specific representations of space and attention systems (e.g., for a particular sector of space, "near" peripersonal versus "far" extrapersonal), whose derangements may give rise to the manifestations of neglect summarized in Table 19.1. The precise anatomic correlates of such a system are poorly known. A recent study, which investigated in humans the cerebral areas conjointly activated by vestibular stimulation and transcutaneous mechanical vibration of the neck muscles (two treatments that may improve or worsen spatial neglect), suggests the parieto-insular vestibular cortex as a main component part (15).

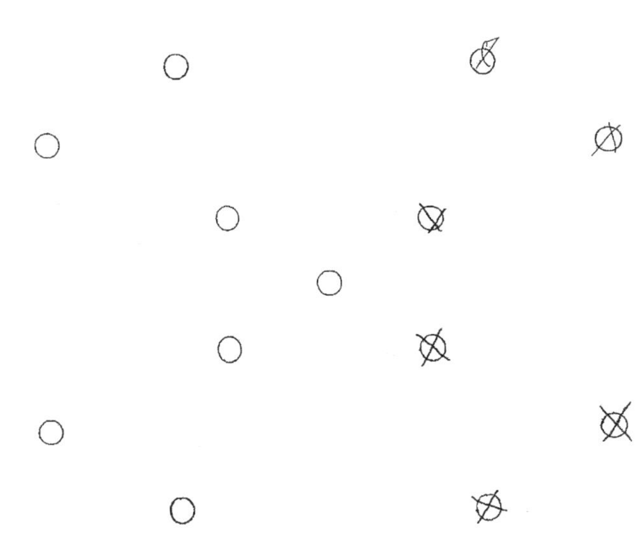

FIG. 19.1. Circle cancellation in visuospatial unilateral neglect:target omission on the left-hand side of the display.

Although the spectrum of clinical manifestations of neglect is wide (see Table 19.1), current knowledge about its neuroanatomic correlates concerns mainly a single, better known and investigated aspect of the syndrome: *visual extrapersonal neglect.* In most anatomic-clinical correlation studies, visual extrapersonal neglect was assessed by target cancellation, line bisection, and other tasks, such as reading, drawing, and copying. In cancellation tasks, patients are required to cross out targets, which may or may be not interspersed among distracters. Figure 19.1 shows an example of left neglect in a circle crossing visuomotor exploratory task, with omission errors in the left-hand side of the display. Figure 19.2 shows a rightward error in setting the midpoint of a horizontal line, committed by a patient with left neglect. In both tasks, the deficit (target omission and rightward bisection error) can not be explained by elementary sensory and motor deficits. Patients are free to move their head and eyes, and use the unaffected right hand.

Extinction to double simultaneous stimulation is a well-known neurologic deficit, frequently, although not necessarily, associated with spatial unilateral neglect: Patients with damage to the central nervous system fail to report one of two stimuli in conditions of double simultaneous stimulation, whereas perception of single stimuli is preserved. In the majority of patients with unilateral cerebral damage, extinction affects the stimulus

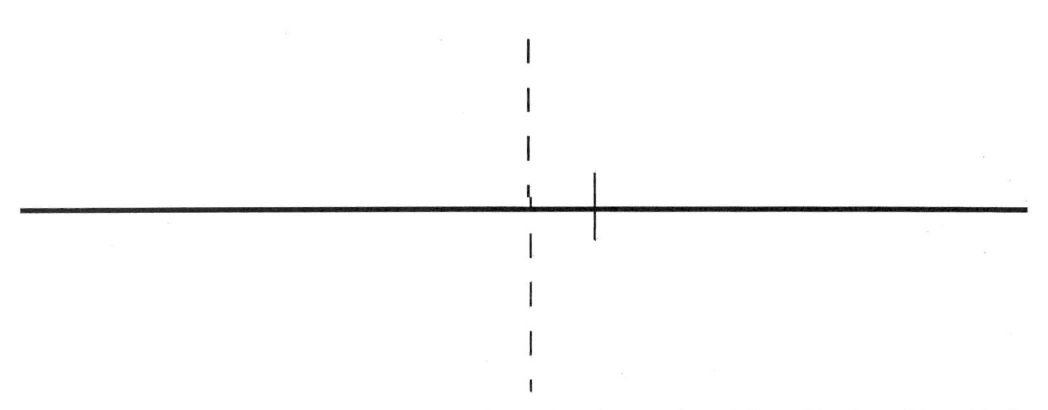

FIG. 19.2. Line bisection in left visuospatial unilateral neglect:setting of the subjective midpoint to the right of the objective center of the segment.

contralateral to the side of the lesion. Extinction phenomena have been reported in different sensory modalities (visual, tactile, auditory, olfactory), and these modality-specific forms may occur independent of one another. Stimuli may be extinguished both within (intramodal) and across (crossmodal extinction) sensory modalities (16–18).

The precise pathologic mechanisms underlying the manifold manifestations of the neglect syndrome are not discussed in this chapter (5,19). There is consensus, however, around the opinion that spatial unilateral neglect can not be interpreted as a deficit concerning lower levels of the sensorimotor neural machinery, namely, reduced or disordered processing of current sensory input, defective exploration through eye or limb movements, owing to the impairment of relatively peripheral motor mechanisms, or a combination of the two types of disorders. Current views consider neglect as a higher-order cognitive deficit, affecting the multiple components of spatial directed attention, or the spatial representations, underpinning perceptual and action awareness.

Extinction phenomena, by contrast, may depend on dysfunction at different levels of neural processing of sensory information. On the one hand, a dysfunction involving the peripheral stages of sensory processing is likely to be responsible for extinction because of spinal cord lesion (20,21). Spatial neglect, conversely, is not associated with such a lower-level damage to the nervous system. On the other hand, extinction of left-sided visual stimuli within the right visual field (22–24), or of left-sided somatosensory stimuli within the right side of the body of right-brain–damaged patients (25) indicates a much higher-order level of dysfunction. In line with this view, right-hemisphere–damaged patients may show extinction of stimuli presented to the contralesional left nostril on olfactory double simultaneous stimulation (26). Because the olfactory pathways are not crossed, this pattern of impairment is likely to reflect defective representation of the contralesional side of space, or impaired orientation of attention toward it. The alternative interpretation in terms of attenuated sensory input instead predicts extinction of olfactory stimuli presented to the right nostril. A further demonstration of high-level dysfunction underlying some extinction phenomena is the finding that the rate of somatosensory extinction of right-brain–damaged patients on bimanual stimulation decreases when the hands are crossed, so that the left (affected) hand is positioned on the ipsilesional right side, with respect to the trunk's sagittal midplane (27). Finally, the observation that the sensory stimulations that improve spatial unilateral neglect may also reduce left-sided contralesional extinction associated with right hemispheric damage suggests underlying pathologic mechanisms shared by extinction and unilateral neglect (28,29).

Hemispheric Differences

Spatial unilateral neglect is more frequent and severe after damage to the right cerebral hemisphere in right-handed patients. This hemispheric difference has been found by many studies using different tests (visuomotor exploration, bisection, drawing, and copying) to assess neglect (19,30,31). Although right-sided neglect is less frequent and severe, right-sided neglect associated with left-brain damage is behaviorally similar to the left-sided deficit.

The hemispheric asymmetry of extinction to double simultaneous stimulation is less definite than that of spatial unilateral neglect. Extinction has been found in both left- and right-hemisphere–damaged patients (32), although suggestions of a tendency toward a greater incidence in right-hemisphere–damaged patients have been made (16,33). The higher rate of contralesional (left-sided) tactile extinction after right-sided intracarotid sodium Amytal injection, compared to the effects of a left-sided injection, indicates an hemispheric asymmetry similar to the one that characterizes neglect, although less pronounced (34).

The hemispheric asymmetry of spatial unilateral neglect, and, to a lesser degree, of extinction, may be explained by the assumption that the right cerebral hemisphere possesses a largely bilateral representation of space, and may readily direct spatial attention toward either side of space, although with a contralesional bias, preference, or more effective processing ability. The left hemisphere, by contrast, is mainly concerned with the contralesional right side of space, with a minor representation of the ipsilesional side, and ability to direct spatial attention toward that side (19,35,36). Neuroimaging activation studies in normal subjects (37–42) and experiments in split-brain patients (43,44) support this view. A model of the hemispheric asymmetry in spatial representation and attention is shown in Figure 19.3.

THE ANATOMIC BASIS OF EXTRAPERSONAL VISUOSPATIAL UNILATERAL NEGLECT: THE ROLE OF PARIETAL DAMAGE

Anatomic-Clinical Correlations Studies in Brain-Damaged Patients

Studies in Large Series of Brain-Damaged Patients

Since its discovery as a neurologic deficit, spatial unilateral neglect has been regarded as a symptom with a remarkable localizing value, which indicates a lesion of the parietal lobe (9,17,45,46). Since the late 1970s, the availability of noninvasive radiologic techniques has confirmed these early observations in larger series of patients, showing also that both lesions located outside the parietal lobe, and purely subcortical damage sparing the

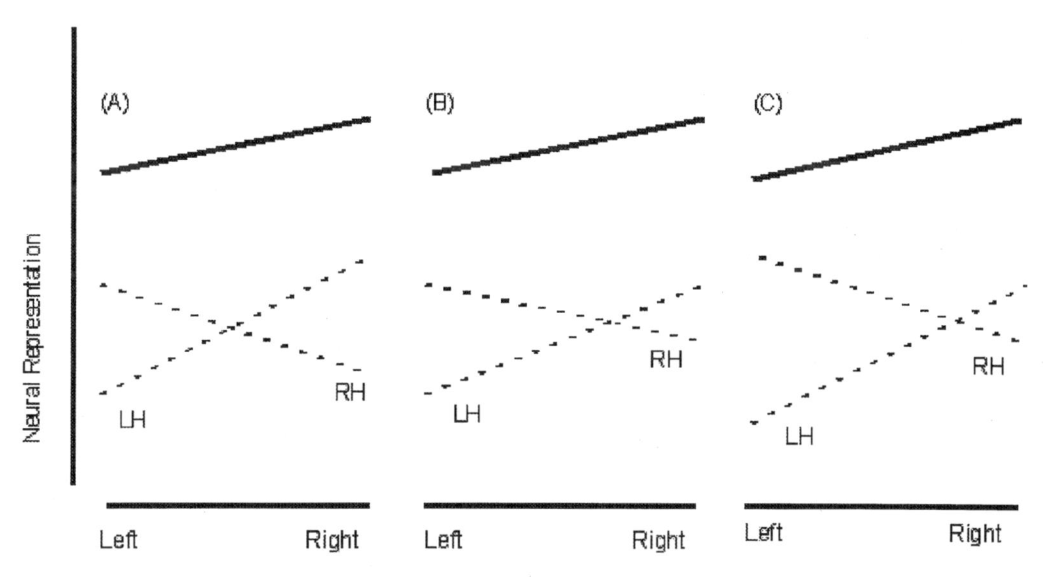

FIG. 19.3. Three possible patterns of left (LH) and right (RH) hemisphere contributions *(dashed lines)* to the overall neural representation of egocentric space *(continuous line),* which is assumed to be skewed rightward in humans. **A:** The neural representations of the contralateral and ipsilateral hemispaces are greater in the LH and RH, respectively, in such a way that the two hemispheres provide an equal contribution to the overall neural representation. **B:** The contralateral hemispaces are equally represented in the two hemispheres, but the ipsilateral representation is greater in the RH, which then provides the major contribution to the overall neural representation. **C:** The representations of both the contralateral and ipsilateral hemispaces are greater in the RH. (Reprinted from: Bisiach E, Vallar G. Unilateral neglect in humans. In: Boller F, Grafman J, Rizzolatti G, eds. *Handbook of neuropsychology,* 2nd ed. Amsterdam: Elsevier Science, 2000:459–502, with permission.)

cortex, may produce extrapersonal spatial unilateral neglect. The anatomic-clinical correlation studies reviewed in this section considered visuospatial extrapersonal neglect as a unitary disorder, assessed by a range of tasks including cancellation, line bisection, drawing, and copying.

Group studies comprising right-brain–damaged patients both with and without evidence of extrapersonal visuospatial unilateral neglect have consistently shown that the deficit occurs much more frequently after parietal damage, although instances of extrapersonal visual unilateral neglect associated with lesions confined to the frontal lobe have been reported repeatedly (47,48–50). This conclusion is even stronger if one considers that the anatomic-clinical group studies, far from being homogeneous, differ in a number of important respects, such as the etiology of the cerebral lesion, the duration of disease, and the tests used to assess neglect (51). In 41 right-brain–damaged patients with space-occupying lesions, a battery including figure-ground discrimination, constructional, copying, drawing from memory, geographical location, and visuospatial tasks was used to assess neglect: A 25% incidence of visual neglect after anterior damage, and a 54% incidence after posterior lesions was found (52). No association between neglect and frontal damage was detected in a series of 179 patients (53). These early data subsequently were confirmed by computed tomography (CT) clinical correlation studies. In 40 right-brain–damaged patients, a study that used a battery including drawing, copying, and cancellation tasks found unilateral neglect in 66% of patients with posterior vascular and neoplastic lesions, but only in 20% of patients with anterior lesions (32). In one study, neglect was assessed by a circle cancellation task (see Fig. 19.1) in 59 right-brain–damaged stroke patients, examined in the early phase of their disease. Eighty percent of right-brain–damaged patients with extensive anterior-posterior lesions and 63% of patients with posterior lesions showed evidence of neglect, but this percentage fell to only 8% in patients with anterior damage (50). In 120 right-hemisphere–damaged patients a more recent CT and single-photon emission tomography (SPET) study confirmed the association of left visual neglect, assessed by a battery including copying, drawing, cancellation, and bisection tasks, with damage in the right posterior parietal cortex and in the posterior white matter fiber bundles (54). An analysis using SPET only showed a major hypoperfusion in the right temporo-parietal junction, and in an extensive set of regions, including the visual association cortex (Brodmann's area, BAs 18 and 19), the temporal lobe (BAs 22, 21, and 37), and the inferior parietal lobule (BAs 39 and 40) (55). The perfusion data, together with the lesion data that have been discussed, suggest that an extensive neural network, centered in the high-order association areas of the inferior parietal lobule, at the temporo-parietal junction, underpins spatial representation and attention.

The close association of visuospatial unilateral neglect with damage involving the right posterior-inferior parietal region at the temporo-parietal junction does not appear to be related to etiology at the temporo-parietal junction. In one series of 19 cerebral tumors, eight out of nine patients with visual neglect showed an involvement of the right parietal region, which was spared in eight out of 10 cases without neglect (51).

In a number of anatomic-clinical correlation studies, attempts have been made to specify in more detail the intraparietal localization of the lesion site associated with visual neglect. Studies comprising both patients who underwent precisely localized corticectomy for the relief of epilepsy (56), and patients suffering from natural (neoplastic and vascular) lesions (50,51,54,57) converge on the *supramarginal gyrus of the right inferior parietal lobule* as a crucial region. The role of damage to the underlying temporo-parieto-occipital white matter is also relevant (50,54,58). The subcortical damage may disrupt both posterior lobar, and anterior-posterior connections, as well as interrupting ascending activating pathways (59,60). Lesions involving the right middle (58) and superior

(61) temporal gyri may contribute to bringing about visual neglect.

The specificity of the association of damage to the inferior parietal lobule, at the temporo-parietal junction, with visual neglect is further supported by data suggesting that nonhemianopic patients with reaching disorders in the contralesional half-field have lesions clustering in the upper parietal region (62). In one clinical CT correlation study (63), the lesions of six left- and four right-brain–damaged patients clustered in the upper part of the posterior parietal lobe, around the intraparietal sulcus. In this series, four left- and two right-brain–damaged patients with misreaching showed neither visual neglect nor extinction to double simultaneous stimulation. These anatomic-clinical data indicate a dissociation within the posterior parietal region, with inferior parietal damage bringing about visual neglect, superior parietal damage, a reaching disorder (optic ataxia) (57). Additional support for the association of visual neglect with inferior-posterior parietal damage is provided by the observation that patients with posterior damage and no signs of visual neglect have lesions superimposing in the superior-posterior parietal areas, as well as in the occipital regions (50). Other studies found an association of left neglect with lesions involving the right fronto-temporo-parietal junction, but not with either purely frontal or superior parietal damage (64).

Patients with visual extinction to double simultaneous stimulation but minimal-to-mild signs of visual neglect are disproportionately impaired in orienting visual attention toward contralesional targets, when an (invalid) cue has appeared in the opposite ipsilesional side (65–67). In one study (66) this deficit in the orientation of attention correlated with damage to the superior parietal regions. A successive study in patients with no clinical evidence of extinction and neglect, however, showed that the impairment in the orientation of attention toward contralesional targets was associated with damage to the temporo-parietal junction and not to the superior-posterior parietal region (67). In these experiments (65–67) the defective disengagement from ipsilesional invalid cues was produced by both left- and right-sided hemispheric lesions, although a tendency toward a more severe impairment after right-hemispheric damage was reported (66).

The second main cortical lesion site associated with neglect is the frontal lobe. Damage involving *Brodmann's areas 6, 8, and 44 of the right dorsolateral premotor cortex and the medial frontal regions* (anterior cingulate cortex, supplementary motor area) may bring about visual neglect (47,48,50,54,55,68). One group study suggests a main role of damage to the more ventral premotor cortex (Brodmann's area 44 in the right hemisphere) (68).

The cortical lesion sites more frequently associated with extrapersonal visual unilateral neglect are shown in Figure 19.4, which shows discrete areas of damage. It should be noted, however, that cognitive activity is supported by neural networks (69), including multiple connected brain regions (70,71). More specifically, in patients with neglect, structural damage to specific cerebral regions frequently brings about a more extensive dysfunction in remote connected areas (diaschisis) (72). This, together with the site and size of the lesion, may account in part for the severity of the disorder, because it is a prognostic factor for functional recovery (73–76).

The lesion pattern shown in Figure 19.4 shows that spatial unilateral neglect is associated with damage to a set of higher-order association areas. Damage confined to the primary sensory and motor cortices does not bring about neglect. Accordingly, the disruption of sensory representations of the stimulus, specific for each modality (e.g., *retinotopic* in vision and somatotopic in the tactile domain) is not a primary causal factor of neglect, although impairments at these more peripheral levels of representation may worsen aspects of the disorder.

This is compatible with the view that neglect is a higher-order cognitive deficit, involving spatial representation and attention, in spatial egocentric and allocentric coordinate frames. In egocentric coordinate frames,

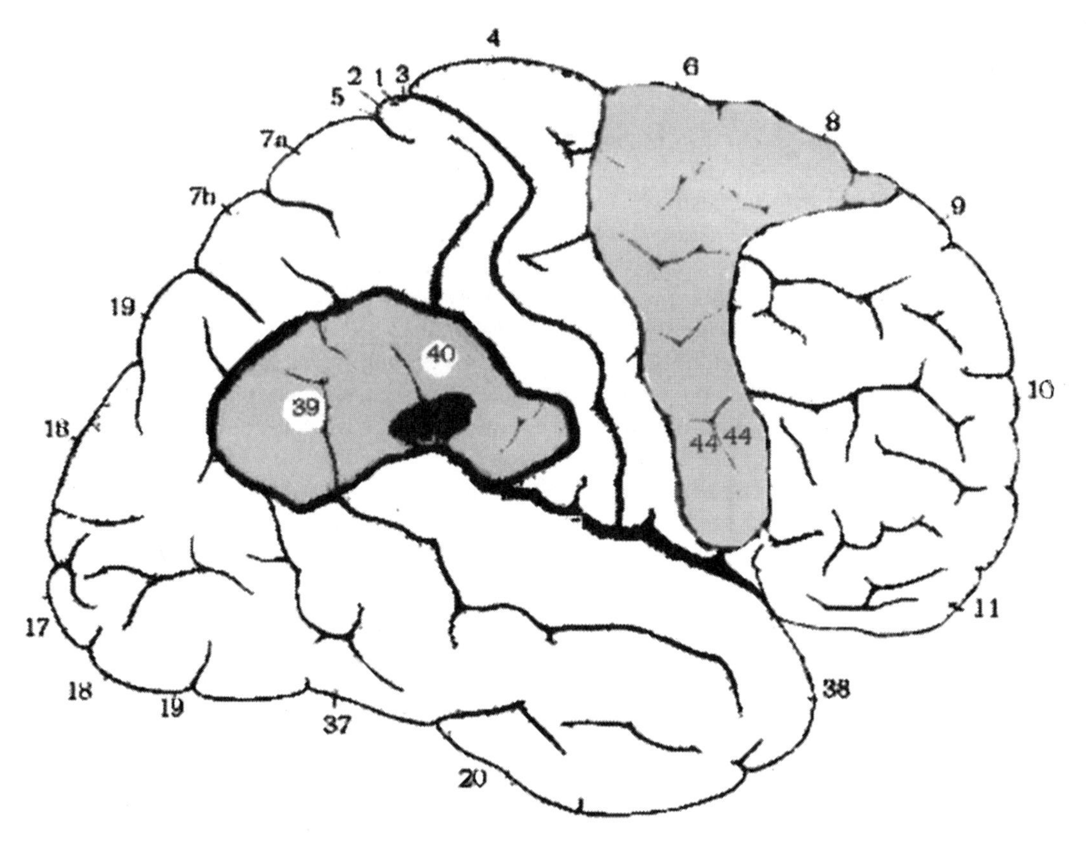

FIG. 19.4. Parietal lesion site in patients with left unilateral neglect. The critical lesion involves the inferior parietal lobule *(dark gray area),* at the temporo-parietal junction *(black area).* Neglect may be associated with damage involving the premotor frontal cortex (Brodmann's areas 44 and 6, gray area) and the superior temporal gyrus also.

the position of the object is coded with reference to the subject's whole body or body parts, giving rise to representations that may be head-centered (in the visual domain, resulting from the combination of the retinotopic map with information about eye position), trunk-centered (based also on information about the position of the head and about posture), arm-centered, and so forth. In *allocentric* coordinate frames, objects are primarily coded with reference to their spatial and configurational properties, such as the relationships between their component parts, and among different objects present in the environment. Egocentric representations may be used for the organization of goal-directed movements, such as reaching a target or avoiding a harmful stimulus. Allocentric representations, encoding the configurational properties of objects and the relationships among them, may be useful for their identification and for navigation in space (77).

The Neural Correlates of Extinction

Damage to the parietal lobe has long been considered not only the main anatomic correlate of spatial unilateral neglect (as discussed in the previous section) but also of extinction to double simultaneous stimulation (16,17). However, left tactile and visual extinction has been reported in individual patients with right-brain damage to the frontal lobe and the subcortical gray nuclei and white matter fiber tracts (47,78–80). A similar pattern of anatomic-clinical correlation holds for auditory extinction (18,78,81). An anatomic-clinical correlation study (82) investigated the anatomic correlates of visual and tactile extinction and of visual neglect in 159 right-brain–damaged patients. Over 50% of the patients with extinction had deep lesions (thalamus, basal ganglia, white matter), which were found in about 25% of the patients with unilateral neglect without extinction. Extensive fronto-temporo-parietal lesions and damage confined to the frontal lobe or the temporo-parietal regions were associated with extinction in the remaining patients. The more frequent anatomic correlates were damage to the paraventricular occipital white matter in patients with visual extinction, and damage to the premotor and rolandic regions in patients with tactile extinction. When visual neglect also was present, the lesions clustered in the inferior parietal lobule. This lesion pattern, with a difference between extinction and unilateral neglect, suggests that the former may include a sensory component, produced by damage to different levels of the ascending pathways. The previously discussed observation of extinction phenomena after lesions to the spinal cord supports this view (20,21).

Studies Exploring the Neural Correlates of Specific Pathologic Mechanisms Underlying Extrapersonal Visuospatial Neglect

Along the processing chain that leads to the production of a motor response from perceptual processing, two main pathologic mechanisms may account for extrapersonal visual neglect, as assessed by target cancellation and bisection tasks: perceptual, or, more generally, input-related, defective processing of the relevant stimulus *(perceptual neglect),* and defective programming and organization of the motor response *(premotor neglect).* Patients showing the latter disorder may manifest absence, delayed initiation, slow execution *(bradykinesia),* reduced amplitude *(hypometria)* and incompleteness or impersistence of motor activity, directed toward the side opposite to the brain lesion. This group of symptoms often is referred to as *directional hypokinesia* (83), which may be caused by a *premotor* (or more generally, an *output-related*) dysfunction.

There is some evidence that the distinction between *perceptual* versus *premotor* or *response* aspects of extrapersonal visuospatial neglect has an anatomic counterpart. The putative premotor versus perceptual dichotomy has been assessed through two main types of paradigms. Some tasks attempted at dissociating the two components by decoupling the direction of the hand movement from the visual control of the display. In one such bisection paradigms, a lateral (leftward or rightward) movement of the hand brought about a movement of the pointer setting the subjective midpoint of the line in the opposite direction (84).

Applying a similar logic to a cancellation task, the direction of the hand movement from the patient's visual control of the display was decoupled through a 90-degree angle mirror (85,86), and by an epidiascope (87). In another paradigm, the direct view of the hand and the target was precluded, with a TV monitor guiding performance (88,89). Under these conditions, patients saw a mirror-reversed display, so that, left-sided targets were seen on the right side in a cancellation task, for instance. Therefore, these paradigms included a *congruent* or *compatible* condition, in which as in a canonic cancellation or bisection paradigm, the patient saw the actual direction of the hand movement in the visual display. In a *noncongruent* or *incompatible* condition, however the movement of the hand took place in a direction opposite to that of the relevant visual stimuli through a device such as a pulley, mirror, or TV monitor, which was seen mirror-reversed. In these paradigms, a premotor directional deficit is characterized by the patient's inability to perform leftward manual movements, in both the congruent and noncongruent conditions. A perceptual deficit, by contrast, brings about a right-sided bias, relative to the display as seen by the patient, independent of the actual direction of the hand movement. Other paradigms investigated the dissociation contrasting verbal versus manual responses (90), or space of vision versus space of manual action (91), however, without introducing any incompatible condition. Taken together, the studies that made use of noncompatible conditions (84–86,88,89) suggest an association of the premotor pathologic mechanisms of extrapersonal visuospatial neglect with damage involving more anterior regions (frontal lobe, basal ganglia), of the perceptual factors with damage involving more posterior (temporo-parieto-occipital regions). A potentially confounding factor in experiments making use of incompatible conditions is the role of some working memory or executive components (92), which may become involved in the more difficult, noncongruent task, and have been associated with frontal lobe function (93). It should be noted, however, that a similar anterior-posterior dissociation has been found in paradigms that made use only of congruent conditions (90,91). A role of the right posterior-inferior parietal area in some aspects of motor programming is suggested by the finding that neglect patients with damage to this region were disproportionately slow in initiating leftward movements toward visual targets in the left side of space; such a deficit was not found in neglect patients with frontal lesions (94,95). Also, the clinical observation that motor neglect may be associated with both anterior frontal (96) and posterior (temporo-parietal) (96,97) damage suggests a parietal contribution to the programming of motor action (see Table 19.1).

The relative contribution of perceptual and response factors to the performance of patients with extrapersonal visuospatial neglect has been assessed by the so-called Landmark task also (98), which requires subjects to compare the length of the two segments of prebisected lines, giving verbal or manual responses (pointing to or naming the color of the shorter or longer segment of the line) (99). On this task, consistent choice of the contralesional segment as shorter (or of the ipsilesional segment as longer) reflects neglect resulting from perceptual bias. Consistent choice of the ipsilesional segment, both when patients are required to indicate the shorter and longer segments, reflect neglect resulting from response bias. The various versions of the Landmark task differ from the paradigms discussed earlier at least in two relevant respects: (a) They do not include incompatible conditions; and (b) the response is confined to a pointing toward the selected segment or a verbal response, rather than involving an actual, more or less complete leftward or rightward movement. A number of studies using the Landmark task, although corroborating the distinction between perceptual and response factors in visuospatial unilateral neglect, do not provide evidence supporting the anatomic dissociation mentioned earlier. In a group study by Bisiach et al. (99), frontal damage was found to be more frequently and definitely associated with a perceptual bias, whereas a response bias was found to be more strongly associated with subcortical damage. These findings, together with the observation that the same particular patient may show perceptual or response biases in different tasks (88,89,99), suggest that the issue of the perceptual versus premotor factors of neglect should be addressed through a fine-grain componential analysis of the tasks used, and of their anatomic correlates in brain-damaged patients with specific patterns of impairment.

The studies discussed in this section attempted to distinguish putative discrete mechanisms of unilateral neglect and their neural correlates on the basis of different behavioral responses in specifically devised tasks. This theoretically based approach contrasts with a recent suggestion, based on the lesion localization in neglect patients with and without visual field deficits, that spatial awareness is a function of the temporal lobe, whereas the posterior parietal lobe codes space for action (61,100). This anatomic-functional distinction, which resembles the perceptual/premotor dichotomy discussed earlier, is unwarranted: In the two groups of patients with and without hemianopia (61), neglect was assessed with the same general tasks (target cancellation, copying, visuomotor exploration) discussed in the preceding, which do not distinguish specific mechanisms underlying visuospatial neglect (101, 102).

Task Dissociations in Visuospatial Unilateral Neglect: Anatomic Correlates

Patients with extrapersonal neglect may show selective impairment in each of the two types of tasks that have been used to establish the anatomic-clinical correlations discussed earlier: target cancellation and line bisection. Right-brain–damaged patients have been described who show left neglect in line bisection, but not in cancellation tasks and vice versa (2,103). These observations did not reveal an anatomic counterpart to the behavioral dissociation. Both patient HD (defective cancellation/preserved bisection) and patient WS (defective bisection/preserved cancellation) had right-sided temporo-parietal damage and left visual-field deficits (2). Patient JB, who was selectively impaired in line bisection, and showed a preserved performance in cancellation tasks, had a posterior subcortical lesion (posterior limb of the internal capsule and thalamus) and full visual fields (103). In one group study (104), where lesions were localized according to standard templates (105), right-brain–damaged patients with abnormal line bisection had more posterior damage, clustering in the inferior parietal lobule, in the posterior and middle temporal gyrus, and in the anterior-lateral occipital lobe. By contrast, patients with cancellation deficits and normal line bisection had lesions clustering in the precentral and premotor frontal regions. In a large series of right-brain–damaged patients, the double dissociation between impairments in line bisection and cancellation tasks was confirmed, but no specific anatomic correlates were found. The lesions, classified by a neurologist with no reference to standard templates, were equally likely in the temporo-parietal region, in the dorso-lateral frontal cortex, and in the deep frontal structures (106).

In one study, left hemianopia was more frequent in patients with defective line bisection (four out of 11), than in patients with a normal performance (one out of 10) (104). In line with these results, two more recent investigations (107,108) found that right-brain–damaged patients with left extrapersonal visuospatial neglect, left hemianopia documented by event-related potentials, and lesions (mapped with reference to a standard atlas) (105), extending posteriorly to the occipital lobe, made a rightward error in line bisection greater than that committed by patients with visuospatial neglect without hemianopia, and more anterior lesions. The two groups of neglect patients were comparably impaired in cancellation tasks. Earlier data are in line with these findings (109,110).

These results suggest that lesions extending to the occipital lobe bring about a more severe contralesional error in line bisection, but not in other tasks assessing visuospatial neglect, such as cancellation. Cancellation and line bisection tasks may differ in that the latter, requiring a more panoramic appreciation of the single stimulus (the segment), involves a more substantial contribution of retinotopic representations, supported by the occipital regions. By contrast, the retinotopic factor may be less relevant in the more serial cancellation tasks, which require the successive crossing out of targets relatively small in size. The view that different underlying pathologic mechanisms are involved in the patient's defective performance in these two types of tasks is further supported by the association of the perceptual bias on the Landmark task with performance on line bisection, of the response bias with performance on cancellation tasks (111).

However, the observations discussed earlier suggest dissociation between impairments in two tasks widely used to assess left neglect in right-brain–damaged patients: target cancellation and line bisection. The anatomic correlates of these discrete patterns of behavioral impairment are less clear cut. Nevertheless, there is some evidence that visual field defects (associated with more posterior lesions extending to the occipital regions) may further increase the rightward error in line bisection.

These additive effects of neglect and visual field defects also may be related to particular aspects of the tasks used. It has long been known that patients with hemianopia without neglect make a contralesional error (toward the blind field) in line bisection and in tasks in which subjects are required to set the straight ahead (109,112–114). In one study using a straight ahead task (115), the opposite directional errors of right-brain–damaged patients with left hemianopia without neglect (left contralesional deviation) and of patients with left neglect without hemianopia (right ipsilesional deviation) canceled out in patients with associated impairments, namely, in right-brain–damaged patients with both left neglect and left hemianopia, the subjective straight ahead was within the normal range. Other studies, however, found a marked ipsilesional rightward deviation of the subjective straight ahead in right-brain–damaged patients with both left neglect and hemianopia (116,117).

Productive Manifestations in Visuospatial Unilateral Neglect: Anatomic Correlates

The anatomic-clinical correlation studies considered in the previous section were all concerned with the better known, and more frequently investigated, defective manifestations of spatial unilateral neglect: omission of left-sided details in drawing and copying, of left-sided targets in cancellation tasks, rightward error in line bisection (see Figs. 19.1 and 19.2). The neglect syndrome also includes so-called productive manifestations (see Table 19.1), characterized by the presence of pathologic behavioral patterns, which add to the defective exploration of the contralesional side of space and to the defective orientation of spatial attention toward contralesional targets.

Definite anatomic information concerning these productive manifestations is available for perseveration in drawing and cancellation tasks. This pathologic behavior of right-brain–damaged patients with left neglect includes relatively simple perseveration patterns (overscoring lines already drawn, adding crosses or lines in a circle cancellation task, see Fig. 19.5A) and more complex graphic productions, such as adding irrelevant graphic material on the right-hand side of the paper, as shown in Figure 19.5B (86,118,119).

Two recent studies in large series of right-brain–damaged patients have not only confirmed that perseveration in cancellation tasks is closely associated with left neglect, but also have shown that frontal and subcortical damage is a relevant factor in bringing about this pathologic behavior (120,121). Figure 19.6 shows that the severity of left unilateral neglect in cancellation tasks, indexed by omission errors, was comparable after posterior (parietal, occipital), subcortical, and extensive anterior-posterior (frontal, temporal, parietal, occipital) damage. However, perseveration errors were more frequent after lesions including the frontal lobe and subcortical damage. These anatomic findings suggest that repetitive drawing responses during cancellation in the ipsilesional sector of the sheet, where exploration is abnormally confined because of left neglect, are caused by the pathologic release of complex motor behavior, brought about by frontal damage, or by a frontal dysfunction caused by subcortical lesions, through a disconnection mechanism. A distinction has been drawn between two impairments of exploratory behavior, "frontal or magnetic apraxia" and "parietal or repellent apraxia" (122) (although the use of the term "apraxia" with reference to these disorders is questionable) (123): The magnetic exploratory aspect of behavior, involving "perseveration of all contactual reactions," would be managed by the parietal cortex and released by frontal and temporal lesions; the repellent, negative, bias would be determined by the premotor, cingulated, and hippocampal re-

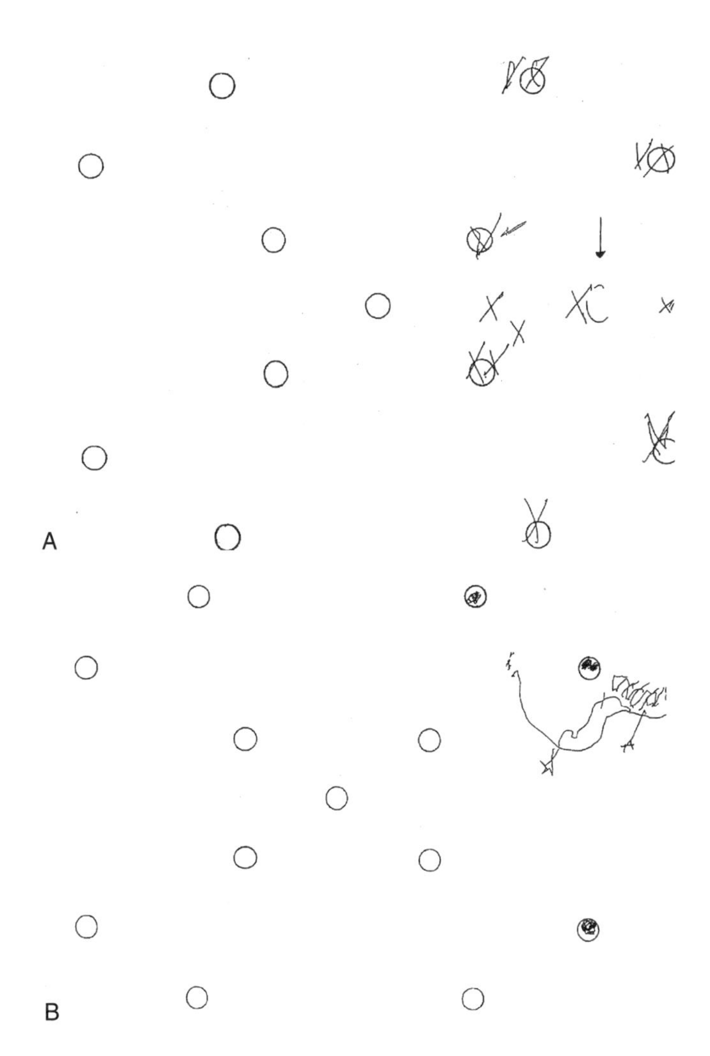

FIG. 19.5. Examples of perseverative errors during the circle cancellation task, committed by right-brain–damaged patients with left spatial neglect. **A:** Spontaneous addition of crosses ("simple perseveration") and of a target-circle subsequently crossed out ("complex" perseveration, *indicated by a vertical arrow).* **B:** Spontaneous drawing of an incomplete hen ("complex" perseveration). (Reprinted from: Rusconi ML, et al. Is the intact side really intact? Perseverative responses in patients with unilateral neglect: a productive manifestation. *Neuropsychologia* 2002;40: 594–604, with permission.)

gions and released by parietal lesions (122). In patients with right frontal damage, the presence of left neglect accounts for the occurrence of perseveration confined to the ipsilesional, right-sided, sector of the sheet. However, when the lesion is confined to the posterior, mainly parietal, regions, the exploratory deficit has no positive, productive components, being limited to a negative bias toward the ipsilesional side. It is the combined effect of the pathologic restriction of the representation of extrapersonal space to the contralesional side, on the one hand, and of the release of repetitive visuomotor activity (drawing, in the context of a cancellation task), on the other, which brings about perseveration behavior in right-brain–damaged patients with left neglect.

Transient Disruption of Brain Activity

In recent years, transcranial magnetic stimulation (TMS) (124,125) has been used in normal individuals to investigate visual cognition (126), with a logic similar to anatomic-clinical correlation studies. Transcranial magnetic stimulation may temporarily disrupt the brain activity of specific brain regions, bringing about behavioral deficits, which may be related to the more long-lasting disorders caused by brain damage. Transcranial magnetic stimulation, in addition, may provide in-

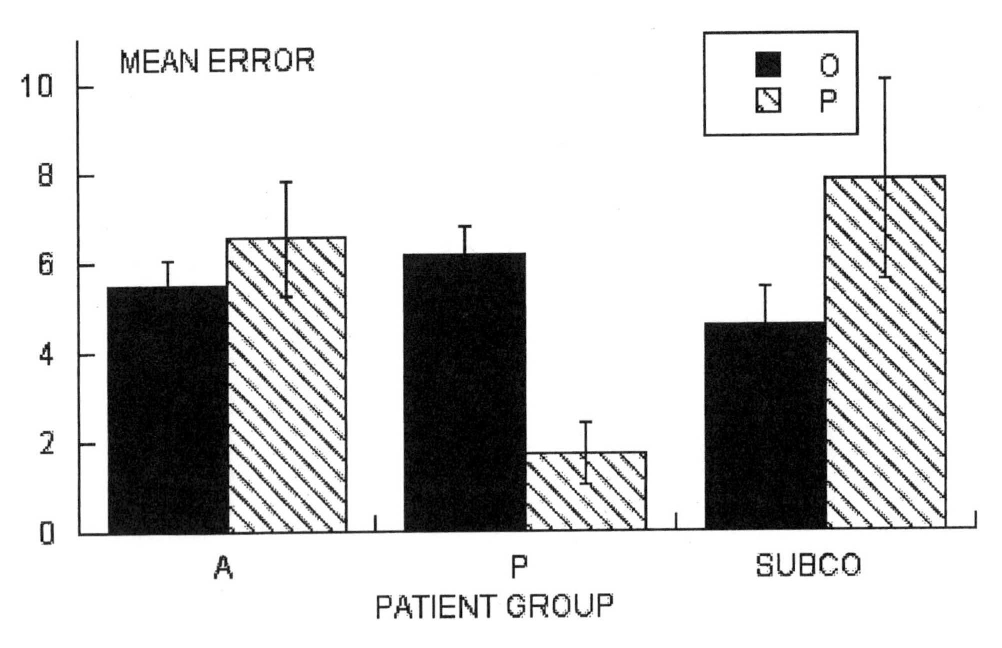

FIG. 19.6. Average number of perseverative (P) and omission (O) errors in 31 right-brain–damaged patients with left spatial neglect, by localization of the hemispheric lesion. (A/P, including/sparing the frontal lobe; SUBCO, confined to subcortical structures.) (Reprinted from: Rusconi ML, et al. Is the intact side really intact? Perseverative responses in patients with unilateral neglect: a productive manifestation. *Neuropsychologia* 2002;40:594–604, with permission.)

formation concerning the timing of cortical activity. Early studies investigated the phenomenon of visual extinction to double simultaneous stimulation, showing that repetitive (r)TMS of the left or of the right parietal lobe inducted visual extinction of contralateral stimuli during double simultaneous stimulation. Occipital rTMS impaired the detection of contralesional single stimuli, affecting more peripheral sensory processing, whereas temporal rTMS had inconsistent effects (127). A recent study confirmed the observation that parietal rTMS brings about contralesional visual extinction, showing also increased detection of ipsilesional stimuli, with the enhancement effect being significant only after right-sided rTMS (128). In right-brain–damaged patients single pulse TMS of the left frontal regions reduced contralesional left tactile extinction (129), suggesting that one mechanism underlying the disorder may be the unbalanced activity of the two sides of the brain, caused by the unilateral damage (130–132). The unbalance would be temporarily reduced by the TMS interference with the postlesional prevailing activity of the undamaged hemisphere.

Recent studies investigated in normal subjects the effects of unilateral rTMS by paradigms used in brain-damaged patients to assess visuospatial unilateral neglect. Right parietal rTMS induced a rightward bias in judgments about the symmetry of prebisected lines, which mimics the rightward error in line bisection, committed by right-brain–damaged patients with left neglect (133). Left parietal rTMS had no detectable effects (134). These results have been confirmed with single pulse TMS, with right parietal stimulation bringing about a rightward bias 150 msec after stimulus presentation (135). In brain-damaged patients with unilateral neglect, parietal rTMS of the unaffected hemisphere reduced the rightward bias in the perceptual judgments discussed earlier (136). As for extinction, these results support the view that a mechanism underlying neglect may be an unbalanced activity of the two cerebral hemispheres, caused by the unilateral lesion (130–132). The modulatory effects of lateralized sensory stimulations, which may temporarily improve or worsen many aspects of the neglect syndrome also may be interpreted in terms of the reduction or increase of the unbalance between the two hemispheres (13,14).

Some animal and human data concerning the effects of bilateral lesions further support the unbalance account of neglect. In the cat, hemianopia and unilateral neglect produced by extensive temporo-occipital damage improve after either a subsequent lesion in the contralateral superior colliculus or splitting the collicular commissure (137). One patient after a right-sided posterior parietal lesion exhibited left neglect and hemianopia, which recovered after a second stroke in the dorsolateral premotor cortex of the left hemisphere (138). Other observations, concerning the effects of symmetrical lesions, suggest, however, cooperative aspects of the interhemispheric interactions. In the monkey unilateral parietooccipital lesions bring about contralesional extinction and the disorder becomes bilateral after a second symmetrical lesion (139). In one patient the first, left-sided, posterior inferior parietal lesion brought about right neglect, which became a bilateral inattention after the second, right-sided, posterior-inferior parietal infarct (140).

Neuroimaging Studies

Functional neuroimaging measures hemodynamic changes–blood flow in the case of positron emission tomography (PET) (141) and blood oxygenation in the case of functional magnetic resonance imaging (fMRI) (142–144). These indexes are used as indirect measures of synaptic activity and neural firing. Positron emission tomography and fMRI have rapidly become a major source of neurophysiologic information on the neural bases of spatial cognition in humans. Neurophysiological research in nonhuman primates, together with evidence from brain damaged patients, provide a challenging research agenda for functional imaging investigations of space representation and attention in humans. The relevant imaging data provide information complementary to the anatomic-clinical correlation findings in patients with spatial unilateral neglect and to neurophysiologic observations in nonhuman primates (8).

Activation Studies in Normal Subjects

Functional Anatomic Studies of Space Representation

Space, perceived as unitary (5,145), is in fact mapped by several systems all of which provide a representation relevant for a given set of actions and for different parts of space itself, including the body (146–148). A constant feature of sensory information processing is a progressive transformation of the receptive field properties of neurons the farther from primary sensory cortices: taking the visual system as an example, the farther from area V1, the more complex the receptive fields of the neurons, with neurons having a receptive field nearly as big as the whole visual field (149). Cortical visual areas may show a retinotopic organization, in that a single neuron responds to visual events arising from a specific part of the retina, or more complex properties as in the case of retinotopic gaze-dependent cells (i.e., neurons with retinotopic receptive fields, whose firing is modulated by the direction of gaze, found in areas V3, V3A, V5/MT, MST, V6, 7a, LIP), or the so-called "real-position" cells (i.e., neurons that respond to stimuli in specific spatial positions, in egocentric—craniocentric—coordinates found in area V6a, in the ventral intraparietal cortex, and in the premotor cortex, area F4) (146,148,150). Both neurophysiologic studies in the monkey and experiments in patients with spatial unilateral neglect have permitted to dissociate elementary somatotopic and retinotopic codes from body-centered coordinate frames (5,19). Neurophysiologic evidence shows that visual space around our body segments has specific neural representations in the ventral premotor and inferior parietal cortices (148,151). The neural response of most of these neural systems is modulated by attentional phenomena (146–148).

Studies in brain-damaged patients with spatial unilateral neglect also provide the challenge of explaining the hemispheric asymmetry in spatial competence, that is only documented, if not only present, for humans. This last aspect makes functional imaging studies of unique import, because hemispheric asymmetry represents a major difference between nonhuman and human neural architecture for spatial cognition.

Together with clear advantages, such as the possibility of permitting the *in vivo* exploration of the activity of the whole human and monkey brain at once, functional imaging has also limitations, that should be kept in mind, and concern temporal and spatial resolution. These make it difficult to track the progressive making of spatial representation from early sensory visual codes to the higher-order ones as precisely as in the monkey. For example, with the current techniques it has not been possible to isolate signals from neuronal populations of cells with a gaze-dependent visual response (146), nor to isolate responses that can be readily interpreted as real-position like (150). There are obvious reasons why this may remain difficult with imaging techniques: Such cells are present in small areas that contain simpler cells as well; in the primate brain, real-position cells coexist in the same area with gaze-dependent cells.

Nonetheless, a substantial part of the detailed information derived from the primate brain is reproducible using functional imaging in humans. A large number of retinotopically organized areas have been now identified outside area V1 and area V2. The existence of separate extrastriate areas specialized in color and motion perception has been demonstrated in the early 1990s by PET studies. These are likely homologs of the monkey areas V4 and V5 (152). A more detailed cartography of at least 10 retinotopically organized extrastriate areas has now been described with fMRI (153). The receptive fields of the neurons within these areas are increasingly complex, with neurons mapping also the ipsilateral visual field starting from area V3a and area MT/V5 (154). An increasing complexity of processing has been also observed: In the ventral stream, for example, whereas posterior regions are preferentially activated by object attributes, like colors or scrambled objects, more anterior regions of the temporal lobe are activated by intact objects or faces (155).

The neural representation of somatosensory space has been explored using a variety of approaches. Simple experiments in which somatosensory stimuli were delivered to the hands or feet have broadly confirmed the well-established notion of a somatotopic organization of area S1. A possible somatotopic organization of area S2 also has been reported (156). In keeping with primate data, receptive fields in S1 appear to be strictly unilateral, whereas bilateral activation of area S2 is observed for stimuli delivered to either sides of the body, suggesting the presence in the human S2 of cells with at least bilateral, if not ipsilateral, receptive fields (157).

The broad dichotomy between two visual streams (the "what" and "where" pathways) (158) has been demonstrated comparing a face matching task (a ventral "what" stream task) and a dot location task (a spatial "where" task) (159,160). Compared to baseline, both experimental tasks showed activation of the lateral occipital cortex. Face discrimination alone, activated a region of the occipito-temporal cortex, whereas spatial location alone activated a region of the lateral superior parietal cortex.

This dichotomy has been replicated and further characterized several times (161), particularly with reference to visual working memory (162). In keeping with electrophysiologic data in the monkey (163), when visual processing involves active maintenance of the spatial location of visual stimuli and delayed response, a dorso-lateral pre-frontal cortex activation is observed, together with posterior parietal activation (162). Consistent again with primate data (164,165), active maintenance of object-oriented visual information involves the more ventral part of lateral prefrontal cortex together with the ventral visual cortex (162). Therefore, the "what" (ventral) and "where" (dorsal) distinction is also present in frontal cortex (166).

The neural correlates of extrapersonal spatial coordinates have been explored in further detail in recent years. This literature is directly relevant for the issue of the neural bases of unilateral neglect, because the experimental tasks used in the activation experiments are similar to the clinical tests used to detect the disorder in patients. Object-centered (allocentric) spatial judgments have been studied in at least three experiments (167–169), concerned with line bisection: the subjects' task was to judge whether or not a short vertical bar divided a horizontal line into two segments equal in length. A metaanalysis of the main activations detected in the right hemisphere in these experiments is shown in Figure 19.7. (The colored areas represent the brain regions usually damaged in patients with spatial unilateral neglect; see also Fig. 19.4.) Activation foci are superimposed on a standard MRI view of the right hemisphere. The activation pattern includes the lateral extrastriate cortex, the most dorsal part of area 40 in the inferior parietal lobule and the intraparietal sulcus, the superior parietal lobule, and the dorsal premotor cortex. This network closely resembles the visuomotor transformation network involved in reaching tasks in the monkey and in humans (170,171). In a largely egocentric task, where subjects judged whether a vertical bar was located to the left or to the right of the subjective midsagittal plane of their body, a similar, although more extensive, bilateral pattern of activation, compared to a baseline condition, emerged in the superior-posterior parietal cortex (mesial and lateral), in the dorsal premotor cortex (bilaterally) and in the left temporo-parietal cortex (168). The premotor cortex, the medial dorsal-parietal and the right intraparietal cortex were more active when the midsagittal plane task was compared with the allocentric line bisection task. Finally, there was a larger activation in the medial ventral extrastriate cortex

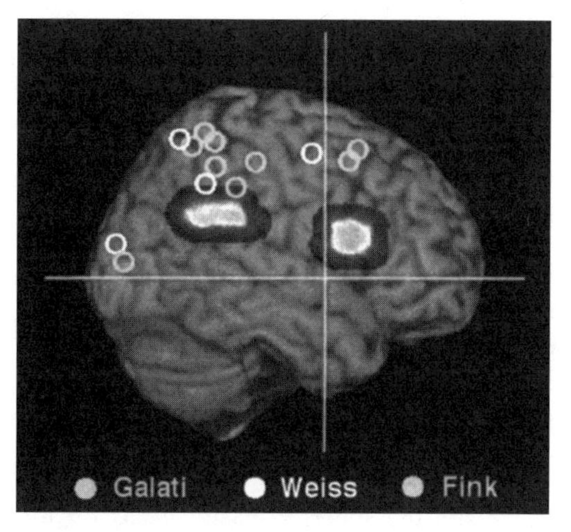

FIG. 19.7. Brain areas associated with spatial neglect (shaded colored areas) and metaanalysis of three functional neuroimaging experiments on line bisection (167–169) The temporoparietal area is redrawn from Vallar and Perani (51), the frontal area from Husain and Kennard (68), and reproduces the regions of maximal overlap of frontal lesions in that patient series. Circles represent the location of activation peaks expressed in stereotactic space in three functional neuroimaging experiments on line bisection. Only the activations on the lateral surface of the right hemisphere are reported. The horizontal line represents the plane passing through the two commissures, the vertical line the coronal plane passing through the anterior commissure. Mesial ventral occipital activations are not shown. (See color plate of this figure after page 186.)

in the allocentric task, compared with the egocentric one (not shown in Fig. 19.7). This latter pattern of activation was also observed in a bisection judgment on a two-dimensional (2D) object (a square), compared with canonical bisection of a one-dimensional (1D) object (a segment) (172). Complementary neuropsychologic data show that patients with left neglect in line bisection may perform normally when marking the center of a square (173,174). Similar to the experiment involving square bisection (172), extensive bilateral activation of the ventral occipital cortex was found in a study that compared a pointing task (including bisection) in far space as opposed to near space. By contrast, comparing activations for near space as opposed to far space revealed a greater left hemisphere activation, in the dorsal occipital and the premotor cortices (169).

A more recent study (175) assessed the neural correlates of setting the midsagittal plane under condition of visual control and in a proprioceptive-somatosensory task. As previous experiments (168,176), the body-centered coding task activated a bilateral fronto-parietal network, although more extensively in the right hemisphere, including posterior parietal regions around the intraparietal sulcus and frontal regions around the precentral and superior frontal sulci, the inferior frontal gyrus and the superior frontal gyrus on the medial wall.

Taken together, the results of the meta-analysis shown in Figure 19.7 (167–169) and other studies (172,176) seem to indicate a weak dissociation in a distributed and partially overlapping system for spatial representation of peripersonal and far extrapersonal space: The visuomotor transformation network involved in the reaching behavior is activated, with greater emphasis on the dorsal parietal and premotor cortex, when the experimental task involves attending spatial locations that are within reaching distance; ventral retinotopically organized and object-based visual cortices are more active when the comparisons of blood flow maps isolate the processing of spatial locations which are either outside reaching distance (far space) (169), or when the stimulus is, comparatively, more object-like (172).

The neural basis of representations for egocentric space also has been explored, by using sensory stimulations. The experimental rationale of this approach is based on the hypothesis that spatial representations are built up through the convergence and integration of different afferent inputs. Visual, vestibular, and proprioceptive-somatosensory stimuli, including those conveyed by the muscle spindles of the neck, play an important role (146,177). Appropriate, direction-specific stimulations of these systems may bring about a lateral spatial bias in normal controls and temporarily reduce, or worsen, many manifestations of spatial unilateral neglect (13,14). In the monkey, a perisylvian network of cortices including the parieto-insular and retroinsular cortex, the ventral premotor cortex, the tip of the intraparietal sulcus, all receive inputs from the vestibule, from the body surface and from muscle spindles, including those of the neck muscles (178). As for the monkey, in humans the perisylvian somatosensory cortex (retroinsular cortex area secondary somatosensory cortex, supramarginal gyrus) are a main site of convergence of these inputs, suggesting that these brain regions contribute to egocentric space representation (15).

Orienting of Attention

In visual space, orienting of attention usually, but not necessarily, implies fast (saccadic) eye movements, which are made to bring the visual stimuli in foveal space (179). Spatial attention can be attracted automatically by the sudden appearance of a stimulus in the visual field or by the perceptual salience of a stimulus among others. These issues have been addressed in a number of functional imaging experiments (41,180,181) and in detailed metaanalyses (182). In brief, both covert (without eye movements) orienting of attention and generation of saccades share neural resources in the regions around the intraparietal sulcus and in the superior parietal lobule, in the dorsal lateral premotor cortex (human homolog of frontal eye fields), and in the mesial dorsal premotor cortex (human supplementary frontal eye fields) (Fig. 19.8).

Therefore, at the level of spatial and temporal resolution of functional neuroimaging, the regions involved in covert orienting attention and saccadic eye movements appear not to be entirely independent. However, according to a recent metaanalysis (182), there is no complete overlap between the two systems. Eye movements are associated with more extensive involvement of regions nearby the primary motor and somatosensory cortices; covert orienting shows a larger involvement of the prefrontal regions, and the more posterior lateral occipito-parietal junctions. The additional areas seen in the saccadic eye move-

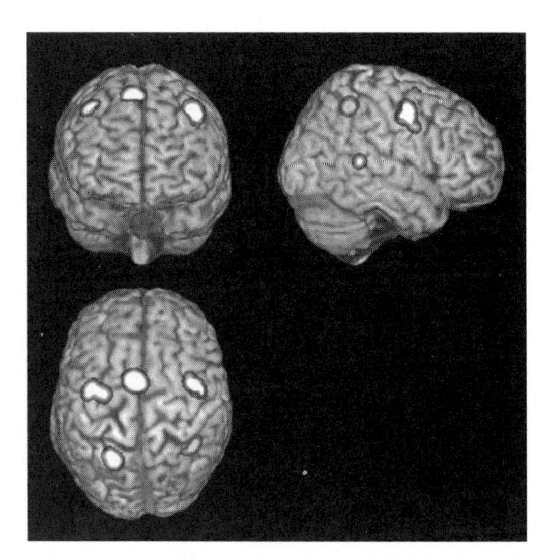

FIG. 19.8. Metaanalysis of two functional magnetic resonance imaging studies (180,181) in which covert orienting of attention and generation of saccades were compared in the same subjects. The metaanalysis represents only brain regions found by both studies. The *areas in yellow* represent brain regions that were activated by both eye movements and covert attention. These include the dorsal premotor cortex, the supplementary motor region, the region of the left intraparietal sulcus/superior parietal lobule, and the right superior temporal sulcus. The right dorsal premotor region and the cortex of the right intraparietal sulcus/superior parietal lobule *(areas in blue)* represent cortical regions more active for covert attention. There was also another such region at the junction between the right ventral mesial extrastriate cortex and the cerebellum (not shown). There was also one area in left ventral occipital cortex and left cerebellum more active for eye movements in both studies (not shown). One study (180) reported separate comparisons of the two experimental tasks against central fixation baseline, and direct comparisons of the two experimental tasks. The other study (181) reported conjunctions of experimental conditions against baseline and differences across conditions (as interaction effects). To accommodate for the different statistical approaches, the metaanalysis was generated as follows: (a) each peak of activation was transformed into a 12-mm diameter sphere in stereotactic space, which approximates spatial resolution of group studies using fMRI; (b) the intersection of attention and eye movement maps was calculated for one study (180) (this makes the data similar to the conjunctions of ref. 181); (c) intersections for three sets of maps from each study were then calculated: (1) areas of shared activation for covert attention and eye movements; (2) areas of greater activations for covert attention; and (3) areas of greater activation for eye movements. (See color plate of this figure after page 186.)

ments can be explained readily by the fact that subjects are actually moving the eyes; the additional areas seen in covert orienting may represent the more cognitive neural counterpart of the task. A metaanalysis (182) also shows that the distribution of the activation during covert orienting of visual attention, although maintaining central fixation, is independent of the specific requirements of the visuospatial task (i.e., detecting a stimulus, discriminating, or tracking its movement). A similar discrepancy, between a motor task and a more controlled task, has been observed also in other cognitive domains. For example, comparison of articulation of a stereotyped string of digits and random generation of the same digits shows a greater involvement of more anterior prefrontal regions (183).

In line with these findings, clinical studies have shown that disorders of ocular movements are not always associated with spatial unilateral neglect (184,185), even though right-brain–damaged patients with unilateral neglect may show abnormal patterns of ocular movements toward the left contralesional side (186–189). However, saccade deficits and unilateral visuospatial neglect after damage to the posterior parietal cortex may occur independently; the oculomotor abnormalities resulting from unilateral neglect should not be confused with disorders specifically affecting the programming of single saccades (190).

Functional Imaging Data on Unilateral Neglect and Related Disorders

Few functional imaging studies have been devoted to unilateral neglect. A variety of subjects have been approached such as the physiopathology of subcortical neglect (73,74, 191), the neural mechanisms of recovery from spatial neglect (75,76,192), the neural correlates of some neglect-related symptoms such as extinction and attentional hemianesthesia (193). Early studies using SPET (74) and PET (194) have shown that unilateral neglect following subcortical damage is associated with hypoperfusion of the cerebral cortex, with behavioral recovery being correlated with increase of cortical blood perfusion (73,75). Recovery from neglect was studied in two right-brain–damaged patients using [18F]-fluorodeoxyglucose ([^{18}F]FDG) PET resting state measurements in the recent acute and in a chronic phase. A reduction of hypometabolism in both the ipsilateral and contralateral hemispheres was found in one patient with a subcortical lesion, who recovered; by contrast, one patient with a massive hemispheric lesion did not show recovery at both the behavioral and the brain metabolic levels (76).

Resting state studies have the limitation that no explicit correlation between behavioral recovery and metabolic recovery can be made because the task in hand (resting state) has nothing to do with spatial exploration. This caveat is particularly relevant for single case studies. In the case of group studies, some correlation can be made between behavioral recovery and metabolic recovery. These correlations become much more explicit in the case of activation studies. One such PET activation study was performed in three right-brain–damaged patients with left neglect, before and after rehabilitation. Regional cerebral blood flow was measured during a visuospatial task requiring the detection of visual targets moving on a computer screen toward the left (neglected) side (192). The cerebral activation was compared with a baseline task in which subjects responded to a dot flashing in the unaffected side of visual space. For all three patients, the regions more active after recovery were almost exclusively found in right-hemisphere–undamaged cortical areas, and largely overlapped with those observed in a group of four normal subjects performing the same task. These regions included the extrastriate visual cortex, primarily at the parieto-occipital junction, and the superior parietal lobule. The only areas that were active in recovered patients, but not in normal subject engaged in the same task, were mainly in the right, damaged, hemisphere, and may be possibly considered as functionally related, although not primarily involved in space representation and exploration. A similar within-system effect was observed in a right-brain–damaged patient showing short-term recovery from left tactile hemi-inattention (195) induced by caloric vestibular stimulation: the brain areas modulated by vestibular stimulation (right insula and putamen) may be considered component parts of the neural system involved in touch perception (193).

Functional imaging and event-related potential studies on extinction, a symptom frequently associated with spatial neglect, provide information about the stages of cortical processing of stimuli in relation to conscious perception. One fMRI study in one right-brain–damaged patient showed that extinguished stimuli can activate extrastriate visual cortex (196). Therefore, this cortical processing may not translate into conscious perception, providing a neural basis for implicit processing in patients with neglect (197–200). Conversely, in one right-brain–damaged patient with left-sided extinction to double simultaneous stimulation, a study visual event-related potentials demonstrated that, following bilateral presentation of the stimuli, the early attention-sensitive P1 (80 to 120 ms) and N1 (140 to 180 ms) components of the event-related potential response were absent in those trials when the contralesional left-sided stimulus was extinguished (201). These results provide information as to the electrophysiologic correlates of conscious vision.

Functional Neuroimaging Data on Spatial Attention, Spatial Processing, and the Anatomy of Spatial Neglect: Commonalities and Differences

Taken together, the functional imaging data on the neural basis of spatial attention and representation discussed previously present some commonalities with the published anatomy of spatial neglect (4,5). A posterior parietal-frontal premotor network for spatial cognition is revealed by both data sets.

A finer-grained look to the data also shows some nontrivial discrepancies, however. Consider the distribution of the lesions observed in patients with left neglect (Fig. 19.4, the shaded areas in Fig. 19.7). The pattern of activation reported for the task of line bisection (a sensitive tool to assess spatial neglect) involves brain regions of the parietal lobe, which are overall more dorsal. On the other hand, the more ventral regions receive signals from the vestibular system and the neck muscle spindle system (15,202). This is a relevant finding, given the well known ability of vestibular and neck signals to recalibrate spatial misperception and representation in neglect patients (14).

A similar remark on the topography of parietal and premotor activations applies to the anatomy of covert orienting of attention (see Fig. 19.8), which appears overall more dorsal than the neuroanatomy of spatial neglect. The limited overlap of the two anatomic patterns supports the view that spatial unilateral neglect involves more than the mere defective contralesional orienting of attention (19). A further discussion of the distinction between a more dorsal (parts of the intraparietal cortex, superior frontal cortex) and a more ventral (inferior parietal lobule and temporo-parietal junction) with reference to the neuroanatomy of spatial neglect may be found in reference 203.

A further possible discrepancy between functional imaging data and the anatomy of spatial unilateral neglect is in the lateralization of brain activations, compared to the comparatively more clear-cut predominance of left unilateral neglect after right hemispheric lesions. It should be first pointed out that hemispheric asymmetries are seldom tested formally, as task by hemisphere interaction effects. In other words, asymmetries are usually reported qualitatively rather than assessed formally. A larger activation of the right hemisphere was found in some studies on spatial attention (41), in studies on sustained attention (37), and in experiments on line bisection and setting the subjective straight ahead (168,175,176). Metaanalyses of experiments on covert orienting of attention show patterns that are substantially bilateral and symmetric in the dorsal premotor, and in the superior parietal and intraparietal cortices (155,182). The functional predominance of the left hemisphere for near space reported in one PET experiment (169) needs to be tested in brain-damaged patients. Taken together, these findings suggest that the hemispheric asymmetry revealed by spatial unilateral neglect may be a matter of degree in a largely bilateral system. Individual differences may be relevant also. Compatible with this view, contralesional, right, neglect has been repeatedly reported after left brain damage, although the deficit is usually less severe and less frequent (19,30,51).

Mismatch between functional imaging and lesion data has been seen already before for important areas of cognitive neuroscience, such as episodic memory (204). The resolution of these discrepancies in spatial neurocognition may contribute to a better understanding of spatial unilateral neglect, of normal human spatial processing and attention, and of the nature of the right hemispheric specialization for spatial cognition. To achieve this, further studies are required on the neuroanatomy of spatial unilateral neglect and on spatial processing in normal subjects and in patients with acquired cerebral lesions. The neuroanatomy of unilateral neglect needs to be mapped by recruiting patients using a wider range of tasks, and looking for groups of patients showing dissociations of behavioral deficits. Indeed, as noted, the lesion data available so far have been collected primarily by recruiting patients on the basis of single tasks, or test batteries (mainly cancellation, bisection, and drawing), which cover spatial exploration in peripersonal space alone.

ACKNOWLEDGMENT

This work was supported in part by MIUR grants to GV and EP.

REFERENCES

1. Barbieri C, De Renzi E. Patterns of neglect dissociation. *Behav Neurol* 1989;2:13–24.
2. Halligan PW, Marshall JC. Left visuospatial neglect: a meaningless entity? *Cortex* 1992;28:525–535.
3. Heilman KM. Neglect and related disorders. In: Heilman KM, Valenstein E, eds. *Clinical neuropsychology.* New York: Oxford University Press, 1979:268–307.
4. Vallar G. Extrapersonal visual unilateral spatial neglect and its neuroanatomy. *NeuroImage* 2001;14:S52–S58.
5. Vallar G. Spatial hemineglect in humans. *Trends Cog Sci* 1998;2:87–97.
6. Bisiach E, Perani D, Vallar G, et al. Unilateral neglect: personal and extrapersonal. *Neuropsychologia* 1986;24:759–767.
7. Guariglia C, Antonucci G. Personal and extrapersonal space: a case of neglect dissociation. *Neuropsychologia* 1992;30:1001–1009.
8. Vallar G. The methodological foundations of human neuropsychology: studies in brain-damaged patients. In: Boller F, Grafman J, eds. *Handbook of neuropsychology,* 2nd ed. Amsterdam: Elsevier, 2000:305–344.
9. Brain WR. Visual disorientation with special reference to lesions of the right cerebral hemisphere. *Brain* 1941;64:244–272.
10. Marshall CR, Maynard FM. Vestibular stimulation for supranuclear gaze palsy: a case report. *Arch Phys Med Rehabil* 1983;64:134–136.
11. Silberpfennig J. Contributions to the problem of eye movements. III. Disturbances of ocular movements with pseudohemianopsia in frontal lobe tumors. *Conf Neurol* 1941;4:1–13.
12. Rubens AB. Caloric stimulation and unilateral visual neglect. *Neurology* 1985;35:1019–1024.
13. Rossetti Y, Rode G. Reducing spatial neglect by visual and other sensory manipulations: non-cognitive (physiological) routes to the rehabilitation of a cognitive disorder. In: Karnath H-O, Milner AD, Vallar G, eds. *The cognitive and neural bases of spatial neglect.* Oxford, UK: Oxford University Press 2002:375–396.
14. Vallar G, Guariglia C, Rusconi ML. Modulation of the neglect syndrome by sensory stimulation. In: Thier P, Karnath H-O, eds. *Parietal lobe contributions to orientation in 3D space.* Heidelberg: Springer-Verlag, 1997:555–578.
15. Bottini G, Karnath HO, Vallar G, et al. Cerebral representations for egocentric space: functional-anatomic evidence from caloric vestibular stimulation and neck vibration. *Brain* 2001;124:1182–1196.
16. Critchley M. The phenomenon of tactile inattention with special reference to parietal lesions. *Brain* 1949;72:538–561.
17. Critchley M. *The parietal lobes.* New York: Hafner, 1953.
18. De Renzi E, Gentilini M, Pattacini F. Auditory extinction following hemisphere damage. *Neuropsychologia* 1984;22:733–744.
19. Bisiach E, Vallar G. Unilateral neglect in humans. In: Boller F, Grafman J, Rizzolatti G, eds. *Handbook of neuropsychology,* 2nd ed. Amsterdam: Elsevier, 2000:459–502.
20. Shapiro MF, Feldman DS. Double simultaneous stimulation phenomena in spinal cord disease. *Neurology* 1952;2:509–513.
21. Bender MB. Extinction and precipitation of cutaneous sensations. *Arch Neurol Psychiatr* 1945;54:1–9.
22. Di Pellegrino G, De Renzi E. An experimental investigation on the nature of extinction. *Neuropsychologia* 1995;33:153–170.
23. Rapcsak SZ, Watson RT, Heilman KM. Hemispace-visual field interactions in visual extinction. *J Neurol Neurosurg Psychiatry* 1987;50:1117–1124.
24. Bisiach E, Geminiani G. Anosognosia related to hemiplegia and hemianopia. In: Prigatano GP, Schacter DL, eds. *Awareness of deficit after brain injury.* New York: Oxford University Press, 1991:17–39.
25. Moscovitch M, Behrmann M. Coding of spatial information in the somatosensory system: evidence from patients with neglect following parietal lobe damage. *J Cogn Neurosci* 1994;6:151–155.
26. Bellas DN, Eskenazi B, Wasserstein J. The nature of unilateral neglect in the olfactory sensory system. *Neuropsychologia* 1988;26:45–52.
27. Smania N, Aglioti S. Sensory and spatial components of somaesthetic deficits following right brain damage. *Neurology* 1995;45:1725–1730.
28. Vallar G, Bottini G, Rusconi ML, et al. Exploring somatosensory hemineglect by vestibular stimulation. *Brain* 1993;116:71–86.
29. Nico D. Effectiveness of sensory stimulation on tactile extinction. *Exp Brain Res* 1999;127:75–82.
30. De Renzi E. *Disorders of space exploration and cognition.* Chichester, UK: Wiley, 1982.
31. Ogden JA. The "neglected" left hemisphere and its contribution to visuospatial neglect. In: Jeannerod M, ed. *Neurophysiological and neuropsychological aspects of spatial neglect.* Amsterdam: Elsevier, 1987:215–233.
32. Ogden JA. Anterior-posterior interhemispheric differences in the loci of lesions producing visual hemineglect. *Brain Cog* 1985;4:59–75.

33. Schwartz AS, Marchok PL, Kreinick CJ, et al. The asymmetric lateralization of tactile extinction in patients with unilateral cerebral dysfunction. *Brain* 1979; 102:669–684.
34. Meador KJ, Loring DW, Lee GP, et al. Right cerebral specialization for tactile attention as evidenced by intracarotid sodium Amytal. *Neurology* 1988;38: 1763–1766.
35. Mesulam M-M. Functional anatomy of attention and neglect: from neurons to networks. In: Karnath H-O, Milner AD, Vallar G, eds. *The cognitive and neural bases of spatial neglect.* Oxford: Oxford University Press 2002:33–45.
36. Mesulam MM. *Principles of behavioral and cognitive neurology.* Oxford, UK: Oxford University Press; 2000.
37. Pardo JV, Fox PT, Raichle ME. Localization of a human system for sustained attention by positron emission tomography. *Nature* 1991;349:61–64.
38. Corbetta M, Miezin FM, Shulman GL, et al. A PET study of visuospatial attention. *J Neurosci* 1993;13: 1202–1226.
39. Corbetta M, Shulman GL, Miezin FM, et al. Superior parietal cortex activation during spatial attention shifts and visual feature conjunction. *Science* 1995;270: 802–805.
40. Gitelman DR, Alpert NM, Kosslyn S, et al. Functional imaging of human right hemispheric activation for exploratory movements. *Ann Neurol* 1996;39:174–179.
41. Nobre AC, Sebestyen GN, Gitelman DR, et al. Functional localization of the system for visuospatial attention using positron emission tomography. *Brain* 1997; 120:515–533.
42. Kim Y-H, Gitelman DR, Nobre AC, et al. The large scale neural network for spatial attention displays multi-functional overlap but differential asymmetry. *NeuroImage* 1999;9:269–277.
43. Proverbio AM, Zani A, Gazzaniga MS, et al. ERP and RT signs of a rightward bias for spatial orienting in a split-brain patient. *NeuroReport* 1994;5:2457–2461.
44. Mangun GR, Luck SJ, Plager R, et al. Monitoring the visual world. Hemispheric asymmetries and subcortical processes in attention. *J Cogn Neurosci* 1994;6: 267–275.
45. Jewesbury ECO. Parietal lobe syndromes. In: Vinken PJ, Bruyn GW, eds. *Handbook of clinical neurology.* Amsterdam: North Holland, 1969:680–699.
46. Adams RD, Victor M, Ropper AH, eds. *Principles of neurology,* 6th ed. New York: McGraw-Hill, 1997.
47. Heilman KM, Valenstein E. Frontal lobe neglect in man. *Neurology* 1972;22:660–664.
48. Heilman KM, Watson RT, Valenstein E. Localization of lesions in neglect and related disorders. In: Kertesz A, ed. *Localization and neuroimaging in neuropsychology.* San Diego: Academic Press, 1994:495–524.
49. Vallar G. The anatomical basis of spatial hemineglect in humans. In: Robertson IH, Marshall JC, eds. *Unilateral neglect: clinical and experimental studies.* Hove, UK: Erlbaum, 1993:27–59.
50. Vallar G, Perani D. The anatomy of unilateral neglect after right hemisphere stroke lesions. A clinical CT/scan correlation study in man. *Neuropsychologia* 1986;24:609–622.
51. Vallar G, Perani D. The anatomy of spatial neglect in humans. In: Jeannerod M, ed. *Neurophysiological and neuropsychological aspects of spatial neglect.* Amsterdam: North-Holland, 1987:235–258.
52. Battersby WS, Bender MB, Pollack M, et al. Unilateral "spatial agnosia" ("inattention") in patients with cerebral lesions. *Brain* 1956;79:68–93.
53. Hécaen H. *Introduction à la neuropsychologie.* Paris Larousse, 1972.
54. Leibovitch FS, Black SE, Caldwell CB, et al. Brain-behavior correlations in hemispatial neglect using CT and SPECT: the Sunnybrook Stroke Study. *Neurology* 1998;50:901–908.
55. Leibovitch FS, Black SE, Caldwell CB, et al. Brain SPECT imaging and left hemispatial neglect covaried using partial least squares: the Sunnybrook Stroke Study. *Hum Brain Mapp* 1999;7:244–253.
56. Hécaen H, Penfield W, Bertrand C, et al. The syndrome of apractognosia due to lesions of the minor cerebral hemisphere. *Arch Neurol Psychiatr* 1956;75: 400–434.
57. Perenin M-T. Optic ataxia and unilateral neglect: clinical evidence for dissociable spatial functions in posterior parietal cortex. In: Thier P, Karnath H-O, eds. *Parietal lobe contributions to orientation in 3D space.* Heidelberg: Springer-Verlag, 1997:289–308.
58. Samuelsson H, Jensen C, Ekholm S, et al. Anatomical and neurologic correlates of acute and chronic visuospatial neglect following right hemisphere stroke. *Cortex* 1997;33:271–285.
59. Heilman KM, Schwartz HD, Watson RT. Hypoarousal in patients with the neglect syndrome and emotional indifference. *Neurology* 1978;28:229–232.
60. Robertson IH, Mattingley JB, Rorden C, et al. Phasic alerting of neglect patients overcomes their spatial deficit in visual awareness. *Nature* 1998;395:169–172.
61. Karnath H-O, Ferber S, Himmelbach M. Spatial awareness is a function of the temporal not the posterior parietal lobe. *Nature* 2001;411:950–953.
62. Ratcliff G, Davies-Jones GAB. Defective visual localization in focal brain wounds. *Brain* 1972;95:49–60.
63. Perenin MT, Vighetto A. Optic ataxia: a specific disruption in visuomotor mechanisms. I. Different aspects of the deficit in reaching for objects. *Brain* 1988; 111:643–674.
64. Kertesz A, Dobrowolsky S. Right-hemisphere deficits, lesion size and location. *J Clin Neuropsychol* 1981;3: 283–299.
65. Posner MI, Walker JA, Friedrich FA, et al. How do the parietal lobes direct covert attention? *Neuropsychologia* 1987;25:135–145.
66. Posner MI, Walker JA, Friedrich FJ, et al. Effects of parietal injury on covert orienting of attention. *J Neurosci* 1984;4:1863–1874.
67. Friedrich FJ, Egly R, Rafal RD, et al. Spatial attention deficits in humans: a comparison of superior parietal and temporal-parietal junction lesions. *Neuropsychology* 1998;12:193–207.
68. Husain M, Kennard C. Visual neglect associated with frontal lobe infarction. *J Neurol* 1996;243:652–657.
69. Buxhoeveden DP, Casanova MF. The minicolumn hypothesis in neuroscience. *Brain* 2002;125:935–951.
70. Mesulam MM. From sensation to cognition. *Brain* 1998;121:1013–1052.
71. Mesulam MM. Spatial attention and neglect: parietal, frontal and cingulate contributions to the mental rep-

resentation and attentional targeting of salient extrapersonal events. *Phil Trans R Soc Lond* 1999;B354: 1325–1346.

72. Feeney DM, Baron J-C. Diaschisis. *Stroke* 1986;17: 817–830.
73. Hillis AE, Wityk RJ, Barker PB, et al. Subcortical aphasia and neglect in acute stroke: the role of cortical hypoperfusion. *Brain* 2002;125:1094–1104.
74. Perani D, Vallar G, Cappa S, et al. Aphasia and neglect after subcortical stroke. A clinical/cerebral perfusion correlation study. *Brain* 1987;110:1211–1229.
75. Vallar G, Perani D, Cappa SF, et al. Recovery from aphasia and neglect after subcortical stroke. *J Neurol Neurosurg Psychiatry* 1988;51:1269–1276.
76. Perani D, Vallar G, Paulesu E, et al. Left and right hemisphere contribution to recovery from neglect after right hemisphere damage—an [^{18}F]FDG pet study of two cases. *Neuropsychologia* 1993;31:115–125.
77. Vallar G. Spatial disorders. In: *Encyclopedia of cognitive science.* London: Macmillan 2003:125–131.
78. Watson RT, Heilman KM. Thalamic neglect. *Neurology* 1979;29:690–694.
79. Stein S, Volpe BT. Classical "parietal" neglect syndrome after subcortical right frontal lobe infarction. *Neurology* 1983;33:797–799.
80. Cambier J, Elghozi D, Strube E. Lésion du thalamus droit avec syndrome de l'hémisphère mineur. Discussion du concept de négligence thalamique. *Rev Neurol (Paris)* 1980;136:105–116.
81. Ferro JM, Kertesz A, Black S. Subcortical neglect: quantitation, anatomy, and recovery. *Neurology* 1987; 37:1487–1492.
82. Vallar G, Rusconi ML, Bignamini L, et al. Anatomical correlates of visual and tactile extinction in humans: a clinical CT scan study. *J Neurol Neurosurg Psychiatry* 1994;57:464–470.
83. Heilman KM, Bowers D, Coslett HB, et al. Directional hypokinesia: prolonged reaction times for leftward movements in patients with right hemisphere lesions and neglect. *Neurology* 1985;35:855–859.
84. Bisiach E, Geminiani G, Berti A, et al. Perceptual and premotor factors of unilateral neglect. *Neurology* 1990;40:1278–1281.
85. Tegnér R, Levander M. Through a looking glass. A new technique to demonstrate directional hypokinesia in unilateral neglect. *Brain* 1991;114:1943–1951.
86. Bisiach E, Tegnèr R, Làdavas E, et al. Dissociation of ophthalmokinetic and melokinetic attention in unilateral neglect. *Cereb Cortex* 1995;5:439–447.
87. Nico D. Detecting directional hypokinesia: the epidiascope technique. *Neuropsychologia* 1996;34:471–474.
88. Adair JC, Na DL, Schwartz RL, et al. Analysis of primary and secondary influences on spatial neglect. *Brain Cog* 1998;37:351–367.
89. Na DL, Adair JC, Williamson DJ, et al. Dissociation of sensory-attentional from motor-intentional neglect. *J Neurol Neurosurg Psychiatry* 1998;64:331–338.
90. Bottini G, Sterzi R, Vallar G. Directional hypokinesia in spatial hemineglect. *J Neurol Neurosurg Psychiatry* 1992;55:431–436.
91. Coslett HB, Bowers D, Fitzpatrick E, et al. Directional hypokinesia and hemispatial inattention in neglect. *Brain* 1990;113:475–486.
92. Baddeley AD. Exploring the central executive. *Q J Med Exp Psychol* 1996;49A:5–28.
93. Fink GR, Marshall JC, Halligan PW, et al. The neural consequences of conflict between intention and the senses. *Brain* 1999;122:497–512.
94. Mattingley JB, Husain M, Rorden C, et al. Motor role of human inferior parietal lobe revealed in unilateral neglect patients. *Nature* 1998;392:179–182.
95. Husain M, Mattingley JB, Rorden C, et al. Distinguishing sensory and motor biases in parietal and frontal neglect. *Brain* 2000;123:1643–1659.
96. Castaigne P, Laplane D, Degos J-D. Trois cas de négligence motrice par lésion frontale pré-rolandique. *Rev Neurol (Paris)* 1972;126:5–15.
97. Triggs WJ, Gold M, Gerstle G, et al. Motor neglect associated with a discrete parietal lesion. *Neurology* 1994;44:1164–1166.
98. Milner AD, Harvey M, Roberts RC, et al. Line bisection errors in visual neglect: misguided action or size distortion? *Neuropsychologia* 1993;31:39–49.
99. Bisiach E, Ricci R, Lualdi M, et al. Perceptual and response bias in unilateral neglect: two modified versions of the Milner landmark task. *Brain Cog* 1998; 37:369–386.
100. Karnath H-O. New insights into the functions of the superior temporal cortex. *Nat Rev Neurosci* 2001;2: 568–576.
101. Marshall JC, Fink GR, Halligan PW, et al. Spatial awareness: a function of the posterior parietal lobe? *Cortex* 2002;28:253–257.
102. Karnath H-O, Himmelbach M. Strategies of lesion localization. Reply to Marshall, Fink, Halligan and Vallar. *Cortex* 2002;38:258–260.
103. Marshall JC, Halligan PW. Within- and between-task dissociations in visuospatial neglect: a case study. *Cortex* 1995;31:367–376.
104. Binder J, Marshall R, Lazar R, et al. Distinct syndromes of hemineglect. *Arch Neurol* 1992;49:1187–1194.
105. Damasio H, Damasio AR. *Lesion analysis in neuropsychology.* New York: Oxford University Press, 1989.
106. McGlinchey-Berroth R, Bullis DP, Milberg WP, et al. Assessment of neglect reveals dissociable behavioral but not neuroanatomical subtypes. *J Int Neuropsychol Soc* 1996;2:441–451.
107. Doricchi F, Angelelli P. Misrepresentation of horizontal space in left unilateral neglect. Role of hemianopia. *Neurology* 1999;52:1845–1852.
108. Daini R, Angelelli P, Antonucci G, et al. Exploring the syndrome of spatial unilateral neglect through an illusion of length. *Exp Brain Res* 2002;144:224–237.
109. D'Erme P, De Bonis C, Gainotti G. Influenza dell'emiinattenzione e dell'emianopsia sui compiti di bisezione di linee nei pazienti cerebrolesi. *Arch Psicol Neurol Psichiatr* 1987;48:193–207.
110. Halligan PW, Marshall JC, Wade DT. Do visual field deficits exacerbate visuospatial neglect? *J Neurol Neurosurg Psychiatry* 1990;53:487–491.
111. Bisiach E, Ricci R, Neppi Mòdona M. Visual awareness and anisometry of space representation in unilateral neglect: a panoramic investigation by means of a line extension task. *Conscious Cog* 1998;7:327–355.
112. Kerkhoff G. Displacement of the egocentric visual midline in altitudinal postchiasmatic scotomata. *Neuropsychologia* 1993;31:261–265.
113. Kerkhoff G, Artinger F, Ziegler W. Contrasting spatial hearing deficits in hemianopia and spatial neglect. *NeuroReport* 1999;10:3555–3560.

114. Axenfeld D. Eine einfache Methode Hemianopsie zu constatiren. *Neurol Centralblatt* 1894;13:437–438.
115. Ferber S, Karnath HO. Parietal and occipital lobe contributions to perception of straight ahead orientation. *J Neurol Neurosurg Psychiatry* 1999;67:572–578.
116. Farnè A, Ponti F, Ladavas E. In search for biased egocentric reference frames in neglect. *Neuropsychologia* 1998;36:611–623.
117. Rossetti Y, Rode G, Pisella L, et al. Prism adaptation to a rightward optical deviation rehabilitates left hemispatial neglect. *Nature* 1998;395:166–169.
118. Gainotti G, Tiacci C. Patterns of drawing disability in right and left hemispheric patients. *Neuropsychologia* 1970;8:379–384.
119. Mark VW, Kooistra CA, Heilman KM. Hemispatial neglect affected by non-neglected stimuli. *Neurology* 1988;38:1207–1211.
120. Rusconi ML, Maravita A, Bottini G, et al. Is the intact side really intact? Perseverative responses in patients with unilateral neglect: a productive manifestation. *Neuropsychologia* 2002;40:594–604.
121. Na DL, Adair JC, Kang Y, et al. Motor perseverative behavior on a line cancellation task. *Neurology* 1999; 52:1569–1576.
122. Denny-Brown D. The nature of apraxia. *J Nerv Ment Dis* 1958;126:9–32.
123. De Ajuriaguerra J, Tissot R. The apraxias. In: Vinken PJ, Bruyn GW, eds. *Handbook of clinical neurology.* Amsterdam: North-Holland, 1969:48–66.
124. Walsh V, Rushworth M. A primer of magnetic stimulation as a tool for neuropsychology. *Neuropsychologia* 1999;37:125–135.
125. Pascual-Leone A, Grafman J, Cohen LG, et al. Transcranial magnetic stimulation: a new tool for the study of higher cognitive functions in humans. In: Boller F, Grafman J, eds. *Handbook of neuropsychology.* Amsterdam: Elsevier, 1997:267–290.
126. Walsh V, Cowey A. Magnetic stimulation studies of visual cognition. *Trends Cog Sci* 1998;2:103–110.
127. Pascual-Leone A, Gomez-Tortosa E, Grafman J, et al. Induction of visual extinction by rapid-rate transcranial magnetic stimulation of parietal lobe. *Neurology* 1994;44:494–498.
128. Hilgetag CC, Théoret H, Pascual-Leone A. Enhanced visual spatial attention ipsilateral to rTMS-induced 'virtual lesions' of human parietal cortex. *Nat Neurosci* 2001;4:953–957.
129. Oliveri M, Rossini PM, Traversa R, et al. Left frontal transcranial magnetic stimulation reduces contralesional extinction in patients with unilateral right brain damage. *Brain* 1999;122:1731–1739.
130. Kinsbourne M. Hemi-neglect and hemisphere rivalry. In: Weinstein EA, Friedland RP, eds. *Hemi-inattention and hemispheric specialization.* New York: Raven Press, 1977:41–49.
131. Kinsbourne M. Mechanisms of unilateral neglect. In: Jeannerod M, ed. *Neurophysiological and neuropsychological aspects of spatial neglect.* Amsterdam: North-Holland, 1987:69–86.
132. Kinsbourne M. Orientational bias model of unilateral neglect: evidence from attentional gradients within hemispace. In: Robertson IH, Marshall JC, eds. *Unilateral neglect: clinical and experimental studies.* Hove, UK: Erlbaum, 1993:63–86.
133. Bisiach E, Bulgarelli C, Sterzi R, et al. Line bisection and cognitive plasticity of unilateral neglect of space. *Brain Cog* 1983;2:32–38.
134. Fierro B, Brighina F, Oliveri M, et al. Contralateral neglect induced by right posterior parietal rTMS in healthy subjects. *NeuroReport* 2000;11:1519–1521.
135. Fierro B, Brighina F, Piazza A, et al. Timing of right parietal and frontal cortex activity in visuospatial perception: a TMS study in normal individuals. *NeuroReport* 2001;12:2605–2607.
136. Oliveri M, Bisiach E, Brighina F, et al. rTMS of the unaffected hemisphere transiently reduces contralesional visuospatial hemineglect. *Neurology* 2001;57: 1338–1340.
137. Sprague J. Interaction of cortex and superior colliculus in mediation of visually guided behavior in the cat. *Science* 1966;153:1544–1547.
138. Vuilleumier P, Hester D, Assal G, et al. Unilateral spatial neglect recovery after sequential strokes. *Neurology* 1996;46:184–189.
139. Lynch JC, McLaren JW. Deficits of visual attention and saccadic eye movements after lesions of parietooccipital cortex in monkeys. *J Neurophisiol* 1989; 61:74–90.
140. Pierrot-Deseilligny C, Gray F, Brunet P. Infarcts of both inferior parietal lobules with impairment of visually guided eye movements, peripheral visual inattention and optic ataxia. *Brain* 1986;109:81–97.
141. Raichle ME. Circulatory and metabolic correlates of brain function in normal humans. In: Mouncastle VB, Plum F, Geiger SR, eds. *Handbook of physiology. The nervous system. Higher functions of the brain.* Bethesda, MD: American Physiological Society, 1987: 643–674.
142. Ogawa S, Lee T, Nayak A, et al. Oxygenation-sensitive contrast in magnetic resonance image of rodent brain at high magnetic fields. *Magn Reson Med* 1990; 14:68–78.
143. Turner R, Le Bihan D, Moonen C, et al. Echo-planar time course MRI of cat brain oxygenation changes. *Magn Reson Med* 1991;22:159–166.
144. Logothetis NK, Pauls J, Augath M, et al. Neurophysiological investigation of the basis of the fMRI signal. *Nature* 2001;412:150–157.
145. Rizzolatti G, Fadiga L, Fogassi L, et al. The space around us. *Science* 1997;277:190–191.
146. Andersen RA, Snyder LH, Bradley DC, et al. Multimodal representation of space in the posterior parietal cortex and its use in planning movements. *Ann Rev Neurosci* 1997;20:303–330.
147. Colby CL, Goldberg ME. Space and attention in parietal cortex. *Ann Rev Neurosci* 1999;22:319–349.
148. Rizzolatti G, Berti A, Gallese V. Spatial neglect: neurophysiological bases, cortical circuits and theories. In: Boller F, Grafman J, Rizzolatti G, eds. *Handbook of neuropsychology,* 2nd ed. Amsterdam: Elsevier, 2000: 503–537.
149. Zeki S. *A vision of the brain.* Oxford, UK: Blackwell, 1993.
150. Galletti C, Fattori P. Posterior parietal networks encoding visual space. In: Karnath HO, Milner AD, Vallar G, eds. *The cognitive and neural bases of spatial neglect.* Oxford, UK: Oxford University Press, 2002:59–69.
151. Graziano MS, Gross CG. The representation of extrapersonal space: a possible role for bimodal, visual-tactile neurons. In: Gazzaniga MS, ed. *The cognitive neu-*

rosciences. Cambridge, MA: MIT Press, 1995: 1021–1034.
152. Zeki S, Watson J, Lueck C, et al. A direct demonstration of functional specialization in human visual cortex. *J Neurosci* 1991;11:641–649.
153. Sereno MI, Dale AM, Reppas JB, et al. Borders of multiple visual areas in humans revealed by functional magnetic-resonance-imaging. *Science* 1995;268: 889–893.
154. Tootell RB, Mendola JD, Hadjikhani NK, et al. The representation of the ipsilateral visual field in human cerebral cortex. *Proc Natl Acad Sci USA* 1998;95: 818–824.
155. Kastner S, Ungerleider LG. Mechanisms of visual attention in the human cortex. *Ann Rev Neurosci* 2000; 23:315–341.
156. Ruben J, Schwiemann J, Deuchert M, et al. Somatotopic organization of human secondary somatosensory cortex. *Cereb Cortex* 2001;11:463–473.
157. Paulesu E, Frackowiak RSJ, Bottini G. Maps of somatosensory systems. In: Frackowiak RSJ, Friston KJ, Frith CD, et al., eds. *Human brain function.* San Diego: Academic Press, 1997:183–242.
158. Ungerleider LG, Mishkin M. Two cortical visual systems. In: Ingle DJ, Goodale MA, Mansfield RJW, eds. *Analysis of visual behavior.* Cambridge, MA: MIT Press, 1982:549–586.
159. Haxby JV, Grady CL, Horwitz B, et al. Dissociation of object and spatial visual processing pathways in human extrastriate cortex. *Proc Natl Acad Sci USA* 1991; 88:1621–1625.
160. Ungerleider LG, Haxby JV, Grady CL, et al. 'What' and 'where' in the human brain. Dissociation of object and spatial visual processing pathways in human extrastriate cortex. *Curr Opin Neurobiol* 1994;4:157–165.
161. Haxby JV, Horwitz B, Ungerleider LG, et al. The functional organization of human extrastriate cortex: a PET-rCBF study of selective attention to faces and locations. Dissociation of object and spatial visual processing pathways in human extrastriate cortex. *J Neurosci* 1994;14:6336–6353.
162. Courtney S, Ungerleider L, Keil K, et al. Object and spatial visual working memory activate separate neural systems in human cortex. *Cereb Cortex* 1996; 6:39–49.
163. Goldman-Rakic PS. The prefrontal landscape: implications of functional architecture for understanding human mentation and the central executive. *Phil Trans R Soc Lond* 1996;B351:1445–1453.
164. Wilson FAW, O'Scalaidhe SP, Goldman-Rakic PS. Dissociation of object and spatial processing domains in primate prefrontal cortex. *Science* 1993;260: 1955–1958.
165. Chelazzi L, Miller EK, Duncan J, et al. A neural basis for visual search in inferior temporal cortex. *Nature* 1993;363:345–347.
166. Ungerleider LG, Courtney SM, Haxby JV. A neural system for human visual working memory. *Proc Natl Acad Sci USA* 1998;95:883–890.
167. Fink GR, Marshall JC, Shah NJ, et al. Line bisection judgments implicate right parietal cortex and cerebellum as assessed by fMRI. *Neurology* 2000;54: 1324–1331.
168. Galati G, Lobel E, Vallar G, et al. The neural basis of egocentric and allocentric coding of space in humans: a functional magnetic resonance study. *Exp Brain Res* 2000;133:156–164.
169. Weiss PH, Marshall JC, Wunderlich G, et al. Neural consequences of acting in near versus far space: a physiological basis for clinical dissociations. *Brain* 2000;123:2531–2541.
170. Wise SP, Boussaoud D, Johnson PB, et al. Premotor and parietal cortex: corticocortical connectivity and combinatorial computations. *Ann Rev Neurosci* 1997; 20:25–42.
171. Jeannerod M. *The cognitive neuroscience of action.* Oxford: Blackwell, 1997.
172. Fink GR, Marshall JC, Weiss PH, et al. 'Where' depends on 'what': a differential functional anatomy for position discrimination in one- versus two-dimensions. *Neuropsychologia* 2000;38:1741–1748.
173. Tegner R, Levander M. The Influence of stimulus properties on visual neglect. *J Neurol Neurosurg Psychiatry* 1991;54:882–887.
174. Halligan PW, Marshall JC. Figural modulation of visuospatial neglect: a case-study. *Neuropsychologia* 1991; 29:619–628.
175. Galati G, Committeri G, Sanes JN, et al. Spatial coding of visual and somatic sensory information in body-centered coordinates. *Eur J Neurosci* 2001;14: 737–746.
176. Vallar G, Lobel E, Galati G, et al. A fronto-parietal system for computing the egocentric spatial frame of reference in humans. *Exp Brain Res* 1999;124: 281–286.
177. Andersen RA, Snyder LH, Li C-S, et al. Coordinate transformations in the representation of spatial information. *Curr Opin Neurobiol* 1993;3:171–176.
178. Guldin WO, Grüsser O-J. Is there a vestibular cortex? *Trends Neurosci* 1998;21:254–259.
179. Umiltà C. Mechanisms of attention. In: Rapp B, ed. *The handbook of cognitive neuropsychology.* Philadelphia: Psychology Press, 2001:135–158.
180. Corbetta M, Akbudak E, Conturo TE, et al. A common network of functional areas for attention and eye movements. *Neuron* 1998;21:761–773.
181. Nobre AC, Gitelman DR, Dias EC, et al. Covert visual spatial orienting and saccades: overlapping neural systems. *NeuroImage* 2000;11:210–216.
182. Corbetta M. Frontoparietal cortical networks for directing attention and the eye to visual locations: identical, independent, or overlapping neural systems? *Proc Natl Acad Sci USA* 1998;95:831–838.
183. Petrides M, Alivisatos B, Meyer E, et al. Functional activation of the human frontal cortex during the performance of working memory tasks. *Proc Natl Acad Sci USA* 1993;90:878–882.
184. Hécaen H. Clinical symptomatology in right and left hemispheric lesions. In: Mouncastle VB, ed. *Interhemispheric relations and cerebral dominance.* Baltimore: The Johns Hopkins Press, 1962.
185. Kumral E, Evyapan D. Associated exploratory-motor and perceptual-sensory neglect without hemiparesis. *Neurology* 1999;52:199–202.
186. Chedru F, Leblanc M, Lhermitte F. Visual searching in normal and brain-damaged subjects (contribution to the study of unilateral inattention). *Cortex* 1973;9:94–111.
187. Girotti F, Casazza M, Musicco M, et al. Oculomotor disorders in cortical lesions in man:the role of unilateral neglect. *Neuropsychologia* 1983;21:543–553.

188. Ishiai S, Furukawa T, Tsukagoshi H. Visuospatial processes of line bisection and the mechanisms underlying unilateral spatial neglect. *Brain* 1989;112: 1485–1502.
189. Barton JJ, Behrmann M, Black S. Ocular search during line bisection. The effects of hemi-neglect and hemianopia. *Brain* 1998;121:1117–1131.
190. Pierrot-Deseilligy C, Müri R. Posterior parietal cortex control of saccades in humans. In: Thier P, Karnath H-O, eds. *Parietal lobe contributions to orientation in 3D space.* Heidelberg: Springer-Verlag, 1997:135–148.
191. Bogousslavsky J, Miklossy J, Regli F, et al. Subcortical neglect: neuropsychological, SPECT, and neuropathologic correlations with anterior choroidal artery infarction. *Ann Neurol* 1988;23:448–452.
192. Pizzamiglio L, Perani D, Cappa SF, et al. Recovery of neglect after right hemispheric damage. (H_2O)-O^{15} positron emission tomographic activation study. *Arch Neurol* 1998;55:561–568.
193. Bottini G, Paulesu E, Sterzi R, et al. Modulation of conscious experience by peripheral stimuli. *Nature* 1995;376:778–781.
194. Baron JC, D'Antona R, Pantano P, et al. Effects of thalamic stroke on energy metabolism of the cerebral cortex. A positron tomography study in man. *Brain* 1986;109:1243–1259.
195. Heilman KM, Watson RT, Valenstein E. Neglect and related disorders. In: Heilman KM, Valenstein E, eds. *Clinical neuropsychology,* 3rd ed. New York: Oxford University Press, 1993:279–336.
196. Rees G, Wojciulik E, Clarke K, et al. Unconscious activation of visual cortex in the damaged right hemisphere of a parietal patient with extinction. *Brain* 2000;123:1624–1633.
197. Berti A, Rizzolatti G. Visual processing without awareness: evidence from unilateral neglect. *J Cogn Neurosci* 1992;4:345–351.
198. Marshall JC, Halligan P. Blindsight and insight in visuospatial neglect. *Nature* 1988;336:766–767.
199. Làdavas E, Paladini R, Cubelli R. Implicit associative priming in a patient with left visual neglect. *Neuropsychologia* 1993;31:1307–1320.
200. McGlinchey-Berroth R, Milberg WP, Verfaellie M, et al. Semantic processing and orthographic specificity in hemispatial neglect. *J Cogn Neurosci* 1996;8: 291–304.
201. Marzi CA, Girelli M, Miniussi C, et al. Electrophysiological correlates of conscious vision: evidence from unilateral extinction. *J Cogn Neurosci* 2000;12: 869–877.
202. Bottini G, Paulesu E, Frith CD, et al. Functional anatomy of the human vestibular cortex. In: Collard M, Jeannerod M, Crysten Y, eds. *Le cortex vestibulaire.* Paris: Irvinn, 1996:27–48.
203. Corbetta M, Shulman GL. Control of goal-directed and stimulus-driven attention in the brain. *Nat Rev Neurosci* 2002;3:215–229.
204. Dolan RJ, Paulesu E, Fletcher P. Human memory systems. In: Frackowiak RSJ, Friston KJ, Frith CD, et al., eds. *Human brain function.* San Diego: Academic Press, 1997:367–404.

20

Hemispheric Asymmetries in the Parietal Lobes

Mary K. Colvin, Todd C. Handy, and Michael S. Gazzaniga

Department of Psychological and Brain Sciences and Center for Cognitive Neuroscience, Dartmouth College, Hanover, New Hampshire

INTRODUCTION

Historically, the parietal lobes have eluded a concise functional definition. Although there is clear evidence that parietal cortex serves as a key interface between sensation and action, it also serves a role in memory, language, mathematic abilities, spatial navigation, and object perception. To complicate matters, the left and right parietal lobes are differentially involved in many of these cognitive tasks. Although early descriptions of patients with unilateral parietal lobe damage led to a loose dissociation between "visuospatial" abilities associated with right parietal function and "visuoconstructive" abilities associated with left parietal function, this distinction fails to capture the true range of processing operations subserved by parietal cortex.

In this chapter, we first review the general role of parietal cortex as a sensorimotor interface directing goal-specific action, focusing on the complementary specialization of the left and right parietal lobes for motor attention and spatial attention, respectively. We then discuss the evolutionary origins of this specialization in relation to the intrahemispheric networks linking the left and right parietal cortices to the frontal and superior temporal areas. We stress the role of each network's parietal lobe as a convergence site between "top-down" and "bottom-up" processes. We then conclude by proposing that lateralized parietal involvement in many higher level cognitive functions is a direct result of these hemisphere-specific, fronto-parieto-temporal networks.

THE COMPLEMENTARY ROLES OF THE RIGHT AND LEFT PARIETAL LOBES IN ATTENTION

The parietal lobes have long been considered critical for the selection and maintenance of attention to the self and the environment (1). As such, the most widely understood functional asymmetry of the two parietal lobes concerns processes associated with selective attention. Selective attention can be characterized as the "filtering" of information in the brain, an ability that allows us to select or isolate what events in the sensory world we want to process, and what responses we want to make to those events (2). In this section, we discuss how parietal-mediated selective attention interacts with functional differences between the left and right hemispheres.

Left Parietal Specialization for Motor Attention

Damage to the left parietal lobe can result in deficits of sequencing motor actions as well as allocating attention to motor acts (motor attention). Two neuropsychological conditions illustrate this specialized function. Optic ataxia, defined as an inability to make accurate movements to objects with one's contralesional hand, can follow damage to either the left or right superior parietal lobe (3). However, optic ataxia patients with left parietal damage make errors with their right hand throughout the entire visual field, whereas optic ataxia patients with right parietal damage only make errors with their left hands in the left visual field (4). Thus, the left parietal lobe attends to motor acts throughout visual space, even though this duplicates right parietal function.

Apraxia, defined as a deficit in performing learned movements, results from left inferior parietal damage and is often observed in both hands (5,6). There are two common forms of apraxia. Ideomotor apraxia, characterized by an inability to execute movements in response to a verbal command despite normal spontaneous execution (7), has been associated with damage to the supramarginal gyrus (8). Conceptual apraxia, characterized by an inability to sequence a series of actions necessary to execute a complex, goal-directed action involving the use of tools, has been associated with damage to the left parieto-occipital or temporoparietal regions. Note that conceptual apraxia is a restricted form of what has traditionally been called ideational apraxia; we have chosen to limit the definition because ideational apraxia is often masked by co-occurring language impairments (aphasia) (8,9). Thus, although the neural substrates of optic ataxia and the apraxias are functionally dissociable (10), all result from left parietal damage and are deficits in motor attention throughout the visual field. This evidence supports a left parietal specialization for execution and attention to sequences of motor acts.

To precisely characterize the nature of the motor attention subserved by the left parietal lobe, Rushworth and colleagues (11) designed a motor-orienting paradigm analogous to established spatial orienting paradigms. In the traditional spatial orienting task, subjects see a cue as to the most likely location of an upcoming target. In Rushworth et al.'s (11) study, unilateral (left or right) parietal patients were required to make a motor movement at the onset of a target; the target itself indicated the specific movement to make. Critically, prior to the onset of the target, patients were cued as to the most likely movement they were going to have to make; the cue was predictive of the pending target. Both left and right parietal patients were proficient in performing the appropriate motor movement when the cue correctly predicted the target identity. However, compared to right parietal patients, left parietal patients experienced more difficulty in their ability to perform the appropriate movement when the cue incorrectly predicted the target. Rushworth et al. (11) proposed that left parietal patients were unable to disengage motor attention from the cued movement.

Converging evidence of a left parietal specialization for motor attention has come from the domain of neuroimaging in normal populations. In a recent positron emission tomography (PET) experiment, participants were asked to make finger movements in response to the presentation of a visual cue (12). In one condition, the cue indicated which of two fingers to move (choice task). In the other condition, the same finger was moved regardless of the cue type (simple task). When activation patterns were compared between these two conditions, there was increased activity in left but not right intraparietal cortex during the choice task relative to the simple task. Critically, the effect was independent of which hand was used for the task. Given that the choice task uniquely required a response selection between two alternative choices, the PET data were taken as evidence for a direct role of left parietal cortex in motor attention. In a similar PET experiment, activation was compared between a condition where participants covertly attended to planned movements in their left hand, and a condition that was identical with the exception that there was no covert attention to planned motor movements (13). These data again revealed left-lateralized parietal activity associated with motor attention, specifically, in the left supramarginal gyrus and adjacent anterior intraparietal sulcus.

Right Parietal Specialization for Spatial Attention

The right hemisphere's dominant role in allocating attention across space (spatial attention) was first discovered from studies of neglect patients with right parietal damage, particularly to the inferior parietal lobule (IPL) and the temporo-parietal junction (14–18). Visuospatial neglect patients do not orient or respond to objects presented to their left visual space (14,19). Although this account describes a deficit coding egocentric spatial representation, the condition can extend to other reference frames (20,21; for a review, see ref. 22), directly reflecting the parietal lobes' ability to code multiple spatial reference frames (23,24; for review, see ref. 25). Critically, patients with lesions in homologous regions of left parietal cortex rarely show deficits in right-oriented processing analogous to those associated with damage to right parietal cortex. This led researchers to propose that the right parietal cortex is responsible for left-related spatial processing, as well as right-related spatial processing. Thus, following left parietal brain damage, there is a redundant, right parietal mechanism for coding the right side of space. In contrast, no such redundancy exists in right parietal cortex and the ability to attend to the left side of space is irrevocably impaired following right parietal damage. Although this attention deficit has been primarily studied within the visual modality, neglect has recently been demonstrated for auditory (26) and somatosensory stimuli (27) as well, reflecting the parietal lobes role in integrating primary sensory input across all modalities.

Traditionally, neglect patients' deficits have been described in terms of spatial attention, including both exogenous and endogenous orienting in contralesional space (28). In a classic study of spatial attentional orienting, patients with unilateral (left or right) parietal cortex lesions were presented with visual cues at fixation signaling the most likely location of a pending target, which would be in the left or right visual field (29). All patients' responses were greatly slowed when targets were presented in the uncued (i.e., unlikely) and contralesional visual space. However, right parietal patients were significantly more impaired than left parietal patients. This was taken as evidence for specialized right hemisphere control of spatial attention (30,31). This notion has been supported by studies of attentional orienting in split-brain patients. For example, Mangun et al. (32) reported that the left hemisphere could only orient attention to the right visual field, whereas the right hemisphere was equally capable of orienting attention to both the left and right visual fields.

Convergent findings have come from neuroimaging. In a seminal study using PET, Corbetta and colleagues examined the network of cortical areas activated during shifts of spatial attention (33). In one condition, subjects made voluntary attentional movements to the left or right of fixation throughout the scanning period while engaged in a target detection task. In a second condition, subjects were required to detect the onset of a target at fixation while ignoring nonfoveal probe stimuli that matched the sequence of parafoveal stimuli in the attentional switching condition. When patterns of cortical activation were compared between these two conditions, they found that the superior parietal lobule was selectively active when attention was shifted between locations within the visual field. Moreover, there was an asymmetry in parietal activation such that the right superior parietal lobule was activated during shifts of attention to both visual fields, but the left superior parietal lobule was only activated during shifts of attention to the right visual field.

This basic finding of superior parietal activation during covert shifts of spatial attention has now been replicated and extended by a number of neuroimaging studies (34–38). These studies have converged on the position that although there is evidence of some left parietal activation during attentional orienting, the dominant role is played by right parietal cortex. However, the evidence for right parietal involvement in attentional shifting has not been unanimous. In a recent PET study, parietal activity was examined in an attentional orienting task as a function of whether there were a low or high number of attentional shifts during the scanning period (39). Surprisingly, parietal activity did not vary as a function of the frequency of attentional shifting, a finding that appears to run counter to the model that parietal cortex plays a specific role in shifting spatial attention (40). Regardless, the collective data across domains converge on the notion that although left parietal cortex has some spatial capacities, spatial selection—and the orienting of spatial attention, in particular—is strongly correlated with right parietal function.

Testing the Parietal Asymmetry Hypothesis

Recently, Rushworth and colleagues (41) directly tested the hypothesis that the left and right parietal cortices demonstrate complementary specialization for motor and spatial selective attention. Using transcranial magnetic stimulation (TMS), they disrupted cortical function in normal participants performing either a motor attention or spatial selective attention task. Performance on the motor task was selectively disrupted by TMS to the left supramarginal gyrus of parietal cortex, whereas performance on the spatial task was selectively disrupted by TMS to the right angular gyrus of parietal cortex. Furthermore, evidence from optic ataxia and patients with unilateral parietal damage indicates that the left parietal lobe governs motor attention throughout visual space as well as spatial attention in the right visual field. Similarly, the right parietal lobe governs spatial attention throughout visual space, as well as motor attention in left visual field.

ORIGINS OF PARIETAL ASYMMETRIES AND INTRAHEMISPHERIC CORTICAL NETWORKS

What are the roots of parietal asymmetries in attention? In this section, we propose that parietal asymmetries reflect more general asymmetries between left and right hemispheres. As such, we begin with a discussion of the evolutionary origins of general hemispheric asymmetries. We then suggest that the left parietal specialization for motor attention is a reflection of an underlying left hemisphere specialization for temporal processing and possibly object-based attention. The right hemisphere is specialized for visuospatial perceptual processes. Given that the parietal asymmetries are rooted within more general hemispheric specializations, we outline the intrahemispheric connections among the parietal, frontal, and superior temporal areas. In the following section, we refer to these networks to account for particular, lateralized higher-level cognitive functions requiring a convergence of input from all areas of the network.

Evolutionary Origins of Hemispheric Asymmetry

Our central premise is that lateralization of parietal function reflects more general asymmetries in hemispheric function. Hemispheric asymmetries are seen throughout many species, including lower vertebrates such as fish, frogs, and chicks (42). However, the prevalence of lateralized cognitive function in the human brain is considerable, even more than nonhuman primates (43). Why has this evolved? Gazzaniga and colleagues have suggested that the addition of new cognitive abilities throughout the course of human evolution required the lateralization of function. As new cognitive capacities were acquired through evolutionary processes, each capacity required a dedicated region in cortex. An obvious adaptive solution was to eliminate bilateral representation of certain functions. In a normal brain, lateralized representations can be quickly transferred to the other hemisphere as needed, creating an efficient, unified system (43,44). Hemispheric asymmetries are typically only observed in overt human behavior when unilateral brain damage is sustained, resulting in a permanent loss of function, or when the corpus callosum is severed. Surgical callosal sections are performed as a treatment for intractable epilepsy, and eliminate the majority of communication between the hemispheres. Thus, callosotomy (split-brain) patients are ideal for studying the specialized functions of the two hemispheres. Here, we review findings from split-brain and other research demonstrating the overarching patterns of hemispheric asymmetry that are reflected in parietal asymmetry.

Relationship Between Left Hemisphere Specialization for Motor and Temporal Processes

The left hemisphere's specialization for motor attention appears to be a uniquely human quality. The majority of nonhuman primates prefer to use the left hand for visually guided reaching movements and use of the right hand to manipulate objects. This context-driven handedness is believed to reflect right hemisphere dominance for visuospatial processing to facilitate exploration, and left hemisphere dominance for coordinating complex movements (45). In contrast, humans demonstrate preferred use of the right hand for both visually guided movements and object manipulation, indicating left hemisphere superiority for movement coordination, regardless of context. Goodale (6) has postulated that the course of human evolution has increasingly demanded accurate and rapid visually guided movements, emphasizing temporal precision and sequencing over visuospatial processing in visually guided movements. As a result, the left hemisphere began to drive all motor sequencing, with indirect visuospatial input from the right hemisphere (6).

Left hemisphere superiority for motor acts is supported by findings from split-brain research. Only the left hemisphere can generate voluntary facial expressions on command, even though both hemispheres are involved in spontaneous facial expressions (46). Furthermore, the left hemisphere is specialized for control and planning of movements involving the proximal musculature (e.g., reaching). However, the left hemisphere's ipsilateral projections to distal musculature of the left hand are not as strong; therefore, movements involving the hands (e.g., grasping) are less left-lateralized. Nevertheless, the majority of movement requires a combination of distal and proximal muscles, leading to a dominant role of the left hemisphere in motor planning and control (47–50).

Recently, Nobre and colleagues have proposed that the left-hemisphere specialization for motor attention reflects a more general specialization for orienting attention in time (36,51,52). In a neuroimaging study comparing attentional orienting to locations in space to particular points in time, differential activations of the right and left parietal cortices, respectively, were observed (36). In a later study, Coull et al. (51) observed overlapping activations between left parietal and prefrontal areas involved in orienting in time and initiating motor control. Thus, the left hemisphere specialization for the coordination of motor actions appears to be linked to a more fundamental specialization for bridging events across time (11,51). As Coull and Nobre (36) discuss, this left hemisphere specialization for temporal orienting also may underlie the left hemisphere specialization for language, given that successful language function requires rapid temporal integration (43,53).

The link between motor and temporal attention also may be related to a left hemisphere specialization for attention to objects or features independent of location. Historically, the role of objects in directing selective attention has been difficult to dissociate from spatial attention. However, several neuroimaging studies have shown increased left frontoparietal activations corresponding to attention to object or feature information (54–57). Left parietal patients also demonstrate object-based attentional deficits (58). Given the left parietal region's strong role in guiding movement in space, it seems plausible that the effectiveness of these movements would be enhanced by a specialized ability to attend specifically to object features. Thus, an apraxic patient's difficulty in formulating an appropriate movement toward an object may represent a combined deficit of appropriate temporal orienting, motor planning, and object feature representation. The relationship among these three left-lateralized attentional roles remains to be explored.

Right Hemisphere Specialization for Visuospatial Perceptual Processes

Split-brain studies have demonstrated that the right hemisphere is specialized for some visual perception tasks but not others. For example, the right hemisphere is superior on tasks of mental rotation (59), amodal boundary completion (60), apparent motion detection (61), spatial matching (62), and mirror image discrimination (63). However, both hemispheres can equally match object identities (62), and perceive anorthoscopic figures (64) and illusory-contours (60). Historically, the complex nature of the visual tasks listed in the preceding has precluded a unified theory of asymmetric perceptual function, making it difficult to predict which visual tasks would elicit right hemisphere superiority. Recently, Corballis et al. (65) tested split-brain patients on simple visual discriminations that were either spatial or nonspatial. Right hemisphere superiority was only found on the spatial discrimination tasks. Corballis et al. (65) propose that the fundamental difference between the perceptual functions of the two hemispheres is that the right hemisphere has a specialized ability to perform spatial discriminations.

As Corballis et al. (65) discuss, the simple discrimination tasks they used could be accomplished in early visual processing, suggesting lateralization of early perceptual processes. However, as the authors note, hemispheric asymmetries appear to be dependent on task demands, not the intrinsic nature of the stimuli. In an earlier study, Corballis et al. (62) asked callosotomy patients to either match spatial location (spatial condition) or object identity (identity condition), using the same stimuli in both conditions (Fig. 20.1). Right hemisphere dominance was only observed in the spatial condition. A similar PET study of normal controls demonstrated bilateral, visual ventral stream activation when subjects matched object identity and right parietal activation when subjects matched spatial location (66). Combined, these data suggest that asymmetric, top-down regulatory mechanisms mediated by the dorsal stream, such as spatial or temporal attention, interact with symmetric visual processes mediated by the ventral stream to produce hemispheric asymmetries in visual perception.

This hypothesis is supported by neuroimaging studies. When a visual stimulus is presented, the frontoparietal areas directing spatial orienting toward that stimulus also modulate the activity of ventral stream areas involved in object analysis (67; for review, see 68). In other words, top-down spatial attention (see the preceding) and spatial working memory (69) can influence bottom-up stimulus identification. One might predict that this interaction between top-down and bottom-up processing is critical in synthesizing a complex sensory scene. Indeed, Treisman (70,71) has proposed that spatial attention is necessary to appropriately bind perceptual features of multiple objects in space. Recent neuroimaging research has supported this view, demonstrating that spatial attention resolves competition between object representations in the ventral pathway (72). In short, when a perceptual process requires attention to and integration of multiple objects across space, one would predict that the specialized right frontoparietal network direct the task.

Sensory Networks and the Parietal Lobes

One of the defining features of the parietal lobes is that they are a convergence point for multiple sensory inputs, including vision, audition, and somatosensation. However, the type of information projected from primary sensory cortex to the parietal lobes is specific; Milner and Goodale (3) propose that it is the information necessary to generate goal-directed action in response to the sensory cue. Multiple studies of nonhuman primates have

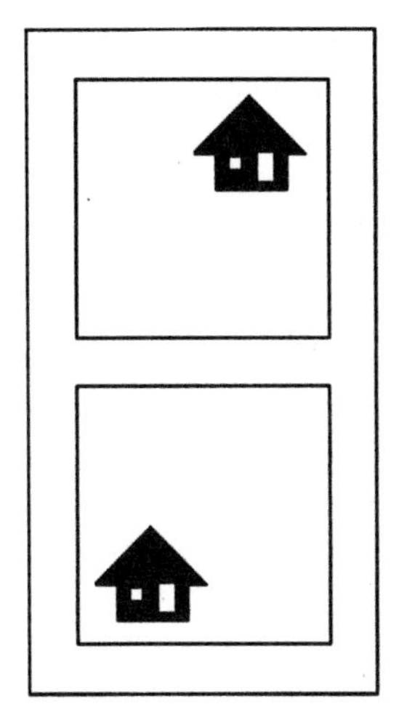

FIG. 20.1. The stimuli used in the split-brain study of Corballis et al. (62). In the identity condition *(left)* the subject's task was to discriminate whether the two objects were the same. In the spatial condition *(right)* the subject's task was to discriminate whether the two objects occurred in the same quadrant of the surrounding square. Data indicated that although the left hemisphere performed marginally better in the identity condition, the right hemisphere clearly performed better in the spatial condition.

shown that the posterior parietal cortex (PPC) receives direct visual input from primary visual cortex (VI) (73,74), direct auditory input from primary auditory cortex (AI) (75,76), and direct somatosensory input from primary sensory cortex (SI) (77). Across these three sensory systems, these anatomic connections are known as the "dorsal stream" and quickly relay specific information necessary for action (3). For example, the visual dorsal stream primarily carries information with low spatial resolution and high contrast sensitivity (78–80). Visual dorsal stream neurons respond to speed and direction of stimulus motion (81,82). This type of input is necessary for the guidance of action toward a sensory stimulus in space; thus, the dorsal visual, auditory, and somatosensory paths are commonly referred to as the "where" pathways (73,75–77). Given that the parietal lobes receive primary sensory input across modalities, it follows that the parietal lobes are involved in the integration of multimodal input, specifically for the guidance of action through space.

Supplementing the sensory "where" pathways, a second pathway emerges from primary sensory cortices and terminates in more ventral or anterior areas. Given the role of these pathways in object identification, they have commonly been called the "what" pathways. The visual and auditory ventral streams terminate in inferotemporal cortex (ITC) (74–76,83). The somatosensory ventral stream projects to lateral somatosensory areas (77). Neurons in these pathways are tuned for specific features of the sensory stimulus that allow for precise characterization. For example, neurons in the ITC selectively respond to complex visual features, such as color, shape, and texture (84). Thus, the two anatomic streams have complementary functions: The dorsal stream allows for orientation and action toward a stimulus in space, whereas the ventral stream subserves fine perception of stimulus features.

The anatomic and functional segregation of the dorsal and ventral visual and auditory streams extends to dorsal prefrontal cortex (DPFC) and ventral prefrontal cortex (VPFC). These areas are associated with spatial memory and object or nonspatial memory, respectively (74–76,83,85–87). However, it seems intuitive that many higher level cognitive functions require the integration of the information necessary for action and the information required for detailed stimulus perception (88). Indeed, in other frontal lobe areas, neurons show sensitivity to both spatial and object information (89), indicating eventual integration of dorsal and ventral stream input. Anatomic studies in nonhuman primates support the connection between the dorsal and ventral frontal areas (90–92). Thus, it is highly probable that the frontal lobes integrate dorsal and ventral stream input, then project this information back to more posterior areas, including the parietal lobes, to guide subsequent behavior (88).

Similarly, the superior temporal sulcus (STS) is a likely site of integration between the dorsal and ventral streams of the visual and auditory systems (93). Physiologic studies have shown that the superior temporal gyrus is a polysensory association area (94–97) and that the STS receives input from both the PPC and ITC (98,99). The STS is so interconnected with the parietal cortex that it has been difficult to distinguish how the two regions differentially contribute to a particular function. For example, Karnath (100) has recently proposed that the STS is the critical area subserving spatial awareness. According to this view, spatial neglect results from right STS damage, not right IPL damage. Although more research needs to be done to explore this possibility, clearly, the inferior parietal and superior temporal areas share close anatomic and functional links.

HIGHER LEVEL COGNITIVE ABILITIES ASSOCIATED WITH THE LEFT AND RIGHT PARIETAL LOBES

How do the left and right parietal lobes' specialized top-down attentional roles interact within the fronto-parieto-temporal network described in the preceding? We propose that the asymmetric top-down attentional mechanisms interact with the largely symmetric and bottom-up object recognition systems, giving rise to multiple higher level cognitive functions associated solely with the right or left parietal lobes.

Role of the Left Parietal Lobe in Language and Calculation

For the majority of people, a distributed neural network subserving language function resides in the left hemisphere. The left temporal lobe has long been associated with language function, as have some regions of the left frontal lobe. Combined with evidence of the dense reciprocal connections discussed in the preceding, one might predict that the left parietal lobe serves as a bridge between object processing and frontally mediated linguistic abilities. Indeed, this is the case. In the following, we review the left parietal lobe's role in linguistic abilities and how this specialization provides a foundation for some mathematic abilities.

The left parietal lobe plays an important role in language comprehension and verbal abilities. At the intersection of the left temporal and parietal lobe, Wernicke's area serves as the linguistic link between object representations and their meanings. Patients with damage to this area suffer from Wernicke's aphasia, characterized by fluent, yet incomprehensible speech. Critically, these patients are unable to understand speech; they can not name or identify common objects presented in any sensory modality (101,102). Based on the characteristics of this deficit, Mesulam (102) has proposed that Wernicke's area is a "transmodal gateway," connecting occipitotemporal sensory representations of word forms and their associated meanings. The left parietal lobe also functions to link word representations to linguistic (verbal or motor) output. Specific damage to the angular gyrus of the left parietal lobe can cause anomia (inability to name objects), alexia (inability to read), and dysgraphia (inability to write) (102). Thus, Wernicke's area and the left angular gyrus of the left parietal lobe are critical to connect ventrally generated word or object representations to linguistic (verbal or motor) output.

Mathematic ability also has long been associated with parietal lobe function of left hemisphere. In addition to the purely linguistic deficits discussed in the preceding, Gerstmann first noticed that damage to the left angular gyrus could cause acalculia (inability to perform arithmetic operations), dysgraphia, finger agnosia (inability to identify one's fingers), and difficulties discerning left and right. This tetrad of cognitive deficits is commonly known as Gerstmann's syndrome (102) and are believed to co-occur because of close anatomic proximity and functional similarity. In general, the form of acalculia following damage to the left inferior parietal lobe, such as that seen in Gerstmann's syndrome, is limited to an inability to perform arithmetic operations. Recent functional magnetic resonance imaging (fMRI) studies of normal subjects performing arithmetic operations have localized more specific activations of the left intraparietal sulcus and a left-lateralized network linking the cingulate, parietal, and frontal lobes (103). As Chochon et al. (103) discuss, this network is closely linked to the dorsal parietal pathway and has been implicated in cognitive tasks involving working memory and visuospatial attention (33,35,104). Presumably, performing arithmetic operations requires the ability to deploy working memory and attentional resources while critically linking to the language and calculation abilities of the left hemisphere. The left inferior parietal lobe is the convergence point for all of these functions.

It should be noted that although calculation abilities have been repeatedly linked to left parietal function, number comparison abilities appear to be bilaterally represented in the parietal lobes. Within the neuropsychological literature, a pattern of impaired calculation and preserved number comparison can be seen following left hemisphere lesions (105,106). In addition, studies of callosotomy (split-brain) patients have demonstrated that the right hemisphere can make magnitude comparisons but can not perform arithmetic operations (107–111). Neuroimaging studies of neurologically normal subjects have confirmed that both the left and right parietal lobes are active during number comparison (103,112,113). One explanation for the involvement of the right parietal lobe in number comparison but not calculation is that comparison entails placing numerical quantities on a mental number line. There is evidence that normal subjects perform this nonverbal, visuospatial conversion automatically (114) and that this ability is supported by the angular gyrus (115).

The bilateral representation of numerical comparison abilities and the left-lateralized representation of calculation abilities can be seen as a combination of converging ventral and dorsal stream processing in the inferior parietal lobes (103) and generalized hemispheric asymmetries. According to Dehaene's triple-code model (116,117), numerical representations can take one of three visual forms: Arabic numerals, number words, or representations of quantity (e.g., circles) that are represented in the occipito-temporal pathway. The left and right parietal lobes support the manipulation of these quantities on a mental number line, enabling comparison operations. However, the left inferior parietal lobe links the occipito-temporal number representations to left-lateralized language function and verbal memory, enabling arithmetic operations.

Role of the Right Parietal Lobe in Conscious Awareness of Implicitly Processed Visual Stimuli

As we have mentioned, patients with right parietal damage, particularly to the inferior parietal and temporo-parietal junction, commonly suffer from the complex disorder of visuospatial neglect. Although this disorder traditionally has been studied to elucidate mechanisms of spatial attention, it has recently shed light on the right parietal lobe's role in conscious awareness of visual percepts. Neglect patients demonstrate normal preattentive perceptual processes, but later fail to attend to these early representations if they are located in the left hemispace. Thus, neglect causes a dissociation between implicit visual processing and explicit spatial awareness.

Studies of visual processing in neglect patients allow categorization of visual processes as either requiring right parietal interaction (attentive) or mediated by earlier ventral visual activity (preattentive). Neglect patients demonstrate residual implicit processing for contralesional stimuli, including figure ground segregation based on luminance contrast (118), shape symmetry (119), and geometric size illusions induced by neglect stimuli (120,121). However, the question remains whether these preattentive processes can be performed by the ventral visual systems of both the left and right hemispheres. Studies of split-brain patients are ideal to answer this question. For example, both split-brain patients and neglect patients have been tested for the perception of illusory contours (Fig. 20.2). Neglect patients can perceive illusory contours (122), as can both the left and right hemispheres (60). Although more convergent evidence from split-brain studies is needed, this research argues that preattentive visual processes can be performed bilaterally by the ventral visual streams. Thus, the right parietal cortex is necessary for awareness of the implicitly formed percepts of objects in contralesional space.

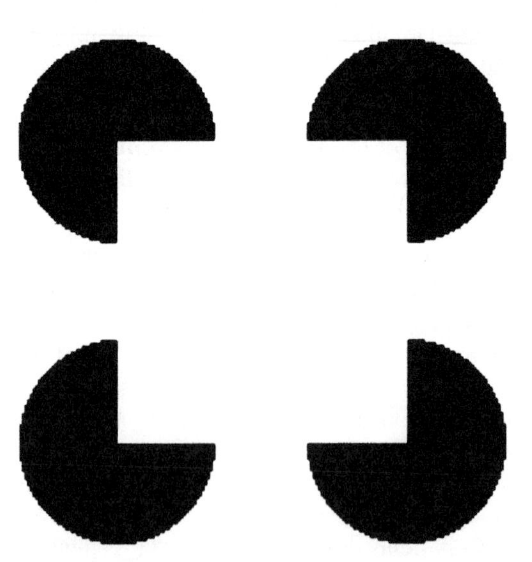

FIG. 20.2. An example of a standard Kanizsa square that can be used to test for the perception of illusory contours. Corballis et al. (60) demonstrated that both hemispheres of split-brain patients are equally able to discriminate between shapes formed by illusory contours. In addition, Vuilleumier et al. (122) demonstrated that neglect patients perceive the square formed by illusory contours, even when the patients are not explicitly aware of the two left pac-men because they are located in the neglected field. Combined, this research argues that both hemispheres can support the preattentive visual processes required for parsing illusory contours.

This view is supported by studies of right parietal patients who demonstrate extinction, or an inability to acknowledge stimuli presented in the left visual field when right visual field stimuli is simultaneously presented (123). Volpe et al. (124) found that patients were able to successfully compare extinguished stimuli to stimuli presented in the opposite visual field, despite having no conscious awareness of the extinguished stimuli. This result is interesting for two reasons. First, it supports the idea that extinguished stimuli are processed although they fail to reach conscious awareness. A recent fMRI study of right parietal patients demonstrated that extinguished stimuli do activate the right hemisphere ventral object-processing stream (125). The second implication is that extinguished stimuli can be used in a comparison task with only the final decision reaching conscious awareness. These findings argue that the right parietal lobe may be necessary to bring an extinguished stimulus into conscious awareness, and that the region links perceptual processing of the stimulus to conscious decision making.

CONCLUSIONS

Traditionally, the left and right parietal lobes have both been linked to selective attention, but only recently have we begun to appreciate their complementary lateralized roles. In this chapter, we have provided evidence that the left parietal lobe is specialized for motor attention. In contrast, the right parietal lobe is specialized for spatial attention. These unique and complementary attentional roles reflect underlying hemispheric asymmetries. The left hemisphere selectively maintains temporal processing, whereas the right hemisphere selectively maintains spatial processing. Thus, in a normal brain, the corpus callosum subserves the integration of attention across time and space. However, we have gone beyond this simple story of attention to propose that the parietal lobes are central to a higher-level cognitive network integrating the frontal, parietal, and temporal areas. The left and right parietal lobes play different roles in higher level cognitive function because of their individual asymmetries as well as general hemispheric asymmetries. The left parietal lobe supports language and calculation, whereas the right parietal lobe links implicitly formed visual percepts to conscious awareness.

ACKNOWLEDGMENTS

We wish to thank Paul Corballis, Margaret Funnell, and Abigail Baird for their helpful comments and suggestions. This work was funded by a graduate research fellowship to MKC from the National Science Foundation and by grants from the National Institutes of Health.

REFERENCES

1. Posner MI, Petersen SE. The attention system of the human brain. *Annu Rev Neurosci* 1990;13:25–42.
2. Handy TC, Hopfinger JB, Mangun GR. Functional neuroimaging of attention. In: Cabeza R, Kingstone A, eds. *Handbook of functional neuroimaging of cognition.* Cambridge, MA: MIT Press, 2001:75–108.
3. Milner AD, Goodale M. *The visual brain in action.* Oxford, UK: Oxford University Press, 1995.
4. Perenin M-T, Vighetto A. Optic ataxia: A specific disruption in visuomotor mechanisms. I. Different aspects of the deficit in reaching for objects. *Brain* 1988; 111:673–674.
5. Kimura D. Left-hemisphere control of oral and brachial movements and their relation to communication. *Philos Trans R Soc Lond* 1982;B298:135–149.
6. Goodale MA. Brain asymmetries in the control of reaching. In: Goodale MA, ed. *Vision and action: the control of grasping.* Norwood, NJ: Ablex, 1990:14–32.
7. Weintraub S. Neuropsychological assessment of mental state. In: Mesulam M-M, ed. *Principles of behavioral and cognitive neurology,* 2nd ed. New York: Oxford University Press, 2000:135–136.
8. Leiguarda RC, Marsden CD. Limb apraxias: higher-order disorders of sensorimotor integration. *Brain* 2000;123:860–879.
9. Ochipa C, Rothi LJ, Heilman KM. Conceptual apraxia in Alzheimer's disease. *Brain* 1992;115:1061–1071.
10. Heilman KM, Mahler LH, Greenwald ML, et al. Conceptual apraxia from lateralized lesions. *Neurology* 1997;49:457–464.
11. Rushworth MF, Nixon PD, Renowden S, et al. The left parietal cortex and motor attention. *Neuropsychologia* 1997;35:1261–1273.
12. Schulter ND, Krams M, Rushworth MFS, et al. Cerebral dominance for action in the human brain: the selection of actions. *Neuropsychologia* 2001;39:105–113.
13. Rushworth MFS, Krams M, Passingham RE. The attentional role of the left parietal cortex: The distinct lateralization and localization of motor attention in the human brain. *J Cog Neurosci* 2001;13:698–710.
14. Mesulam M-M. A cortical network for directed attention and unilateral neglect. *Ann Neurosci* 1981;28: 598–613.
15. Heilman KM, Watson RT, Valenstein E, et al. Localization of lesions in neglect. In: Kertesz A, ed. *Localization in neuropsychology.* New York: Academic Press, 1983:471–492.
16. Vallar G, Perani D. The anatomy of unilateral neglect after right-hemisphere stroke lesions. A clinical/CT-scan correlation study in man. *Neuropsychologia* 1986;24:609–622.
17. Driver J, Mattingley JB. Parietal neglect and visual awareness. *Nat Neurosci* 1998;1:17–22.
18. Mesulam M-M. Spatial attention and neglect: parietal, frontal and cingulate contributions to the mental representation and attentional targeting of salient extrapersonal events. *Philos Trans R Soc Lond* 1999;354: 1325–1346.
19. Heilman KM, Watson RT, Valenstein E. Neglect and related disorders. In: Heilman KM, Valenstein E, eds. *Clinical neuropsychology,* 2nd ed. New York: Oxford University Press, 1985:243–293.
20. Humphreys GW, Romani C, Olson A, et al. Non-spatial extinction following lesions of the parietal lobe in humans. *Nature* 1994;372:357–359.
21. Bisiach E. Unilateral neglect and the structure of space representation. *Curr Dir Psychol Sci* 1996;5:62–65.
22. Behrman M. Spatial reference frames and hemispatial neglect. In: Gazzaniga MS, ed. *The new cognitive neurosciences.* Cambridge, MA: MIT Press, 2000:651–666.
23. Humphreys GW. Neural representation of objects in space: a dual coding account. In: Humphreys GW, Duncan J, Treisman A, eds. *Attention, space, and action.* New York: Oxford University Press, 1999: 165–182.
24. Mesulam M-M. Attentional networks, confusional states, and neglect syndromes. In: Mesulam M-M, ed. *Principles of behavioral and cognitive neurology,* 2nd ed. New York: Oxford University Press, 2000: 174–256.
25. Colby CL, Goldberg ME. Space and attention in parietal cortex. *Annu Rev Neurosci* 1999;22:319–349.
26. Bellmann A, Meuli R, Clarke S. Two types of auditory neglect. *Brain* 2001;124:676–687.
27. Vallar G. Spatial frames of reference and somatosensory processing: a neuropsychological perspective. *Philos Trans R Soc Lond* 1997;352:1401–1409.
28. Husain M, Shapiro K, Martin J, et al. Abnormal temporal dynamics of visual attention in spatial neglect patients. *Nature* 1997;385:154–156.
29. Posner MI, Walker JA, Friedrich FJ, et al. Effects of parietal injury on covert orienting of attention. *J Neurosci* 1984;4:1863–1874.
30. Posner MI, Walker JA, Friedrich FJ, et al. How do the parietal lobes direct covert attention? *Neuropsychologia* 1987;25:135–145.
31. Townsend J, Courchesne E. Parietal damage and narrow "spotlight" spatial attention. *J Cog Neurosci* 1994; 6:220–232.

32. Mangun GR, Luck SJ, Plager R, et al. Monitoring the visual world: hemispheric asymmetries and subcortical processes in attention. *J Cog Neurosci* 1994;6: 265–273.
33. Corbetta M, Miezin FM, Shulman SE, et al. A PET study of visuospatial attention. *J Neurosci* 1993;13: 1202–1226.
34. Corbetta M, Shulman GL, Miezin FM, et al. Superior parietal cortex activation during spatial attention shifts and visual feature conjunction. *Science* 1995;270: 802–805.
35. Nobre AC, Sebestyen GN, Gitelman DR, et al. Functional localization of the system for visuospatial attention using positron emission tomography. *Brain* 1997; 120:515–533.
36. Coull JT, Nobre AC. Where and when to pay attention: the neural systems for directing attention to spatial locations and to time intervals as revealed by both PET and fMRI. *J Neurosci* 1998;18:7426–7435.
37. Culham JC, Brandt SA, Cavanagh P, et al. Cortical fMRI activation produced by attentive tracking of moving targets. *J Neurophysiol* 1998;80:2657–2670.
38. Gitelman DR, Nobre AC, Parrish TB, et al. A large scale distributed network for covert spatial attention: further anatomic delineation based on stringent behavioural and cognitive controls. *Brain* 1999;122: 1093–1106.
39. Vandenberghe R, Duncan J, Arnell KM, et al. Maintaining and shifting attention within left or right hemifield. *Cerebral Cortex* 2000;10:706–713.
40. Robertson IH, Manly T. Sustained attention deficits in time and space. In: Humphreys GW, Duncan J, Treisman A, eds. *Attention, space, and action.* New York: Oxford University Press, 1999:297–310.
41. Rushworth MFS, Ellison A, Walsh V. Complementary localization and lateralization of orienting and motor attention. *Nature* 2001;4:656–661.
42. Vallortigara G, Rogers LJ, Bisazza A. Possible evolutionary origins of cognitive brain lateralization. *Brain Res Rev* 1999;30:164–175.
43. Corballis PM, Funnell MG, Gazzaniga MS. An evolutionary perspective on hemispheric asymmetries. *Brain Cogn* 2000;43:112–117.
44. Gazzaniga MS. Cerebral specialization and interhemispheric communication: does the corpus callosum enable the human condition? *Brain* 2000;123: 1293–1326.
45. MacNeilage PF. Grasping in modern primates: the evolutionary context. In: Goodale MA, ed. *Vision and action: the control of grasping.* Norwood, NJ: Ablex, 1990:1–13.
46. Gazzaniga MS, Smylie CS. Hemispheric mechanisms controlling voluntary and spontaneous facial expressions. *J Cog Neurosci* 1990;2:239–245.
47. Gazzaniga MS, Bogen JE, Sperry RW. Dyspraxia following division of the cerebral commissures. *Arch Neurol* 1967;16:606–612.
48. Milner B, Kolb B. Performance of complex arm movements and facial-movement sequences after cerebral commissurotomy. *Neuropsychologia* 1985;23: 791–799.
49. Johnson SH. Cerebral organization of motor imagery: contralateral control of grip selection in mentally represented prehension. *Psychol Sci* 1998;9:219–222.
50. Johnson SH, Corballis PM, Gazzaniga MS. Roles of the cerebral hemispheres in planning prehension: accuracy of movements selection following callosotomy. *J Cog Neurosci* 1999;11:85.
51. Coull JT, Frith CD, Buchel C, et al. Orienting attention in time: behavioural and neuroanatomic distinction between exogenous and endogenous shifts. *Neuropsychologia* 2000;28:808–819.
52. Nobre AC. Orienting attention to instants in time. *Neuropsychologia* 2001;39:1317–1328.
53. Merzenich MM, Jenkins WM, Johnston P, et al. Temporal processing deficits of language-learning impaired children ameliorated by training. *Science* 1996; 271:77–81.
54. Arrington CM, Carr TH, Mayer AR, et al. Neural mechanisms of visual attention: object-based selection of a region in space. *J Cog Neurosci* 2000;12: SI06–SI17.
55. Fink GR, Dolan RJ, Halligan PW, et al. Space-based and object-based visual attention: shared and specific neural domain. *Brain* 1997;120:2013–2028.
56. Vandenberghe R, Duncan J, Dupont P, et al. Attention to one or two features in left or right visual field: a positron emission tomography study. *J Neurosci* 1997; 17:3739–3750.
57. Nobre AC. The attentive homunculus: now you see it, now you don't. *Neurosci Biobehav Rev* 2001;25: 477–496.
58. Egly R, Driver J, Rafal RD. Shifting visual attention between objects and locations: evidence from normal and parietal lesion subjects. *J Exp Psychol Gen* 1994; 113:501–517.
59. Corballis MC, Sergent J. Imagery in a commissurotomized patient. *Neuropsychologia* 1988;26:13–26.
60. Corballis PM, Fendrich R, Shapley R, et al. Illusory contours and amodal completion: Evidence for a functional dissociation in callosotomy patients. *J Cog Neurosci* 1999;11:459–466.
61. Forster BA, Corballis PM, Corballis MC. Effect of luminance on successiveness discrimination in the absence of the corpus callosum. *Neuropsychology* 2000; 38:441–450.
62. Corballis PM, Funnell MG, Gazzaniga MS. A dissociation between spatial and identity matching in callosotomy patients. *NeuroReport* 1999;10:2183–2187.
63. Funnell MG, Corballis PM, Gazzaniga MS. A deficit in perceptual matching in the left hemisphere of a callosotomy patient. *Neuropsychologia* 1999;37: 1143–1154.
64. Fendrich R, Wessinger CM, Gazzaniga MS. Hemispheric equivalence in anorthoscopic perception. *Cog Neurosci Soc Abs* 1996;3:94.
65. Corballis PM, Funnell MG, Gazzaniga MS. Hemispheric asymmetries for simple visual judgments in the split brain. *Neuropsychologia* 2002;40:401–410.
66. Koehler S, Kapur S, Moscovitch M, et al. Dissociation of pathways for object and spatial vision: a PET study in humans. *NeuroReport* 1995;6:1865–1868.
67. Barcelo F, Suwazono S, Knight ST. Prefrontal modulation of visual processing in humans. *Nat Neurosci* 2000;3:299–303.
68. Corbetta M. Frontoparietal cortical networks for directing attention and the eye to visual locations: identical, independent, or overlapping neural systems? *Proc Natl Acad Sci USA* 1998;95:831–838.
69. Jonides J, Smith EE, Koeppe RA, et al. Spatial work-

ing memory in humans as revealed by PET. *Nature* 1993;363:623–625.
70. Treisman A, Gelade G. A feature integration theory of attention. *Cog Psychol* 1980;12:87–136.
71. Treisman A. Features and objects: the fourteenth Bartlett memorial lecture. *Q J Exp Psychol* 1988; 40:201–237.
72. Kastner S, De Weerd P, Desimone R, et al. Mechanisms of directed attention in the human extrastriate cortex as revealed by functional MRI. *Science* 1998;282:108–111.
73. Ungerleider LG, Mishkin M. Two cortical visual systems. In: Ingle DJ, Goodale MA, Mansfield RJW, eds. *Analysis of visual behavior.* Cambridge, MA: MIT Press, 1982:549–586.
74. Ungerleider LG, Courtney SM, Haxby JV. A neural system for human visual working memory. *Proc Natl Acad Sci USA* 1998;95:883–890.
75. Romanski LM, Tian B, Fritz J, et al. Dual streams of auditory afferents target multiple domains in the primate prefrontal cortex. *Nat Neurosci* 1999;2: 1131–1136.
76. Rauschecker JP, Tian B. Mechanisms and streams for processing of "what" and "where" in auditory cortex. *Proc Natl Acad Sci USA* 2000;97:11800–11806.
77. Romo R, Salinas E. Touch and go: decision-making mechanisms in somatosensation. *Annu Rev Neurosci* 2001;24:107–137.
78. Livingstone MS, Hubel DH. Segregation of form, color, movement, and depth: anatomy, physiology, and perception. *Science* 1988;240:740–749.
79. Ferrera VP, Nealey TA, Maunsell JH. Mixed parvocellular and magnocellular geniculate signals in visual area V4. *Nature* 1992;358:756–761.
80. Ferrera VP, Nealey TA, Maunsell JH. Responses in macaque visual area V4 following inactivation of the parvocellular and magnocellular LGN pathways. *J Neurosci* 1994;14:2080–2088.
81. Newsome WT, Salzman CD. Neuronal mechanisms of motion perception. *Cold Spring Harbor Symp Quant Biol* 1990;55;697–705.
82. Andersen RA, Snyder LH, Bradley DC, et al. Multimodal representation of space in the posterior parietal cortex and its use in planning movements. *Annu Rev Neurosci* 1997;20:303–330.
83. Alain C, Arnott SR, Hevenor S, et al. "What" and "where" in the human auditory system. *Proc Natl Acad Sci USA* 2001;98:12301–12306.
84. Desimone R, Gross G. Visual areas in the temporal cortex of the macaque. *Brain Res* 1979;178:363–380.
85. Wilson FAW, Scalaidhe SPO, Goldman-Rakic PS. Dissociation of object and spatial processing domains in primate prefrontal cortex. *Science* 1993;260: 1955–1958.
86. Kikuchi-Yorioka Y, Sawaguchi T. Parallel visuospatial and audiospatial working memory processes in the monkey dorsolateral prefrontal cortex. *Nat Neurosci* 2000;3:1075–1076.
87. Anourova I, Nikouline VV, Ilmoniemi RJ, et al. Evidence for dissociation of spatial and nonspatial auditory information processing. *NeuroImage* 2001;14: 1268–1277.
88. Passingham RE, Toni I. Contrasting the dorsal and ventral visual systems: guidance of movement versus decision making. *NeuroImage* 2001;14:SI25–SI31.
89. Rao SC, Rainer G, Miller EK. Integration of what and where in the primate prefrontal cortex. *Science* 1997; 276:821–824.
90. Barbas H. Anatomic organization of basoventral and mediodorsal visual recipient prefrontal region in the rhesus monkey. *J Comp Neurol* 1998;276:313–342.
91. Pandya DN, Yeterian EH. Comparison of prefrontal architecture and connections. In: Roberts AC, Robbins TW, Weiskrantz L, eds. *The prefrontal cortex.* Oxford, UK: Oxford University Press, 1998:51–66.
92. Petrides M, Pandya DN. Dorsolateral prefrontal cortex: comparative cytoarchitectonic analysis in the human and the macaque brain and corticocortical connection patterns. *European J Neurosci* 1999;11: 1011–1036.
93. Karnath H-O. New insights into the functions of the superior temporal cortex. *Nat Neurosci Revs* 2001;2: 568–576.
94. Seltzer B, Pandya DN. Afferent cortical connections and architectonics of the superior temporal sulcus and surrounding cortex in the rhesus monkey. *Brain Res* 1978;149:1–24.
95. Bruce C, Desimone R, Gross CG. Visual properties of neurons in a polysensory area in superior temporal sulcus of the macaque. *J Neurophysiol* 1981;46:369–384.
96. Felleman DJ, Van Essen DC. Distributed hierarchical processing in the primate cerebral cortex. *Cerebral Cortex* 1991;1:1–47.
97. Seltzer B, Pandya DN. Parietal, temporal, and occipital projections to cortex of the superior temporal sulcus in rhesus monkey: a retrograde tracer study. *J Comp Neurol* 1994;343:445–463.
98. Morel A, Bullier J. Anatomic segregation of two cortical visual pathways in the macaque monkey. *Vis Neurosci* 1990;4:555–578.
99. Baizer JS, Ungerleider LG, Desimone R. Organization of visual inputs to the inferior temporal and posterior parietal cortex in macaques. *J Neurosci* 1991;11: 168–190.
100. Karnath H-O, Ferber S, Himmelbach M. Spatial awareness is a function of the temporal not the posterior parietal lobe. *Nature* 2001;411:950–953.
101. Damasio AR, Damasio H. Aphasia and the neural basis of language. In: Mesulam M-M, ed. *Principles of behavioral and cognitive neurology,* 2nd ed. New York: Oxford University Press, 2000:294–315.
102. Mesulam M-M. Behavioral neuranatomy: large-scale networks, association cortex, frontal syndromes, the limbic system, and hemispheric specialization. In: Mesulam M-M, ed. *Principles of behavioral and cognitive neurology,* 2nd ed. New York: Oxford University Press, 2000:1–120.
103. Chochon F, Cohen L, van de Moortele PF, et al. Differential contributions of the left and right inferior parietal lobules to number processing. *J Cog Neurosci* 1999;11:617–630.
104. Goldman-Rakic PS. Modular organization of prefrontal cortex. *Trends Neurosci* 1984;7:419–424.
105. Grafman J, Kampen D, Rosenberg J, et al. Calculation abilities in a patient with a virtual left hemispherectomy. *Behav Neurol* 1989;2:183–194.
106. Dehaene S, Cohen L. Two mental calculation systems: a case study of severe acalculia with preserved approximation. *Neuropsychologia* 1991;29:1045–1074.
107. Gazzaniga MS, Hillyard SA. Language and speech ca-

pacity of the right hemisphere. *Neuropsychologia* 1971;9:273–180.
108. Gazzaniga MS, Smylie CE. Dissociation of language and cognition: a psychological profile of two disconnected right hemispheres. *Brain* 1984;107:145–153.
109. Seymour SE, Reuter-Lorenz PA, Gazzaniga MS. The disconnection syndrome: basic findings reaffirmed. *Brain* 1994;117:105–115.
110. Funnell MG, Colvin MK, Gazzaniga MS. Numbers in the brain: studies with a callosotomy patient. 2002.
111. Funnell MG, Colvin MK, Gazzaniga MS. Calculation abilities of the two hemispheres. 2002.
112. Pesenti M, Thioux M, Seron X, et al. Neuroanatomic substrates of arabic number processing, numerical comparison, and simple addition: a PET study. *J Cog Neurosci* 2000;12:461–479.
113. Pinel P, Dehaene S, Riviere D, et al. Modulation of parietal activation by semantic distance in a number comparison task. *NeuroImage* 2001;14:1013–1026.
114. Dehaene S, Spelke E, Pinel E, et al. Sources of mathematic thinking: behavioral and brain-imaging evidence. *Science* 1999;284:970–974.
115. Gobel S, Walsh V, Rushworth MFS. The mental number line and the human angular gyrus. *NeuroImage* 2001;14:1278–1289.
116. Dehaene S. Varieties of numerical abilities. *Cognition* 1992;44:1–42.
117. Dehaene S, Cohen L. Toward an anatomic and functional model of number processing. *Math Cog* 1995;1:83–120.
118. Marshall JC, Halligan PW. The yin and the yang of visuospatial neglect: a case study. *Neuropsychologia* 1994;32:1037–1057.
119. Driver J, Baylis GC, Rafal RD. Preserved figure-ground segmentation and symmetry perception in a patient with neglect. *Nature* 1993;360:73–75.
120. Mattingley JB, Bradshaw JL, Bradshaw JA. The effects of unilateral visuospatial neglect on perception of Muller-Lyer illusory figures. *Perception* 1995;24: 415–433.
121. Ro T, Rafal RD. Perception of geometric illusion in hemispatial neglect. *Neuropsychologia* 1996;34: 973–976.
122. Vuilleumier P, Valenza N, Landis T. Explicit and implicit perception of illusory contours in unilateral spatial neglect: behavioural and anatomic correlates of preattentive grouping. *Neuropsychologia* 2001;39: 597–610.
123. Vuilleumier PO, Rafal RD. A systematic study of visual extinction. Between- and within-field deficits of attention in hemispatial neglect. *Brain* 2000;123: 1263–1279.
124. Volpe BT, Ledoux JE, Gazzaniga MS. Information processing of visual stimuli in an 'extinguished' field. *Nature* 1979;282:722–724.
125. Driver J, Vuilleumier P, Eimer M, et al. Functional magnetic resonance imaging and evoked potential correlates of conscious and unconscious vision in parietal extinction patients. *NeuroImage* 2001;14:S68–S75.

21

Parietal Lobe Epilepsy

Adrian M. Siegel

Department of Neurology, University Hospital of Zurich, Zurich, Switzerland

INTRODUCTION

Neurological diseases that exclusively disturb parietal lobe functioning are rare in clinical practice. Functional deficits of the parietal lobe are mainly evidenced following widespread cortical damage due typically to brain tumors, cerebrovascular diseases, trauma, dementias (e.g., Alzheimer's disease), encephalitis, and so on. Thus, it is not uncommon that a patient with a large stroke in the territory of the middle cerebral artery suffers both from a hemiplegia caused by precentral gyral damage and from parietal symptoms such as visuospatial neglect and sensory deficits. The spectrum of disorders following parietal lobe damage is wide and includes visual or tactile agnosia, apraxias, acalculia, impaired cross-modal matching, contralateral neglect, disorders of body image, and so on (1). More uncommon syndromes also have been associated with parietal lobe damage. Gerstmann's syndrome (the tetrad of symptoms finger agnosia, right-left disorientation, agraphia, and acalculia) is a rare syndrome caused by damage to the left angular gyrus (2). Bilateral posterior parietal damage may result in another rare syndrome first described by Badal in 1888 (3) and later by Balint (4) and Holmes (5). The Balint, or Balint-Holmes, syndrome is a clinical complex of visual attentional disorders including gaze apraxia, optic ataxia, and disturbances in the estimation of distance (4,5).

A rare clinical disorder of the parietal lobe is parietal lobe epilepsy. The nature of the neuropsychological functions associated with the parietal lobe is such that a temporary parietal dysfunction as in partial seizures may go unnoticed (e.g., a temporary visuospatial neglect). Indeed, most forms of parietal lobe epilepsy are clinically silent and may first manifest following spread to other brain regions. Subjective symptoms, such as paresthesias and pain, are characteristic of seizures that remain isolated to the parietal lobe. Although rare, proper treatment of this disorder requires a familiarity with its wide array of associated clinical features.

EPIDEMIOLOGY

Incidence and Prevalence

The available epidemiologic data on parietal lobe epilepsy must be interpreted with caution, because they emanate from retrospective studies with methodologic drawbacks. These suggest, however, that although the parietal lobes account for approximately 25% of brain volume, partial seizures of parietal lobe origin are relatively rare. Gibbs and Gibbs (1952), for example, reported parietal foci in only 5% of patients with focal epilepsies, although not all cases were adequately confirmed (6). Recent data from a comprehensive surgical series also found that parietal lobe seizures constituted at most 5% of all partial seizures (6–8). Still the relative frequency of parietal lobe epilepsy may be underestimated, because often it is initially clinically silent and only manifests following spread to other brain regions.

Age at Onset

Partial seizures of parietal lobe origin generally begin at an early age. Seizure onset occurred in the first year of life in circa 18% of the patients studied by Gibbs and Gibbs (6), although diagnoses were based on scalp electroencephalograph (EEG) only. In a large surgical series of 82 patients with nontumoral parietal lobe epilepsy, the age at seizure onset ranged from 1 to 50 years (mean 14.1 years) (9).

Etiology

Brain tumors are regarded as the most common cause of parietal lobe epilepsy (10). Approximately 10% of large groups of neurosurgery patients with tumoral epilepsy were found to suffer from parietal lobe tumors (11,12). Compared with nonparietal tumors, those of the parietal lobe appear to be highly epileptogenic: in two large surgical series, parietal lobe tumors caused epilepsy in 46% and 68% of patients, respectively (11,12).

Parietal lobe seizures are rarely associated with nontumoral lesions. In a surgical series of nontumoral epilepsy, parietal lobe lesions were found in only 3% of the patients (13). In 82 patients with nontumoral parietal lobe epilepsy, a history of head trauma and birth trauma were identified as risk factors in 43% and 16% of patients, respectively (9). The presumed etiology in 21% of the patients included a history of encephalitis, febrile convulsions, gunshot wounds to the head, forme fruste of tuberous sclerosis, hamartoma, vascular malformations, tuberculoma, arachnoid or porencephalic cysts, microgyria, and posttraumatic thrombosis of the middle cerebral artery. The etiology remained unknown in 20% of the patients (9).

Partial epilepsy is a relatively common consequence of parietal lobe traumas. The incidence of posttraumatic epilepsy is 50.7% following damage to central regions compared to 35.0% following temporal, 33.3% following occipital, and 26.7% following frontal lobe trauma (14). A subsequent study reported posttraumatic epilepsy in 65% of patients with parietal trauma compared to 39% of patients with frontal and 38% with occipital and temporal lobe damage (15).

Modern imaging methods have begun to elucidate the pathogenetic mechanisms involved in parietal lobe epilepsy. These techniques have identified neuronal migration disorders such as heterotopia and focal cortical dysplasia, as well as pachygyria and schizencephalia as frequent underlying structural abnormalities in this disorder (16–18).

CLINICAL FEATURES

Parietal lobe seizures present with a wide clinical spectrum. According to Gibbs and Gibbs (6) seizures of parietal lobe origin include grand mal seizures (in 41%), Jacksonian fits (30%), "focal" seizures (26%), petit mal (17%), and psychomotor seizures (1%; either as the only manifestation or combined with others). Other types of seizures have been found in one third of patients (6). Complex partial seizures with loss of consciousness and automatisms are not observed in patients with seizures restricted to the parietal lobe (19,20), but may appear if the seizure spreads beyond the parietal cortex. In a series of 50 patients with medically refractory epilepsy caused by lesions involving the parietal lobe, the typical features of temporal lobe seizures were experienced by two patients with parietal lesions, two with parietotemporal lesions, and one with a parieto-occipital lesion (21). In contrast to the predominantly nocturnal seizures found in frontal lobe epilepsy, parietal lobe seizures do not predominantly occur during a particular part of the day (22).

Anatomic Classification

Certain clinical characteristics of parietal lobe seizures are indicative of the region of seizure origin. The primary somatosensory area (SI) in the postcentral gyrus is the most common origin of seizures in parietal lobe epilepsy. These are associated with contralateral somatosensory symptoms such as paresthesias, thermal sensations, and pain, sensations that may spread along the sensory homunculus. The localizing and lateralizing significance of such "sensory fits" has been documented recently (23–27). Seizures from the second somatosensory area (SII) along the upper portion of the sylvian fissure may include ipsilateral or bilateral sensory symptoms. Seizures of posterior parietal origin are associated with automatisms, unresponsiveness, and other complex clinical manifestations (28,29). They also have been described as "psychoparetic," implying a psychic aura such as déjà vu or fear followed by an impairment of consciousness and motor arrest (29). Some patients experience elementary visual hallucinations or ictal amaurosis as the initial seizure in "silent" posterior parietal foci spreads to the occipital lobe (28,30). Seizures originating in the temporo-parieto-occipital region also may be associated with complex visual symptoms. Spatial disorientation is a common symptom of seizures originating in the inferior parietal lobe; whereas illusions of unreality may be reported by patients with seizures emanating from the supramarginalis gyrus. Vertiginous sensations are predominantly associated with seizures of temporoparietal origin (9), and seizures in the parietal operculum may be manifested in gustatory sensations. Seizures with origin in the parieto-temporo-occipital junction may be associated with speech arrest, Wernicke's aphasia, gestural, verbal, and dyspraxic automatisms.

Although these clinical–anatomic relationships aid in the classification of parietal seizures, not all symptoms are specific to an anatomic structure. Cephalic auras characterized by a vague head sensation, for instance, may be a parietal symptom (discharges in the postcentral gyrus) but also may be an ictal feature of frontal or temporal discharges (31). Similarly, genital sensations ("orgasmic seizures") have been reported by patients with seizure origins either in the parietal or temporal lobes (32–35).

CLINICAL CLASSIFICATION

Like all partial seizures, parietal lobe seizures consist of subjective and objective components. The subjective symptoms include somatosensory seizures such as paresthesias, thermal and genital sensations, pain, cephalic auras, gustatory hallucinations, the feeling of an inability to move a body part, somatosensory illusions (e.g., kinetic, feelings of distorted body shape, asomatognosia, feelings of unreality), vertigo, and visual illusions or hallucinations (e.g., metamorphopsia, palinopsia). Objective signs mostly occur when discharges extend beyond the parietal lobe, and consist of postural and focal clonic movements, focal paralysis, eye and head deviation, speech disorders, nystagmus, and panic attacks.

Subjective Features

According to Penfield and Jasper (26), the most common subjective sensations or auras are paresthesias. These usually are described as numbness and tingling but may be a sensation of "pins and needles" and, more rarely, of crawling or itching. In a series of 82 patients with parietal lobe seizures, 94% of the patients exhibited auras (9). Ictal paresthesias, however, occurred in less than half of parietal lobe epilepsy patients recently studied by Williamson et al. and Cascino et al., respectively (28,30). Moreover, these somatosensory symptoms were not always lateralized. When lateralized, they were not always contralateral to the side of seizure origin, presumably reflecting activity in secondary sensory systems (30). Figure 21.1 lists the frequencies of experienced epileptic paresthesias.

Seizures originating in the suprasylvian region close to the sensory area may be associated with thermal sensations, that is, usually an unpleasant feeling of cold or heat (26,36). In the series of Maugière and Courjon (27), 11% of 127 patients with somatosensory auras had thermal sensations. Thermal sensations usually occurred with other somatosensory sensations such as paresthesias or pain (27). Like other somatosensory seizures, they were mainly manifested unilaterally and spread along the sensory homunculus, although a bilateral spread also has been expe-

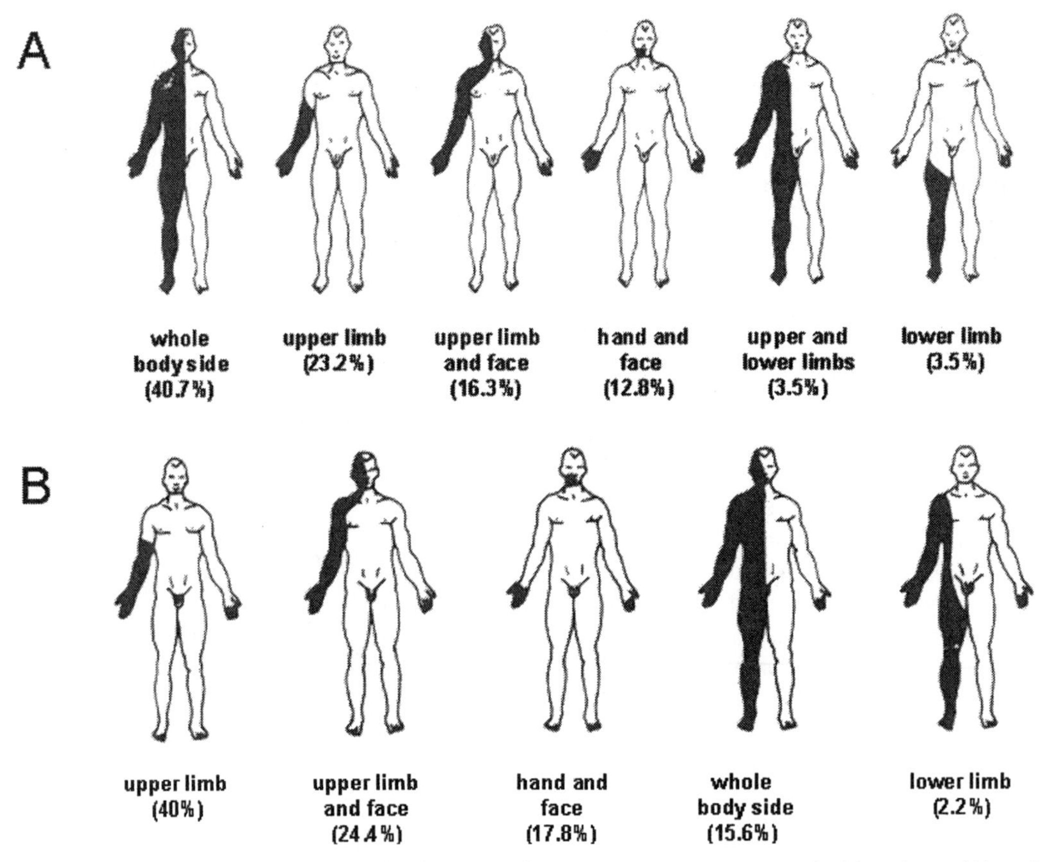

FIG. 21.1. The figure shows the maximal spread of somatosensory symptoms in 86 patients *(A);* and the maximal spread of somatosensory symptoms in 45 patients with seizures beginning in the hand or fingers *(B).* (Modified from: Mauguière F, Courjon J. Somatosensory epilepsy: a review of 127 cases. *Brain* 1978;101:307–332, with permission.)

rienced (37). Rarely, the tongue and mouth are the only body parts involved (19,36).

Another type of somatosensory seizure is manifested as genital sensations. Patients describe paresthesias, numbness, or an unpleasant feeling in the genital region (37). Others report an "orgasmic feeling" (32–35,38). Genital sensations have been reported with seizure origin both in the mesial area of the paracentral lobule, the opercular area of the secondary somatosensory area, and in the temporal lobe. If only half of the genital region is involved, a contralateral seizure focus usually is involved. Similar to genital sensations, somatosensory seizures also may manifest with sensations in the peroneal and rectal regions (35).

The subjective ictal phenomenon of pain has been recognized for over 100 years (39,40). However, because pain rarely occurred in response to electrical stimulation of the parietal lobe (41), Penfield and Jasper (26) questioned whether pain truly constituted a cortical sensation. Subsequent stimulation studies, however, supported the role of the primary somatosensory area (SI; Brodmann's areas 3, 2, and 1) in pain perception. In more recent human studies, the controversy regarding the role of SI in pain perception was again revived. For example, studies measuring regional cerebral blood flow in SI in response to experimental pain, a surrogate marker of its involvement in pain perception, produced conflicting results. Some authors found an increase in cerebral blood flow in SI (42,43), whereas other studies reported negative findings (44), and still others documented decreases in cerebral blood flow during painful stimulation (45). Another structure presumably involved in pain perception is the secondary somatosensory area (SII), which represents both sides of the body with a contralateral predominance (37,46–50). Animal studies (50–52) identified regions containing sparse nociresponsive neurons (i.e., <10% of the cells sampled) contiguous to SII such as area 7b of the parietal lobe, the retroinsular and posterior auditory areas, and the granular portion of the insula. Subsequent human studies confirmed the involvement of SII in pain processing. One possible mechanism for epileptic pain is that seizures in SI or SII cause spreading depression that interferes with control of pain perception (47,53).

The clinical spectrum of epileptic pain sensations is vast. A precise description is often difficult to obtain from the patient and may depend on cultural and emotional influences (54). Penfield and Jasper (26) reported that ictal pain usually was mild, and suggested that the term "pain" might even be a misnomer. However, other authors and many patients have reported very severe pain. Ictal pain may be felt anywhere in the body, but the most common locations are a part or all of one side of the body, the head (cephalic), and the abdomen (39,47,55). Unilateral or cephalic pain is felt in the majority of painful seizures and occurred in one series in 41.7% and 45.8% of patients, respectively (47). Among several large series of patients with epileptic pain, the incidence of abdominal pain varied widely between 12.5% and 38.6% (39,47). Bilateral pain involving all limbs and paroxysmal rectal or genital pain are exceptional (37,56–58).

Painful unilateral sensations of the extremities are most commonly experienced as burning, dull, throbbing, cramplike, or electric-like. Unilateral pain is more often a well-localizable than a diffuse sensation (39,47,59). At the time of seizure onset, it usually involves one arm, especially the hand and fingers, then spreads to the face and leg, and then to the shoulder (47,60,61). The spread of pain usually follows the sensory homunculus, but the trunk sometimes may be spared. There are two possible explanations for this. First, spread from the upper to the lower limb (or vice versa) may be so rapid that the involvement of the trunk is not registered. Second, the lack of sensation may be caused by the limited representation of the trunk in the sensory cortex (62). As with paresthesias, painful seizures may occur ipsilateral to the seizure focus or bilaterally if the second somatosensory area is involved (63).

Abdominal pain is a more severe, sharp ("like a knife") feeling (47,63,64), but also can be oppressive (65). Abdominal pain is most often periumbilical (64,66), but may involve the whole or a quadrant of the abdomen (63,65). Painful abdominal sensations rarely spread across the homunculus to involve other parts of the body (63). Ictal abdominal pain is associated both with parietal and temporal lobe seizure origins (63).

Cephalic pain may be nonspecific, or throbbing and migrainelike (47). It may be diffuse or well localized. In their 11 patients with cephalic ictal pain, Young and Blume (47) reported that three suffered from a diffuse and eight from a localized headache with no location predominating. Epileptic cephalic pain may either secondarily involve the whole head or spread to one side of the body (47,63). Ictal head pain has no localizing value (63).

Seizures associated solely with pain are rare, and only a few patients have been reported in the literature (59,68–70). In their examination of 30 patients with ictal pain, Mauguière and Courjon (27) found only two cases in which pain was the only seizure manifestation. In a large series of somatosensory seizures, almost 40% of the patients had multiple sensory symptoms (27). Unilateral pain is most commonly associated with paresthesias, thermal sensations, and disturbances in body image, whereas motor signs are less frequent (27,47,54,63,67,68,70). Abdominal pain is often accompanied by nausea and diarrhea (66,71).

Cephalic Aura

Cephalic aura, a sudden vague head sensation, also may be experienced by patients with parietal lobe seizures (62). This vague sensation is often a signal symptom. Although its localizing and lateralizing value in parietal lobe epilepsy is poor, a discharge posterior to the central region may be involved. However, cephalic sensations are more common in frontal and, particularly, temporal lobe epilepsy.

Gustatory Hallucinations

Some patients report gustatory hallucinations, usually experienced as an unpleasant sensation of salty, metallic, or bitter taste (72). Although gustatory hallucinations mainly suggest a temporal or frontal lobe seizure origin, parietal lobe seizures with gustatory sensations have been reported (72–75).

The Inability to Rapidly Move or Maintain a Sustained Muscle Contraction

The inability to rapidly move a body part or maintain a sustained muscle contraction without detectable loss of power has been reported as an ictal feature. This sensation has been elicited by electrical stimulation in the suprasylvian postcentral cortex or supplementary motor area (26), in the postrolandic suprasylvian region, and in the inferior frontal area (76,77).

Somatosensory Illusions

Epileptic seizures of parietal lobe origin may be associated with a wide spectrum of somatosensory illusions, such as feelings of changed or distorted body shape, kinetic illusions, asomatognosia, and illusions of unreality. Penfield and Jasper (26) considered somatosensory illusions a rare finding, but Mauguière and Courjon (27) reported such illusions in 10% of their patients with somatosensory auras. In illusions of distorted or changed body shape patients experience the whole body or a part of it as elongated, shortened, shrunk, or swelled. In kinetic illusions, patients experience motions of the whole body or a body part. A typical example is the feeling of being in an elevator. Because it is difficult to distinguish these kinetic illusions from vertiginous sensations, Penfield and Jasper (1954) interpreted vertigo in parietal lobe seizures as a labyrinthine sensation rather than a simple sensory perception (26). Asomatognosia, a feeling of absence of a body part, is another somatosensory illusion that is usually experienced on one side of the body as the absence of a limb or a hemibody. Rarely, the sensation of a phantom limb has been reported. The illusion of unreality with the feeling of being "far away" is a rare somatosensory illusion that has been elicited by electrical stimulation in the supramarginalis gyrus (26,78). Somatosensory illusions suggest a seizure origin in the inferior parietal lobule and the superior part of the postcentral gyrus, more commonly in the nondominant hemisphere (73). Other structures presumably involved in the genesis of these sensations are the nondominant parieto-occipital junction, the posteroinferior frontal lobe, and the posterior parietal lobe (19,62).

Ictal Vertigo

Ictal vertigo is experienced as either rotatory vertigo, the feeling of falling, mostly backward, or as unspecific dizziness. Ictal vertigo suggests a seizure origin in the suprasylvian, parietotemporal, or parieto-occipital region (79,80).

Visual Illusions or Hallucinations

Visual illusions or hallucinations comprise metamorphopsia, micropsia (objects appear smaller), macropsia (objects appear larger), teleopsia (objects appear both smaller and at a distance), palinopsia (visual perseveration), visual allesthesia (combination of palinopsia and transference of images from one hemifield to the other), heautoscopia (perception of mirror images), and complex visual hallucination (19,62,81–86). Visual illusions and hallucinations are associated with a seizure origin in the parieto-occipital region.

Objective Features

Parietal lobe seizures are usually objectively manifested only when the seizures spreads to adjacent brain areas such as the precentral gyrus and temporal lobe. When confined to the parietal lobe, the objective features of the seizure consist mainly of positive or negative motor phenomena such as postural and focal clonic movements, focal paralysis, eye and head deviation, speech disorders, nystagmus, and panic attacks.

Tonic Postural Movements

Tonic postural movements may be involve a limb or consist of tonic eye and head deviation. Rarely, both arms or both legs, or even more infrequently, the entire body may be involved. They may be associated with other clinical features such as paresthesias or clonic movements. Although contralateral postural movements in parietal lobe seizures have been documented (87), their lateralizing and localizing value is limited because they also occur during frontal lobe seizures of primarily supplementary motor area origin.

Focal Clonic Movements

Parietal lobe seizures also may be focal clonic movements. Although contralateral clonic movements are commonly associated with seizure origins in the precentral motor area, electrical stimulations of the postcentral gyrus elicited more motor than sensory responses in about 25% of stimulations. It remains unclear whether these findings are relevant to the clinical practice. However, a contralateral parietal discharge was presumed in several cases with focal motor seizures of the truncal muscles (88).

Ictal Focal Paralysis

Seizures originating in the centroparietal area may be associated with ictal focal paralysis, a rare clinical parietal manifestation (9,10,27,63). This type of seizure also has been called inhibitory motor seizures, somatic inhibitory seizures, atonic partial seizures, hemiparetic seizures, focal akinetic seizures, and ictal paralysis (19).

Versive Eye and Head Movements

Versive eye and head movements may be observed in parietal lobe seizures. These eye and head turning may be ipsilateral or contralateral to the seizure origin (26,63,89,90). Because versive eye and head movements also may be found in frontal, temporal, and occipital lobe seizures, their localizing and even lateralizing value are very poor.

Speech Disorders

Speech disorders in parietal lobe seizures are uncommon (90), and ictal aphasia suggests seizure discharges in the parieto-temporo-occipital junction.

Ictal Nystagmus

Ictal nystagmus has been associated with seizures of various origin, namely the occipital, parieto-occipital, temporooccipital, and frontotemporal regions (91–94).

Episodes that fulfill the diagnostic criteria for panic attacks have been described in two patients with parietal lobe seizure origin (95,96).

The "Silent" Parietal Lobe Seizures

Seizures in most parts of the parietal lobe may be clinically silent, or may only be demonstrable under extraordinary circumstances. For example, while undergoing electrocorticography under local anesthesia, Penfield's and Jasper's patient J St suffered an electrically induced seizure that remained restricted to the parietal lobe (26). During the seizure, two-point discrimination was impaired in the contralateral hand, but returned to normal when the seizure subsided. The patient was unaware of any specific symptoms.

The signs and symptoms of parietal lobe seizures were retrospectively analyzed in two recent studies of patients with parietal lobe epilepsy who all had probable epileptogenic lesions detected with magnetic resonance imaging (MRI) (28,30). Although some signs and symptoms were associated with parietal lobe seizures, most patients displayed no clinical seizure manifestations until after the seizures had spread beyond the parietal lobe. These observations indicate that the parietal lobe is clinically silent in terms of seizures, and consequently, that most patients with parietal lobe seizure origin do not evidence a clinically localizable form of localization-related epilepsy. These conclusions, coupled with misleading scalp EEG findings, almost certainly explain some of the surgical failures in patients with normal neuroimaging findings or those studied with invasive EEG that neglected coverage of the parietal lobe.

Spread Pathways of Parietal Lobe Seizures

Most objective manifestations of parietal lobe seizures reflect seizure spread outside of the parietal lobe: (a) anteriorly into the frontal lobe; (b) inferiorly into the temporal lobe; or (c) posteriorly into the occipital lobe, with seizure characteristics corresponding to the direction of spread. As such, tonic motor activity and/or automatisms may occur in patients with parietal lobe seizures. Asymmetrical tonic posturing was associated with a spread from the parietal lobe to the supplementary motor area (SMA) in a depth electrode study (30), but this finding was not replicated in another study of parietal lobe seizures using invasive monitoring (87).

Seizures that originate in the parietal lobes and spread to medial temporal structures have been well documented using intracranial recording (30,87). The clinical characteristics of this event resemble temporal lobe seizures. Two patients with unsuspected parietal lobe seizure origin underwent unsuccessful temporal lobe surgeries as reported by Williamson et al. (30), as did two patients from the series reported by Ho et al. (29). Parietal lobe seizure origin in the two patients from the Yale series was first considered after parietal lobe lesions were detected using previously unavailable MRI and was verified following successful surgery that removed the lesion and immediately surrounding brain. Ictal single-photon emission computed tomography was employed to determine the parietal lobe seizure origin of the two patients unsuccessfully operated on in Ho et al.'s study (29).

ELECTROENCEPHALOGRAPHIC FINDINGS

In a study of 11 patients with parietal lobe epilepsy documented with intracranial EEG, Williamson et al. (30) reported that scalp EEG findings were normal, nonspecific, or misleading. Scalp EEG clearly localized or lateralized the side of seizure origin in only one out of 11 patients (21). In another scalp EEG study of 66 patients (Salanova), interictal discharges were recorded fronto-centro-parietally in 33%, parieto-posterior-temporally in 14%, parietally in 9%, parieto-occipitally in 9%, fronto-centro-temporally in 4.5%, fronto-temporo-parietally in 4.5%, hemispheric ipsilaterally with a posterior maximum in 9%, and bilaterally in 4.5% of the patients. Electroencephalography was normal in 7.5% and secondary bilateral synchrony was found in 32% of the patients. Salanova et al. (9,10) observed ictal discharges predominantly lateralized with a maximum over the centro-parietal region or over the posterior head region.

Occasionally, the sleep-EEG provides unspecific assistance in the diagnosis of parietal lobe epilepsy. Although extrinsic stimuli during non–rapid eye movement sleep (stage 2) usually induce vertex waves and K-complexes, the same stimuli may trigger seizures from parietal foci.

NEUROPSYCHOLOGICAL EXAMINATION

In a series of 11 patients with parietal lobe epilepsy (30), six had lateralized neuropsychological findings. Of these, the abnormality was congruent with the side of seizure origin in only three patients. Neuropsychological testing did not indicate the hemisphere of seizure origin in five patients. The most common neuropsychological deficits in patients with parietal lobe epilepsy were impairments in spatial ability, difficulties in the reproduction of complex pictorial material, left-right confusion, hemispatial neglect, and so on (1,30).

CONCLUSIONS

An estimated 5% of patients with focal epilepsies have parietal foci, although the prevalence is most likely underestimated. Compared to other brain regions, the parietal lobe appears to be highly epileptogenic, and brain tumors are the most common etiology in parietal lobe epilepsy. Parietal lobe seizure origin should be suspected when symptoms such as lateralized paresthesias or pain occur prominently and early in partial seizures. Most patients with parietal lobe seizures, however, have no symptoms or signs indicating their origin, and spread patterns are unpredictable and can result in false localization (97). Thus, in the absence of detectable epileptogenic lesions and clinical seizure characteristics suggesting a parietal lobe origin, patients may present with very misleading findings, resulting in erroneous localization (97,98).

ACKNOWLEDGMENT

The author gratefully acknowledges the assistance of Kirsten Taylor in the preparation of the manuscript.

REFERENCES

1. Kolb B, Whishaw IQ. *Fundamentals of human neuropsychology.* San Francisco: Freeman, 1980.
2. Gerstman J. Syndrome of finger agnosia, disorientation for right and left, agraphia and acalculia. *Arch Neurol Psychiatry* 1940;44:398–408.
3. Badal J. Contribution à l'étude des cécités psychique. Alexie, agraphie, hemianopsie inférieure. Trouble du sens de l'éspace. *Archives d'Ophtalmologie* 1888;140: 97–117.
4. Balint R. Seelenlähmung des 'Schauens', optische Ataxie, räumliche Störung der Aufmerksamkeit. *Monatsschrift fïr Psychiatrie und Neurologie* 1909; 25:51–81.
5. Holmes G. Disturbances of visual orientation. *Br J Ophtalmol* 1938;2:449-468
6. Gibbs FA, Gibbs EL. *Atlas of electroencephalography,* vol 2. *Epilepsy.* Cambridge, MA: Addison-Wesley, 1952:163–252.
7. Rasmussen T. Surgery for epilepsy arising in regions other than the temporal and frontal lobes. In: Purpura DP, Penry JK, Walter RD, eds. *Neurosurgical management of the epilepsies.* New York: Raven Press, 1975: 207–226.
8. Rasmussen T. Focal epilepsies of nontemporal and nonfrontal origin. In: Wieser HG, Elger CE, eds. *Presurgical evaluation of epilepsies: basics, techniques, implications.* Berlin: Springer-Verlag, 1987:301–305.
9. Salanova V, Andermann F, Rasmussen T, et al. Parietal lobe epilepsy. Clinical manifestations and outcome in 82 patients treated surgically between 1929 and 1988. *Brain* 1995;118:607–628.
10. Salanova V, Andermann F, Rasmussen T, et al. Tumoural parietal lobe epilepsy. Clinical manifestations and outcome in 34 patients treated between 1934 and 1988. *Brain* 1995;118:1289–1304.
11. Penfield W, Erickson TC, Tarlov I. Relation of intracranial tumors and symptomatic epilepsy. *Arch Neurol Psychiatr* 1940;44:300–315.
12. White JC, Liu CT, Mixter WJ. Focal epilepsy: a statistical study of its causes and the results of surgical treatment. I. Epilepsy secondary to intracranial tumors. *NEJM* 1948;238:891–899.
13. Bhatia R, Kollevold T. A follow-up study of 91 patients operated on for focal epilepsy. Epilepsia 1976;17: 61–66.
14. Watson CW. The incidence of epilepsy following cranio-cerebral injury. In: *Epilepsy. Proceedings of the association held jointly with the International League against Epilepsy.* Baltimore: Williams & Wilkins, 1947: 516–528.
15. Russell WR, Whitty CWM. Studies in traumatic epilepsy. I. Factors influencing the incidence of epilepsy after brain wounds. *JNNP* 1952;15:93–98.
16. Barkovich AJ, Kuzniecky RI, Dobyns WB. Radiologic classification of malformations of cortical development. *Curr Opin Neurol* 2001;14:145–149.
17. Palmini A, Andermann F, de Grissac H, et al. Stages and patterns of centrifugal arrest of diffuse neuronal migration disorders. *Dev Med Child Neurol* 1993;35: 331–339.
18. Rodrigues W, Kher A, Rathi S, et al. Recurrent seizures due to pachygyria. *Indian Pediatr* 1998;35:1230–1233.

19. Loiseau P. Parietal lobe epilepsies. In: Meinardi H, ed. *Handbook of neurology,* vol. 73. *The epilepsies,* part II. Amsterdam: Elsevier Science, 2000:97–106.
20. Siegel AM, Williamson PD. Parietal lobe epilepsy. In: Williamson PD, Siegel AM, Roberts DW, eds. *Advances in neurology,* vol. 84. *Neocortical epilepsy.* Philadelphia Lippincott Williams & Wilkins, 2000:189–199.
21. Boon PA, Williamson PD, Fried I, et al. Intracranial, intraaxial, space-occupying lesions in patients with intractable partial seizures: an anatomoclinical, neuropsychological, and surgical correlation. *Epilepsia* 1991;32: 467–476.
22. Janz D. The grand mal epilepsies and the sleeping-waking cycle. *Epilepsia* 1962;3:69–109.
23. Foerster O, Penfield W. The structural basis of traumatic epilepsy and results of radical operation. *Brain* 1930;53: 99–119.
24. Foerster O. The cerebral cortex in man. *Lancet* 1931;2: 309–312.
25. Cushing H. The parietal tumors. Inaugural sensory fits. In: Cushing H. *Meningiomas: their classification, regional behaviors, life history, and surgical end results.* Springfield, IL: Charles C Thomas, 1938:632–656.
26. Penfield W, Jasper H. *Epilepsy and the functional anatomy of the human brain.* London: Churchill, 1954: 773.
27. Mauguière F, Courjon J. Somatosensory epilepsy: a review of 127 cases. *Brain* 1978;101:307–332.
28. Cascino GD, Hulihan JF, Sharbrough FW, et al. Parietal lobe lesional epilepsy: electroclinical correlation and operative outcome. *Epilepsia* 1993;34:522–527.
29. Ho SS, Berkovic SF, Newton MR, et al. Parietal lobe epilepsy: clinical features and seizure localization by ictal SPECT. *Neurology* 1994;44:2277–2284.
30. Williamson PD, Boon PA, Thadani VM, et al. Parietal lobe epilepsy: diagnostic considerations and results of surgery. *Ann Neurol* 1992c;31:193–201.
31. Nair DR, Lüders HO. Cephalic and whole-body auras. In Lüders HO, Noachtar S, eds. *Epileptic seizures. Pathophysiology and clinical semiology.* New York: Churchill Livingstone, 2000:355–360.
32. Erickson TC. Erotomania (nymphomania) as an expression of cortical epileptiform discharge. *Arch Neurol Psychiatry* 1945;53:226–231.
33. Currier RD, Little SC, Suess JF, et al. Sexual seizures. *Arch Neurol* 1971;25:260–264.
34. Calleja J, Carpizo R, Berciano J. Orgasmic epilepsy. *Epilepsia* 1988;29:635–639.
35. O'Donovan CA, Burgess RC, Lüders HO. Autonomic auras. In Lüders HO, Noachtar S, eds. *Epileptic seizures. Pathophysiology and clinical semiology.* New York: Churchill Livingstone, 2000:320–328.
36. Tuxhorn I, Kerdar MS. Somatosensory auras. In Lüders HO, Noachtar S, eds. *Epileptic seizures. Pathophysiology and clinical semiology.* New York: Churchill Livingstone, 2000:286–297.
37. Young GB, Barr HW, Blume WT. Painful epileptic seizures involving the second sensory area. *Ann Neurol* 1986;19:412.
38. Mulder DW, Daly D, Bailey AA. Visceral epilepsy. *Arch Intern Med* 1954;93:481–493.
39. Gowers W. *Epilepsy and other chronic convulsive disorders: their causes, symptoms and treatment.* London: Churchill Livingstone, 1901:29–58.
40. Head H, Holmes G. Sensory disturbances from cerebral lesions. *Brain* 1911;34:102–254.
41. Penfield W, Boldrey E. Somatic motor and sensory representation in the cerebral cortex of man as studied by electrical stimulation. *Brain* 1937;60:389–443.
42. Talbot JD, Marrett S, Evans AC, et al. Multiple representation of pain in human cerebral cortex. *Science* 1991;251:1355–1358.
43. Casey KL, Minoshima S, Berger KL, et al. Positron emission tomographic analysis of cerebral structures activated specifically by repetitive noxious heat stimuli. *J Neurophysiol* 1994;71:802–807.
44. Jones AKP, Brown WD, Friston KJ, et al. Cortical and subcortical localization of response to pain in man using positron emission tomography. *Proc R Soc Lond Biol* 1991;244:39–44.
45. Apkarian AV, Stea RA, Manglos SH, et al. Persistent pain inhibits contralateral somatosensory cortical activity in humans. *Neurosci Lett* 1992;140:141–147.
46. Lüders H, Lesser RP, Dinner DS, et al. The second sensory area. Humans: evoked potential and electrical stimulation studies. *Ann Neurol* 1985;17:177–184.
47. Young GB, Blume WT. Painful epileptic seizures. *Brain* 1983;106:537–554.
48. Lewin W, Phillips CG. Observations on the partial removal of the post-central gyrus for pain. *JNNP* 1952;15: 143–147.
49. Marshall J. Sensory disturbances in cortical wounds with special reference to pain. *JNNP* 1951;14:187–204.
50. Robinson CJ, Burton H. Somatotopographic organization in the second somatosensory area of *M. fascicularis. J Comp Neurol* 1980;192:43–67.
51. Robinson CJ, Burton H. Organization of somatosensory receptive fields in cortical area 7b, retroinsula, postauditory and granular insula of *M. fascicularis. J Comp Neurol* 1980;192:69–92.
52. Robinson CJ, Burton H. Somatic submodality distribution within the second somatosensory (SII), 7b, retroinsular, postauditory, and granular insular cortical areas of M. fascicularis. *J Comp Neurol* 1980;192:93–108.
53. Ajmone-Marsan C, Gumnit RJ. Neurophysiological aspects of epilepsy. In: Magnus O, DeHaas AML, eds. *Handbook of clinical neurology,* vol. 15. *The epilepsies.* Amsterdam: North-Holland, 1974:30–59.
54. Critchley M. *The parietal lobes.* London: Edward Arnold, 1955.
55. Babb RR, Eckman PB. Abdominal epilepsy. *JAMA* 1972;222:65–66.
56. Otani K, Imai K, Futagi Y, et al. Bilateral painful epileptic seizures of the hands. *Dev Med Child Neurol* 1995; 37:933–936.
57. York GK, Gabor AJ, Dreyfus PM. Paroxysmal genital pain: an unusual manifestation of epilepsy. *Neurology* 1979;29:516–519.
58. Schubert R, Cracco JB. Familial rectal pain: a type of reflex epilepsy? *Ann Neurol* 1992;32:824–826.
59. Wilkinson HA. Epileptic pain. An uncommon manifestation with localizing value. *Neurology* 1973;23: 518–520.
60. Trevathan E, Cascino GD. Partial epilepsy presenting as focal paroxysmal pain. *Neurology* 1988;38:329–330.
61. Lancman ME, Ascoape JJ, Penry KT, et al. Paroxysmal pain as sole manifestation of seizures. *Pediatr Neurol* 1993;9:404–406.

62. Sveinbjornsdottir S, Duncan JS. Parietal and occipital lobe epilepsy: a review. *Epilepsia* 1993;34:493–521.
63. Siegel AM, Williamson PD, Roberts DW, et al. Localized pain associated with seizure origin in the parietal lobe. *Epilepsia* 1999;40:845–855.
64. Eschle D, Siegel AM, Wieser HG. Pure amygdalar seizure onset characterized by severe abdominal pain: a depth electrode study. *Mayo Clin Proc* 2002;77:1358–1361.
65. Solana de Lope J, Alarcon Fernandez O, Aguilar Mendoza J, et al. Epilepsia abdominal en el adulto. *Rev Gastroenterol Mex* 1994;59:297–300.
66. Zarling EJ. Abdominal epilepsy: an unusual cause of recurrent abdominal pain. *Am J Gastroenterol* 1984;79: 687–688.
67. Soques A. Diagnostic du siège et de la nature d'une variété de tumeurs cerebrales (psammomes ou sarcomes angiolithiques par la radiographie). *Rev Neurol* 1921; 28:984–986.
68. Marchand L, de Ajuriaguerra J. Crises d'épilepsie a aura douloureuse. *Ann Médico-psychologiques* 1939; 97:794–796.
69. Penfield W, Gage L. Cerebral localization of epileptic manifestations. *Arch Neurol Psychiatry* 1933;30: 709–727.
70. Michelsen JJ. Subjective disturbances of the sense of pain from lesions of the cerebral cortex. *Res Publ Assoc Nerv Ment Dis* 1943;25:86–99.
71. Peppercorn MA, Herzog AG, Dichter MA, et al. Abdominal epilepsy. A cause of abdominal pain in adults. *JAMA* 1978;240:2450–2451.
72. Ebner A, Kerdar MS. Olfactory and gustatory auras. In Lïders HO, Noachtar S, eds. *Epileptic seizures. Pathophysiology and clinical semiology.* New York: Churchill Livingstone, 2000:313–319.
73. Bancaud J. Epilepsies. In: *Système nerveux.* Paris: Encyclopédie Médico-Chirurgicale, 1976:1–22.
74. Swartz BE, Hallgren E, Delgado-Escueta AV, et al. Multidisciplinary analysis of patients with extratemporal complex partial seizures. II. Predictive value of semiology. *Epilepsy Res* 1990;5:146–154.
75. Hauser-Hauw C, Bancaud J. Gustatory hallucinations in epileptic seizures: electrophysiological, clinical and anatomic correlates. *Brain* 1987;110:339–359.
76. Lüders HO, Lesser RP, Dinner DS, et al. Inhibition of motor activity by electrical stimulation of the human cortex. *Epilepsia* 1983;24:519.
77. Lüders HO, Lesser RP, Dinner DS, et al. The second sensory area in humans: evoked potential and electrical stimulation studies. *Ann Neurol* 1985;17:177–184.
78. Blanke O, Ortigue S, Landis T, et al. Neuropsychology: stimulating illusory own-body perceptions. The part of the brain that can induce out-of-body experiences has been located. *Nature* 2002;419:269–270.
79. Elger CE, Brockhaus A. Vertigo as epileptic seizure manifestation. *Nervenheilkunde* 1997;16:98–104.
80. Kluge M, Beyenburg S, Fernandez G, et al. Epileptic vertigo: evidence for vestibular representation in human frontal cortex. *Neurology* 2000;55:1906–1908.
81. Cummings JL, Syndulko K, Goldberg Z, et al. Palinopsia reconsidered. *Neurology* 1982;32:444–447.
82. Vuilleumier P, Despland PA, Assal G, et al. Astral and out-of-body voyages. Heautoscopy, ecstasis and experimental hallucinations of epileptic origin. *Revue Neurologique* 1997;153:115–119.
83. Cummings JL, Miller BL. Visual hallucinations: clinical occurrence and use in differential diagnosis. *West J Med* 1987;146:46–51.
84. Mooney AJ, Carey P, Ryan M, et al. Parasagittal parieto-occipital meningioma with visual hallucinations. *Am J Ophthalmol* 1965;59:197.
85. Jacobs L. Visual allesthesia. *Neurology* 1980;30: 1059–1063.
86. Anand I, Geller EB. Visual auras. In Lüders HO, Noachtar S, eds. *Epileptic seizures. Pathophysiology and clinical semiology.* New York: Churchill Livingstone, 2000:298–303.
87. Geier S, Bancaud J, Talairach J, et al. Ictal tonic postural changes and automatisms of the upper limb during epileptic parietal lobe discharges. *Epilepsia* 1977;18: 517–524.
88. Matsuo F. Parietal epileptic seizures beginning in the truncal muscles. *Acta Neurol Scand* 1984;69:264–270.
89. Landré E, Chassoux F, Devaux B, et al. Mouvements oculaires critiques dans un cas d'épilepsie pariétale. *Epilepsies* 1997;9:125–130.
90. Rasmussen T. Seizures with local onset and elementary symptomatology. In: Vinken PJ, Bruyn GW, eds. *Handbook of clinical neurology,* vol. 15. *The epilepsies.* Amsterdam: North-Holland, 1974:74–86.
91. Garcia-Pastor A, Lopez-Esteban P, Peraita-Adrados R. Epileptic nystagmus: a case study video-EEG correlation. *Epilept Disord* 2002;4:23–27.
92. Gire C, Somma-Mauvais H, Nicaise C, et al. Epileptic nystagmus: electroclinical study of a case. *Epilept Disord* 2001;3:33–37.
93. Kaplan PW. Neurophysiological localization of epileptic nystagmus. *Am J Electroneurodiag Techn* 1999;39: 77–83.
94. White JC. Epileptic nystagmus. *Epilepsia* 1971;12:157.
95. Alemayehu S, Bergey GK, Barry E, et al. Panic attacks as ictal manifestations of parietal lobe seizures. *Epilepsia* 1995;38:824–830.
96. Marsh L, Rao V. Psychiatric complications in patients with epilepsy: a review. *Epilepsy Res* 2002;49:11–33.
97. Siegel AM, Roberts DW, Thadani VM, et al. The role of intracranial electrode reevaluation in epilepsy patients failing initial invasive monitoring. *Epilepsia* 2000;41: 571–580.
98. Siegel AM, Jobst BC, Thadani VM, et al. Medically intractable, localization-related epilepsy with normal MRI: presurgical evaluation and surgical outcome in 43 patients. *Epilepsia* 2001;42:883–888.

22

Surgical Management of Parietal Lobe Epilepsy

Hahnah J. Kasowski, Michael R. Stoffman, Susan S. Spencer*, and Dennis D. Spencer

Departments of Neurosurgery and Neurology, Yale University, New Haven, Connecticut*

INTRODUCTION

The surgical management of parietal lobe epilepsy is challenging yet frequently rewarding. The presurgical investigation, the operation, and resulting pathology shares similarities with extraparietal epilepsy; however, there are unique features in the semiology and spread of the seizures. Most important, the following data suggest that excellent postsurgical seizure outcomes are tenable in these patients and every effort should be made to identify the critical epileptogenic substrate in these patients. In this chapter, the Yale epilepsy protocol, an illustrative case, and our series of 28 patients with parietal lobe epilepsy are presented.

SURGICAL EVALUATION

Patients in the Yale epilepsy program undergo an evaluation which begins with a detailed history, physical examination, and an outpatient electroencephalogram (EEG). The patient is then admitted to the hospital for continuous audiovisual EEG monitoring. During the admission the patient has a magnetic resonance imaging (MRI) scan, which includes special sequences particularly useful in visualizing subtle dysplastic cortex. Positron emission tomography (PET) imaging is also obtained, as are interictal and ictal single-photon emission computed tomography (SPECT) imaging, which are subtracted for a difference image, and a PET/SPECT ratio is determined. These forms of imaging have been found to be helpful in localizing seizure onset, especially in patients with normal MRI scans. Patients also undergo detailed neuropsychological testing. If an operation may impact on language or memory, intracarotid amobarbital testing (WADA) is performed.

All of the collected data are then reviewed by a multidisciplinary epilepsy team. A decision is made either to proceed to surgery for resection or to acquire more information with an intracranial monitoring study. A complete discussion of the rational for this decision is beyond the scope of this chapter; however, some patients with a well-circumscribed lesion (usually a tumor or vascular malformation) and a concordant preoperative evaluation may proceed directly to surgery. If the lesion is in or adjacent to sensorimotor or language cortex, the patient may require either intraoperative awake mapping or an intracranial monitoring study to define the extent of the resection that can be achieved without causing postoperative neurologic deficit.

The patient may be a candidate for intracranial monitoring if discordant data are obtained during the initial seizure work-up, or if no discrete lesion is seen on the imaging studies. Subdural grids, strips, and depth electrodes are used to record the ictal onset and the grids are used for functional mapping of the cortex. When possible, extraoperative mapping is preferred to intraoperative awake mapping because the patient is fully awake and cooperative. There is also no time pressure on the testing because it can be carried out over a day or two. Functional MRI (fMRI) is also obtained to localize sensorimotor and language functions. At this time, however, functional imaging is not used as a replacement for extraoperative intracranial or intraoperative awake mapping.

The patient is then taken to the operating room for resection of the epileptic region. If the epileptogenic region is discrete, such as a tumor, we attempt to resect the entire lesion with a margin of normal brain tissue. Occasionally this is not possible because of the proximity of the lesion to sensorimotor or language cortex, and a subtotal resection must be performed. In patients with suspected cortical dysplasia, we resect cortex subpially to the depth of the white matter protecting the en-passant vessels. Subpial transections are performed on cortex that is thought to be epileptogenic but not resectable because of its location in functional cortex.

THE YALE EXPERIENCE

Patients

The Yale epilepsy database was searched for all patients from 1990 to the present who had surgery involving the parietal lobe by the senior author. Twenty-eight patients were identified, 19 women and nine men. The patients had a mean age of seizure onset of 9 years (range, 2 months to 38 years). The mean age at the time of first surgery was 26 years (range, 18 months to 51 years). Two patients had two operations. Mean duration of seizures before the first surgery was 18 years (range 13 months to 43 years). Two patients had a history of head trauma. Four patients had a history of febrile seizures. One of the patients with febrile seizures had a history of central nervous system infection, as did four other patients.

Seizure Characteristics

Complex partial seizures were most common, with 18 of 28 (64%) patients having this seizure type. Of these, nine patients experienced rare generalization and another two patients experienced generalization during 50% of their seizures. The next most common seizure type was simple partial experienced by five patients (18%). One of these patients had epilepsia partialis continua. Three patients had a mixture of complex partial and simple partial seizures and two patients experienced rapidly generalized seizures. The seizure frequency ranged from three per year to 600 per month, with a median of 32.5 per month.

Most patients (18 out of 28) experienced an aura prior to their seizures. The most common were somatosensory described by 13 (46%) as numbness or tingling. In most patients, this was contralateral to their seizure onset but two patients had somatosensory aura in bilateral extremities and two experienced the sensation ipsilateral to seizure onset. One of these patients experienced tingling in her arm that occasionally progressed to a sharp and excruciatingly painful feeling.

One patient had a cephalic sensation prior to seizures consisting of a "throbbing or vibratory" sensation in her head. Epigastric sensations were seen in three patients prior to seizure onset. Two patients experienced lightheadedness or vertiginous sensation and one experienced fear of impending doom.

Ten patients had focal clonic activity contralateral to the epileptogenic zone. Five had tonic posturing of the extremities. Seven patients had staring, four had head deviation, and four had automatisms.

Interictal EEG was abnormal in 18 out of 25 patients (not available in three patients). All of these patients had epileptiform spiking except two that showed slowing. The distribution of the abnormal spiking is as follows: frontal, one; temporal, four; parietal or central, four; occipital, none; bilateral, five and temporoparietal, one; and temporal-frontal, one. Thus, most patients did not have interictal abnormalities in the parietal lobe. Ictal EEG was available in 25 patients and was lateralizing in nine and localizing in five. Of the patients with lateralizing ictal EEG findings, onset was bilateral in seven, temporal in four, fronto-temporal in two, and fronto-polar in one.

Neuropsychological testing was available in all but two patients in our series. Seventeen patients had lateralized findings suggestive of dysfunction in the hemisphere of seizure onset. Intelligence quotient (IQ) testing revealed a median verbal IQ (VIQ) of 92, a median performance IQ of 93, and a median full-scale IQ (FSIQ) of 94. WADA testing was performed in 16 patients. Left speech dominance was occurred in 14 (88%) patients. Bilateral memory testing was done in 13 patients. Seven were found to have predominantly left memory, two had right memory, and memory was equal on both sides in four patients.

Imaging

Magnetic resonance imaging results were available in all patients. Only five patients had a normal MRI scan. Magnetic resonance imaging findings, when present, localized the area of abnormality in all but one patient. This patient had a vascular malformation in the hemisphere opposite to the side of seizure onset.

Ictal HMPAO Single-Photon Emission Computed Tomography

Seventeen (59%) of the patients underwent ictal SPECT. In nine (53%) patients, the SPECT localized to the epileptogenic focus. SPECT was considered to be localizing if hypoperfusion (seven patients) or hyperperfusion (two patients) correlated with the other preoperative tests and confirmed on surgical pathology.

Positron Emission Tomography

Fourteen (48%) of the patients underwent PET scanning. In seven (50%) of these patients, hypometabolism was demonstrated in the epileptogenic focus and confirmed on surgical pathology to be gliosis, cortical dysplasia, or low-grade tumor. In the remaining seven patients, either no metabolic defect was illustrated or a focus distant from the surgical resection was documented.

Surgical Procedure

As illustrated in Figure 22.1, 15 patients underwent lesionectomy. Seven of the patients

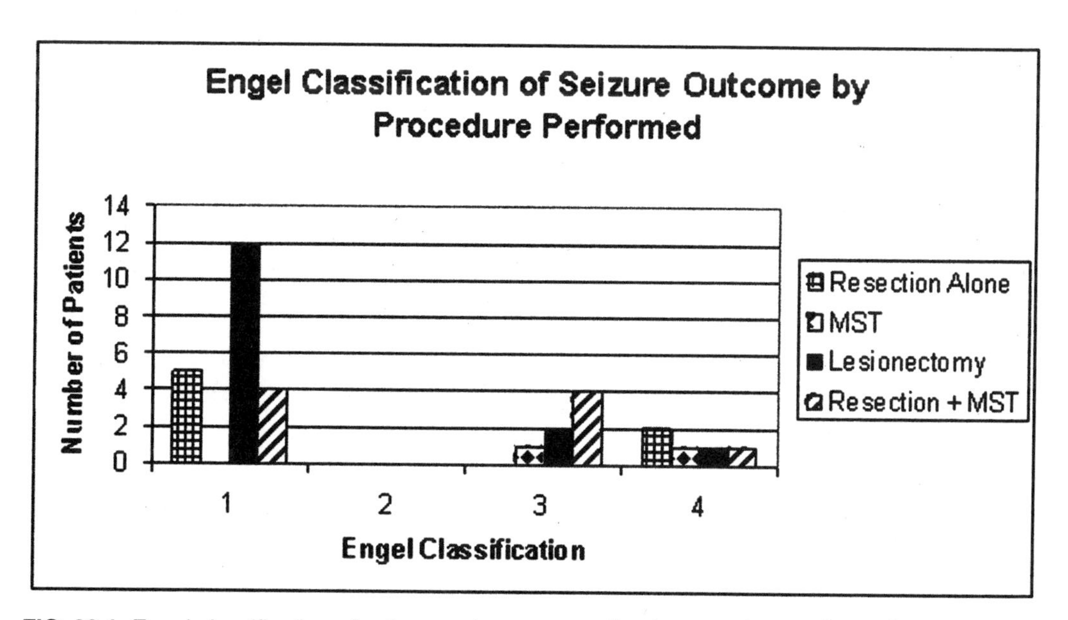

FIG. 22.1. Engel classification of seizure outcome according to procedure performed.

underwent resection only. Eight of the patients underwent resection and multiple subpial transections (MST). Two patients underwent MST alone. MST was performed when an area of cortex could not be resected because of its proximity to sensorimotor or language function. When frozen section intraoperative pathologic diagnosis suggested low-grade glioma (astrocytoma or ganglioglioma), an attempt was made to produce "clear" margins. Tumor was resected until intraoperative pathology revealed an absence of tumor cells. Twenty-seven patients (93%) had one surgical procedure and two (7%) patients had two surgical procedures.

Histopathology

Results of histopathology are the following: Gliosis was present in 12 patients, low-grade tumors in eight, malformations of cortical development in six, a vascular malformation in one, viral infection in one, and laminar necrosis with inflammation in one.

Surgical Morbidity and Mortality

There was no mortality with a median follow-up of 6.2 years (range 3 months to 15 years). There were five transient neurologic deficits (defined as lasting <6 months). These included paraphasia, foot weakness, difficulty reading, decreased hand proprioception, and right hemiparesis. There were six permanent neurologic deficits, which included nondominant hand weakness, anomia, hemiparesis, homonymous hemianopsia, decreased proprioception, and right-left disorientation. One bone flap infection occurred requiring delayed cranioplasty.

Outcome

With a median follow-up of 6.2 years, 16 (55%) patients were seizure and aura free, and 13 (45%) patients were not seizure free. The Engel classification was also used as an outcome scale. A class I result occurred in 19 (66%) patients, class II in no patients, class III in six (21%) patients, and class IV in four (13%). Variables affecting seizure outcome were analyzed. As illustrated in Figures 22.1 to 22.4, Engel class I outcome was most common in patients harboring low-grade gliomas. Low-grade gliomas included World Health Organization (WHO) grade II astrocytoma, grade II oligodendroglioma and oligoastrocytoma. Gangliogliomas were analyzed separately. However, the two patients with gangliogliomas have also had an Engel class 1 outcome. Six patients (67%) with MCD had a class I outcome and the remaining three (33%) had a class III outcome. Four (36%) patients with gliosis had a class I outcome. The remaining seven patients had either a class III or IV outcome. The one patient with tuberous sclerosis had a class IV outcome.

Correlation between tumor or nontumor pathology and outcome was also analyzed. Eighty-six percent of class III and 100% of class IV outcomes occurred in nontumoral cases. Thus, the majority of poor outcomes occurred in nontumor cases. However, 12 out of a total of 23 (52%) patients with nontumoral pathology had a class I outcome.

Procedure type was also analyzed as a factor affecting outcome. In this retrospective cohort, it is difficult to determine the precise effect of procedure. However, only four (40%) patients who underwent MST in addition to resection had a class I outcome. This is likely related to bias because MST was performed when pathology was present in cortex where it was felt that removal would result in unacceptable morbidity.

Last, secondary generalization was analyzed. As illustrated in Figure 22.4, this variable did not determine outcome. Even patients who generalized more than 50% of the time, had a class I outcome. Additionally, patients who never or rarely generalized (clinically or on EEG) had a class III or IV outcome.

Illustrative Case

JW is a 23-year-old right-handed woman who began having febrile seizures at age two. Initially, her seizures were controlled with phenobarbital but subsequently she developed

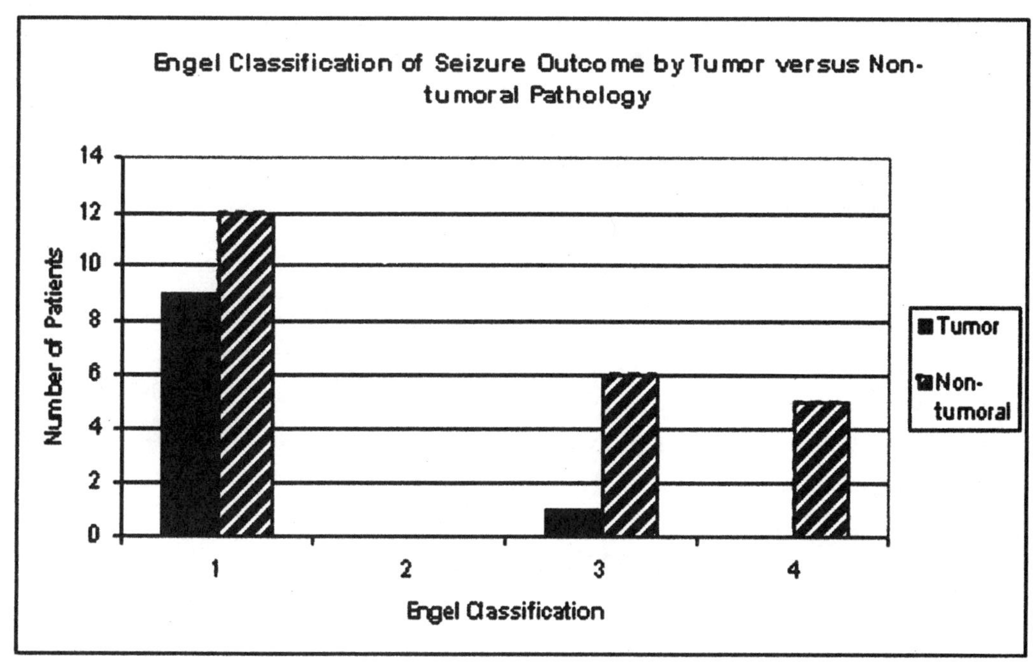

FIG. 22.2. Engel classification of seizure outcome according to pathologic diagnosis.

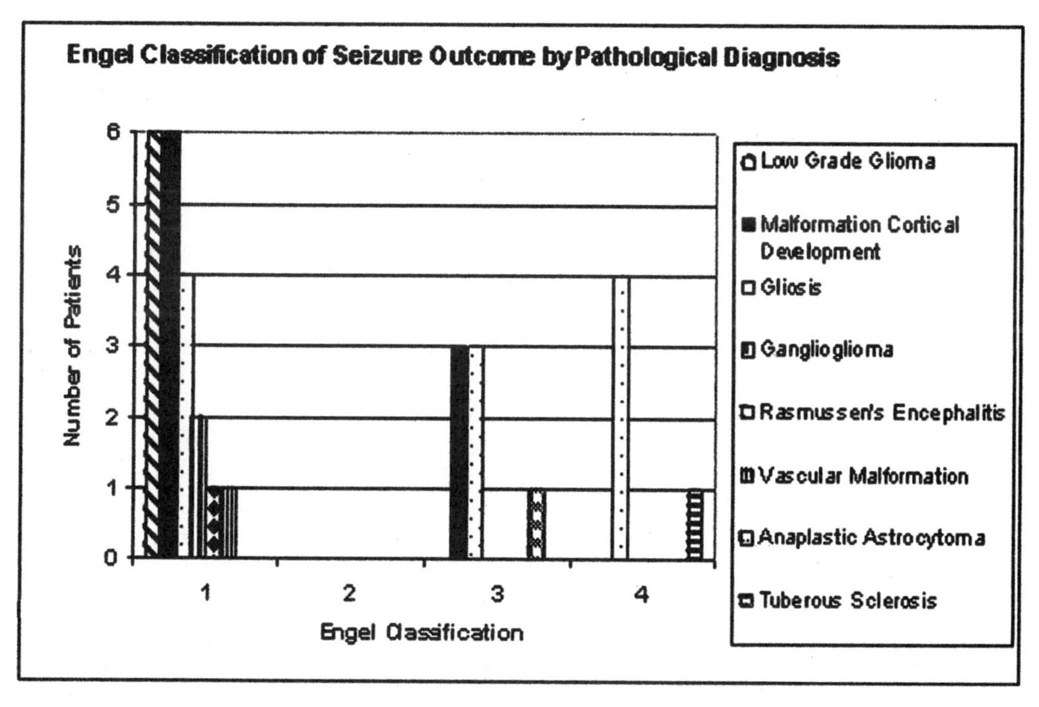

FIG. 22.3. Engel classification of seizure outcome according to tumoral versus nontumoral pathology.

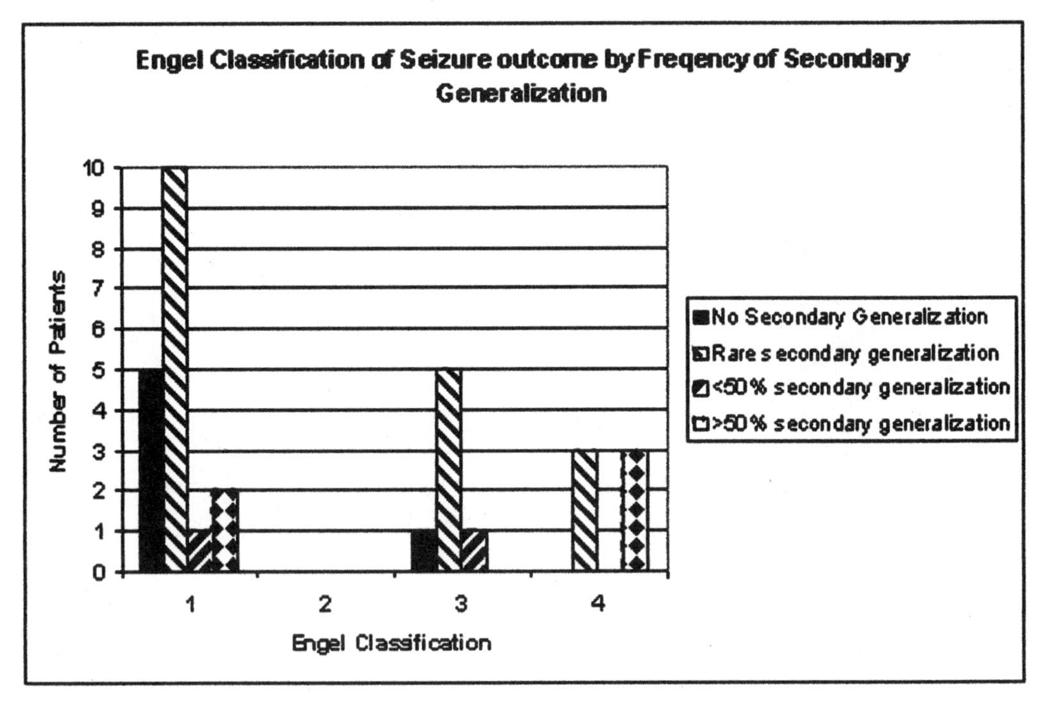

FIG. 22.4. Engel classification of seizure outcome according to presence versus absence of secondary generalization.

afebrile seizures. The patient describes an aura of something feeling "funny, like a vibration" in her head. Then, anticipating a seizure, she sits down or calls her parents. This is followed by loss of contact and half of the time proceeds to a generalized clonic-tonic seizure. She does not experience postictal weakness or aphasia, although she is drowsy. She has tried many medications, including phenobarbital, felbamate, Neurontin, Gabitril, and Depakote. Currently, she is taking Trileptal and dilantin and has a seizure every 1 to 2 months.

In addition to a history of febrile seizures, there is a family history of febrile seizures in her mother and grandmother. The patient has been otherwise healthy. She had a normal birth and delivery and denies history of head trauma. Although she had high fevers at age two, there is no clear history of meningitis or encephalitis. She graduated from college and works as a substitute teacher. Physical and neurologic examinations were unremarkable.

The patient was admitted for video-EEG monitoring. Interictal EEG showed right temporal spikes and ictal EEG showed right temporal neocortical onset. Magnetic resonance imaging of the brain was normal. Interictal SPECT was normal and ictal subtraction SPECT was nonlocalizing. Positron emission tomography scan showed right temporal hypometabolism.

Neuropsychologic evaluation showed subtle nondominant neocortical dysfunction. WADA testing showed left dominance for language, left memory 10/10, and right memory 3/10.

The patient was discussed by the epilepsy team and it was felt that the patient should have a right hemisphere study widely based on the temporo-parietal region. She was subsequently taken to the operating room for right hemisphere subdural grids and strips with a right hippocampal depth electrode. Antiepileptic medications were weaned and several stereotyped seizures were recorded.

The focus was localized to the superior parietal lobule. Extraoperative mapping then was performed for language and sensorimotor function.

The patient then was brought back to the operating room and the strips and depth electrodes were removed. The superior parietal lobule then was resected and multiple subpial transections were performed on the inferior parietal lobule. Pathology revealed gliosis.

Postoperatively, the patient's neurologic examination remained intact. She has not experienced any seizures since surgery. She remains on antiepileptic medications.

DISCUSSION

Somatosensory auras have long been recognized as the most common preceding parietal lobe seizures (1–4). The majority of patients in this series presented with complex partial seizures and most had a preceding aura. In nine of these 13 patients, the sensation was felt in the limb contralateral to the seizure focus. In another four patients, the sensation was bilateral or ipsilateral to seizure onset. Patients most commonly experience paresthesias, although painful seizures are also known to occur with parietal lobe seizures (3,5,6), which occurred in one patient in our study. Painful sensations are thought to arise from secondary somatosensory areas posterior to the postcentral gyrus and rarely have been evoked by cortical stimulation of these areas (7–9).

Two patients described a feeling of lightheadedness or vertigo prior to their seizures. Vertigo also has been reported in parietal lobe epilepsy (4,10). Although cortical representation of vertigo is not fully understood, it is thought to be located near the temporo-parietal border near the auditory cortex (11).

Other patients described auras that are more commonly seen in temporal lobe seizures. Three patients had epigastric sensations prior to seizure onset. Although this has been reported in other series, this represents a somewhat higher percentage of patients than had been reported previously (4,10,12). Epigastric sensations are most often seen with temporal lobe seizures, and likely this aura suggests rapid spread from the ictal onset to the temporal lobe, thus imitating temporal lobe seizures. It has been demonstrated that epileptic zones in association cortex can rapidly spread to cause initial symptoms at a distant cortical area (13,14).

Although a variety of clinical seizure patterns were observed, they can be grouped by general characteristics into three groups. As has been described, both in a previous report from this institution, and in other studies, parietal lobe seizures often present as asymmetric tonic seizures with or without clonic features and with seizures causing a loss of contact and automatisms (4,12,15). It is surmised that these two different seizure patterns represent spread from the ictal focus either into the supplementary motor area in the frontal lobe or into the temporal lobe and limbic structures (16,17). In our study, seven patients demonstrated tonic posturing and four demonstrated loss of contact and automatisms. The other general group of seizures observed in this study was focal motor clonic activity not associated with tonic posturing from involvement of the central area. Seven patients exhibited this type of seizure; the second most common type of seizure observed.

Interictal scalp EEG was notable in that it was commonly abnormal (17 of 25 patients). However, the abnormality was suggestive of a parietal lobe lesion in only four of these patients. Ictal EEG was similarly often abnormal. The ictal EEG was suggestive of a parietal lobe onset in six patients, but was only helpful in lateralizing the hemisphere of onset in seven patients. Seven additional patients had bilateral ictal EEG changes. Ictal scalp EEG seems to be useful in lateralizing the ictal onset, as has been shown, but is not useful in further localizing the lesion (4,12,15).

Neuropsychiatric testing suggested a lateralized abnormality in 20 patients. The abnormality lateralized correctly to the hemisphere of ictal onset in 17 of these. Other groups have reported that a significant number of patients have abnormalities on neuropsychiatric testing (4). However, the percentage of correlation seen in this study is higher than reported in our previous study (12). Intelligence quotient testing revealed average IQ values in the low-normal range. It is interesting that despite the high number of patients with abnormalities on neuropsychiatric testing, IQ scores were preserved near normal. WADA testing revealed most patients had left-hemisphere–dominant speech and memory.

Magnetic resonance imaging proved to be very useful in localizing the area of ictal onset. All of the patients in this study had MRIs and the abnormality was correctly identified by MRI in 18 patients. Of the patients in whom an MRI did not localize the lesion, six were normal, two indicated bilateral abnormalities, and two revealed a lesion opposite to the side of ictal onset. This is in agreement with other studies that have demonstrated the importance of adequate MRI imaging (12,15,18). Several of the patients referred from other institutions had initial MRI scans that were read as normal. These patients received repeat MRI imaging at Yale using special sequences that are effective in showing subtle cortical abnormalities or signal changes in the white matter. Often, when an initial MRI was interpreted as normal, an area of cortical dysplasia was identified in thin cut imaging.

Over half of the current patients underwent an ictal HMPAO SPECT. The majority of patients with concordant examinations demonstrated hypoperfusion. Ho (19) demonstrated hyperperfusion on ictal-SPECT in all 14 patients examined with parietal lobe epilepsy. However, Avery (20) demonstrated that four out of five patients examined with ictal-SPECT had hypoperfusion localizing to an epileptogenic focus. Two of these patients also had areas of hyperperfusion. The current series confirms that epileptogenic focus ictally has areas of increased and decreased perfusion. In isolation, ictal-SPECT has difficulty localizing the focus, but it adds useful information for seizure localization in combination with other noninvasive tests.

With respect to the type of operation performed, we analyzed the effect of performing MST in addition to a standard resection for a lesion. Not surprisingly, over 80% of the class I patients underwent resection alone, but only 50% of the class III or IV patients underwent resection alone. However, four (40%) patients who underwent resection and MST had a class I outcome. This retrospective cohort was not designed to predict differences in outcome according to procedure. Therefore, it is difficult to draw any conclusions with regard to the precise effect of performing MST in addition to a resection. There is also a source of selection bias in this series. Patients were not randomized to receive resection with or without MST. This was determined by examining preoperative investigation and determining the likelihood of producing neurologic morbidity with resection. For example, no attempt was made to perform a complete resection in lesions located within functionally active cortex, as demonstrated on invasive electrical studies or fMRI. In this situation, a small biopsy was performed to assess the histology and MST then was performed. In a meta-analysis (21) of 211 patients who underwent MST with resection or MST alone, similar postoperative seizure control rates were found. This issue needs to be addressed in a larger multicenter prospective randomized controlled trial.

The histopathologic diagnosis was important in determining outcome. All patients undergoing resection of a low-grade glioma (defined in results) or ganglioglioma had a class I outcome. Sixty-seven percent of the patients with malformation of cortical development also had a class I outcome. Interestingly, no patient had a class II outcome. Thirty-three percent of patients with MCD had a class III outcome. This suggests that these patients often have diffuse cerebral pathology not amenable to resection. These data are supported by other groups as well (22).

Four (36%) patient with gliosis had a class I outcome. The remaining patients with gliosis had a class III or IV outcome. The pathologic diagnosis gliosis is an ambiguous term often given to tissue that appears normal. The fact that any patients with this pathologic diagnosis had a class I outcome suggests that these patients have a more subtle diagnosis, such as microdysgenesias. Using an elegant three-dimensional cell counting method, Thom et al. (23) demonstrated increased neuronal densities in temporal lobe surgical specimens. Neuronal density correlated positively with postresection seizure outcome. This is one of the first papers to rigorously compare with an appropriate control group cell densities in white matter. Some groups have refuted that microdysgenesias even exists (24–29). However, Thom's paper provides strong evidence for microdysgenesias. We propose that microdysgenesias may have been present in our patients with "gliosis" but an Engel class IA outcome.

In fact, the surgical morbidity and mortality in this cohort is consistent with previously published data (22). There has been no mortality with a median follow-up of 6.9 years. There were only six permanent neurologic deficits (defined as lasting >6 months). These were mainly expected and clearly related to the resection, and patients were prepared for functional changes. In general, these primarily sensory association changes did not affect daily activities. One patient developed a contralateral mild hemiparesis, objective power testing 4/5 (Medical Research Council scale) throughout both distal and proximal upper and lower extremity. The other five permanent neurologic deficits were more subtle findings only demonstrable on sensory cortical testing.

SUMMARY

Parietal lobe seizure foci are difficult to localize unless there is an MRI lesion or contralateral sensory aura. Rapid network projection often makes scalp EEG and semiology misleading. However, seizure control can be achieved with reasonable success when concordant information guides the physician to a parietal ictal onset. Perhaps the most important messages that this small surgical series provides is that of neurologic outcome. The parietal lobe is a highly convergent cortical region and a major network way station. Except for primary sensory phenomena and language, one cannot temporarily ablate parietal cortical association area within a presumed epileptogenic region and predict the visuospatial, cognitive, and neurologic outcome. Therefore, data demonstrating that one can resect regions of parietal cortex and not cause serious dysfunction are helpful. The mild morbidity encountered in this group of patients would not be necessarily predicted if the same region of normal parietal lobe was resected. Therefore, one must consider cortical plasticity and functional redistribution as possible reasons for this, particularly when most of these substrates are of developmental origin.

REFERENCES

1. Cushing H. The parietal tumors. Inaugural sensory fits. In: Cushing H. *Meningiomas: their classification, regional behaviours, life history, and surgical end results.* Springfield, IL: Charles C Thomas, 1938:632–656.
2. Ajmone-Marsan C, Goldhammer L. Clinical ictal patterns and electrographic data in cases of partial seizures of frontal-central-parietal origin. In: Brazier MAB, ed. *Epilepsy: its phenomena in man.* New York: Academic Press, 1973:235–258.
3. Mauguière F, Courjon J. Somatosensory epilepsy: a review of 127 cases. *Brain* 1978;101:307–322.
4. Salanova V, Andermann F, Rasmussen T, et al. Tumoural parietal lobe epilepsy: clinical manifestations and outcome in 34 patients treated surgically between 1934 and 1988. *Brain* 1995;118:1289–1304.
5. Russell WR, Whitty CWM. Studies in traumatic epilepsy. 2. Focal motor and somatic sensory fits: a study of 85 cases. *Neurol Neurosurg Psychiatry* 1953; 16:73–97.
6. Young GB, Blume WT. Painful epileptic seizures. *Brain* 1983;106:537–554.
7. Foerster O. The cerebral cortex in man. *Lancet* 1931:2; 309–312.
8. Penfield W, Gage L. Cerebral localization of epileptic manifestations. *Arch Neurol Psychiatry* 1933;30: 709–727.
9. Penfield W, Boldrey E. Somatic motor and sensory representation in the cerebral cortex of man as studied by electrical stimulation. *Brain* 1937;60:398–443.
10. Sveinbjornsdottir S, Duncan JS. Parietal and occipital lobe epilepsy: a review. *Epilepsia* 1993;34:493–521.
11. Penfield W, Rasmussen T. Epileptic seizure patterns. In: *The cerebral cortex of man.* New York: Macmillan, 1951.
12. Williamson PD, Boon PA, Thadani VM, et al. Parietal lobe epilepsy: diagnostic considerations and results of surgery. *Ann Neurol* 1992;31:193–201.
13. Lüders HO, Awad IA. Conceptual considerations. In Lüders HO, ed. *Epilepsy surgery.* New York: Raven Press, 1991:51–62.

14. Williamson PD, Spencer SS. Clinical and EEG features of complex partial seizures of extratemporal origin. *Epilepsia* 1986;27:S46–S63.
15. Cascino GD, Hulihan JF, Sharbrough FW, et al. Parietal lobe lesional epilepsy: electroclinical correlation and operative outcome. *Epilepsia* 1993;34:522–527.
16. Bancaud J, Talairach J, Bonus A, et al. *La stereo electroencephalographie dans l'epilepsy.* Paris: Masson, 1965.
17. Geier S, Bancaud J, Talairach J, et al. Ictal tonic postural changes and automatisms of the upper limb during epileptic parietal lobe discharges. *Epilepsia* 1977;18:517–524.
18. Olivier A, Boling W Jr. Surgery of parietal and occipital lobe epilepsy. *Adv Neurol* 2000;84:533–575.
19. Ho SS, Berkovic SF, Newton MR, et al. Parietal lobe epilepsy: clinical features and seizure localization by ictal SPECT. *Neurology* 1994;44:2277–2284.
20. Avery RA, Zubal IG, Stokking R, et al. Decreased cerebral blood flow during seizures with ictal SPECT injections. *Epilepsy Res* 2000;40:53–61.
21. Spencer SS, Schramm J, Wyler A, et al. Multiple subpial transection for intractable partial epilepsy: an international meta-analysis. *Epilepsia* 2002;43:141–145.
22. Salanova V, Andermann F, Rasmussen T, et al. Parietal lobe epilepsy: clinical manifestations and outcome in 82 patients treated surgically between 1929 and 1988. *Brain* 1995;118:607–628.
23. Thom M, Sisodiya S, Harkness W, et al. Microdysgenesis in temporal lobe epilepsy. A quantitative and immunohistochemical study of white matter neurones. *Brain* 2001;124:2299–2309.
24. Meencke HJ, Veith G. The relevance of slight migrational disturbances (microdysgenesis) to the etiology of the epilepsies. *Adv Neurol* 1999;79:123–131.
25. Engel J Jr, ed. *Surgical treatment of the epilepsies,* 2nd ed. New York: Raven Press, 1993.
26. Lüders HO. *Epilepsy surgery.* New York: Raven Press, 1992.
27. Siegel AM, Williamson PD. Parietal lobe epilepsy. *Adv Neurol* 2000;84:189–199.
28. Siegel AM, Williamson PD, Roberts DW, et al. Localized pain associated with seizure origin in the parietal lobe. *Epilepsia* 1999;40:845–855.
29. Williamson P, Spencer DD, Spencer SS, et al. Complex partial seizures of frontal lobe origin. *Ann Neurol* 1985;18:497–504.

Subject Index

R